Hydrogen in Semiconductors II

SEMICONDUCTORS
AND SEMIMETALS
Volume 61

Semiconductors and Semimetals

A Treatise

Edited by R. K. Willardson
   Consulting Physicist
   Spokane, Washington

Eicke R. Weber
Department of Materials Science
and Mineral Engineering
University of California
at Berkeley

Hydrogen in Semiconductors II

SEMICONDUCTORS
AND SEMIMETALS

Volume 61

Volume Editor

NORBERT H. NICKEL

HAHN-MEITNER-INSTITUT BERLIN
BERLIN, GERMANY

ACADEMIC PRESS
San Diego   London   Boston
New York   Sydney   Tokyo   Toronto

This book is printed on acid-free paper.

COPYRIGHT © 1999 BY ACADEMIC PRESS

ALL RIGHTS RESERVED.
NO PART OF THIS PUBLICATION MAY BE REPRODUCED OR TRANSMITTED IN ANY FORM OR BY ANY MEANS, ELECTRONIC OR MECHANICAL, INCLUDING PHOTOCOPY, RECORDING, OR ANY INFORMATION STORAGE AND RETRIEVAL SYSTEM, WITHOUT PERMISSION IN WRITING FROM THE PUBLISHER.
The appearance of the code at the bottom of the first page of a chapter in this book indicates the Publisher's consent that copies of the chapter may be made for personal or internal use of specific clients. This consent is given on the condition, however, that the copier pay the stated per-copy fee through the Copyright Clearance Center, Inc. (222 Rosewood Drive, Danvers, Massachusetts 01923), for copying beyond that permitted by Sections 107 or 108 of the U.S. Copyright Law. This consent does not extend to other kinds of copying, such as copying for general distribution, for advertising or promotional purposes, for creating new collective works, or for resale. Copy fees for pre-1999 chapters are as shown on the title pages; if no fee code appears on the title page, the copy fee is the same as for current chapters. 0080-8784/99 $30.00

ACADEMIC PRESS
525 B Street, Suite 1900, San Diego, CA 92101-4495, USA
http://www.apnet.com

ACADEMIC PRESS
24–28 Oval Road, London NW1 7DX, UK
http://www.hbuk.co.uk/ap/

International Standard Book Number: 0-12-752170-4
International Standard Serial Number: 0080-8784

PRINTED IN THE UNITED STATES OF AMERICA
99 00 01 02 03 EB 9 8 7 6 5 4 3 2 1

Contents

PREFACE xi
LIST OF CONTRIBUTORS xiii

Chapter 1 Introduction to Hydrogen in Semiconductors II 1
Norbert H. Nickel

Chapter 2 Isolated Monatomic Hydrogen in Silicon 13
Noble M. Johnson and Chris Van de Walle
I. INTRODUCTION 13
II. THEORY 14
    1. Hydrogen in the Positive Charge State: H^+ 14
    2. Hydrogen in the Neutral Charge State: H^0 16
    3. Hydrogen in the Negative Charge State: H^- 17
    4. Relative Stability of Different Charge States and
       Negative-U Character 17
III. EXPERIMENT 18
    1. Donor Level 18
    2. Acceptor Level 20
    3. Equilibrium Densities 21
IV. CONCLUSIONS 21
    REFERENCES 22

**Chapter 3 Electron Paramagnetic Resonance Studies of Hydrogen
and Hydrogen-Related Defects in Crystalline Silicon** 25
Yurij V. Gorelkinskii
I. INTRODUCTION 25
II. INTERSTITIAL (BC) HYDROGEN IN SILICON 28
    1. Experimental Procedure 28
    2. EPR Spectrum of Bond-Centered Hydrogen 29
    3. Stress-Induced Alignment of Bond-Centered Hydrogen 37
    4. Thermally Activated Annealing 42
III. EPR OF HYDROGEN-RELATED COMPLEXES IN SILICON 45

		1. Hydrogen-Intrinsic Defect Complexes	45
		2. EPR of Platinum-Hydrogen and Sulphur-Hydrogen Complexes	47
		3. ENDOR Spectra of Si—H Bonds at the (111) Si Surface	48
		4. ENDOR of Hydrogen in the Oxygen Thermal Donor (NL10)	49
IV.	HYDROGEN-INDUCED EFFECTS IN SILICON		51
		1. Hydrogen-Associated Shallow Donors in Hydrogen-Implanted Silicon	51
		2. Stress-Induced Alignment of the AA1 Spectrum	59
		3. Neutral Charge State of Hydrogen-Associated Donor	65
		4. EPR Evidence of Hydrogen-Enhanced Diffusion of Al in Silicon	69
V.	SUMMARY AND CONCLUSIONS		75
	REFERENCES		77

Chapter 4 Hydrogen in Polycrystalline Silicon 83

Norbert H. Nickel

I.	INTRODUCTION	83
II.	EXPERIMENTAL TECHNIQUES	85
	1. Sample Preparation and Characterization	85
	2. Hydrogen Passivation	86
	3. Characterization	86
III.	HYDROGEN DIFFUSION	87
	1. Hydrogen Diffusion from a Plasma Source	87
	2. Hydrogen Diffusion from a Silicon Layer	98
	3. Hydrogen Density of States	102
IV.	HYDROGEN PASSIVATION OF GRAIN-BOUNDARY DEFECTS	110
V.	METASTABILITY	124
	1. Light-Induced Defect Generation	125
	2. Metastable Changes in the Electrical Conductivity	132
VI.	HYDROGEN-INDUCED DEFECTS DURING PLASMA EXPOSURE	143
	1. Generation of Acceptor-Like Defects	144
	2. Platelets	152
VII.	SUMMARY AND FUTURE DIRECTIONS	159
	REFERENCES	161

Chapter 5 Hydrogen Phenomena in Hydrogenated Amorphous Silicon 165

Wolfhard Beyer

I.	INTRODUCTION	165
II.	MATERIAL CHARACTERIZATION BY HYDROGEN EFFUSION AND INFRARED ABSORPTION	166
	1. Measurement Techniques and Material Preparation	166
	2. Hydrogen Effusion Data	167
	3. Infrared Absorption Data	174
III.	EXPERIMENTAL HYDROGEN DIFFUSION AND SOLUBILITY DATA	182
	1. Hydrogen Diffusion Data	182
	2. Hydrogen Solubility Data	192
IV.	HYDROGEN DIFFUSION AND EFFUSION EFFECTS	199
	1. Hydrogen Diffusion Processes	199
	2. Hydrogen Density of States Distribution and Hydrogen Chemical Potential	206
	3. Temperature Shift of Hydrogen Chemical Potential and	

  Meyer-Neldel Rule of Hydrogen Diffusion 212
 4. *Time Dependence of Hydrogen Diffusion Coefficient* 217
 5. *Deviations from Error-Function Diffusion Profiles* 220
 6. *Plasma In-Diffusion Versus Layer Diffusion* 225
 7. *Relation Between SIMS Diffusion Data and Hydrogen Effusion Data* 226
 8. *Interrelation Between IR Absorption Spectra and Effusion Transients* 229
V. HYDROGEN SOLUBILITY EFFECTS 230
 1. *Solubility in Compact Material* 231
 2. *Hydrogen-Related Void Formation* 234
VI. CONCLUSIONS 235
 REFERENCES 236

Chapter 6 Hydrogen Interactions with Polycrystalline and Amorphous Silicon—Theory 241

Chris G. Van de Walle

I. INTRODUCTION 241
 1. *Role of Hydrogen in Amorphous and Polycrystalline Silicon* 241
 2. *General Features of Hydrogen in Silicon* 242
 3. *Computational Approaches* 245
II. HYDROGEN INTERACTIONS WITH AMORPHOUS SILICON 248
 1. *Hydrogen Motion – Introduction* 248
 2. *Hydrogen Interactions with Dangling Bonds* 253
 3. *Hydrogen Interactions with Overcoordination Defects* 256
 4. *Hydrogen Interactions with Weak Si—Si Bonds* 261
 5. *Simulations of Amorphous Networks* 262
 6. *Hydrogen Diffusion and Metastability – Discussion* 264
 7. *Hydrogen Versus Deuterium for Passivation of Dangling Bonds* 269
III. HYDROGEN IN POLYCRYSTALLINE SILICON 271
 1. *Grain Boundaries* 271
 2. *Hydrogen Interactions with Grain Boundaries* 272
 3. *Hydrogen-Induced Generation of Donor-Like Metastable Defects* 273
 4. *Hydrogen-Induced Generation of Acceptor-Like Defects* 277
IV. CONCLUSIONS AND FUTURE DIRECTIONS 277
 REFERENCES 278

Chapter 7 Hydrogen in Polycrystalline CVD Diamond 283

Karen M. McNamara Rutledge

I. INTRODUCTION 283
II. SOLID-STATE CHARACTERIZATION TECHNIQUES 286
 1. *Fourier Transform Spectroscopy* 286
 2. *Nuclear Magnetic Resonance* 288
 3. *Electron Paramagnetic Resonance* 290
 4. *Other Analysis Techniques* 293
III. RESULTS OF SOLID-STATE ANALYSIS 294
 1. *Covalent Bonding Environments* 294
 2. *Quantitative Hydrogen Concentrations* 297
 3. *Local Hydrogen Distribution* 298
 4. *Proximity to Paramagnetic Defects* 302
 5. *Macroscopic Hydrogen Distributions* 304

IV.	EFFECTS OF HYDROGEN ON OBSERVED PROPERTIES	305
	1. Infrared Transmission	305
	2. Thermal Conductivity	306
V.	SUMMARY	308
	REFERENCES	309

Chapter 8 Dynamics of Muonium Diffusion, Site Changes and Charge-State Transitions 311

Roger L. Lichti

I.	INTRODUCTION	311
II.	EXPERIMENTAL TECHNIQUES	316
	1. Transverse-Field Methods	318
	2. Longitudinal-Field Methods	319
III.	IDENTIFICATION AND CHARACTERIZATION OF MUONIUM STATES	328
	1. Neutral Paramagnetic Centers	329
	2. Charged Diamagnetic Centers	334
IV.	DYNAMICS OF MUONIUM TRANSITIONS	341
	1. Silicon: The Basic Model	342
	2. Germanium	354
	3. Gallium Arsenide	359
	4. Other III-V Materials	364
V.	RELEVANCE TO HYDROGEN IMPURITIES	366
	REFERENCES	369

Chapter 9 Hydrogen in III-V and II-VI Semiconductors 373

Matthew D. McCluskey and Eugene E. Haller

I.	INTRODUCTION	373
II.	HYDROGEN IN III-V SEMICONDUCTORS	376
	1. Hydrogen in GaAs	376
	2. Hydrogen in AlAs	390
	3. Hydrogen in InP	396
	4. Hydrogen in GaP	402
	5. Hydrogen in AlSb	412
	6. Hydrogen in GaN	422
	7. Hydrogen in Other III-V Semiconductors	426
III.	HYDROGEN IN II-VI SEMICONDUCTORS	426
	1. Hydrogen in ZnSe	426
	2. Hydrogen in CdTe	433
IV.	SUMMARY AND FUTURE DISCUSSION	434
	REFERENCES	436

Chapter 10 The Properties of Hydrogen in GaN and Related Alloys 441

S. J. Pearton and J. W. Lee

I.	INTRODUCTION	441
II.	HYDROGEN IN AS-GROWN NITRIDES	442
	1. Doped Material	442
	2. Sources of Hydrogen	445
	3. Diffusion	446

III.	Dopant Passivation	450
	1. Calcium	451
	2. Carbon	454
	3. Summary	455
IV.	Diffusion and Reactivation Mechanism	456
	1. Alloys	456
	2. In-Containing Nitrides	461
	3. Mechanisms	462
	4. Heterostructures	466
V.	Role of Hydrogen During Processing	469
	1. Implant Isolation	470
	2. Wet Processing	470
	3. Deposition and Etching	471
VI.	Theory of Hydrogen in Nitrides	472
VII.	Summary and Conclusions	476
	References	477

Chapter 11 Theory of Hydrogen in GaN 479

Jörg Neugebauer and Chris G. Van de Walle

I.	Introduction	479
II.	Method	481
	1. Defect Concentrations and Solubility	481
	2. Energetics, Atomic Geometries, and Electronic Structure	482
III.	Monatomic Hydrogen in GaN	483
	1. Atomic Geometries and Stable Positions	484
	2. Migration Path and Diffusion Barriers	486
	3. Formation Energies and Negative-U Effect	486
IV.	Hydrogen Molecules in GaN	489
V.	Hydrogen-Acceptor Complexes in GaN	489
	1. The Mg-H Complex	490
	2. Other H-Acceptor Complexes	492
VI.	Complexes of H with Native Defects	492
	1. Hydrogen Interacting with Nitrogen Vacancies	493
	2. Hydrogen Interacting with Gallium Vacancies	494
VII.	Role of Hydrogen in Doping GaN	495
	1. Doping in the Absence of Hydrogen	495
	2. Doping in the Presence of Hydrogen	497
	3. Activation Mechanism of the Dopants	498
VIII.	General Criteria for Hydrogen to Enhance Doping	499
IX.	Conclusions and Outlook	500
	References	501

Index	503
Contents of Volumes in This Series	509

Preface

The extensive research devoted to the physics of hydrogen in semiconductors during the past decade has led to a more complete understanding of the properties of hydrogen in semiconductors. This progress was driven mainly by technological interests since most preparation and processing steps introduce hydrogen inadvertently into the material. This can lead to undesirable changes of electrical and optical properties. On the other hand, for some device fabrication processes hydrogen is deliberately incorporated into semiconductors for the purpose of passivating defects or dopants. In either case, a fundamental understanding of the properties of hydrogen in semiconductors is extremely important from both the scientific and engineering points of view.

This second volume on the topic of *Hydrogen in Semiconductors* reviews the latest experimental and theoretical studies on the properties of hydrogen in crystalline and disordered semiconductors. The eleven chapters, written by recognized authorities representing industrial and academic institutions, cover thoroughly the most recent phenomena related to hydrogen in semiconductors. All chapters were reviewed. Because the chapters may be read independently, the editor retained some overlapping among the chapters. Moreover, related discussions among the chapters may not necessarily agree with each other. The authors were encouraged to adopt a tutorial format in order to make the volume as useful as possible both to graduate students and to scientists from other disciplines, as well as to active participants in the exciting arena of semiconductor research. I trust that this volume will prove to be an important and timely contribution to the semiconductor literature.

NORBERT H. NICKEL

List of Contributors

Numbers in parentheses indicate the pages on which the authors' contribution begins.

WOLFHARD BEYER (165), *Forschungszentrum Jülich, Institut für Schicht-und Ionentechnik, Jülich, Germany*

YURIJ V. GORELKINSKII (25), *The Institute of Physics and Technology of the Ministry of Science, Academy of Sciences of the Republic Kazakstan, Almaty, Kazakstan*

EUGENE E. HALLER (373), *Lawrence Berkeley National Laboratory and University of California at Berkeley, Berkeley, California*

NOBLE M. JOHNSON (13), *Xerox Palo Alto Research Center, Palo Alto, California*

J. W. LEE (441), *PlasmaTherm, St. Petersburg, Florida*

ROGER L. LICHTI (311), *Department of Physics, Texas Tech University, Lubbock, Texas*

MATTHEW D. MCCLUSKEY (373), *Xerox Palo Alto Research Center, Palo Alto, California*

KAREN M. MCNAMARA RUTLEDGE (283), *Department of Chemical Engineering, Worcester Polytechnic Institute, Worcester, Massachusetts*

JÖRG NEUGEBAUER (479), *Fritz-Haber-Institut der Max-Planck Gesellschaft, Berlin, Germany*

NORBERT H. NICKEL (1, 83), *Department of Applied Physics, Hahn-Meitner-Institut Berlin, Berlin, Germany*

STEPHEN J. PEARTON (441), *Department of Materials Science and Engineering, University of Florida, Gainesville, Florida*

CHRIS G. VAN DE WALLE (13, 241, 479), *Xerox Palo Alto Research Center, Palo Alto, California*

CHAPTER 1

Introduction to Hydrogen in Semiconductors II

Norbert H. Nickel

DEPARTMENT OF APPLIED PHYSICS
HAHN-MEITNER-INSTITUT BERLIN
BERLIN, GERMANY

In the past decade there has been a considerable increase in understanding the properties of hydrogen in semiconductors. This has been fueled mainly by the fact that most semiconductor fabrication and processing steps can give rise to the introduction of hydrogen. Many fabrication processes use hydride gases such as AsH_3, SiH_4, and NH_3, which commonly cause the incorporation of vast amounts of hydrogen into the growing semiconductor. Another common source of hydrogen is residual water vapor in vacuum systems. In plasma deposition systems, the water molecules dissociate into H and OH on electron impact (Bozzelli and Barat, 1983), and the H atom can easily diffuse into the semiconductor (Nickel *et al.*, 1994). The amount of hydrogen introduced into a specimen is difficult to control and depends strongly on the condition and history of the deposition system. In order to study the properties of hydrogen in semiconductors, it is often useful to introduce hydrogen in a controllable way. Therefore, a number of hydrogenation techniques such as plasma and ion-beam hydrogenations, anneals in molecular hydrogen, and electrochemical techniques have been developed. A comprehensive review of hydrogenation processes by Seager (1991) was published previously in a book entitled, *Hydrogen in Semiconductors* (Pankove and Johnson, 1991).

One of the most prominent properties of hydrogen is its ability to passivate deep and shallow defects. In silicon, the passivation of shallow donors and acceptors was demonstrated experimentally by Johnson *et al.* (1986) and Pankove *et al.* (1984), respectively. A significant impact on the field of hydrogen in semiconductors was made by the observation that the properties of amorphous silicon improved vastly on incorporation of hydrogen (Paul et al., 1976). This was due to the elimination of preexisting silicon dangling bonds that act as recombination centers (Pankove, 1978). The so-called device-grade amorphous silicon, grown by decomposing siliane in a plasma discharge, contains up to 10 at.% hydrogen. In the 1980s it was shown that hydrogen passivates a vast number of defects such as surface states, grain boundaries, dislocations, and implantation-induced defects (Pankove, 1991; Kamins and Marcoux, 1980). For this purpose, hydrogen also could be introduced in a posthydrogenation step.

Without the presence of hydrogen, amorphous silicon–based devices such as solar cells, thin-film transistors, large-area two-dimensional scanners, and flat panel displays would not be possible. However, the presence of hydrogen also gives rise to the formation of new defects. Already in 1977 Staebler and Wronski (1977) observed that prolonged illumination with visible light of hydrogenated amorphous silicon resulted in a decrease of the photo- and dark conductivity that is reversible on annealing. It is widely accepted that these changes are due to the formation of silicon dangling bonds (Dersch et al., 1981). Although the microscopic origin of the Staebler-Wronski effect is still an unsolved problem, the participation of hydrogen in this effect was established previously (Nickel et al., 1993).

Hydrogen-induced gap states also have been detected by deep-level transient spectroscopy performed on c-Si Schottky-barrier diodes (Johnson et al., 1987). In addition, hydrogenation of n-type silicon performed at moderate temperatures results in the formation of extended two-dimensional defects commonly known as *platelets* (Johnson et al., 1987). The observation of hydrogen-related defects is not confined to silicon but has been observed in a variety of semiconductors such as germanium, carbon-related semiconductors, and compound semiconductors (Haller, 1991; Zhou et al., 1996; Chevallier et al., 1991).

The discovery of a variety of new phenomena related to the presence of hydrogen and the following profound impact on technology caused a veritable explosion in research activities. A review of the many studies on hydrogen in semiconductors that were performed mainly in the 1980s can be found in the book entitled, *Hydrogen in Semiconductors* (Pankove and Johnson, 1991). Recent developments in the field of hydrogen in crystalline, polycrystalline, and amorphous semiconductors are reviewed in the present volume. In the remainder of this introduction, a brief summary of each of the chapters is provided.

In Chapter 2, Johnson and Van de Walle summarize experimental and theoretical investigations that have led to current understanding of the different charge states and associated energy levels of hydrogen in silicon. In the positive (H^+) and neutral (H^0) charge states, hydrogen is most stable at the bond-center site. A three-center bond is formed between hydrogen and the neighboring silicon atoms, resulting in a displacement of the silicon atoms by 0.41 and 0.45 Å for H^+ and H^0, respectively. The associated defect level corresponds to a donor level located in the upper part of the band gap at $E_C - E_D = 0.2$ eV. The donor level for isolated hydrogen in silicon was first observed with deep-level transient spectroscopy (DLTS) experiments and labeled E3 (Kimmerling *et al.*, 1979). Subsequent electron paramagnetic resonance (EPR) and DLTS studies established that this center is a hydrogen-related defect with an activation energy of 0.16 eV (Gorelkinskii and Nevinnyi, 1987; Holm *et al.*, 1991).

In the negative charge state hydrogen prefers a region of low electron charge density and therefore can be found at the tetrahedral interstitial site. H^- is the preferred charge state in *n*-type silicon, and the associated acceptor level is located at $E_C - E_A = 0.6$ eV. Experimental evidence for H^- in silicon was inferred from the field-drift experiments of H (Tavendale *et al.* 1990; Zhu *et al.*, 1990). Johnson and Herring (1992) showed that negatively charged hydrogen is stable when the Fermi level is positioned 0.3 eV above midgap and possibly at lower positions. The position of the acceptor level was obtained by observing the kinetics of the charge-state transitions of monatomic H with time-resolved capacitance transient measurements (Johnson *et al.*, 1994).

The stability of the charge state depends on the Fermi level. H^+ is most stable in *p*-type silicon. H^0 is less stable than H^+ because the donor level is occupied by an electron, and hence H^0 is independent of the Fermi level. H^- is most stable in *n*-type silicon. Under equilibrium conditions, the energy of H^0 is always higher than that of H^+ and H^-, which is characteristic of a defect center with a negative correlation energy U. In silicon, the correlation energy is $U \approx -0.2$ eV. The negative-U behavior has been shown to be a general characteristic of H in semiconductors. For example, negative-U behavior for H in GaN is discussed in Chapter 11 (Neugebauer and Van de Walle, 1995).

In Chapter 3, Gorelkinskii reviews EPR and ENDOR results of hydrogen and hydrogen-related complexes in silicon. The EPR studies unambiguously confirm the existence of bond-centered hydrogen [BC(H)] and show that the silicon-silicon distance increases on insertion of a hydrogen atom into the BC site.

Hydrogen-impurity complexes such as PtH_2 and $(SH)^0$ were observed, and exact microscopic models of these complexes were extracted from resolved hyperfine interactions with protons. Using ENDOR measurements, hydrogen was found in the structure of shallow oxygen thermal donors. The oxygen-related NL10 center was identified as a singly passivated double donor. Moreover, it was shown that hydrogen plays a dominant role in the formation of these defects. A second EPR center, due to a hydrogen-induced shallow donor, was observed with properties similar to the oxygen thermal donor NL10. From EPR measurements it was concluded that the molecular structure of the hydrogen-induced thermal donor is different from the NL10 center. Gorelkinskii suggests that the NL10 center arises from accumulation of hydrogen-induced shallow donors.

The presence of hydrogen enhances not only the migration rate of oxygen but also the diffusivity of aluminum at temperatures below 200 K. This becomes evident from EPR measurements. In order to create Al-Al pairs, the presence of radiation defects and hydrogen is required. The diffusion takes place in the form of some radiation-induced interstitial $(Al + H)_i$ complex. The role of hydrogen is most likely to decrease the thermal barrier for the migration of Al atoms, which is similar to hydrogen-enhanced oxygen diffusion (Newman and Jones, 1994).

In Chapter 4, Nickel discusses the properties of hydrogen in polycrystalline silicon (poly-Si). He shows that the properties extend well beyond the simple passivation of silicon dangling bonds at grain boundaries. Although hydrogen is readily introduced into poly-Si, the diffusion process is complex and depends on the microstructure of the material. Hydrogen transport is consistent with a simple two-level model used to describe diffusion in amorphous silicon. In solid-state crystallized poly-Si, shallow and deep H traps are located at 0.5 and 1.5–1.7 eV below the transport states. The hydrogen chemical potential was found to reside about 1.2 eV below the transport sites. In response to an increase in the hydrogen concentration from mid 10^{17} to mid 10^{21} cm^{-3}, the hydrogen chemical potential was found to increase abruptly by roughly 0.2 eV.

The most prominent deep hydrogen trap in poly-Si, the silicon dangling bond, is effectively passivated by hydrogen, resulting in the well-known improvements in the electrical properties of the material. Defect passivation is diffusion-limited, and model calculations of the defect passivation show that the amount of H necessary to passivate Si dangling bonds exceeds the defect concentration by a factor larger than unity. The excess H atoms give rise to metastability. Two types of metastable defect formation are reported: (1) light-induced defect generation and (2) a cooling-rate-dependent change in the electrical dark conductivity. In the first effect, the participation of hydrogen manifests itself by the ability to rejuvenate the light-induced defect generation simply by reexposing the sample to monatomic H. The latter effect is clearly due to the formation and dissociation of a hydrogen donor complex.

Nickel also reviews hydrogen-induced generation of defects during plasma exposures of poly-Si. The defects reported are not due to plasma damage, since for all plasma treatments an optically isolated remote system was used. At high temperatures ($>300°C$), posthydrogenation caused the formation of acceptor-like states leading to conductivity-type conversion. On the other hand, at low temperatures ($<300°C$), hydrogen-stabilized platelets were observed. As in c-Si, the platelets appear within a depth of $\sim 0.1\,\mu$m below the sample surface and are oriented along $\{111\}$ crystallographic planes.

In Chapter 5, Beyer deals with a multitude of hydrogen-related phenomena in hydrogenated amorphous silicon (a-Si:H). He reviews and confirms previous concepts regarding the structural information in infrared (IR) absorption spectra and the role of voids in hydrogen stability. Hydrogen diffusion in compact a-Si:H was investigated, and the data are interpreted in terms of a quasi-equilibrium process including the rupture of Si—H and subsequent reconstruction of Si—Si bonds. The hydrogen diffusion process is described in terms of an energy-band model with the hydrogen chemical potential located near the upper edge of a hydrogen density-of-states distribution that determines the concentration of mobile hydrogen atoms in the transport states. The time dependence of the diffusion coefficient is attributed to microstructural effects rather than to H diffusion in a distribution of states.

Theoretical investigations of hydrogen in polycrystalline and amorphous silicon are discussed by Van de Walle in Chapter 6. The calculations were performed using a state-of-the-art first-principles approach based on density-functional theory in the local density approximation (Hohenberg and Kohn, 1964; Kohn and Sham, 1965). In disordered silicon, hydrogen interactions with silicon dangling bonds are of great interest because they influence H diffusion and metastability directly. Van de Walle discusses the structure and energy of Si—H bonds in a crystalline and amorphous environment. The binding energy to remove one hydrogen from a fully hydrogenated vacancy was found to be 2.1 eV. With increasing size of the vacancy, the energy required to remove the H atom and place it in an interstitial site increases to 2.5 eV. The difference is attributed to H—H repulsion effects. The influence of strain in the backbonds on the binding energy of Si—H was investigated. The three Si atoms to which the Si atom with the dangling bond is bonded were moved outward. The displacement was chosen to be in the plane of these three atoms and had a magnitude of 0.13 Å. This caused a decrease in the binding energy by only 0.05 eV.

The difference between hydrogen and deuterium passivation of dangling bonds at surfaces and at the Si/SiO_2 interface is discussed. Desorption of hydrogen from a $Si(100)(2 \times 1)$ surface was observed by irradiation with electrons emitted from a scanning tunneling microscope tip (Lyding *et al.*, 1994). When the experiment was repeated with deuterium, the desorption yield was about two orders of magnitude lower (Avouris *et al.*, 1996). A similar strong isotope effect was observed in hot carrier degradation experiments in metal oxide–semiconductor transistors. This effect is attributed to coupling of the Si—D bending mode to the bulk TO phonon states resulting in an efficient channel for deexcitation of Si—D.

Hydrogen atoms inserted into a Si—Si bond center site can be stabilized for temperatures up to room temperature if the Si—Si bond is subjected to strain. Commonly, bond-stretching and bond-angle distortions occur at grain boundaries in poly-Si. For bond-length distortions up to 0.3 Å, the formation energy of the Si—H—Si bond will be lowered by ~ 0.4–$0.5\,\text{eV}$ per 0.1 Å of distortion in the Si—Si bond. Finally, remaining problems are identified, and future directions are suggested.

In Chapter 7, McNamara Rutledge reviews the properties of hydrogen in polycrystalline chemical vapor deposited (CVD) diamond. Like many CVD processes for depositing thin films, the growth precursor for diamond deposition contains hydrogen. The overall hydrogen content estimated from nuclear magnetic resonance (NMR) is typically $\leq 0.2\,\text{at.\%}$. This value correlates quantitatively with the CH-stretch absorption in IR spectra. Typical proton NMR spectra consist of a Gaussian and a Lorentzian component. The low temperature NMR spectra reveal a broadening of the Lorentzian component, which is indicative of reduced molecular motion (McNamara et al., 1994). Since the motion of H_2 would not be reduced significantly at 100 K, it is concluded that this peak is not due to trapped hydrogen molecules, as observed in silicon (Boyce et al., 1992). The Gaussian component remains unchanged at 100 K, indicating rigidly held hydrogen. Although the average H concentration is low, the proton homonuclear dipolar line broadening amounts to $\sim 60\,\text{kHz}$, indicating that locally the hydrogen concentration is high. This observation suggests that H segregates in polycrystalline diamond (McNamara et al., 1992). A small fraction of the hydrogen observed is associated with covalently bound groups such as methyl groups, and only $5 \times 10^{-4}\,\text{at.\%}$ of the total H content is associated with dangling-bond defects. EPR measurements suggest that all the dangling bonds are located near highly distorted defects found at grain boundaries and surfaces.

It is extremely difficult to study isolated hydrogen directly. However, an analogous impurity exists in muonium for which obtaining the relevant properties is much more straightforward. Muonium is a very light, short-lived pseudoisotope of hydrogen with a mass of 0.1125 amu. In Chapter 8, Lichti gives a brief overview of the experimental techniques and reviews the dynamics of muonium diffusion, site changes, and charge-state transitions in a variety of semiconductors including silicon, germanium, and GaAs. A significant contribution of muonium research to the field of impurities in semiconductors was the identification of the bond-center site as the ground-state configuration for Mu^0 in silicon (Estle et al., 1986; Estreicher, 1987) and the experimental determination of the physical and electronic structure of this center (Kiefl et al., 1988). Comparison of the Mu_{BC}^0 hyperfine parameters with those of the AA9 EPR center for isolated hydrogen (Gorelkinskii et al., 1987) showed the complete equivalence of the Mu and H electronic structures in the only case where similar data exist for hydrogen and Mu impurities. Detailed comparison of the available hydrogen and muonium experimental results from Kreitzman et al. (1995) and Lichti et al. (1995) lead to the conclusion that the muonium results should be considered as a good quantitative estimate for hydrogen. Based on the larger zero-point energy for muonium, barriers for motion between sites should be somewhat higher for H than for Mu. On the other hand, electronic energies are expected to be very similar. For example, DLTS measurements give an energy of 0.164 eV associated with the H_{BC}^0 to H_{BC}^+ transition (Holm et al., 1991). When correlated for electric field effects, this yields an ionization energy of 0.18 eV, somewhat below the present value of about 0.2 eV for ionization of Mu_{BC}^0, which is within the range of results from various muonium-spin resonance techniques.

At least two results obtained from muonium studies are directly relevant to models of hydrogen diffusion. The first is the likely identification of the fast-diffusing state in silicon, namely, metastable H^0, which can easily hop among tetrahedral interstitial sites directly analogous to Mu_T^0 motion. Quantum tunneling behavior of Mu_T^0 at low temperatures should be less relevant for hydrogen, and above 100 K, Mu_T^0 and H motion should be very similar, with some adjustment to the barrier height. Thus the muonium experiments identify a good candidate for the fast-diffusion species in the T-site neutral center that does not require trapping and detrapping steps, as commonly assumed in hydrogen diffusion models. Muonium charge-cycle results suggest that diffusion could occur as repeatedly interrupted motion even without formation and dissociation of H-impurity complexes, even though there is no doubt that such complexes play a dominant role. The details of these transitions were measured in silicon, germanium, and GaAs. The results suggest that diffusion of charged species may very well be dominated by charge-state transitions into the mobile H^0 center rather than motion of H^+ or H^- centers themselves. Transfer of the muonium results to hydrogen implies that transitions among the various isolated hydrogen states should occur extremely rapidly on the time scale of most of the measurement techniques applied to the study of H impurities.

The properties of hydrogen in III-V and II-VI semiconductors are reviewed in Chapter 9. McCluskey and Haller investigated the microscopic structure of hydrogen-related complexes and their interaction with the host lattice using IR and Raman spectroscopy. The microscopic structures of hydrogen-related complexes follow two general trends: (1) In donor-hydrogen complexes, hydrogen resides in an antibonding position due to repulsion of the H^- ion by the high electron density at the bond center. (2) In acceptor-hydrogen complexes, hydrogen resides in a bond-centered position, since it is attracted by the electron density in the covalent bond. However, in semiconductors with strongly ionic bonds such as GaN, the microscopic structure of hydrogen-related complexes is still debated. Magnesium, for example, is the most common p-type dopant in GaN. As a result of hydrogen passivation during growth, as-grown Mg-doped GaN is semi-insulating. Theoretical studies of the Mg-H complex predict that the hydrogen atom binds to a host nitrogen in an antibonding configuration (Neugebauer and Van de Walle, 1996). Although experimental studies show conclusively that hydrogen binds to nitrogen, it is not known whether the hydrogen resides in an antibonding or bond-centered position.

Hydrogen-related defects also interact with the host lattice. As a result of the compression of the P—H bond, group II acceptor-hydrogen complexes in GaP show a higher stretch-mode frequency than in InP. In GaAs:Zn and ZnSe:As, hydrogen binds to an As atom in a bond-centered position adjacent to a Zn atom. To probe interactions between hydrogen complexes and the host lattice and to determine the microscopic structure of hydrogen complexes, variable-pressure or variable-temperature spectroscopy was used. Two models are described that explain the temperature-dependent frequency shift and line-width broadening of local vibrational modes as a consequence of anharmonic coupling to acoustical phonons. A new interaction between local vibrational modes and extended-lattice phonons in AlSb is reported. The resonant interaction results from a near degeneracy of the Se—H stretch mode with a wag-phonon combination mode. A distinct anticrossing was observed with varying temperature or pressure.

In Chapter 10, Pearton and Lee review the properties of hydrogen in GaN. In as-grown GaN, there are numerous potential sources of hydrogen. Vast amounts of hydrogen incorporated during growth easily passivate dopants such as Mg, Zn, Ca (Lee et al., 1996), C (Pearton et al., 1994), and Cd (Burchard et al., 1997), resulting in highly resistive material. The most prominent acceptor is Mg, which can be activated by a postgrowth thermal anneal or by annealing under minority-carrier injection conditions. In the ladder case, the reactivation follows a second-order kinetics process in which the $(MgH)^0$ complex is stable up to 450°C in thin, lightly doped GaN samples. In thicker, more heavily doped layers, trapping of H at Mg acceptors is more prevalent, and annealing temperatures up to 700°C may be required to achieve full activation of the Mg. Hydrogen, however, remains in the material until much higher temperatures and can accumulate in defective regions.

Deuterium diffusion experiments show that in *p*-type material deuterium follows an indiffusion profile where it is trapped at Mg atoms or defects. In addition, in the near-surface region, a very high D concentration suggests the presence of deuterium clusters. On the other hand, in *n*-type GaN there is no measurable D concentration due to the plasma exposure. Optical absorption measurements performed on the ternary alloys AlGaN and AlN reveal substantial band tailing. The origin of the band tails may be related to fluctuations in local electric fields due to fluctuations in the density of charged defects. Posthydrogenation reduces the concentration of these defects, reducing the near-bandedge absorption.

The theory of hydrogen in GaN is reviewed in Chapter 11 by Neugebauer and Van de Walle. The calculations were performed using a state-of-the-art first-principles approach based on density-functional theory in the local density approximation (Hohenberg and Kohn, 1964; Kohn and Sham, 1965). Many features of hydrogen in GaN are similar to the behavior of H in other semiconductors. However, a number of important differences occur. The computational studies show that hydrogen and acceptor-hydrogen complexes prefer unusual geometries. Positively charged hydrogen prefers the N antibonding site, which is in striking contrast to the behavior of H^+ in GaAs or silicon, where the bond-centered site was found to be energetically most stable (Pavesi and Giannozzi, 1992; Van de Walle *et al.*, 1989). Negatively charged hydrogen prefers the Ga antibonding site. At this site the distance to the neighboring Ga atom is maximized, and the charge density of the bulk crystal has a local minimum.

Positively charged hydrogen is the energetically most stable species for Fermi-level positions below $E_C - E_F = 2.1\,\text{eV}$. For Fermi levels higher in the gap, H^- is more stable, whereas neutral hydrogen is never stable in GaN. This is characteristic for a negative-*U* system. The value of *U*, given by the difference between acceptor and donor levels, is $U \approx -2.4\,\text{eV}$, which is unusually large compared with any measured or predicted value for hydrogen in any other semiconductor.

Besides the passivation of dopants (Mg, Be, Zn, etc.), hydrogen also can interact with native defects. This can lead either to the passivation of defects or to the conversion from acceptors to donors. An important class of native defects is vacancies. The interactions of hydrogen with nitrogen and gallium vacancies and their influence on the properties of GaN are reviewed.

For device applications, ternary alloys such as AlGaN are crucial. Although qualitatively similar behavior of hydrogen in these alloys is expected, the larger band gap and the stronger ionicity of AlGaN compared with GaN may affect the properties of interstitial hydrogen. It is proposed to extend research toward these ternary alloys.

References

Avouris, P., Walkup, R. E., Rossi, A. R., Shen, T.-C., Abeln, G. C., Tucker, J. R., and Lyding, J. W. (1996). *Chem. Phys. Lett.*, **257**, 148.
Boyce, J. B., Johnson, N. M., Ready, S. E., and Walker, J. (1992). *Phys. Rev. B*, **46**, 4308.
Bozzelli, J. W., and Barat, R. B. (1983). *Plasma Chem. Plasma Proc.*, **8**, 293.
Burchard, A., Deicher, M., Forkel-Wirth, D., Haller, E. E., Magerle, R., Prospero, A., and Stotzler, R. (1997). *Mater. Res. Soc. Symp. Proc.*, **449**, 961.
Chevallier, J., Clerjaud, B., and Pajot, B. (1991). In *Hydrogen in Semiconductors*, **34**, J. I. Pankove and N. M. Johnson, eds. San Diego: Academic Press, p. 447.
Dersch, H., Stuke, J., and Beichler, J. (1981). *Appl. Phys. Lett.*, **38**, 456.
Estle, T. L., Estreicher, S. K., and Marynick, D. S. (1986). *Hyperfine Ineract.*, **32**, 637.
Estreicher, S. K. (1987). *Phys. Rev. B*, **36**, 9122.
Gorelkinskii, Y. V., and Nevinnyi, N. N. (1987). *Sov. Technol. Phys. Lett.*, **13**, 45.
Haller, E. E. (1991). In *Hydrogen in Semiconductors*, **34**, J. I. Pankove and N. M. Johnson, eds. San Diego: Academic Press, p. 351.
Hohenberg, P., and Kohn, W. (1964). *Phys. Rev.*, **136**, B864.
Holm, B., Nielsen, K. B., and Nielsen, B. B. (1991). *Phys. Rev. Lett.*, **66**, 2360.
Johnson, N. M., and Herring, C. (1992). *Phys. Rev. B*, **46**, 15554.
Johnson, N. M., Herring, C., and Chadi, D. J. (1986). *Phys. Rev. Lett.*, **56**, 769.
Johnson, N. M., Herring, C., and Van de Walle, C. G. (1994). *Phys. Rev. Lett.*, **73**, 130.
Johnson, N. M., Ponce, F. A., Street, R. A., and Nemanich, R. J. (1987). *Phys. Rev. B*, **35**, 4166.
Kamins, T. I., and Marcoux, P. J. (1980). *IEEE Electron. Device Lett.*, **EDL-1**, 159.
Kiefl, R. F., Celio, M., Estle, T. L., Kreitzman, S. R., Luke, G. M., Riseman, T. R., and Ansaldo, E. J. (1988). *Phys. Rev. Lett.*, **60**, 224.
Kimmerling, L. C., Blood, P., and Gibson, W. M. (1979). *Inst. Phys. Conf. Ser.*, **46**, 273.
Kohn, W., and Sham, L. J. (1965). *Phys. Rev.*, **140**, A1133.
Kreitzman, S. R., Hitti, B., Lichti, R. L., Estle, T. L., and Chow, K. H. (1995). *Phys. Rev. B*, **51**, 13117.
Lee, J. W., Pearton, S. J., Zolper, J. C., and Stall, R. A. (1996). *Appl. Phys. Lett.*, **68**, 2102.
Lichti, R. L., Schwab, C., and Estle, T. L. (1995). *Mater. Sci. Forum*, **196–201**, 831.
Lyding, J. W., Shen, T. C., Hubacek, J. S., Tucker, J. R., and Abeln, G. C. (1994). *Appl. Phys. Lett.*, **64**, 2010.
McNamara, K. M., Gleason, K. K., and Robinson, C. J. (1992). *J. Vac. Sci. Technol. A*, **10**, 3143.
McNamara, K. M., Williams, B. E., Gleason, K. K., and Scruggs, B. E. (1994). *J. Appl. Phys.*, **76**, 2466.
Neugebauer, J., and Van de Walle, C. G. (1995). *Phys. Rev. Lett.*, **75**, 4452.
Neugebauer, J., and Van de Walle, C. G. (1996). *Appl. Phys. Lett.*, **68**, 1829.
Newman, R. C., and Jones, R. (1994). In *Oxygen in Silicon*, **42**, F. Shimura, ed. San Diego: Academic Press, p. 289–352.
Nickel, N. H., Johnson, N. M., and Jackson, W. B. (1993). *Phys. Rev. Lett.*, **71**, 2733.
Nickel, N. H., Yin, A., and Fonash, S. F. (1994). *Appl. Phys. Lett.*, **65**, 3099.
Pankove, J. I. (1978). *Appl. Phys. Lett.*, **32**, 812.
Pankove, J. I. (1991). In *Hydrogen in Semiconductors*, **34**, J. I. Pankove and N. M. Johnson, eds. San Diego: Academic Press, p. 35.
Pankove, J. I., and Johnson, N. M. (eds.). (1991). *Hydrogen in Semiconductors*. San Diego: Academic Press.
Pankove, J. I., Wance, R. O., and Berkeyheiser, J. E. (1984). *Appl. Phys. Lett.*, **45**, 1100.

Paul, W., Lewis, A. J., Connell, G. A. N., and Moustakas, T. D. (1976). *Solid-State Commun.*, **20**, 969.
Pavesi, L., and Giannozzi, P. (1992). *Phys. Rev. B*, **46**, 4621.
Pearton, S. J., Abernathy, C. R., and Ren, F. (1994). *Electron. Lett.*, **30**, 527.
Seager, C. H. (1991). In *Hydrogen in Semiconductors*, **34**, J. I. Pankove and N. M. Johnson, eds. San Diego: Academic Press, p. 17.
Staebler, D. L., and Wronski, C. R. (1977). *Appl. Phys. Lett.*, **31**, 292.
Tavendale, A. J., Pearton, S. J., and Williams, A. A. (1990). *Appl. Phys. Lett.*, **56**, 949.
Van de Walle, C. G., Denteneer, P. J., Bar-Yam, Y., and Pantelides, S. T. (1989). *Phys. Rev. B*, **39**, 10791.
Zhou, X., Watkins, G. D., McNamara Rutledge, K. M., Messmer, R. P., and Chawla, S. (1996). *Phys. Rev. B*, **54**, 7881.
Zhu, J., Johnson, N. M., and Herring, C. (1990). *Phys. Rev. B*, **41**, 12354.

CHAPTER 2

Isolated Monatomic Hydrogen in Silicon

Noble M. Johnson and Chris G. Van de Walle

XEROX PALO ALTO RESEARCH CENTER
PALO ALTO, CA

I. INTRODUCTION 13
II. THEORY 14
   1. *Hydrogen in the Positive Charge State: H^+* 14
   2. *Hydrogen in the Neutral Charge State: H^0* 16
   3. *Hydrogen in the Negative Charge State: H^-* 17
   4. *Relative Stability of Different Charge States and Negative-U Character* . . . 17
III. EXPERIMENT 18
   1. *Donor Level* 18
   2. *Acceptor Level* 20
   3. *Equilibrium Densities* 21
IV. CONCLUSIONS 21
   REFERENCES 22

I. Introduction

A classic study in the field of hydrogen in semiconductors was published by Van Wieringen and Warmholtz in 1956 on the solubility and diffusivity of hydrogen in silicon. From measurements of hydrogen permeation through silicon at high temperatures (1092–1200°C), they determined the following Arrhenius dependences for the solubility s and diffusivity D (Herring and Johnson, 1991):

$$s = 4.96 \times 10^{21} \exp(-1.86\,\text{eV}/kT) \text{ atoms per cm}^3 \quad (1)$$

at 1 atm;

$$D = 9.67 \times 10^{-3} \exp(-0.48\,\text{eV}/kT) \text{ cm}^2/\text{sec} \quad (2)$$

where k is Boltzmann's constant and T is absolute temperature, and the activation energies were estimated to be accurate to 10%. Further, from the square-root dependence of the permeation rate on H_2 pressure, Van Wieringen and Warmoltz concluded that hydrogen diffuses as a monatomic species. The measurement was a major experimental achievement that has never been duplicated in silicon, and the results, particularly for the diffusivity, remain impressively consistent with those which have been obtained more recently by other methods and over other temperature regimes. However, what was not considered in this early study was the possibility that the monatomic hydrogen was in other than a neutral charge state.

Today we know that one of the most important characteristics of interstitial hydrogen in silicon, and in semiconductors in general, is the fact that it can assume different charge states. Hydrogen is an amphoteric impurity; that is, it behaves both as a donor and as an acceptor with two levels in the bandgap, and the electron occupancy determines the charge state. The charge state, in turn, determines the most favorable location of hydrogen in the lattice. Both theory and experiment have contributed to the quantitative determination of these energy levels. In this chapter we summarize the key studies that have led to our current understanding of these energy levels for isolated hydrogen in silicon.

II. Theory

A great deal of information about hydrogen in semiconductors has been obtained from first-principles calculations. The most extensive calculations have been performed using the density-functional-pseudopotential approach (Van de Walle et al., 1989), but other techniques have been used as well (for a review, see Estreicher, 1995). For hydrogen in silicon, complete total-energy surfaces were mapped out by placing the hydrogen atom in a variety of positions in the crystal and allowing the host atoms to relax. The resulting energies as a function of hydrogen position form an adiabatic total energy surface and include information about diffusion barriers.

We note that most of the features described here apply not just to H in silicon but to H in other semiconductors as well.

1. HYDROGEN IN THE POSITIVE CHARGE STATE: H^+

In the positive charge state (H^+), the impurity is essentially a proton. It thus can be expected to seek out regions of high electronic charge density. In most covalent semiconductors, this maximum charge density is found at the bond-center (BC) site, and in Si and GaAs, H^+ is indeed known to be located at this site.

When H^+ is located at the BC site, a three-center bond is formed between the hydrogen and the neighboring silicon atoms. These silicon atoms have to move outward to accommodate the hydrogen (Fig. 1a). The energy cost

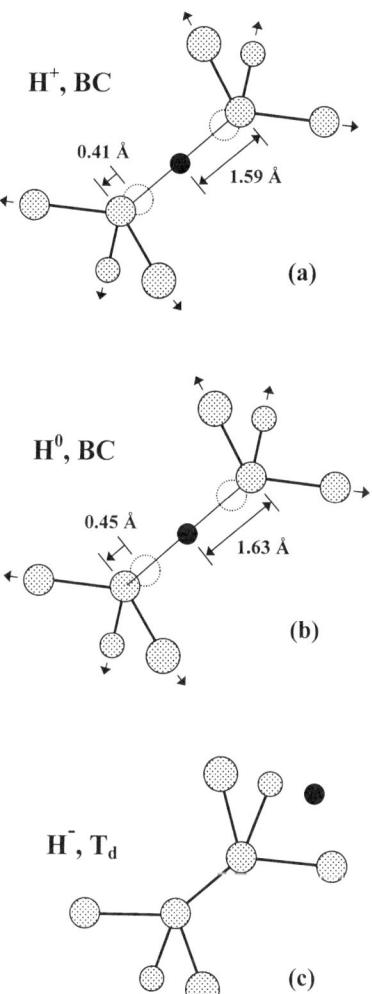

FIG. 1. Schematic illustration of the location of (a) H^+, (b) H^0, and (c) H^- in the silicon lattice. The relaxations of the Si atoms (based on calculations in Van de Walle et al., 1989) are indicated.

associated with these displacements is more than compensated by the energy gained in the formation of the three-center bond. The strength of this bond can be understood within a simple molecular-orbital or tight-binding picture (Van de Walle, 1991), in which the hydrogen 1s orbital overlaps with hybrid orbitals on the neighboring silicon atoms. This picture then shows that the hydrogen-induced defect level actually corresponds to an antibonding combination of Si orbitals; hydrogen is located at a node of the corresponding defect wave function. In the neutral charge state, the defect level would be occupied with one electron; in the positive charge state, it is empty. The defect level therefore corresponds to a donor level; it is located in the upper part of the bandgap.

The stability of each charge state depends on the Fermi level in the material, which is determined by doping. For hydrogen to acquire a positive charge state, it needs to shed an electron. This electron is placed in the reservoir for electrons, which has an energy determined by the Fermi level. The higher the Fermi energy, the more energy it will cost to put an electron in the reservoir, and the higher the formation energy of H^+ will be. It is therefore easy to see that H^+ will be most favorable (lowest formation energy) in p-type material, where the Fermi level is low in the gap.

The proton diffuses easily through the lattice, moving from bond center to bond center, with a migration barrier of less than 0.5 eV. The characteristics of isolated interstitial hydrogen also allow us to predict how hydrogen will interact with shallow dopants. Hydrogen always counteracts the electrical activity of the dopants. The presence of hydrogen in p-type material leads to formation of H^+, a donor, that compensates acceptors. H^+ is coulombically attracted to ionized acceptor impurities; when H^+ forms a *complex* with the acceptor impurity, we say that the acceptor is *passivated* (or *neutralized*). Passivation is characterized by the presence of neutral complexes (which cause little scattering of carriers), whereas *compensation* involves donors and acceptors that are spatially separated (and hence contribute separately to ionized impurity scattering); the difference can be observed in carrier mobilities.

2. Hydrogen in the Neutral Charge State: H^0

The arguments applied earlier to the stability of the three-center bond also apply to neutral hydrogen at the BC. Its stability is slightly less than that of H^+ at the BC because the defect level in the upper part of the bandgap is now occupied by an electron. The BC site is therefore relatively less stable. The energy of the neutral species is, of course, independent of the

Fermi energy. The atomic configuration is very similar to that for H^+ (Fig. 1b), with only small differences in the relaxation of the Si atoms.

3. HYDROGEN IN THE NEGATIVE CHARGE STATE: H^-

In the negative charge state (H^-), hydrogen prefers regions of low electronic charge density, in which its distance to the host atoms is maximized. In semiconductors with the zinc-blende structure, H^- therefore can be found at the tetrahedral interstitial site (Fig. 1c). The extra electron in H^- is acquired by transfer of an electron from the Fermi level; the energy of H^- decreases as the Fermi level rises in the gap. H^- is therefore the preferred charge state in n-type material.

The calculated diffusion barrier for H^- in Si is only slightly larger than that for H^+ (Van de Walle et al., 1989). Just as H^+ passivates acceptors, we expect H^- to passivate donors.

4. RELATIVE STABILITY OF DIFFERENT CHARGE STATES AND NEGATIVE-U CHARACTER

Our discussion so far indicates that hydrogen is an amphoteric impurity; that is, it behaves both as a donor and as an acceptor. In silicon, as in all semiconductors that have been theoretically investigated, the energy of H^0 is always higher than that of H^+ or H^-; that is, the neutral charge state is never the most stable state, under equilibrium conditions. This is the defining characteristic of a *negative-U center*. The magnitude of U is determined by the difference between the donor and acceptor levels, with the donor level being located above the acceptor level in the case of a negative-U center. This behavior can occur because hydrogen assumes different configurations in the lattice as the charge state changes.

First principles calculations (Van de Walle et al., 1989) placed the donor level (i.e., the transition level between the positive and neutral charge states) about 0.2 eV below the conduction band and the acceptor level (the transition level between the neutral and the negative charge states) roughly at midgap. These positions have been observed experimentally, as reviewed in Section III.

The negative-U behavior has been shown to be a general characteristic of hydrogen in any semiconductor; it can be understood by analyzing the total-energy surfaces in detail, as discussed by Neugebauer and Van de Walle (1995).

III. Experiment

Experimental studies have provided identification and quantitative determination of the transition energies of isolated hydrogen in silicon. These results are summarized in Fig. 2 with a schematic energy-band diagram for silicon and the transition energies for the donor level ε_D and acceptor level ε_A. This section highlights the experimental studies that have contributed to these identifications.

1. Donor Level

Several experimental studies have contributed to the identification and characterization of the donor level for isolated hydrogen in silicon. The level was first reported in 1978, with deep-level transient spectroscopy (DLTS), as one of several deep levels introduced into specimens of n-type silicon by proton bombardment at low temperatures (e.g., 45 K) (Kimerling et al., 1979). In the cataloging of defects generated by the implantation, the particular level of interest was labeled E3 and observed to anneal out at 120 K, although in this earliest report neither was the thermal ionization energy quoted nor was the level associated with isolated hydrogen.

A subsequent DLTS study (Irmscher et al., 1984), also on n-type silicon implanted with protons (at 80 K), reported several new observations for this level, now labeled the E3' center. The areal density of the level was found to coincide with the proton implantation dose, which suggested that the level is associated with a hydrogen-related defect. The electron thermal emission

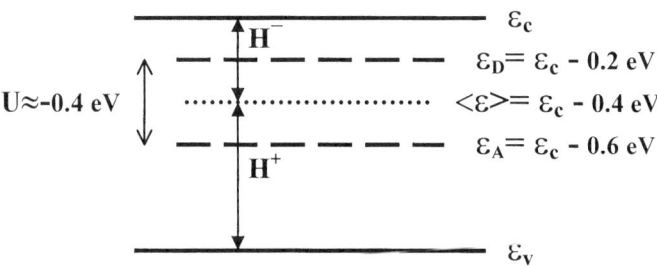

FIG. 2. Schematic energy-band diagram for silicon indicating the experimental assignments for the donor level ε_D and acceptor level ε_A of isolated hydrogen. The inverted order of the acceptor and donor levels defines a negative effective correlation energy U, and the mean energy $\langle \varepsilon \rangle \equiv (\varepsilon_D + \varepsilon_A)/2$ is also indicated.

rate was found to depend on the (average) magnitude of the applied electric field in the depletion layer of a Schottky diode. This phenomenon is termed the *Poole-Frenkel effect* and arises from the electric-field–induced lowering of a Coulomb energy barrier for thermal emission (Frenkel, 1938). Observation of the Poole-Frenkel effect thus established that the E3' center is donor-like. By applying a correction for the effect, the zero-field activation energy was estimated to be 0.2 eV. Finally, the authors included in their speculations on the possible origins of this center the suggestion that the level might arise from a hydrogen atom at a single interstitial site.

A third DLTS study (Holm *et al.*, 1991) further established that E3' is a hydrogen-related defect, reported an activation energy of 0.16 eV after a correction for the Poole-Frenkel effect, and quantitatively characterized the annealing stages. In particular, it was observed that the annealing temperature depends on the presence or absence of free electrons.

The application of electron paramagnetic resonance (EPR) techniques has contributed decisively to the microscopic identification of the E3' center. Studies of silicon implanted with hydrogen at low temperatures revealed a characteristic paramagnetic defect that was designated the AA9 center and assigned to bond-centered hydrogen (Gorelkinskii and Nevinnyi, 1991). The AA9 center is the hydrogenic analogue of anomalous muonium in silicon, which has been studied extensively (Kiefl and Estle, 1991).

Subsequent studies by Bech Nielsen and coworkers (Holm *et al.*, 1991; Bech Nielsen *et al.*, 1994) clearly established that the paramagnetic center AA9 and the deep-level defect center E3' are identical. For example, they observed that the change between plus and zero charge states occurs for both centers at the same donor energy, $\varepsilon_D \approx \varepsilon_C - 0.16\,\text{eV}$, and that the centers display the same annealing behavior.

The E3' center also has been observed in silicon that was hydrogenated without proton implantation (Johnson and Herring, 1994). The hydrogen was introduced with a remote plasma reactor by diffusion into *n*-type, *p*-doped silicon and stabilized at room temperature by the formation of PH complexes. Subsequently, the hydrogen was released from the PH complexes within the depletion layer of Schottky diodes by the injection of minority carriers (holes). By quenching a diode to low temperature during hole injection, the isolated hydrogen was effectively immobilized at an interstitial lattice site. Conventional DLTS clearly revealed the E3' center with its donor-like character. Its observation without low-temperature proton implantation eliminates the possible involvement of displacement damage in the E3' center.

Accurate determination of the hydrogen donor transition energy by DLTS is somewhat problematic. The emission rate depends on the local electric field (i.e., the Poole-Frenkel effect), which varies throughout the depletion layer, and the two methods just described (i.e., implantation and remote-plasma hydrogenation) for introducing hydrogen tend to produce highly nonuniform spatial depth distributions of isolated monatomic hydrogen. The analysis of emission-rate data to obtain the zero-field activation energy also depends on the choice of model for the field dependence. From such considerations, the H^0/H^+ transition energy in silicon is estimated to be 0.2 eV, although the actual value may be slightly smaller than this.

The identity of the AA9 and E3' centers was further strengthened by the work of Gorelkinskii and Nevinnyi (1996) on the effects of uniaxial stress on the AA9 center. Their results indicate that H^+ as well as H^0 occupies BC sites and that the insertion of an H^+ at a BC site is accompanied by an outward relaxation of the two Si nearest neighbors. These conclusions are in complete accord with the computational results summarized in Fig. 1.

2. ACCEPTOR LEVEL

Experimental evidence for negatively charged hydrogen H^- in silicon originally was inferred from the field drift of hydrogen released by the thermal dissociation of PH complexes (Tavendale et al., 1990; Zhu et al., 1990). It was shown subsequently (Johnson and Herring, 1992) that H^- is the stable form when the Fermi level is positioned 0.3 eV above the midgap energy ε_m, and possibly also at lower positions, which established that the hydrogen acceptor level ε_A is below $\varepsilon_m + 0.3$ eV. Since the donor level lies within 0.2 eV of the conduction band (Sec. III.1), that is, $\varepsilon_D \geq \varepsilon_m + 0.35$ eV, the donor level is higher than the acceptor level. This experimentally identifies hydrogen as a negative-U impurity in silicon, for which two H^0's can lower their energy by changing to H^+ and H^- (Anderson, 1975), as had been predicted theoretically and discussed in Section II.4.

With the qualitative information that hydrogen is a negative-U impurity, the experimental observation that H^- is stable for $\varepsilon_F = \varepsilon_m + 0.3$ eV imposes a tighter bound on the location of the acceptor level than would apply for the positive-U case (Johnson et al., 1994). Namely, it is the mean energy $\langle \varepsilon \rangle \equiv (\varepsilon_D + \varepsilon_A)/2$ that is located below $\varepsilon_m + 0.3$ eV, which implies that $\varepsilon_A \leq \varepsilon_m + 0.25$ eV.

A direct quantitative determination of ε_A was obtained by observing the kinetics of the charge-state transitions of monatomic hydrogen with time-resolved capacitance transient measurements (Johnson et al., 1994). A pulse of holes was introduced into the depletion layer of hydrogenated n-type Schottky diodes to generate isolated hydrogen by minority carrier–enhanced dissociation of the PH complexes. Diode bias conditions were used to controllably convert the free hydrogen between H^+ and H^-.

Changes in the diode capacitance with time reflected redistribution of space charge due to the following processes: (1) drift and diffusion of mobile charged hydrogen, (2) charge change of monatomic hydrogen due to emission or absorption of electron or holes, and (3) recapture of hydrogen by dopants or other defects. The analysis fixes ε_A relative to the conduction-band minimum at approximately $\varepsilon_C - 0.6\,\text{eV}$.

3. Equilibrium Densities

In equilibrium, the donor and acceptor energy levels of monatomic hydrogen in silicon determine the dependence of the relative densities of H^+, H^0, and H^- on the Fermi level ε_F via the equations (Herring and Johnson, 1991)

$$n_-/n_0 = (v_- Z_-/v_0 Z_0)\exp[(\varepsilon_F - \varepsilon_A)/kT] \tag{3}$$

$$n_+/n_0 = (v_+ Z_+/v_0 Z_0)\exp[(\varepsilon_D - \varepsilon_F)/kT] \tag{4}$$

where n_j is the density of the jth species, v_j is the number of possible states for this species in a primitive cell, and Z_j is the associated vibrational partition function. With the partition functions estimated to be close to unity (e.g., near room temperature), the factors $v_- Z_-$, $v_0 Z_0$, and $v_+ Z_+$ in Eqs. (3) and (4) are determined by the v_j's, which equal to 2, 8, and 4, respectively (Herring and Johnson, 1991). Because $U < 0$, there is no position of ε_F at which H^0 predominates: n_0 is always $\ll n_-$ or n_+. The dominant species changes quickly from H^+ to H^- as ε_F passes above the mean value

$$\langle \varepsilon \rangle \equiv (\varepsilon_D + \varepsilon_A)/2 \approx \varepsilon_C - 0.4\,\text{eV} \tag{5}$$

IV. Conclusions

Both theory and experiment demonstrate that isolated hydrogen is an amphoteric impurity in silicon with two charge-state transition energies, donor and acceptor levels, in the bandgap. The current best estimates of these transition energies are $\varepsilon_D \approx \varepsilon_C - 0.2\,\text{eV}$ and $\varepsilon_A \approx \varepsilon_C - 0.6\,\text{eV}$, which yields a negative value for the effective correlation energy, $U \approx -0.4\,\text{eV}$. Because of the inverted order of the levels (i.e., $U < 0$), in equilibrium there is no position of ε_F at which H^0 predominates; the density of H^0 is always much less than that of the predominant species (either H^+ or H^-). The dominant species changes quickly from H^+ to H^- as ε_F passes above the mean value $\langle \varepsilon \rangle \approx \varepsilon_C - 0.4\,\text{eV}$.

The now-known amphoteric properties of isolated hydrogen in silicon provide new insight into the classic study by Van Wieringen and Warmoltz (1956). Their solubilities and diffusivities (Eqs. 1 and 2) actually relate predominantly to H^+ rather than to H^0, since at their elevated measurement temperatures ($\geq 1092^\circ C$) the material was intrinsic, and therefore, the Fermi level was substantially below $\langle \varepsilon \rangle$.

Only for hydrogen in silicon do we have direct quantitative information from experiments on the charge-state transition energy levels. Despite the wealth of information available on migration and complex formation of hydrogen in other technologically important semiconductors, notably germanium (Haller, 1991) and gallium arsenide (Chevallier et al., 1991), and despite the fact that computational studies indicate the general property that isolated hydrogen forms a negative-U system in semiconductors with both transition energies in the bandgap (Neugebauer and Van de Walle, 1995), experimental verification of the negative-U character is still lacking in all semiconductors other than silicon.

ACKNOWLEDGMENTS

We are pleased to acknowledge helpful discussions with C. Herring, W. B. Jackson, J. Neugebauer, N. H. Nickel, S. T. Pantelides, and R. A. Street.

REFERENCES

Anderson, P. W. (1975). *Phys. Rev. Lett.*, **34**, 953.
Bech Nielsen, B., Bonde Nielsen, K., and Byberg, J. R. (1994). *Mater. Sci. Forum*, **143–147**, 909.
Chevallier, J., Clerjaud, B., and Pajot, B. (1991). In *Semiconductors and Semimetals*, **34**, J. I. Pankove and N. M. Johnson, eds. San Diego: Academic Press, chap. 13.
Estreicher, S. K. (1995). *Mater. Sci. Engr. Reps.*, **14**, 319.
Frenkel, J. (1938). *Phys. Rev.*, **54**, 647.
Gorelkinskii, Yu V., and Nevinnyi, N. N. (1991). *Physica B*, **170**, 155.
Gorelkinskii, Yu V., and Nevinnyi, N. N. (1996). *Mater. Sci. Engr. B*, **36**, 133.
Haller, E. E. (1991). In *Semiconductors and Semimetals*, **34**, J. I. Pankove and N. M. Johnson, eds. San Diego: Academic Press, chap. 11.
Herring, C., and Johnson, N. M. (1991). In *Semiconductors and Semimetals*, **34**, J. I. Pankove and N. M. Johnson, eds. San Diego: Academic Press, chap. 10.
Holm, B., Bonde Nielsen, K., and Bech Nielsen, B. (1991). *Phys. Rev. Lett.*, **66**, 2360.
Irmscher, K., Klose, H., and Maass, K. (1984). *J. Phys. C: Solid-State Phys.*, **17**, 6317.

Johnson, N. M., and Herring, C. (1992). *Phys. Rev. B*, **46**, 15554.
Johnson, N. M., and Herring, C. (1994). *Mater. Sci. Forum*, **143–147**, 867.
Johnson, N. M., Herring, C., and Van de Walle, C. G. (1994). *Phys. Rev. Lett.*, **73**, 130.
Kiefl, R. F., and Estle, T. L. (1991). In *Semiconductors and Semimetals*, **34**, J. I. Pankove and N. M. Johnson, eds. San Diego: Academic Press, chap. 15.
Kimerling, L. C., Blood, P., and Gibson, W. M. (1979). *Inst. Phys. Conf. Ser.*, **46**, 273.
Neugebauer, J., and Van de Walle, C. G. (1995). *Phys. Rev. Lett.*, **75**, 4452.
Tavendale, A. J., Pearton, S. J., and Williams, A. A. (1990). *Appl. Phys. Lett.*, **56**, 949.
Van de Walle, C. G. (1991). *Physica B*, **170**, 21.
Van de Walle, C. G., Denteneer, P. J. H., Bar-Yam, Y., and Pantelides, S. T. (1989). *Phys. Rev. B*, **39**, 10791.
Van Wieringen, A., Warmoltz, N. (1956). *Physica*, **22**, 849.
Zhu, J., Johnson, N. M., and Herring, C. (1990). *Phys. Rev. B*, **41**, 12354.

CHAPTER 3

Electron Paramagnetic Resonance Studies of Hydrogen and Hydrogen-Related Defects in Crystalline Silicon

Yurij V. Gorelkinskii

THE INSTITUTE OF PHYSICS AND TECHNOLOGY OF THE MINISTRY OF SCIENCE
ACADEMY OF SCIENCES OF THE REPUBLIC KAZAKSTAN
ALMATY, KAZAKSTAN

I. INTRODUCTION 25
II. INTERSTITIAL (BC) HYDROGEN IN SILICON 28
   1. Experimental Procedure 28
   2. EPR Spectrum of Bond-Centered Hydrogen 29
   3. Stress-Induced Alignment of Bond-Centered Hydrogen 37
   4. Thermally Activated Annealing 42
III. EPR OF HYDROGEN-RELATED COMPLEXES IN SILICON 45
   1. Hydrogen-Intrinsic Defect Complexes 45
   2. EPR of Platinum-Hydrogen and Sulfur-Hydrogen Complexes 47
   3. ENDOR Spectra of Si—H Bonds at the (111) Si Surface 48
   4. ENDOR of Hydrogen in the Oxygen Thermal Donor (NL10) 49
IV. HYDROGEN-INDUCED EFFECTS IN SILICON 51
   1. Hydrogen-Associated Shallow Donors in Hydrogen-Implanted Silicon . . . 51
   2. Stress-Induced Alignment of the AA1 Spectrum 59
   3. Neutral Charge State of Hydrogen-Associated Donor 65
   4. EPR Evidence of Hydrogen-Enhanced Diffusion of Al in Silicon 69
V. SUMMARY AND CONCLUSIONS 75
   REFERENCES 77

I. Introduction

Hydrogen is the simplest atom among the elements of the periodic table and, as such, represents the simplest atomic impurity in a crystalline lattice. As a result, there has been a great deal of effort made to understand its role in solids. In particular, its small size, high diffusivity, and extremely high reactivity enable it to interact with other impurities and defects at relatively low temperatures. These interactions can, in turn, strongly influence the optical, electrical, and magnetic properties of semiconductors (Pearton et al., 1987, 1992; Pankove and Johnson, 1991).

One of the most exciting aspects of semiconductor materials science has been the discovery of hydrogen-induced passivation (deactivation of electrical activity) of shallow acceptors (Sah et al., 1983; Johnson et al., 1986; Pankove et al., 1983) and donors in silicon (Johnson et al., 1991; Bergman et al., 1988). Hydrogen can be introduced easily into the silicon crystal in a variety of ways, some of which could easily be accidental in nature, such as boiling in water (Gale et al., 1983; Gorelkinskii et al., 1994) or exposure to water vapor or hydrogen gas at elevated temperatures (Velorisoa et al., 1991; McQuaid et al., 1991; Nickel et al., 1995). Since hydrogen exists in one chemical form or another at almost every stage of device processing in modern semiconductor device technology, investigation of its undesirable role, as well as its useful influence, continues to be one of the hot topics of applied and basic semiconductor physics (Pearton et al., 1987, 1992; Pankove and Johnson, 1991; Lyding et al., 1996).

The phenomenon of hydrogen-enhanced diffusion of impurities in silicon most pronounced for oxygen (McQuaid et al., 1995; Newman, 1996) is also a new, exciting aspect of applied and basic semiconductor physics. These processes are evidently dependent on the ability of hydrogen to move through the crystal, which provides special motivation for the study of its diffusion mechanisms and its possible stable and metastable configurations in the silicon crystal lattice (Blöchl et al., 1990; Pankove and Johnson, 1991; Boucher and Deleo, 1994; Van de Walle and Street, 1995).

In the last several years, concerted effort by both theory and experiment has resulted in much progress in understanding the behavior and properties of interstitial hydrogen (Pankove and Johnson, 1991; Johnson et al., 1994; Johnson and Herring, 1994; Van de Walle and Nickel, 1995), as well as its complexes with doped impurities in crystalline semiconductors (Denteneer et al., 1990; Pearton et al., 1992; Van de Walle, 1994; Stalova et al., 1995; Zhou et al., 1995; Zheng and Stalova, 1996). However, a host of unanswered questions remains concerning isolated hydrogen—especially its complexes in semiconductors (Myers et al., 1992).

Most of the direct experimental information concerning isolated hydrogen states in semiconductors has been obtained from muon spin rotation (μSR) and muon level-crossing resonance (μLCR) experiments on muonium, a light pseudoisotope of hydrogen (Patterson, 1988; Kiefl and Estle, 1991; Kreitzman et al., 1995). A comparison of μLCR data (Kiefl et al., 1988) and theoretical findings (Cox and Symons, 1986; Estle et al., 1987) has established that of the two neutral states, the anisotropic center Mu* (anomalous muonium) is the ground state in Si. In Mu* the muonium lies at the center of a stretched Si—Si bond, i.e., in a bond-centered (BC) configuration (C_{3v} symmetry). The unpaired electron is predominantly localized on the two nearest-neighbor silicon atoms; the associated nonbonding orbital has a node at the muon position. The equivalence of the $Mu^0(BC)$ and $H^0(BC)$ electronic structures has been demonstrated experimentally by comparison of the μLSR results with the hyperfine interaction parameters of the Si-AA9 electron paramagnetic resonance (EPR) center (Kiefl and Estle, 1991; Gorelkinskii and Nevinnyi, 1991). Theoretical investigations (Van de Walle et al., 1989; Chang and Chadi, 1989a) suggest that H^+ (or Mu^+) in BC configuration should be stable and should have a structural configuration very similar to the neutral state, since the electron is removed from the nonbinding orbital.

The isotropic neutral Mu state detected by μSR is a rapidly diffusing state associated with the tetrahedral interstitial T_d site. EPR of H^0 in the T_d site was not observed in silicon so far. Metastable neutral Mu^0 (T_d) has a hyperfine parameter about half that of the vacuum Mu atom, indicating that its donor level should be within the valence band (Kiefl and Estle, 1991).

Recent investigations of transition dynamics among muonium states for both n- and p-type silicon have been carried out by radiofrequency (rf) μSR techniques at temperatures ranging from 10–500 K (Kreitzman et al., 1995). The features are assigned to seven separate transition processes and a single set of parameters, providing an excellent description over the full doping range. These rf μSR results correlate well with the few analogous hydrogen measurements by DLTS (Irmscher et al., 1984; Holm et al., 1991) and EPR (Gorelkinskii and Nevinnyi, 1991, 1996).

Theoretical investigations (Van de Walle et al., 1989; Chang and Chadi, 1989b) suggest that H^- in the T_d $(0/-)$ level lies lower in energy than the H in BC $(+/0)$ level, leading to negative-U electronic behavior for H or Mu in Si. The negative-U property ($U \approx -0.36\,eV$) of hydrogen has been confirmed experimentally (Johnson et al., 1994). It is interesting to note that in these experiments a hydrogen species has been introduced into the sample at room temperature or above, while basic parameters of the monoatomic hydrogen have been extracted from capacitance transients on dissociation of hydrogen species.

Although many splendid results concerning isolated muonium (and hence hydrogen) states in semiconductors have been obtained by the μSR or μLCR methods, important information concerning the stable and long-lifetime metastable states of hydrogen are either ambiguous or cannot be extracted in principle because the short lifetime ($\sim 2.2\,\mu sec$) of muonium prevents the use of the muon techniques on these time scales. There are particular problems of this kind in the investigation of stable hydrogen-related complexes.

Among the various experimental techniques, electron paramagnetic resonance (EPR) is capable of providing the most detailed direct information concerning isolated hydrogen and its complexes in semiconductors. In the case of EPR, there is no limitation for study of stable or long-lifetime metastable states of hydrogen and its complexes. Important, detailed information concerning atomic structure and associated electronic wave function can be obtained from the measured hyperfine parameters. Thus, in the last several years, some remarkable EPR and ENDOR investigations have led to direct identification of hydrogen in defect structures. In particular, platinum-hydrogen (Uftring et al., 1995; Höhne et al., 1994a) and sulfurhydrogen (Zevenbergen et al., 1995a) complexes have been discovered, hydrogen has been identified directly on the surface of porous silicon (Bratus et al., 1994), and hydrogen was identified in the structure of an oxygen thermal shallow donor (Martynov et al., 1995a).

The next section will briefly describe experimental procedures for BC hydrogen in silicon. It will further review the peculiarities of the EPR spectrum for the AA9 defect identified with BC hydrogen, the studies of uniaxial stress alignment of H^+ and H^0 in the BC site, and the data on its annealing kinetics. In Section III the EPR spectra in silicon arising from hydrogen-related complexes will be described; these reveal the hyperfine interaction with the proton(s). Section IV is devoted to EPR centers that do not reveal hyperfine interaction with hydrogen but for which hydrogen's role in defect formation is critical. This section reviews previous and current observations of hydrogen-induced shallow donors in hydrogen-implanted silicon. Very recent EPR observations of hydrogen-enhanced diffusion of aluminum impurities in silicon at low temperature are also described. Section V describes controversial questions and suggestions for future investigations.

II. Interstitial (BC) Hydrogen in Silicon

1. EXPERIMENTAL PROCEDURE

Electron paramagnetic resonance (EPR) measurements were performed in a Q-band (37-GHz) spectrometer at 77 K primarily in the absorption mode. A cylindrical microwave cavity of TE_{011} mode was used in these experiments. The magnetic field could be rotated in a {011} plane of the sample and modulated at 100 or 10 kHz for EPR detection.

Samples with typical dimensions $14 \times 1.1 \times 0.5 \, mm^3$ (the long axis being $\langle 110 \rangle$) were prepared from high-purity silicon grown either by the floating-zone (Fz) method ($\rho \approx 3\text{--}10\,\text{k}\Omega \cdot \text{cm}$) or by the Czochralski (Cz) method (n- or p-type, $\rho \approx 10\,\Omega \cdot \text{cm}$). For hydrogen implantation, silicon samples were bombarded through an aluminum absorber (of corresponding thickness) with protons (deuterons) at a starting energy in the range from 7–30 MeV. The implantation was repeated on a given sample using several different absorber thicknesses (typically 3–4 absorbers) in order to produce a homogeneous distribution of hydrogen and defects. The implantation was carried out using a cyclotron at an ion current density of 0.05–0.08 $\mu\text{A}/\text{cm}^2$ and a sample temperature of 80 K. It is important to note that for the successful generation of the BC interstitial hydrogen (AA9 EPR spectrum), the average power density of the proton beam should be no more than $\sim 1\,\text{W}/\text{cm}^2$.

The sample could be illuminated from a tungsten-halogen bulb (outside the cryostat) through a light guide and optical window in the cavity. In order to reduce heating of the sample under illumination, a silicon infrared (IR) filter ($\sim 300\,\mu\text{m}$ thickness) was placed between the bulb and the light guide. The average intensity of illumination on the sample was about 0.2–0.5 W/cm^2, with essentially all this light having energy larger than the Si bandgap.

Stress was applied to these hydrogen-implanted samples outside the EPR cavity in the special cryostat. Uniaxial stress of 150–200 MPa was applied to a $\langle 110 \rangle$ axis of the sample either in the dark or under bandgap illumination; the temperature was varied from 180–77 K. After the sample was cooled to 77 K under stress, the stress was removed, and the sample was placed (without warmup) into the EPR cavity for measurements at 77 K. It should be noted that the theory of stress-induced alignment and the reorientation kinetics, as well as experimental methods, were developed previously for defects and impurities in silicon (see, for example, Lee and Corbett, 1974; Watkins, 1975; Trombetta et al., 1997).

2. EPR Spectrum of Bond-Centered Hydrogen

a. Condition of Observation and g-Tensor

Following high-energy hydrogen implantation, the AA9 spectrum (Gorelkinskii and Nevinnyi, 1987) associated with BC hydrogen is observed only under bandgap illumination of the sample in either Cz-grown or Fz-grown silicon (Fig. 1). However, the best conditions for the AA9 spectrum analysis are realized for high-purity zone-refined silicon; in Cz silicon, proton implantation produced oxygen-related defects that have EPR spectra overlapping the AA9 spectrum.

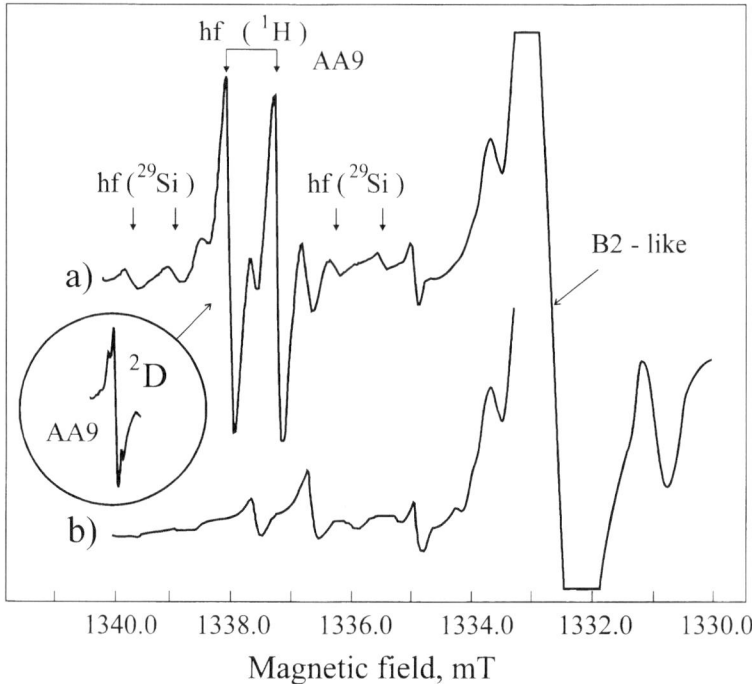

FIG. 1. EPR spectrum of the Si-AA9 center in Fz-Si implanted at 80 K with protons, observed at 77 K with the magnetic field $H\|\langle 100\rangle$, $v_0 = 37.50$ GHz. The spectra were recorded (a) under illumination of the sample and (b) in the dark. The inset shows the EPR spectrum of a sample implanted with deuterons.

Note that for high-purity silicon the EPR signals of the AA9 center are partly saturated at low-level (<1 mW) microwave power even at 77 K. Therefore, the dispersion mode is preferable for the measurement. However, an annealing (~10 min) of the sample in the dark at ~160 K greatly reduces the spin-lattice time relaxation of the AA9 center, and its EPR signal can be taken at 77 K in absorption mode. This effect appears to be due to transformation and annihilation of intrinsic radiation defects that are introduced by proton implantation at 77 K and partly annealed out near 160 K (Watkins, 1994). The AA9 spectrum itself disappears (irreversibly) after annealing the sample in the dark (i.e., H^+ in the BC site) at temperatures about 200 K (Gorelkinskii and Nevinnyi, 1991). Under illumination (i.e., H^0 in the BC site), the spectrum disappears (partly reversibly) at a temperature range from 100–110 K (Holm et al., 1991; Nielsen et al., 1994).

It can be seen from the angular dependence of the AA9 spectrum (Fig. 2) that each line represents a doublet splitting. When hydrogen atoms (nuclear spin $I_N = 1/2$) are replaced by deuterium atoms ($I_N = 1$), the doublet splitting changed into a poorly resolved triplet (inset to Fig. 1). These splittings are evidently due to hyperfine (hf) magnetic interaction from a single hydrogen (deuterium) atom that has been incorporated into the molecular structure of the AA9 defect.

There is a strong-intensity central group of Zeeman lines and weaker pairs of satellites on the spectrum AA9 (Fig. 1). These satellites have been identified as arising from a strong hf interaction with a ^{29}Si nucleus ($I_N = 1/2$, 4.7% naturally abundant) in the immediate vicinity of the defect. The intensity ratio of the hf line satellite to the corresponding Zeeman line is ~ 0.05 and thus clearly corresponds to a ^{29}Si atom in one of the two equivalent sites.

The spectrum of the AA9 center can be described as arising from an anisotropic center that has four equivalent orientations in the cubic silicon lattice by means of the spin Hamiltonian

$$H = \mu_B \mathbf{S g H} + \sum [\mathbf{SA} - (\mu_j/I_j)\mathbf{H}]\mathbf{I}_j \qquad (1)$$

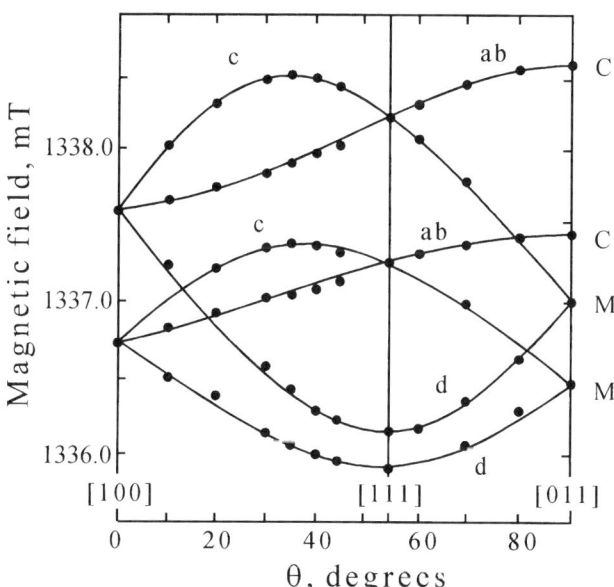

FIG. 2. Angular dependence of the Si-AA9 spectrum with **H** in the $(0\bar{1}1)$ plane at $T = 77$ K and $v_0 = 37.412$ GHz. The pairs of lines of the hf splitting from the ^{1}H are denoted by c, d, and ab. They also denote the corresponding equivalent site in Fig. 3. The intensities of the lines denoted C and M are equal at $\mathbf{H} \| [011]$. (Adapted from Gorelkinskii and Nevinnyi, 1987.)

with a spin $S = 1/2$. The first term in the Hamiltonian describes the electron Zeeman interaction; the second term describes the hyperfine and nuclear Zeeman interactions. The principal values of the **g**- and **A**-tensors and their axes are given in Table I for one of the equivalent center orientations in Fig. 3; the table entries were deduced from angular dependence of the spectrum (Fig. 2). The results show that the tensors of the hf magnetic interaction with the proton and ^{29}Si nucleus as well as the **g**-tensor of AA9 center are axially symmetric with respect to the $\langle 111 \rangle$ axis of the silicon lattice, clearly indicating trigonal (C_{3v}) symmetry of the defect.

Analysis of the AA9 **g**-tensor ($g_\parallel = 2.0011$, $g_\perp = 1.9983 \pm 0.0003$) shows that shift of its g-values from the free-electron value $g_0 = 2.0023$ correspond to the region $\Delta g_\parallel \leq 0$ and $\Delta g_\perp < 0$. According to the classification of paramagnetic defects in silicon on their g-shifts (Lee and Corbett, 1973; Siverts, 1983), the AA9 center can be applied to interstitial impurity defects.

b. *^1H and ^{29}Si Hyperfine Interactions*

Analysis of experimental data shows that the isotropic (Fermi-contact) hf interaction with the ^{1}H nucleus $[a = 1/3\mathrm{Tr}(A)]$ is $a = \pm 23.0$ MHz. The dipole part of the hf interaction, corresponding to $b = 1/3(A_\parallel - A_\perp)$, is

TABLE I

HYPERFINE PARAMETERS OF THE AA9 EPR CENTER AND THE ANOMALOUS MUONIUM (Mu*) IN SILICON

Center	Nucleus	$A_\parallel$ (MHz)	$A_\perp$ (MHz)	a (MHz)	b (MHz)	α_j^2	β_j^2	η_j^2
AA9[a]	^{1}H	6.2	31.4	23.0	8.4	0.016	—	—
	^{29}Si	139.0	72.9	95.3	22.3	0.11	0.89	0.21
Mu*[b]	Mu	5.5	29.3	21.37	7.93	0.0151	—	—
	^{29}Si	137.5	73.9	95.1	21.2	0.114	0.886	0.203
Theory[c]	Mu*(^{1}H)	4.05	25.35	11.15	7.10	0.008	—	—
F.P.C.	^{29}Si	128.0	63.5	85.0	91.5	0.10	0.90	0.20

Hyperfine constants of muonium are divided by the ratio of the free muonium and hydrogen hf constant (3.142) for convenience of comparison. Spin-density parameters (η, α, and β) were calculated using Eqs. (2) and (3).
[a]Gorelkinskii and Nevinnyi, 1987, 1991.
[b]Kiefl et al., 1988; Kiefl and Estle, 1991.
[c]Van de Walle and Blöchl, 1993.

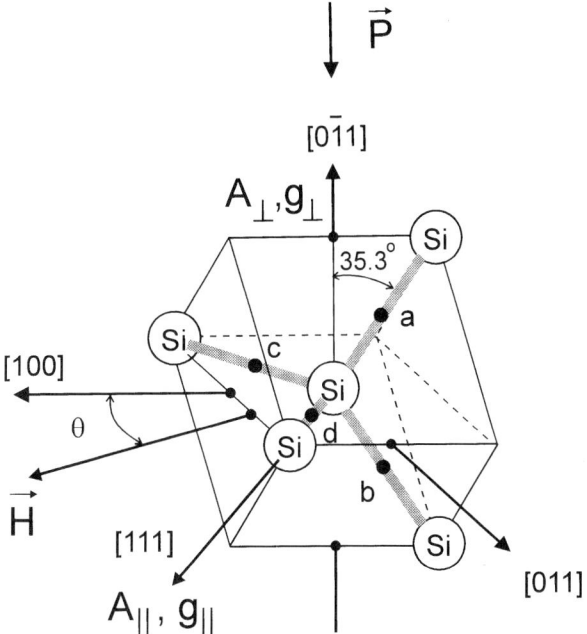

FIG. 3. Fragment of the silicon lattice. a, b, c, and d denote possible equivalent sites of hydrogen atoms in BC positions and their positions relative to the stress direction (P) in the $(0\bar{1}1)$ plane.

$b = \mp 8.4$ MHz; the sign is the opposite that of the isotropic part—a. Using the constant value of hf interaction (1420.4 MHz) of the free hydrogen atom, it was found that only ~1.6% (23.0 MHz/1420.4 MHz ≈ 0.016) of the resonant wave function of the defect belongs to the 1s state of the hydrogen atom (Gorelkinskii and Nevinnyi, 1987). The slight, anisotropic hf interaction from the proton may be due to a small admixture of an excited p-state of the hydrogen atom at the resonant wave function of the AA9 center. This situation cannot be analyzed correctly in terms of the LCAO approximation (Stallinga, 1994).

The ^{29}Si hf interaction of the AA9 center was analyzed, as has been done previously on other defects (Watkins, 1975; Lee et al., 1976), in the terms of one-electron linear-combination of ($3s$, $3p$) atomic orbital (LCAO) resonant wave function: $\psi = \Sigma \eta_j(\alpha_j \varphi_j^{3s} + \beta_j \varphi_j^{3p})$, with $\alpha_j^2 + \beta_j^2 = 1$ and $\Sigma_j \eta_j^2 = 1$, where η_j^2, α_j^2, and β_j^2 refer to a fraction of the total wave function of the s

the p wave function at the jth nuclear site, respectively. The isotropic (a) and anisotropic (b) parts of hf tensor can be described as

$$a_j = 8/3\pi g_0 g_N \mu_B \mu_N \alpha_j^2 \eta_j^2 |\varphi_{3s}(0)|_j^2 \qquad (2)$$

$$b_j = 2/5 g_0 g_N \mu_B \mu_N \beta_j^2 \eta_j^2 \langle r_{3p}^{-3}\rangle_j \qquad (3)$$

Using $g_n = -1.1106$, $|\varphi_{3s}(0)|^2 = 31.5 \times 10^{24}\,\text{cm}^{-3}$, and $\langle r_{3p}^{-3}\rangle = 16.1 \times 10^{24}\,\text{cm}^{-3}$ for the neutral ^{29}Si atom (Lee et al., 1976) and the experimental values of the isotropic part of the hf interaction, $a = 95.3$ MHz and dipole one $b = 22.3$ MHz, it was deduced that $\sim 21\%$ of the resonant wave function of the AA9 center belongs to each of two equivalent silicon atoms. The resonant wave function of the AA9 center is definitely of a p nature: 89% 3p and only 11% 3s. The excellent consistency between experimental data for distribution of the resonant wave function of the anomalous muonium (Mu*) and BC hydrogen (AA9) is shown in Table I.

c. *Comparison with µLSR and Theory Data*

A comparative analysis of both µLSR and EPR experimental data (see Table I) shows that the parameters of the AA9 center measured by EPR are in remarkable agreement with those of the Mu* state. Coincidence of hf constants for AA9 and Mu* (taking into account the ratio of their magnetic moments) and the symmetry of the hyperfine interactions for muon and proton suggest that two equivalent structures exist in the silicon lattice (Mu* and AA9) and differ only in the mass of the hydrogen atom (Kiefl and Estle, 1991; Gorelkinskii and Nevinnyi, 1991). The exact $\langle 111 \rangle$ axial symmetry of hf interaction tensors and the g-tensor, the presence of two equivalent Si atoms in the structure of center, and the fact that the signal intensity of AA9 is independent of the carbon and oxygen concentrations in the sample impose some severe limitations in the choice of a model for the AA9 (Mu*) center.

A comparison of µLCR, EPR data (Kiefl and Estle, 1991; Gorelkinskii and Nevinnyi, 1991) and theoretical calculations (Cox and Symons, 1986; Estle et al., 1987; Van de Walle, 1991) established that the anisotropic center AA9 (Mu*) is the ground state in Si and that the hydrogen (Mu) atom location lies near the center of a stretched Si—Si bond, i.e., in a bond-centered (BC) configuration. The paramagnetic electron is predominantly localized on the two equivalent nearest-neighbor Si atoms and occupies a nonbonding orbital with a node at the hydrogen position (Fig. 4) (Van de Walle, 1991).

The AA9 center has been identified by means of deep-level transient

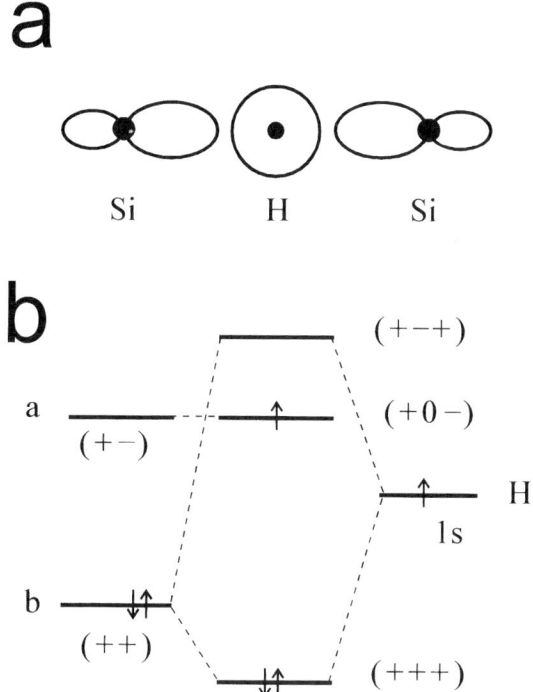

FIG. 4. Formation of a three-center bond for hydrogen in a bond-center site. (a) Schematic illustration of orbitals in the bond-centered configuration. (b) Corresponding energy levels obtained from simple molecular-bonding (or tight-binding) arguments for an elemental semiconductor. b and a indicate bonding and antibonding states. The paramagnetic electron occupies the antibonding state which has a node on the H atom at the bond center. (From Van de Walle, 1991.)

spectroscopy (DLTS) as a deep donor (denoted E3′) (Irmscher et al., 1984), with the ionization energy of the E3′ level measured to be $E_C - 0.16$ (± 0.01) eV (Holm et al., 1991). IR absorption studies (Stein, 1979) reveal an Si—H stretching mode at 1990 cm^{-1} in hydrogen-implanted silicon at 80 K, which is in agreement with the value of 1945 ± 100 cm^{-1} calculated for H^0 in the BC site (Van de Walle et al., 1988).

Theoretical calculations of the hf tensors in paramagnetic states of neutral hydrogen at the BC and T_d sites in silicon or diamond lattice also showed qualitative agreement with the EPR and μLCR data (Estreicher, 1987; Kuten et al., 1988; Deák et al., 1988).

Weak distinction between hf interaction values of hydrogen and muonium in BC configuration (see Table I) has been explained in a theoretical investigation based on cluster calculations in the Hartree-Fook approximation. Using methods of configuration interaction and density functional theory (Paschedag et al., 1993; Paschedag, 1996), different masses of the meson and proton result in different ground states of wave functions for these particles. This effect is independent of nucleus magnetic property and is called the *residual isotope effect* (Paschedag, 1996). The ratio of theoretical values of isotropic hf interaction of $a(Mu^*)/a(H) = 0.91$ is in excellent agreement with the same experimental data where $a(Mu^*)/a(H) = 0.93$, while calculated absolute values of the hf parameters of the BC configuration have shown only their qualitative agreement with the experimental data (Estreicher, 1987; Deák et al., 1988; Paschedag et al., 1993; Paschedag, 1996). However, recent theoretical investigations reported by Van de Walle and Blöchl (1993) demonstrate progress in calculation of the hf parameters of hydrogen and other impurities in semiconductors. They have used first-principle calculations based on spin-density functional theory and pseudopotentials. It is important to note that this approach allowed calculation of isotropic and anisotropic interactions of the resonate wave function for different defect systems (including hydrogen) in various semiconductors. In particular, their calculations of the ^{1}H (Mu) and ^{29}Si hf parameters for BC configuration in silicon are in satisfactory agreement with the experimental data (see Table I).

Previous theoretical and experimental investigations have shown that the position of atomic hydrogen in the silicon lattice (and other semiconductors) depends on its charge state (Pankove and Johnson, 1991; Myers et al., 1992). A number of theoretical calculations (Boucher and Deleo, 1994; Van de Walle et al., 1989; Chang and Chadi, 1989a, 1989b) agree that the minimum-energy configuration for H^0 and H^+ has the hydrogen at the BC site (with outward relaxation of neighboring silicon atoms), while the H^0 in T_d is a metastable state (Van de Walle et al., 1988). Most recently it has been shown experimentally (Johnson et al., 1994) that monatomic hydrogen in silicon has a large negative correlation energy, $U \approx -0.36\,\text{eV}$, with the donor level ($H^0$ in the BC) situated $\sim 0.20\,\text{eV}$ below the conducted band minimum and the acceptor level coincidentally located (essentially) at midgap. The activation energy for electron emission from $H^-(T_d)$ was shown to be $\sim 0.84\,\text{eV}$, which implies a configurational energy barrier of $\sim 0.3\,\text{eV}$ for electron capture by H^0. These experimental findings are in agreement with theoretical investigations (Van de Walle et al., 1989) and the few experimental reports (Irmscher et al., 1984; Holm et al., 1991) for the H(BC) structure.

Since $H^+(Mu^+)$ is not paramagnetic, alternative direct experimental evidence of existing H^+ in the BC configuration is most important. As expected, this evidence can be obtained from study of uniaxial stress alignment of the AA9 EPR center.

3. STRESS-INDUCED ALIGNMENT OF BOND-CENTERED HYDROGEN

The AA9 EPR spectrum is observed only under illumination of the sample by bandgap light. This fact may reflect either that H(BC) is in a diamagnetic state (H^+) or that the H^0 has trapped an additional electron from a radiation defect level and is in the tetrahedral site (H^-). In the latter case, photoionization of the H^- (and conversion to a neutral charge state) would cause a jump to a BC position, leading to observation of the AA9 spectrum. This transition cannot be excluded because the activation energy for electron emission from the $H^-(T_d)$ was measured to be $\sim 0.84\,eV$ (Johnson et al., 1994), i.e., within the bandgap of Si. In this respect it is important to note that a μLSR investigation (Kadono et al., 1994) has shown that the $Mu^0(T_d)$ center can be switched rapidly to the BC site [Mu^0(BC)] by illumination at low temperature (8 K).

For defects with lower than cubic symmetry, uniaxial stress can induce alignment at elevated temperatures. Therefore, if at nonparamagnetic states of the AA9 center a hydrogen atom is transferred to the cubic symmetry T_d position, a response on uniaxial stress must be absent, while for trigonal symmetry of H^+ or H^0 in the BC configuration we can expect a change to normal populations of equivalent defect orientations. As expected, these experiments can give unambiguous answers to these questions (Gorelkinskii and Nevinnyi, 1996).

a. Positive Charge State

After hydrogen implantation at 80 K, the sample was transferred without warmup into the EPR spectrometer cryostat; under bandgap light illumination, a strong AA9 spectrum appeared. As can be seen from the angular dependence of the AA9 spectrum (Fig. 2), each line of trigonal (C_{3v}) symmetry center is a doublet due to the hyperfine interaction from a single hydrogen atom ($I_N = 1/2$). The defect has four equivalent orientations in the cubic silicon lattice; under normal equilibrium conditions and for a magnetic field direction in the ($1\bar{1}0$) plane, three pairs of spectral lines are found with EPR with a relative intensity of 1:1:2. Possible equivalent sites of the atomic hydrogen at BC positions in the silicon lattice are shown in Fig. 3.

For $H \parallel [011]$ in the ($0\bar{1}1$) plane, two pairs of spectral lines with equal intensity are observed (see Figs. 2 and 5b). Uniaxial stress of 200 MPa was applied to a [$0\bar{1}1$] axis (see Fig. 3) of the sample (in the dark) at 170 K for 5 min; the sample then was cooled under stress to 77 K for 50 min. The stress was then removed and the sample placed in the EPR cavity for observation at 77 K.

It is important to note that stressing of the sample was performed for the nonparamagnetic state of the AA9 center, i.e., for H^+ or H^- states. For H^- or H^0 in the (cubic symmetry) T_d site, there should be no response of the AA9 spectrum to uniaxial stress. For H^+ in the (trigonal symmetry) BC position, we do expect a stress effect, as just discussed.

Illumination of the sample produced the AA9 EPR spectrum; a preferential alignment of the defect due to uniaxial stress was found to have been frozen in at 77 K for both the H^+ and H^0 states, as measured by the relative intensities of corresponding spectral lines of the AA9 spectrum (Fig. 5c). This preferential alignment of the nonparamagnetic state of the AA9 center is experimental evidence that H^+ in the BC position indeed exists (Gorelkinskii and Nevinnyi, 1996). The defect whose $\langle 111 \rangle$ axes were perpendicular to the stress directions increased in intensity, while the defects whose $\langle 111 \rangle$ axis was 35.3° off the [0$\bar{1}$1] stress directions decreased (see Figs. 3 and 5). This indicates that the energy of the defect is increased if the defect is compressed along its $\langle 111 \rangle$ axis. The defect, therefore, tends to align with its $\langle 111 \rangle$ axis perpendicular to the compressed direction.

The recovery from the stress-induced alignment versus 10 min of isochronal annealing in the dark is shown in Fig. 6 along with irreversible annealing of the AA9 center. The recovery kinetics from the alignment also were studied by a series of isothermal anneals. The results are shown in Fig. 7. The activation energy required for the reorientation process was determined to be 0.43 ± 0.02 eV with a preexponential factor of $1/\tau = 2.3 \times 10^{12}$ s^{-1}. The preexponential factor has the magnitude for vibrational frequency of $\sim kT/h$, as expected for a local rearrangement process. It is important to note that the activation energy for reorientation is very close to that for H^+ migration (long-range diffusion) during annealing. DLTS investigation gives an activation energy value for the disappearance of the H^+ of $E_a = 0.44 \pm 0.01$ eV (Holm et al., 1991). In addition, nonreversible annealing of the AA9 EPR center (in the dark) is characterized by $E_a = 0.48 \pm 0.04$ eV (Gorelkinskii and Nevinnyi, 1991). However, the preexponential factor in this case is $\sim 10^4 - 10^5$ times smaller than for reorientation. From these experimental findings it was concluded that the energy barrier for the reorientation of H^+ from one to another equivalent position is very close to that for H^+ long-range migration. This is reasonable for monatomic interstitial defects (e.g., interstitial boron; Watkins, 1975) when the same limiting mechanism can be involved on both anneal and reorientation.

As done previously for other trigonal symmetry defects, stress alignment of the H^+ (BC) center can be analyzed in terms of piezospectroscopic tensors **B** (Lee and Corbett, 1974; Watkins, 1975). Uniaxial stress of ~ 170 MPa was applied along the [0$\bar{1}$1] axis (see Fig. 3) of the sample for

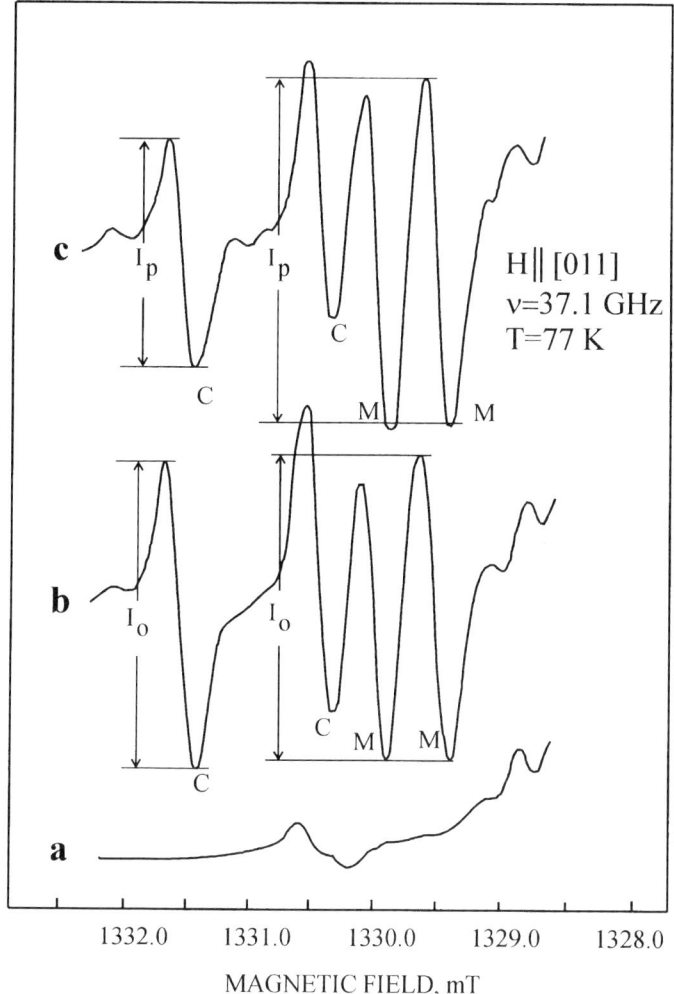

FIG. 5. EPR spectrum of the Si-AA9 center. $H\|[001]$ in $(0\bar{1}1)$ plane. $\nu_0 = 37.233$ MHz. (a) EPR spectrum without illumination of the sample. (b) EPR spectrum of the AA9 center under illumination of the sample. The intensities of the C and M lines, corresponding to two independent equivalent orientations of the defect, are equal ($I_{oc} \approx I_{om}$). (c) EPR spectrum of the AA9 center under illumination after applying $[0\bar{1}1]$ stress in the dark (~ 200 MPa, $T = 135$ K). The alignment corresponds to ($I_{om}/I_{oc} \approx 1.5$). (Adapted from Gorelkinskii and Nevinnyi, 1996.)

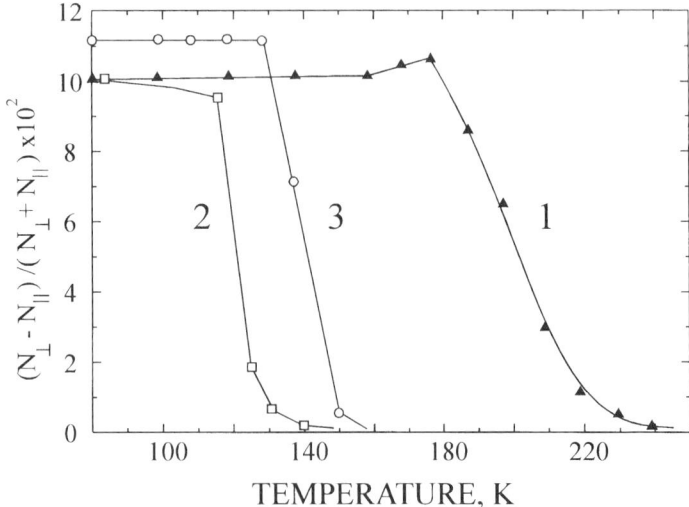

FIG. 6. A 10-min isochronal annealing of the AA9 center (relative intensity): (1) in the dark (H^+ in BC); (2) under illumination of the sample (H^0 in BC). Curve 3 shows the recovery after a 10-min isochronal annealing from the stress-induced alignment of the H^+ in the BC form. The alignment was achieved initially by applying a 150-MPa stress at 170 K (in the dark) and cooling to 77 K under stress. [Curves (1) and (3) adapted from Gorelkinskii and Nevinnyi, 1991, 1996.]

2 hours at 140 K; the crystal was then rapidly cooled to 77 K with stress applied. As can be estimated from the data in Fig. 7, this time period at a temperature of 140 K is sufficient to establish an equilibrium alignment. The energy of the defect in an applied strain field is given by

$$E = \sum_{ij} B_{ij}\varepsilon_{ij} \qquad (4)$$

where ε_{ij} are the strain tensor components and B_{ij} are the components of a symmetric second-rank piezospectroscopic tensor **B** (Kaplyanskii, 1964). The number of the independent components (B_{ij}) depends on the symmetry of the center. For trigonal symmetry, the matrix **B** has only two independent parameters having the form (Watkins, 1975)

$$B = \begin{bmatrix} -B + B_0 & 0 & 0 \\ 0 & -B + B_0 & 0 \\ 0 & 0 & +2B + B_0 \end{bmatrix} \qquad (5)$$

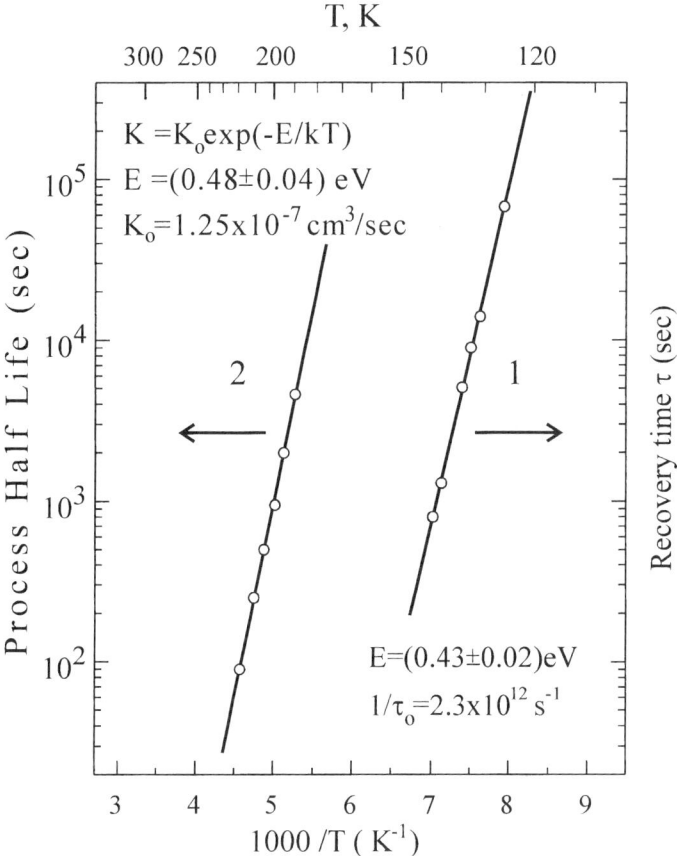

FIG. 7. Plots of recovery time of the AA9 EPR spectrum after ⟨011⟩ stress-induced alignment (1) and lifetime of the H^+ in the BC center versus temperature (2). (Adapted from Gorelkinskii and Nevinnyi, 1991, 1996.)

We equate the alignment in Eq. (4) to a Boltzmann distribution

$$N_\perp/N_\| = \exp[-(E_\perp - E_\|)/kT] \tag{6}$$

where $E_\|$ and $E_\perp$ are given by Eq. (4) for each defect orientation, and T is the temperature of the equilibrium alignment.

For stress $\sigma(0\bar{1}1)$ along the $[0\bar{1}1]$ direction, Eqs. (4) and (5) lead to

$$B = [E_\perp - E_\|]/S_{44}\sigma(0\bar{1}1) \tag{7}$$

where S_{44} is the shear elastic modulus of the silicon lattice ($S_{44} = 12.56 \times 10^{-13}$/MPa) (Watkins, 1975). With $\sigma = 170$ MPa, $N_\perp/N_\parallel \approx 1.4$, and $T = 140$ K, we get

$$B \approx -2.0 \, \text{eV/(unit strain)}$$

The negative sign means that the defect energy increases if compressed along its $\langle 111 \rangle$ trigonal axis, i.e., along Si—H—Si. This is in perfect agreement with well-known theoretical calculations, which predicted the minimum-energy configuration for both H^0 and H^+ in the $\langle 111 \rangle$ bond-centered position only with outward silicon atom relaxation (Van de Walle et al., 1989). We note again that alignment by stress (in the dark) of the nonparamagnetic state of the AA9 center is strong experimental evidence that the H^+ is situated in the BC position on the $\langle 111 \rangle$ axis, while the negative sign of **B**-tensor component is confirmation that Si atoms relax outward from hydrogen.

b. *Neutral Charge State*

The neutral state of H^0(BC), i.e., the AA9 EPR center, also was studied under stress (200 MPa) in the temperature range from 130–77 K (under illumination). No detectable alignment of the H^0(BC) state was observed for temperatures between 100 and 130 K. Previous DLTS studies (Holm et al., 1991) have shown that the E3' center (H^0 in BC) is not stable at temperatures above ~ 110 K. On the other hand, we noted previously that significant alignment of H^0 in the BC remained at 77 K when stress was applied in the dark (i.e., to H^+) at elevated temperature (Fig. 5c). It is important to emphasize that irreversible annealing of the AA9 center (after its alignment in the dark) in the temperature range 110–130 K leads to a decrease of its intensity; however, the degree of alignment is unaffected (Gorelkinskii and Nevinnyi, 1996). Furthermore, the AA9 EPR spectrum at sample temperatures of ~ 130 K indicates that an interpretation in terms of four equivalent sites remains valid; i.e., jumping of the hydrogen atom between these equivalent sites is absent even during annealing of the AA9 center. This finding is strong evidence that, in contrast to the behavior of H^+(BC), H^0(BC) disappears by direct transition to a T_d site without a stage of atomic reorientation between the equivalent sites.

4. THERMALLY ACTIVATED ANNEALING

Studying the annealing kinetics of the atomic hydrogen EPR signal (AA9 center) can give important additional information on the diffusion process of atomic hydrogen in the silicon lattice at a low temperature. Isochronal annealing of the sample for 15 min was performed in the dark, i.e., for H^+ in the BC site. After each annealing period, the AA9 spectrum was regenerated by illumination of the sample at 77 K. The data of isochronal annealing, shown in Fig. 6 (curve 1), demonstrate that H^+ in the BC position appears to be thermally stable for temperatures up to at least 200 K. This is in agreement with the results obtained from isochronal annealing of the ionized E3' donor inferred from DLTS (Holm et al., 1991).

The disappearance of H^+(BC) was studied by a series of isothermal anneals of the sample in the dark (Gorelkinskii and Nevinnyi, 1991). After implantation of hydrogen at 80 K, all samples were first annealed at 195 K for 40 min. This anneal removes some mobile intrinsic defects induced by hydrogen ion implantation, in particular the single vacancy (Watkins, 1975). Other radiation defects such as divacancy or tetravacancy are thermally stable (and presumably immobile) even above room temperature (Watkins, 1994). We note that these experiments used high-purity ($\rho \approx 3$–$10\,\text{k}\Omega\cdot\text{cm}$) silicon samples with low content of oxygen and carbon impurities.

Analysis of isothermal annealing indicates that the disappearance of H^+ in BC is characterized by simple second-order reaction kinetics, $d[N]/dt = -K[N]^2$, which can be solved to give $[N_0 - N]/N_0 N = -Kt$, where N_0 and N are initial and current concentrations of the AA9 center, respectively, and K is a preexponential factor $[K = K_0 \exp(-E/kT)]$. Figure 8 shows the results of isothermal anneals of the AA9 center (for its nonparamagnetic state) for a set of similarly prepared samples. Analysis of the isothermal annealing data for the temperature range 190–220 K indicates that the disappearance of the nonparamagnetic state of the AA9 (H^+ in the BC) is characterized by an activation energy $E_a = 0.48 \pm 0.04$ eV with a second-order preexponential factor $K = (1.25 \pm 2.5) \times 10^{-7}\,\text{cm}^3/\text{sec}$. The results (along with recovery time from stress alignment of the H^+ in the BC) are shown in Fig. 7. This value, within experimental error, agrees with the value of 0.44 ± 0.01 eV obtained from the ionized E3' donor annealing study by the DLTS method (Irmscher et al., 1984; Holm et al., 1991). Recent careful rf μSR investigations give the activation energy of 0.38 ± 0.6 eV for transition from Mu^+ in the BC to Mu^0 in T_d (Kreitzman et al., 1995). Within the experimental error, all these measurements (EPR, DLTS, and rf μSR) come close to overlapping. The lower value of rf μSR measurements is consistent with the expected increase in barrier height for H motions relative to barrier for the lighter Mu as a consequence of their different zero-point energies (Kreitzman et al., 1995).

It is important to note that EPR experiments were performed on hydrogen-implanted samples with hydrogen concentrations of $\sim (0.5$–$5) \times 10^{16}\,\text{cm}^{-3}$, which is five orders of magnitude greater than the hydrogen

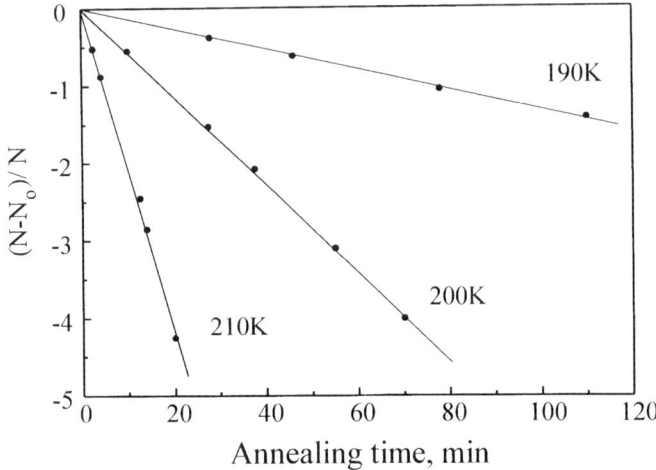

FIG. 8. Isothermal annealing behavior (in the dark) of the AA9 center. Plots demonstrate simple second-order kinetics. N_0 and N are initial and current intensities of the AA9 spectrum, respectively.

concentration in the DLTS experiments (Holm et al., 1991). Moreover, a high concentration of defect states (at least two orders of magnitude more than the concentration of hydrogen atoms) is introduced into the forbidden gap of silicon by proton implantation (Gorelkinskii et al., 1976). Some of this implanted hydrogen will interact with intrinsic defects to form complexes. Immediately after implantation, some of the hydrogen may be situated in the $H^-(T_d)$ configuration, having trapped a second electron from another defect, and thus hydrogen in high concentrations can markedly change the Fermi level in the gap. Approximately 20% of the implanted hydrogen belongs to the AA9 EPR spectrum. Therefore, the annealing kinetics of $H^+(BC)$ from EPR studies strongly differ from the annealing kinetics from DLTS studies, where the disappearance process of the ionized E3' donor is characterized by first-order reaction kinetics and a preexponential factor of $(1.3 \pm 0.6) \times 10^8 \sec^{-1}$ (Holm et al., 1991). If we roughly extrapolate the second-order preexponential factor (extracted from EPR study) to obtain a first-order one, by multiplying the second-order factor by a typical concentration of $\sim (5-8) \times 10^{15} cm^{-3}$ for the AA9 center, a value $K_0 \approx 10^9 \sec^{-2}$ close to the prefactor from DLTS measurements is obtained.

Although one can conceive of various mechanisms for the thermally activated disappearance of atomic hydrogen (AA9 center), the kinetic equation for diamagnetic molecular hydrogen ($d[H]/dt = -K[H]^2$) has been found to be consistent with experimental data. The absolute value of the atomic hydrogen (AA9) concentration cannot be determined exactly due to the uncertainty of spatial distribution in the sample. Nonetheless, a value for the diffusion constant has been roughly estimated at $D_0 \approx 10^{-3} \div 10^{-2}$ cm^2/sec for the temperature range 190–220 K (Gorelkinskii and Nevinnyi, 1991). Unexpectedly, our measurements of activation energy (Fig. 7) and diffusion constant at low temperatures are very close to earlier experimental data on hydrogen diffusion at high temperatures (970–1200°C) by van Wieringen and Warmoltz (1956), to theoretical predictions (Deák et al., 1988; Boucher and Deleo, 1994) for the value of the activation energy of hydrogen diffusion via the BC sites, and to recent experimental data for diffusion of hydrogen in the room-temperature region (Seager and Anderson, 1988, 1990).

Under illumination, annealing of AA9 (H^0 in the BC) occurs in the temperature region of 110–130 K (Fig. 6). The rate of the EPR signal intensity loss strongly depends on illumination intensity and hydrogen implant dose. Therefore, a quantitative measurement of this process by means of EPR was not performed. DLTS studies of this process have, however, allowed extraction of the activation energy for annealing of the neutral E3′ donor; a value 0.293 $\pm$ 0.003 eV was obtained, along with a preexponential factor $v_H = (3.0 \pm 0.5) \times 10^{12}$ sec^{-1} (Holm et al., 1991). The activation energy is very close to that obtained in recent rf μSR data, where the transition of Mu^0(BC) to Mu$^-$(T$_d$) is characterized by an activation energy of 0.34 $\pm$ 0.01 eV (Kreitzman et al., 1995). Presumably, the difference in activation energy may be due to the different zero-point energies of Mu and proton.

III. EPR of Hydrogen-Related Complexes in Silicon

1. HYDROGEN-INTRINSIC DEFECT COMPLEXES

A number of intrinsic defects are created in the silicon lattice on implantation of hydrogen at room temperature. There are a number of paramagnetic defects in proton-implanted silicon that are also found for neutron (Brower, 1976; Lee et al., 1972) and electron (Watkins, 1994) irradiation. However, in a hydrogen-implanted layer, the EPR is dominated by a spectrum with a trigonally symmetric **g**-tensor; this was denoted initially as the S1 spectrum (Lütgemeier and Schnitzke, 1967). The S1 **g**-tensor is close to that of B2 (Daly, 1969), which is found after low-dose, low-energy ($\sim$100 keV) nitrogen or phosphor ion implantation. The S1(B2) center is stable up to $\sim$300°C. Kleinhens et al. (1979) have suggested that the S1(B2) spectrum arises from the negative charge state of a dangling bond in a vacancy cluster; a strong hf interaction from a single ^{29}Si isotope was observed: $A_\parallel[111] = 419.2$ MHz and $A_\perp = 259.6$ MHz at 77 K. It is important to note that an S1(B2) spectrum was not created after helium implantation in pure Fz or Cz silicon (Gorelkinskii et al., 1987). Later, using a Q-band spectrometer (at 77 K), it was found (Gorelkinskii and Nevinnyi, 1991) that the S1(B2) spectrum arises from a sum of different defects with slightly different g-tensors of C_{1h} and C_{3v} symmetry; the differences between the g-values is within ± 0.0003. Two different spectra have been found in Fz silicon and five in a Cz sample.

Recent EPR studies (Stallinga et al., 1993, 1994a) of hydrogen-implanted pure Fz silicon ($\rho \approx 3$ k$\Omega \cdot$cm) at a sample temperature of ~ 6 K have revealed a dominant EPR spectrum with clearly resolved hf splitting from two nearly equivalent protons. The spectrum, labeled NL52, is stable up to $\sim 300°$C, has a trigonal symmetry g-tensor with $g_\parallel = 2.0007$, and $g_\perp = 2.0095$) (± 0.0002), and has hf interaction from two nearly equivalent protons with the same symmetry (with $A_\parallel = 12.1$ MHz and $A_\perp = -5.0$ MHz; Stallinga et al., 1993). This hf splitting is not completely resolved when hydrogen atoms are replaced by deuterium, in agreement with the nuclear moment ratio $\mu_D/\mu_H \approx 0.16$. A strong hf splitting, apparently due to a single ^{29}Si nucleus, also has an A-tensor with trigonal symmetry. This would be typical for dangling bonds on vacancy-type defects; the principal values for the tensor are $A_\parallel = 436$ MHz and $A_\perp = 246$ MHz. A possible model for NL52 is a hydrogen quasi-molecule situated at a vacancy-type defect.

A single vacancy-hydrogen complex (VH)^0 is also a fundamental defect in silicon that has been discovered by EPR (Nielsen et al., 1997). The (VH)^0 EPR spectrum derives from the neutral charge state of the defect that forms in Fz silicon on implantation of hydrogen (deuterium) at room temperature and is stable up to $\sim 250°$C. The (VH)^0 EPR center displays a g-tensor of C_{1h} symmetry at sample temperatures below 65 K and C_{3v} symmetry above 100 K. The hf interaction with protons is very weak ($Ax = -3.3$ MHz, $Ay = -4.6$ MHz, $Az = 8.5$ MHz), but principal values have been extracted by simple EPR. It is interesting to note that the parameters of the g-tensor and the hf parameters with a single ^{29}Si nucleus of the (VH)^0 center are very similar to those of a single vacancy-phosphorus pair, i.e., to the G8 EPR center (Watkins and Corbett, 1964). From these experiments it was concluded that electronic properties of the (VH)^0 defect are determined almost completely by the silicon dangling bond and that the Si—H fragment in the vacancy may be regarded as a "pseudo-group V impurity" (Nielsen et al., 1997).

An important conclusion of these investigations is also the direct evidence that a single hydrogen trapped by a vacancy elevates the thermal stability of the defect significantly (up to 250°C).

3 HYDROGEN AND HYDROGEN-RELATED DEFECTS IN CRYSTALLINE SILICON 47

It should be noted that only three EPR defects, the AA9, NL52, and $(VH)^0$, are so far the hydrogen–intrinsic defect complexes that clearly reveal an hf interaction from proton(s); however, a number of IR bands associated with Si—H bonds have been identified in hydrogen-implanted pure Fz silicon (Stein, 1975; Gerasimenko *et al.*, 1978; Mukashev *et al.*, 1989; Tokmoldin *et al.*, 1990; Nielsen *et al.*, 1994).

2. EPR OF PLATINUM-HYDROGEN AND SULFUR-HYDROGEN COMPLEXES

In the last several years, a method of introducing hydrogen into silicon by means of heat treatments at elevated temperature ($\sim 1250°C$) in the presence of hydrogen gas or water vapor has been used widely (Velorisoa *et al.*, 1991; McQuaid *et al.*, 1991; Lightowlers, 1995). A concentration of hydrogen-related species of about 1×10^{16} cm^{-3} or greater can be reached in this way.

The involvement of one or more hydrogen atoms in the microscopic structure of defects has been established through spectroscopic studies such as IR local vibrational mode spectroscopy and luminescence (Velorisoa *et al.*, 1991; McQuaid *et al.*, 1991; Lightowlers, 1995). In addition, some remarkable EPR and ENDOR investigations with direct identification of hydrogen atoms in defect structure have been carried out. We shall take a quick look at properties of these complexes.

The PtH_2 complex has been found in Pt-doped *n*-type silicon after high-temperature H-indiffusion treatment (William *et al.*, 1994; Uftring *et al.*, 1995). The complex introduces two levels in the Si bandgap: the first level lies near $E_c - 0.1$ eV, and the second is estimated to lie near midgap. There are three pairs of H vibrational bands, one pair for each of the charge states of the PtH_2 center. It has been argued that the PtH_2 center is a double acceptor and that the charge states are PtH_2^{2-}, PtH_2^- (paramagnetic), and PtH_2^0. The EPR spectrum of this defect has an effective spin $S = 1/2$ and a **g**-tensor with C_{2v} symmetry. Hyperfine interactions with a single ^{195}Pt nucleus ($I = 1/2$, 33% abundant) and two equivalent protons ($I_N = 1/2$, 100% abundant) were detected. The value of the isotropic part of the proton hf interaction is $a = 8.7$ MHz. Consequently, only $\sim 0.6\%$ of the resonant wave function is located on the hydrogen atom. The anisotropic hyperfine interaction, $b = 1.08$ MHz, can be used to estimate the Pt–H distance. Assuming an electron concentrated at the Pt atom, this *b* value implies $r \approx 4.2$ Å. A model for this defect has been suggested (Uftring *et al.*, 1995) in which the Pt atom is displaced off-center toward two of its Si neighbors, and the remaining Si bonds are terminated by hydrogen atoms at bonding or back-bonding sites. Thus, using EPR and IR absorption data, as well as the response of the defect to uniaxial stress, the exact molecular model and electronic structure of the PtH_2 center has been established.

The second Pt-hydrogen-related EPR spectrum, labeled NL53 (Höhne *et al.*, 1994a, 1994b), has an effective spin $S = 1/2$ and trigonal symmetry **g**-tensor. It has been identified as three substitutional platinum atoms with hydrogen atoms passivating dangling bonds on the nearest silicon neighbors.

Two distinct EPR spectra, NL54 and NL55, are detected in hydrogenated sulfur-doped silicon for both Fz or Cz material (Zevenbergen *et al.*, 1995a, 1995b). The spectra displayed a clearly resolved hf interaction of the proton and of the ^{33}S nucleus (nuclear spin $I_N = 3/2$). Additional evidence of the involvement of hydrogen in these defects was found by deuterium doping of the sample, along with exact ENDOR measurements of hf interaction constants with the proton or deuteron. The measurements indicate that both centers are sulfur-hydrogen pairs and have **g**- and **A**-tensors with $\langle 111 \rangle$ axial symmetry. Evidently, the sulfur atom occupies a substitutional position in the lattice; the hydrogen atom is situated along a $\langle 111 \rangle$ axis of the crystal with respect to sulfur. Three possible positions for the hydrogen nucleus would be (1) the bond center, (2) antibonded on a sulfur nucleus, and (3) antibonded on a silicon nucleus. Both NL54 and NL55 centers are identified as singly passivated substitutional sulfur donors, i.e., $(S-H)^0$ pairs (Zevenbergen *et al.*, 1995a, 1995b, 1996). Most recently, using the local spin-density pseudopotential method, calculations have been carried out on 87-atom clusters containing S and H (Torres *et al.*, 1996). Two nearly degenerate configurations were found, both with S in a substitutional site. In both configurations, the hydrogen atom is located at a BC site (or an antibonding position) with one of the four Si neighbors of sulfur. In both configurations, the H atom bonds strongly with an Si neighbor to S, and both defects give a deep donor level with almost the same energy. Localization of resonant wave function on the proton corresponds to 0.3% and 0.1% of that free hydrogen for BC and AB configurations, respectively (Torres *et al.*, 1996). Results of the calculations are in good agreement with the experimental value, at about 0.4% (Zevenbergen *et al.*, 1996). The detailed models and the exact identification of the origin of the apparent distinction between the two centers remain to be seen.

3. ENDOR SPECTRA OF Si—H BONDS AT THE (111) Si SURFACE

ENDOR investigations of the P_b center in porous silicon have revealed hf interactions with the nuclei of hydrogen atoms that passivate the dangling bonds at the (111) Si—SiO_2 interface (Bratus *et al.*, 1994; Hofmann *et al.*,

1995). ENDOR measurements of the P_b center reveal a broad line corresponding to the Larmor frequency of the proton with typical width at half maximum equal to 250 kHz (Bratus et al., 1994, 1995). Under appropriate experimental conditions, a set of lines, labeled the N-spectrum, with a width about 40 kHz has been obtained. The lines of the N-spectrum are apparently symmetric about the Larmor frequency of the proton and depend on the orientation of the magnetic field. After annealing the porous sample at 550°C (when hydrogen effuses from SiH species), the ENDOR signal is decreased significantly and the N-spectrum is no longer detectable, while the EPR signal of the P_b center is increased. The shape of the broad line remains unchanged up to 850°C, indicating near independence of the nearby hydrogen environment of the P_b center (Hofmann et al., 1995; Bratus et al., 1995). Analysis of the total ENDOR spectrum has shown that the N-spectrum is caused by hf interaction with hydrogen nuclei of different nearest-neighbor shells; the broad line is due to the more distant shells (Bratus et al., 1995). The constants of isotropic and anisotropic hf interaction have been estimated for these shells. ENDOR results are in good agreement with the current model for the P_b center where the nearest-neighbor shell is created by oxygen atoms (Caplan et al., 1979; Brower, 1983).

One important conclusion of these investigations is confirmation that the ENDOR spectrum of the impurity atom in silicon (in this case of hydrogen) can be clearly detected, while the impurity atom is not involved into the structure of paramagnetic defect.

4. ENDOR OF HYDROGEN IN THE OXYGEN THERMAL DONOR (NL10)

Heat treatment of Cz silicon in the temperature range 350–500°C leads to oxygen interstitial diffusion and formation of oxygen agglomerates that act as donor defects (Jones, 1996). The oxygen donor acts as a near-ideal, effective-mass helium-like double donor, which produces two states in the silicon gap. DLTS (Benton et al., 1983; Kimerling, 1986) and IR absorption (Kimerling, 1986; Stalova and Lee, 1986; Binns et al., 1996; Wagner et al., 1987; Markevich et al., 1994) data confirm the presence of two states in the silicon gap, i.e., a deep level, $E_C - 0.15$ eV and a shallow one, $E_C - (0.035 \div 0.07)$ eV. Two distinct EPR spectra appear in heat-treated Cz silicon. The EPR spectrum labeled NL8 is attributed to a singly ionized thermal donor (TDD$^+$) with a spin $S = 1/2$ and C_{2v} symmetry of the **g**-tensor (Muller et al., 1978; Michel et al., 1986). It is important to note that the **g**-tensor of the NL8 is slightly changed during annealing, which reflects unresolved contributions of different donors (Muller et al., 1978; Trombetta et al., 1997). The

NL10 spectrum, created at longer annealing times than NL8 and its associated electrical level, is very similar to that of the shallow phosphorus donor (Bekman et al., 1988). ENDOR studies have suggested that the cores of both the NL8 and NL10 centers are very similar, having almost identical bonded oxygen atoms and identical symmetry of the ^{29}Si hf interaction. Since the NL8 center has been reliably identified as a positively charged double-donor (TDD$^+$) state (Wagner et al., 1987), its neutral shallow state should have a spin $S = 0$ or 1. The NL10 spectrum clearly has a spin $S = 1/2$ (Gregorkiewicz et al., 1988). The spectrum is poorly resolved, presenting a practically isotropic single line in the X-band (Meilwes et al., 1994), although EPR in the K-band can be described as arising from a C_{2V} symmetry defect (Gregorkiewicz et al., 1987).

Recently, taking into consideration the facts that hydrogen can be introduced easily in silicon as a contaminant during crystal growth (Pearton et al., 1987, 1992; Pankove and Johnson, 1991), and in view of the significant hydrogen enhancement of oxygen diffusion and of the rate of TDD formation (McQuaid et al., 1991, 1995), it has been decided to investigate the possible involvement of hydrogen in the structure of NL10 (Martynov et al., 1995a, 1995b).

In a sample with a *strong* NL10 EPR signal, the ENDOR hf spectrum from a nearby proton was indeed detected. It is important to emphasize that hydrogen has not been introduced intentionally into this material. In subsequent investigations, four kinds of Cz samples (doped with either B, Al, P, or N) have been subjected to heat treatment in the presence of water vapor at 1250°C, followed by a quench to room temperature. The samples were then annealed at 470°C to create the thermal donor (Martynov et al., 1995a). In the samples diffused with hydrogen, production of the NL10 spectrum was significantly enhanced, while the formation of NL8 was suppressed. In all the samples diffused with hydrogen, an ENDOR hf spectrum of the proton has been observed for the NL10 center. However, not even a trace of hydrogen on the NL8 ENDOR spectrum has been detected. Therefore, it has been concluded that hydrogen is not involved in the NL8 defect formation (Martynov et al., 1995a, 1995b). The localization of the resonant wave function on protons of the NL10 center has been estimated to be $\eta \approx 0.006\%$ (Martynov et al., 1995a). This value is comparable with η on oxygen of NL10, but it is about 10 times less than the total localization of the resonant wave function on first-shell Si (which was previously extracted from ENDOR of ^{29}Si; Bekman et al., 1989). The analysis of the hf interaction with protons (deuterons) and Al of the NL10 confirms the multispecies character of this defect (Michel et al., 1989). Although the presence of oxygen is a principal condition for creation of both NL10 and NL8, for paramagnetic states of NL10, impurities such as hydrogen or Al are required. Thus the NL10 center has been identified as a neutral thermal double donor with one of its two electrons passivated by hydrogen. The most recent investigations of IR electronic spectra from annealed hydrogenated Cz samples with different concentrations of the NL10 have shown a good correlation between EPR and IR data. The IR measurements have confirmed that the NL10 center can indeed be a TDD center partially passivated by hydrogen (deuterium), with an ionization energy of ~ 36 meV (Newman et al., 1996).

The authors suggest (Martynov et al., 1995a, 1995b) that the long-standing problem of the interrelation of the NL8 and NL10 EPR spectra and the thermal double donor in silicon appears to be resolved this way. A second important conclusion of these ENDOR investigations (Martynov et al., 1995a) is the direct evidence that hydrogen species as a rule are present in concentrations in the range of 10^{14}–10^{15} cm^{-3} in commercially available silicon.

IV. Hydrogen-Induced Effects in Silicon

1. HYDROGEN-ASSOCIATED SHALLOW DONORS IN HYDROGEN-IMPLANTED SILICON

   a. Conditions of the Donor Formation

The hydrogen-associated shallow donor (HSD) is a fundamental species of hydrogen-related defects (Schwuttke et al., 1970; Ohmura et al., 1972, 1973; Gorelkinskii et al., 1974; Gorelkinskii and Nevinnyi, 1983a, 1983b) known to arise in proton-implanted silicon, in neutron-irradiated Si grown in H_2 atmosphere (Wang and Lin, 1984; Meng, 1991), or in H plasma-treated silicon (Hartung and Weber, 1993, 1995). The HSD dramatically changes the electrical properties of the crystal, although the electronic and molecular structure of HSD continues to be a mystery. The HSD is produced in silicon by hydrogen implantation at room temperature and subsequent annealing of the sample in the temperature range of 300–500°C for 10–20 min. It is important to note that the HSD center is not created if hydrogen implantation is performed at sample temperatures of 300–350°C (Gorelkinskii et al., 1980). The HSD formation rate does not depend strongly on the ion energy, and there is no dramatic difference in the occurrence of these centers between Cz and high-purity Fz silicon (whose oxygen contents differ greatly (Ohmura et al., 1972, 1973; Gorelkinskii et al., 1974; Gorelkinskii and Nevinnyi, 1983a, 1983b). Studies of the depth profile of HSD centers have shown that HSDs are localized only at the end of the proton range (Ohmura et al., 1973; Gorelkinskii et al., 1974, 1980). Initial measurements of the HSD level in silicon performed by Hall effect measurements on a sample implanted with 100- to 300-keV protons have shown a donor level of $E_C - 26$ meV, where E_C is the energy at the bottom of the conduction band (Ohmura et al., 1972). High-energy (7–30 meV) proton implantation also produced the HSD center whose concentration increased proportional to the implantation dose up to $\phi \approx 1 \times 10^{17}$ H/cm^2 (10^{18}–10^{19} H/cm^3), beyond which overlapping of defect clusters takes place. Two donor levels with ionization energies of $E_{C1} - 0.035$ eV and $E_{C2} - 0.16$ eV were found in Hall effect measurements of silicon implanted at high energies of protons (Gorelkinskii and Nevinnyi, 1983a). A shallow donor with ionization energy of $E \approx 0.037$ eV also was obtained after annealing of neutron transmutation-doped Fz silicon grown in hydrogen atmosphere (Wang and Lin, 1984; Meng, 1991), as well as after hydrogen plasma treatment of the neutron-irradiated Fz silicon (Hartung and Weber, 1993, 1995).

Comparative studies using hydrogen and helium implantation (Gorelkinskii et al., 1974; Gorelkinskii and Nevinnyi, 1987) indicate that shallow donors are absent in helium-implanted silicon and EPR spectra characteristic for electron (Watkins, 1975, 1994), neutron (Lee et al., 1972; Lee and Corbett, 1974; Brower, 1976), or helium (Gorelkinskii et al., 1987) irradiation are created on annealing in the temperature range of 30–600°C.

In general terms, the formation mechanism of the hydrogen-associated shallow donor can be understood by the following simple scheme. In fact, the dominant defects that are stable at room temperature in proton (or neutron) irradiated Fz-silicon are the divacancy (Watkins and Corbett, 1965), pentavacancy (Lee and Corbett, 1973), and self-interstitial complexes, such as Si-P6 (Lee et al., 1976), A5 (Lee et al., 1972), and B3 (Brower, 1976). However, in a hydrogen-implanted layer (as well as when hydrogen was introduced in the Fz sample before (Wang and Lin, 1984; Meng, 1991) or after (Hartung and Weber, 1993) neutron irradiation, strong Si—H bonds between some dangling-bond defects and hydrogen are created. In particular, it has been found that some of the hydrogen-associated IR bands in silicon are stable up to 600–650°C (Stein, 1975; Gerasimenko et al., 1978; Mukashev et al., 1989; Tokmoldin et al., 1990; Meng, 1991). Theoretical calculations also predicted the existence of strong, bonded Si—H complexes (Van de Walle and Street, 1994). On heating the sample above $\sim 300°$C, most free divacancies (predominantly acceptor-type defects) are annealed. Strong Si—H bonds neutralize the electrical activity of dangling bonds of the vacancy-type defects and prevent their annihilation by self-interstitial defects. Hence the presence of strong Si—H bonds should promote the clustering (aggregation) of self-interstitial atoms during the annealing process. Since broken bonds on vacancy-type defects are neutralized by hydrogen, while isolated self-interstitial complexes (like B3; Brower, 1976) are characterized by an abundance of the electron density, a self-interstitial complex is the most probable candidate for the HSD core.

Early EPR studies showed that these HSD centers give rise to a single isotropic EPR line of conduction electrons with $g = 1.9987$ at 77 K (Gorelkinskii et al., 1974). Low-temperature EPR measurements in an X-band spectrometer (Gorelkinskii and Nevinnyi, 1991) indicate that the isotropic line intensity is strongly decreased for sample temperature lower than ~ 30 K, while later it was found that the isotropic signal can be partially restored by bandgap illumination even at 4.2 K. Thus the X-band EPR spectrum of HSD (as well as of the oxygen thermal donor NL10; Michel et al., 1986) in its neutral charge state is an isotropic line without any hf structure. The strong dependence of the HSD intensity on illumination of the sample suggests that its ground state is diamagnetic, with a spin $S = 0$ (Gorelkinskii and Nevinnyi, 1991).

b. *Bistability of the Hydrogen-Associated Donor*

Previous experiments (Gorelkinskii and Nevinnyi, 1981a, 1983a, 1983b) also suggested bistability of the electronic properties of the HSD center. The concentration of HSD can be reversibly changed as a function of the quenching temperature of the sample below $\sim 250°C$ (for samples with $\sim 10^{17}$ e/cm^2) after creation of HSD at 400–500°C (Fig. 9). The temperature dependence of the equilibrium concentration of HSD can roughly be written as a pseudomolecular reversible chemical reaction: $dC/dt = -K_1 C + K_2 B$, where C and B denote the HSD and the deep-state concentrations, respectively, and K_1 and K_2 are the reaction rate constants (Fig. 10). At thermal equilibrium ($dC/dt = 0$), the HSD concentration is $C^e = (K_2/K_1)B^e$. Far from saturation (i.e., below $\sim 180°C$ for a sample with an HSD concentration of about 1×10^{17} cm^{-3}), the temperature dependence of the equilibrium concentration of HSD centers corresponds to an Arrhenius relation:

$$C^e = 2.3 \times 10^{22} \exp(-\Delta E_{eq}/kT) \tag{8}$$

From experimental data, ΔE_{eq} was determined to be 0.51 eV. Note that $\Delta E_{eq} = E_A - E_D$, where E_A and E_D are activation energies for transitions from the deep state to HSD and from HSD to the deep state, respectively.

The kinetics of the HSD concentration change on transition from one equilibrium state to another and can be described using the exponential form $C(t) = C^e + [C(0) - C^e] \exp(-t/\tau)$, where $C(0)$ and $C(t)$ are the initial

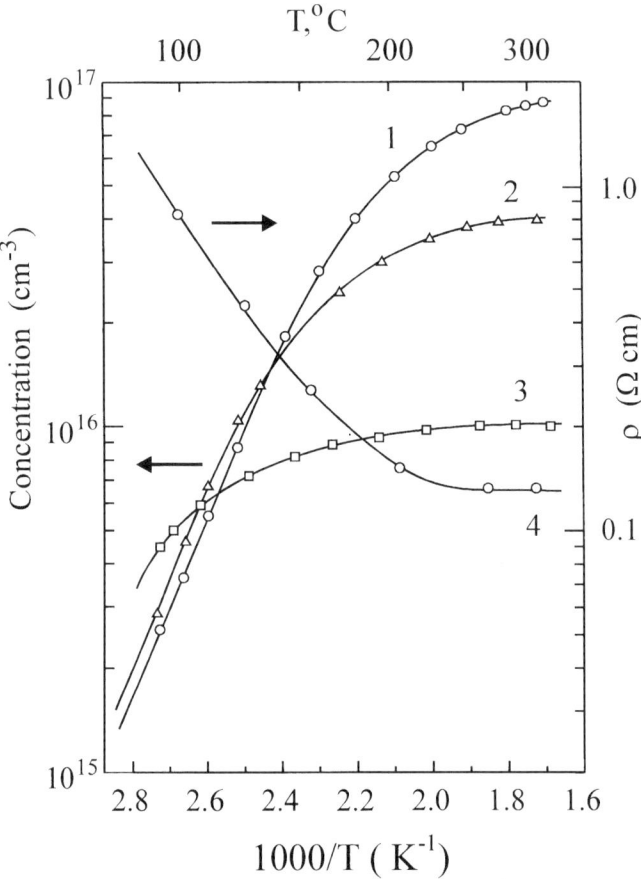

FIG. 9. Plots of the equilibrium concentration of shallow donors versus quenching temperature in Fz-Si doped with 10^{14} boron/cm^3. The shallow donors were formed after H implantation and annealing at 450°C for 20 min. The concentration was determined from the EPR line of conduction electrons at 77 K. The total dose of protons corresponds to concentrations: (1) 8×10^{18} H$^+$/cm^3, (2) 4×10^{18} H$^+$/cm^3, (3) 1.3×10^{18} H$^+$/cm^3. Curve 4 indicates the change in electrical resistivity of sample 1 measured at room temperature. (Adapted from Gorelkinskii and Nevinnyi, 1983a, 1983b.)

and final HSD concentrations, respectively, and $\tau = 1/(K_1 + K_2)$ is the lifetime of the HSD state. The temperature dependence of the lifetime τ is given in Fig. 11. The thermal activation barrier for the transition of the HSD to a deep state is characterized by an activation energy $E_D = (1.23 \pm 0.05)$ eV with the preexponential factor of $1/\tau_0 \approx 10^{13}$ sec^{-1}. The activation energy

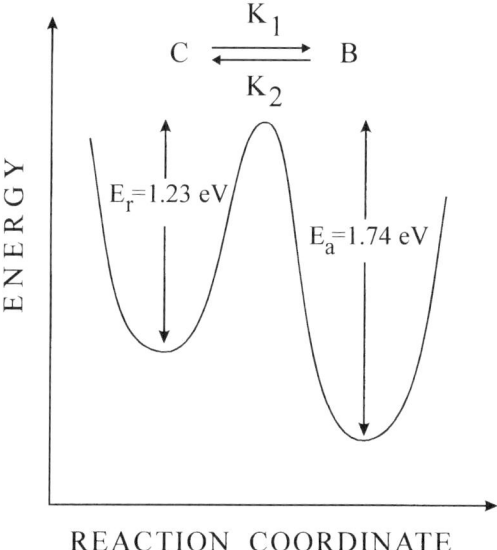

FIG. 10. Energy-reaction coordinate diagram. E_S and E_D are the energy barriers for transitions to shallow and deep donor states, respectively.

barrier for the transition from deep state to HSD was found to be $E_A = E_D + \Delta E_{eq} = 1.74$ eV (Fig. 10). We note that the preexponential factor depends slightly on the preliminary sample treatment (i.e., on the hydrogen ion dose and the annealing temperature), while the activation energy E_D does not vary. The E_D and τ_0 values were obtained for samples with significantly different doses of hydrogen implantation (Fig. 11).

As can be seen in Fig. 11, the EPR data (which were obtained from measurements of HSD concentration after the sample was heated to different temperatures in the range from 80–130°C and subsequently quenched to room temperature) and also direct measurements of the sample conductivity at the current temperature (Gorelkinskii and Nevinnyi, 1981b) give the same value for the activation energy and very similar values for the preexponential factor (Fig. 11). This correlation between EPR data and electrical measurements is an additional confirmation that the isotropic EPR line is indeed caused by the HSD center. The equilibrium concentration of shallow levels (0.035 eV) is also correlated with the signal intensity of the isotropic EPR line (Gorelkinskii and Nevinnyi, 1983a). We note that the equilibrium concentration of HSD, for example, at 100°C is $\sim 3 \times 10^{15}$ cm^{-3} (from Eq. 8) and $\tau \approx 15$ min (Fig. 11). That is, for an H-implanted sample containing $\sim 3 \times 10^{15}$ cm^{-3} of HSD centers, the

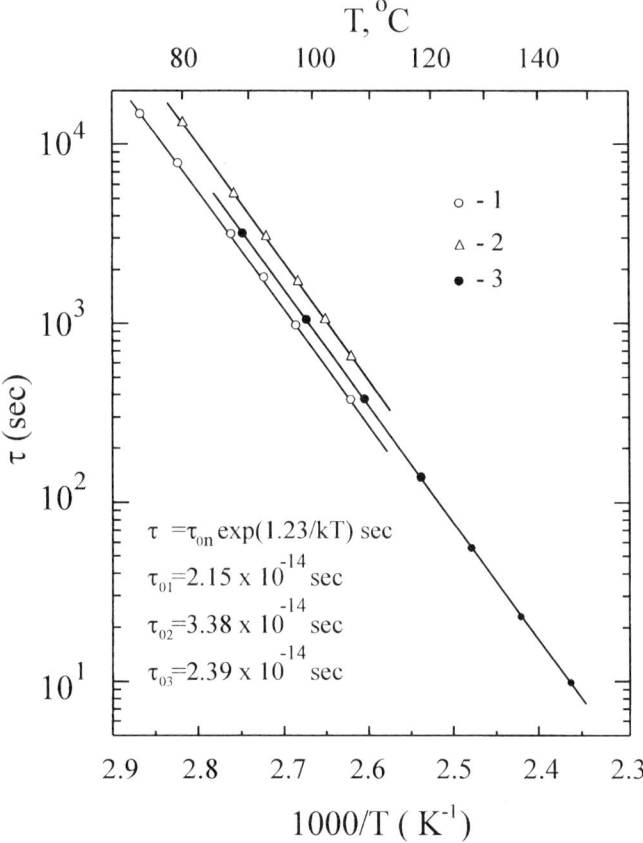

FIG. 11. Temperature dependence of the transition rate ($1/\tau$) of the shallow donor state to the deep state. Fz-Si samples doped with 10^{14} boron/cm^3 were implanted with protons at room temperature. The total dose of protons corresponds to (1) 8×10^{18} H$^+$/cm^3, (2) 4×10^{17} H$^+$/cm^3, (3) $\sim 8 \times 10^{18}$ H$^+$/cm^3. Plots (1) and (2) have been obtained from EPR data at 77 K. Plot (3) is a result of the direct measurement of the sample conductivity at current temperature (150–90°C). (Data sets (1) and (2) are adapted from Gorelkinskii and Nevinnyi, 1983a.)

reversible transitions should be observed in the temperature range below 100°C only, with a characteristic time greater than 15 min.

It should be noted that the conversion from n-type to p-type sample conductivity (at decreasing HSD equilibrium concentration from $\sim 10^{17}$ to $\sim 10^{15}$) was not observed, although a B-doped silicon sample starting with $\rho \approx 0.1\ \Omega \cdot$cm has been used for hydrogen implantation. That is, the solubility of the HSD center is increased with increasing acceptor concentration.

The preexponential factor of $\sim 10^{13}\,\text{sec}^{-1}$ for HSD reversible transitions has a reasonable value for a local rearrangement process, being of the order of kT/h. It is important to note that the preexponential factor of $\sim 10^{13}\,\text{sec}^{-1}$ is a characteristic vibrational frequency for simple point defects or monatomic impurities (Watkins, 1975, 1994) and cannot be caused by long-range migration of defects, where $v_0 \approx 10^7\text{--}10^8\,\text{sec}^{-1}$ (Watkins, 1975). Thus we can conclude that reversible transitions from shallow to deep states (and inversely) are an atomic reorientation processes for the HSD structure itself. On the other hand, the transition times are very short (e.g., $\tau \approx 10\,\text{sec}$ at 150°C; see Fig. 11), while simple defects such as a single vacancy or a single self-interstitial should be annealed completely at 400–500°C (Watkins, 1994) (during HSD creation). High concentrations of Si—H bonds are observed in hydrogen-implanted silicon (Stein, 1975; Gerasimenko et al., 1978; Mukashev et al., 1989; Tokmoldin et al., 1990). Therefore, it is reasonable to infer that hydrogen rebonding to the defect structure occurs in the reversible transformation of the HSD states. Note that bandgap illumination (up to $\sim 10\,\text{W/cm}^{-2}$) of samples containing HSD at temperatures in the range from 30–100°C does not enhance the reversible transitions rates. This fact also confirms that a thermally activated *atomic* process is involved in reversible transformation of the HSD center.

The most intense Si-H IR stretching band observed in hydrogen-implanted silicon with lines at $2162\,\text{cm}^{-1}$ (Stein, 1975; Gerasimenko et al., 1978), $2122\,\text{cm}^{-1}$, $2107\,\text{cm}^{-1}$ (Mukashev et al., 1989), $2222\,\text{cm}^{-1}$ (Tokmoldin et al., 1990), $2122\,\text{cm}^{-1}$, $2107\,\text{cm}^{-1}$, and $2222\,\text{cm}^{-1}$ exhibits the same annealing behavior as the HSD center. However, reversible changes of the intensity of the Si-H-related IR bands in hydrogen-implanted silicon were not observed so far.

c. Role of Oxygen in the Formation of the Hydrogen-Associated Donor

The AA1 EPR spectrum appears at an early stage of HSD formation in the temperature range $\sim 260\text{--}350°C$ during annealing of a hydrogen-implanted sample with $\sim 10^{18}\,\text{H/cm}^3$ (Gorelkinskii and Nevinnyi, 1983a, 1991). After annealing of the sample below $\sim 300°C$, the AA1 spectrum is observed only under sample illumination. On subsequent annealing at 320–350°C for 10 min, the intensity of the AA1 spectrum is increased and can be observed without illumination. With annealing of the sample at temperatures above $\sim 350°C$, the shallow state is created quickly, while the AA1 spectrum disappears. Therefore, it has been proposed that the AA1 spectrum arises from the positive charge state of the HDD center (Gorelkinskii and Nevinnyi, 1991). A comparison of the AA1 spectrum parameters with those of the oxygen-related thermal donor NL8 (Muller et al., 1978; Michel et al., 1986; Bekman et al., 1988) reveals their strong resemblance (Gorelkinskii and Nevinnyi, 1991). In particular, both the AA1 and the NL8 defects are double donors in positive charge states with C_{2V} symmetry of the g-tensor. The principal values of their g-tensors are very close. There is no resolved hf structure in the AA1 and NL8 spectra. Both the AA1 and NL8 centers have two donor levels with similar ionization energies (Benton et al., 1983; Gorelkinskii and Nevinnyi, 1983a). This great resemblance between spectra suggests that their nature should be very similar. The principal distinction lies in the fact that, in general, no difference in HSD formation is observed between Cz and high-purity Fz silicon (Ohmura et al., 1972, 1973; Gorelkinskii et al., 1974; Gorelkinskii and Nevinnyi, 1983a; Wang and Lin, 1984; Meng, 1991; Hartung and Weber, 1993). The oxygen concentrations for these two types of Si are very different, although oxygen is critical for thermal (TSD or TDD) donor formation.

In view of the dramatically hydrogen-enhanced diffusion of oxygen that has been observed (McQuaid et al., 1991; Newman and Jones, 1994), measurements of the HSD concentration in pure Fz samples ($\rho \approx 3\,\mathrm{k\Omega \cdot cm}$) with low oxygen concentration ($\leq 10^{16}\,\mathrm{cm}^{-3}$) and in Cz samples (n- and p-type, $\rho \approx 20\,\Omega \cdot \mathrm{cm}$) were performed. Hydrogen implantation and subsequent annealing were carried out (simultaneously) under identical conditions. The measurements were carried out mainly with an X-band EPR spectrometer at 77 K using the HSD isotropic line signal. Phosphorus-doped silicon ($1 \times 10^{17}\,\mathrm{cm}^{-3}$) was used to calibrate the spin concentration. After high dose implantation (corresponding to $\sim 1 \times 10^{19}\,\mathrm{H/cm^3}$) and subsequent 20-min annealing at 470°C, the HSD concentration for a pure Fz sample was determined to be $(6-8) \times 10^{16}\,\mathrm{cm}^{-3}$ (i.e., the concentration of HSD is higher than the concentration of oxygen), while for the Cz sample the HSD concentration is 30–40% less than for pure Fz. The same measurements for low-dose implantation (corresponding to $\leq 10^{17}\,\mathrm{H/cm^3}$) have shown that the production rate of HSD is far less (3–5 times) in Cz samples than in pure Fz silicon.

It is important to note that HSD centers were not created if hydrogen implantation was carried out at sample temperatures in the range of 300–350°C (Gorelkinskii et al., 1980); hydrogen plasma treatments of Cz silicon at $\sim 300°C$ strongly enhance TDD or STD defect creation (Newman and Jones, 1994).

Taking into consideration these findings and the fact that the HSD concentration in pure Fz silicon is more than that of oxygen, we have to conclude that only intrinsic defects and hydrogen atom(s) can be involved in the structure of the HSD center. However, we are by no means ignoring the existence of the oxygen thermal shallow donor in the structure where

3 HYDROGEN AND HYDROGEN-RELATED DEFECTS IN CRYSTALLINE SILICON 59

hydrogen is also involved [as has been reported by Martynov et al. (1995a) and Newman et al. (1996a)].

2. STRESS-INDUCED ALIGNMENT OF THE AA1 SPECTRUM

Important information about molecular defect structure can be extracted from the uniaxial stress response. Therefore, uniaxial stress-induced alignment of the AA1 center was performed. High-purity zone-refined silicon crystals ($\rho \approx 3$–$6 \,\text{k}\Omega \cdot \text{cm}$) were used in these experiments as starting material. Implantation was carried out through an aluminum absorber (3.6–4.0 mm) at room temperature with a starting proton energy of 30 MeV. The straggling of 30-MeV protons is around their stopping point of ~ 150–$100 \,\mu\text{m}$, so a homogeneous distribution of implanted hydrogen in the sample was obtained. In these experiments the samples were implanted with a relatively low dose of protons, corresponding to $\sim 1 \times 10^{17} \,\text{H/cm}^3$. Moreover, in order to provide a positive charge state of HSD, compensating defects also were introduced by proton irradiation of the same samples (before annealing) without the Al absorber and using a dose of $\sim 1 \times 10^{17}$ protons/cm^2. The measurements were performed using a Q-band EPR spectrometer at 77 K in absorption.

The anisotropic AA1 EPR spectrum appeared on annealing of the sample at 400°C for 20 min (Fig. 12a). The intensity of the AA1 spectrum increases strongly under bandgap illumination of the sample. The AA1 spectrum and the Si-P1 pentavacancy (Lee and Corbett, 1973) are the predominant spectra in samples prepared as just described (Gorelkinskii et al., 1998). The AA1 spectrum (Fig. 12) can be described as arising from a defect with a C_{2V} symmetry g-tensor with the principal values $g_1[110] = 1.9947$, $g_2[\bar{1}10] = 2.0001$, and $g_3[001] = 1.9997(\pm 0.0003)$. Apparently, the AA1 g-tensor parameters are very close to those of the NL8 spectrum (Muller et al., 1978; Michel et al., 1986; Bekman et al., 1988).

It is important to note that Cz silicon samples, prepared under conditions identical with those of the pure Fz samples (implantation, annealing), do not reveal even a trace of the AA1 or NL8 spectra, and only oxygen-dependent spectra (P2, P4, P5, and A16; Lee and Corbett, 1973, 1976) along with the P1 were observed.

Figure 12a shows the AA1 EPR spectrum observed after annealing at 400°C for 20 min of an Fz sample that was implanted with hydrogen as described above. Here, with the magnetic field H//[110], the spectrum consists of three components (A, B, and C) that are the three nonequivalent orientations of a C_{2V} symmetry defect stressed along the [$\bar{1}$10] direction, as illustrated in Fig. 12c. Under normal thermodynamic conditions, the relative

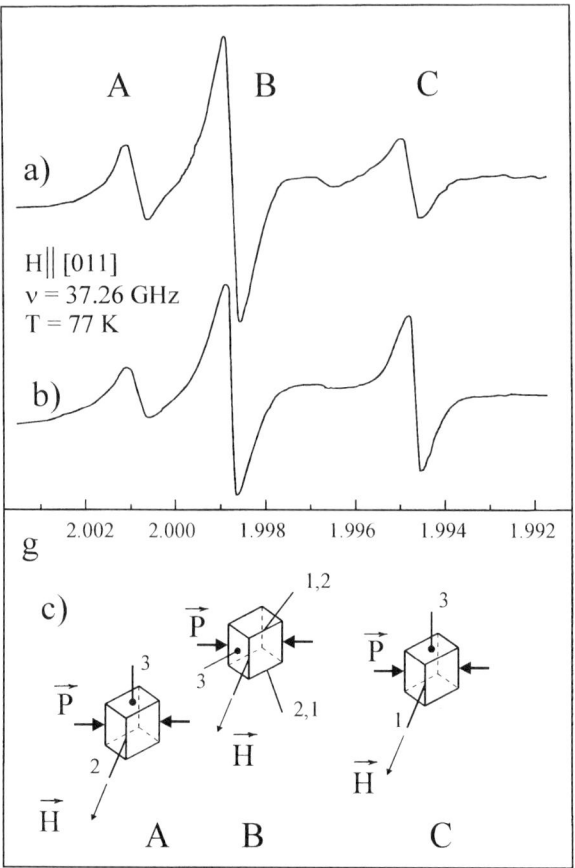

FIG. 12. EPR spectrum of the Si-AA1 center in Fz-Si implanted with hydrogen and annealed at 400°C for 20 min. (a) "As annealed" sample. (b) Additionally annealed at 320°C for 30 min under ⟨011⟩ stress of 200 MPa and cooled to room temperature under stress. The spectra were measured at 77 K under bandgap illumination. (c) The insets show defect orientations associated with each of the resolved EPR lines along with stress and magnetic field directions. (From Gorelkinskii et al., 1998.)

intensities of the three components are $A:B:C = 1:4:1$. Under stress along the [$\bar{1}$10] direction, the component intensities provide a direct quantitative measure of the defect alignment, given by $A:B:C = a:4b:c$, where a, b, and c are the probabilities for each defect orientation (Trombetta et al., 1997).

After creation of the AA1 spectrum (400°C, 20 min), an uniaxial stress of 200 MPa was applied to the [$\bar{1}$10] axis of the sample at 320°C for 20 min.

The sample was then cooled in about 60 min to room temperature under stress. The stress was then removed and the EPR spectrum of the sample was measured at 77 K. As can be seen in Fig. 12b, a significant alignment has been achieved at these conditions. As with NL8 (Trombetta *et al.*, 1997; Watkins, 1996), the dominant effect for AA1 is the tendency for the defect to align with its g_2 [$\bar{1}$10] axis along the stress direction. However, for the AA1 center, the probability a (as well as b) is decreased; this contrasts with NL8, where both the a and c probabilities are increased under [$\bar{1}$10] stress direction (Trombetta *et al.*, 1997). Thus the dominant effect is the tendency of the AA1 center to align with the g_1[110] and g_3[001] axes perpendicular to the compressed direction, as reflected in the probability values (Fig. 12b):

$$a = 0.61 \quad b = 0.62 \quad c = 1.77 \tag{9}$$

(in the unstressed crystal, $a = b = c = 1$). Complete recovery of the AA1 spectrum components A, B, and C (Fig. 12a) is observed on annealing the sample at 320°C for 30 min without stress. The alignment of the AA1 center under stress and then its recovery without stress at 320°C can be repeated many times, suggesting that the defect is actually reorienting. Expressions describing the recovery kinetics of C_{2v} symmetry defects have been established previously (Trombetta *et al.*, 1997) and were used in these experiments for the analysis of the AA1 recovery process:

$$A(t) - C(t) = [A(0) - C(0)] \exp(-t/\tau) \tag{10}$$

$$B(t) - 2A(t) - 2C(t) = [B(0) - 2A(0) - 2C(0)] \exp(-3t/2\tau) \tag{11}$$

We noted that the recovery of component B in the AA1 spectrum is uniformly 1.5 times faster throughout the decay than the recovery between A and C. As has been shown previously, this fact strongly supports the reorientation model (Trombetta *et al.*, 1997). It was found that the recovery process of the AA1 center is characterized by a simple exponential law, suggesting that the AA1 spectrum arises from defects with equal (or very similar) reorientation times (and in contrast to NL8; Trombetta *et al.*, 1997). The temperature dependence of the characteristic time for AA1 atomic reorientation was found to be (Fig. 13)

$$\tau = 3.2 \times 10^{-17} \exp(2.3 \pm 0.1 \text{ eV}/kT) \text{ sec} \tag{12}$$

The magnitude of the preexponential factor ($\sim 10^{16}$ sec^{-1}) of the AA1 recovery process is typical for extended defects in silicon like thermal donors (Trombetta *et al.*, 1997). It is important to note, however, that the charac-

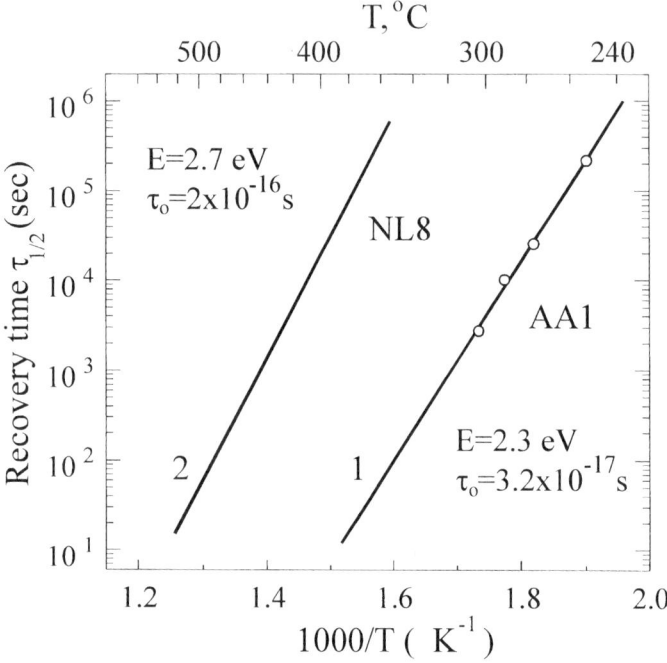

FIG. 13. Recovery time ($\tau_{1/2}$) of the atomic reorientation for the Si-AA1 center versus temperature (1). For comparison, the reorientation decay time $\tau_{1/2}$ for the Si-NL8 center is shown (2). Plot (2) is the result of a calculation using the reorientation constants of the Si-NL8 from Trombetta et al. (1997). (Adapted from Gorelkinskii et al., 1998.)

teristic time of recovery from stress-induced alignment of the AA1 is dramatically ($\sim 10^4$ times) shorter than that of the NL8 defect (Fig. 13).

The activation energy of the AA1 atomic reorientation is very large (~ 2.3 eV) and cannot be related to simple point defects that are efficiently introduced by proton implantation. Among a variety of radiation defects in proton- or neutron-irradiated Fz silicon there is only one predominant defect B3 (Daly, 1971) that has an activation energy for the atomic reorientation comparable with that of the AA1 defect and is stable up to $\sim 500°$C. Si-B3 is a secondary defect that has been identified previously as either a $\langle 001 \rangle$ Si di-interstitial or a $\langle 001 \rangle$ Si split interstitial (Brower, 1976).

As done previously for other C_{2v} symmetry defects (Lee and Corbett, 1974; Watkins, 1996), we may analyze stress alignment of the AA1 center in terms of a piezospectroscopic tensor **B** (Kaplyanski, 1964) (Eq. 4). We have used basic equations for the analysis of the experiments with stress-induced alignment from Wagner et al. (1987), and the traceless part of the **B**-tensor in the defect 1,2,3 is a principal-axis system as for NL8 (Trombetta et al., 1997).

Uniaxial stress of 200 MPa was applied along the [$\bar{1}$10] axis (Fig. 12c) of the sample for 40 min at 320°C. The crystal was then quickly cooled (over ~60 sec) to room temperature with stress applied. As can be seen from Fig. 13, 40 min at this temperature ($\tau \approx$ 10 min) is sufficient to establish equilibrium alignment for the AA1 center. The AA1 alignment (Fig. 12b) with probabilities a, b, and c, from Eq. (9), yields an effective C_{2v} strain-coupling tensor: $B_1 = -12.7$ eV, $B_2 = 30.6$ eV, and $B_3 = -17.9$ eV, where the labeling of the principal axes is shown in Fig. 12c and corresponds to the convention for the g-value component labeling. The negative sign and magnitude of the B_3 component indicate that the core of the AA1 defect as well as the NL8 (Trombetta et al., 1997) produces a strong compressional strain field along its C_{2v}[001] 3 axis that causes the defect to align with this axis perpendicular to the applied compressional strain. However, in contrast to NL8, the B_1 component of the AA1 also has a negative sign. This means a large compressional strain field produced by the AA1 defect along the [110] 1 axis, and hence the defect prefers to align only with the [$\bar{1}$10] 2 axis along the applied compressional strain. In the competition between the 1 and 3 axes, the compressional strain field produced by the AA1 defect along its [001] 3 axis is considerably greater than along its [110] 1 axis. A negative sign of the B_1 component also has been obtained for samples prepared from different silicon ingots and for different sample quenching temperatures. This fact indicates a significant difference between the molecular structures of the AA1 and NL8 defects (Gorelkinskii et al., 1998).

We suggest a [110] chain of the two [001] Si split or di-interstitials (defects that are like B3; Brower, 1976) as a possible defect configuration of the AA1 core. This configuration, as expected, may produce a large compressional strain field along its C_2 [001] 3 axis and less along the [110] 1 axis and is in agreement with the C_{2v} symmetry of the AA1 center.

In summary, it has been demonstrated experimentally that the AA1 spectrum, arising from a C_{2v} symmetry defect with principal **g**-tensor parameters very close to those of NL8 (oxygen thermal donor in positive charge state), really is a defect with a distinct molecular structure. The greatest differences are (1) the B_1 component of the piezospectroscopic tensor has a negative sign for the AA1 defect in contrast to the NL8, (2) the characteristic time of the AA1 recovery from stress-induced alignment is about four orders of magnitude less than that for NL8, and (3) the presence of oxygen decreases the AA1 creation rate. These facts cannot be related to any enhancement of oxygen atom diffusion and can be caused only by differing molecular structures of the defects.

On the other hand, the strong resemblance of the AA1 and NL8 spectra, it seems, cannot be accidental. We can propose that their cores are very similar while the shells are different. The B3 spectrum arises from a $\langle 001 \rangle$ symmetry (Si-Si)$_i$ complex that is most prominent in neutron (Brower, 1976) or proton (Gorelkinskii et al., 1980) irradiated p-type Fz silicon and cannot be ignored at formation of defect complexes in hydrogen-implanted silicon (at least, by no means less than oxygen). The B3 defect prefers to align with its $g_\parallel$ axis perpendicular to the stress direction; atomic reorientation is characterized by an activation energy of $\sim 2.3\,\mathrm{eV}$ (Brower, 1976). The response of the AA1 defect on uniaxial stress is in good agreement with a defect structure such as the $\langle 110 \rangle$ chain of the $\langle 001 \rangle$ Si split interstitials. Therefore, the AA1 core is most probably a self-interstitial complex ($\langle 110 \rangle$ chain of the $\langle 001 \rangle$ Si interstitial complexes, like the B3 defects), while hydrogen atoms (or oxygen atoms in the case of TDD) are located mainly in the shell and influence strongly atomic reorientation and formation kinetics. It is in agreement with an early suggestion for the thermal donor structure where its core is the aggregate of self interstitials (Newman, 1985). These results also demonstrate that the NL8 EPR spectrum does not have a unique nature (as well as the NL10; Michel et al., 1989); that is, the defects with different microscopic structures give rise to very similar EPR signals.

3. Neutral Charge State of Hydrogen-Associated Donor

High-purity Fz silicon ($\rho \approx 3\text{--}6\,\mathrm{k\Omega \cdot cm}$) was used as a starting material for the experiments. Samples were bombarded with protons at room temperature (with an initial energy of 30 MeV) through an aluminum absorber of 3.9–4.1 mm thickness. Hydrogen-implanted samples with two very different concentrations of hydrogen ($\sim 5 \times 10^{16}\,\mathrm{H/cm^3}$ and $\sim 8 \times 10^{18}\,\mathrm{H/cm^3}$) were studied. Measurements were performed on superheterodyne EPR spectrometers with an operating frequency of 9.2 or 23.3 GHz using low-frequency field modulation (12.3 and 910 Hz) in dispersion mode and a sample temperature of $\sim 6\,\mathrm{K}$ (Stallinga et al., 1994b).

After annealing the sample containing the low hydrogen concentration at 340°C for 20 min, vacancy-like defects anneal out, and the EPR signal is almost completely absent. Only a trace EPR signal with a g-value of conduction electrons was observed. It is important to note that annealing strongly decreased the sample resistivity. The starting material (before implantation) had a resistivity of $\rho \approx 10^3\,\mathrm{\Omega \cdot cm}$, which declined to $\rho \approx 8\,\mathrm{\Omega \cdot cm}$ on annealing (it was measured at 300 K, corresponding to a concentration of shallow donors of $\sim 8 \times 10^{14}\,\mathrm{cm^{-3}}$). Under illumination

3 HYDROGEN AND HYDROGEN-RELATED DEFECTS IN CRYSTALLINE SILICON 65

by bandgap light, an intense EPR signal appears. With the X-band EPR spectrometer, one observes an isotropic line with a g-value of conduction electrons (as for the NL10 spectrum measured in the X-band; Meilwes et al., 1994; Michel et al., 1989). Note that we have referred to this EPR line as an HSD center above and that it is the predominant intensity spectrum in samples with low concentrations of implanted hydrogen. No trace of oxygen-dependent EPR spectra was obtained in these samples.

On subsequent annealing of the same sample at 380–400°C for ~20 min, the intensity of the isotropic spectrum (HSD) is partially decreased, and an ($S = 1$) spectrum labeled Si-NL51 (Stallinga et al., 1994b) emerges (Ammerlaan et al., 1992) (Fig. 14a). This latter spectrum can be observed only when the sample is illuminated with bandgap light. The NL51 spectrum can

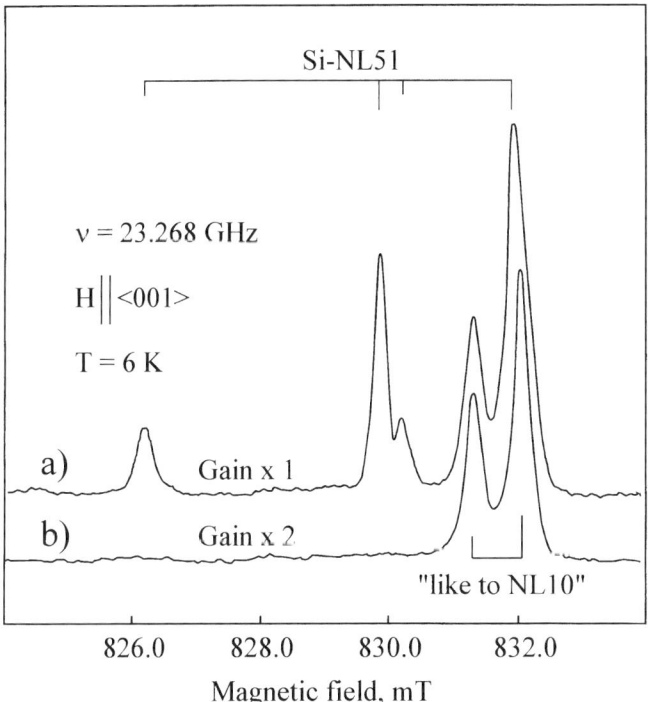

FIG. 14. NL51 and "NL10-like" EPR spectra in hydrogen-implanted Fz silicon after annealing at 520°C for 20 min: (a) under bandgap illumination; (b) in the dark. (From Ammerlaan et al., 1992.)

be described as arising from a tetragonal symmetry defect with the spin Hamiltonian

$$H = \mu_B \mathbf{B g S} + \mathbf{S D S} \quad (13)$$

with an effective spin $S = 1$. The first term describes the electron Zeeman interaction, and the second term corresponds to the fine structure. The spin Hamiltonian constants of the NL51 spectra were $g_{\parallel}[001] = 2.0071(\pm 0.0002)$, $g_{\perp} = 2.0007$, $D_{\parallel}[001] = -37.9$ MHz, and $D_{\perp} = 19.0$ MHz (Stallinga et al., 1994b). Moreover, in this sample, the B3 spectrum (Brower, 1976) is also observed (only under illumination), although its intensity is much smaller than that of the HSD or NL51 spectrum. It is important to emphasize that after annealing the sample at 420°C for 25 min, the intensity of the NL51 is about two times higher than the intensity of the HSD spectrum.

In the second sample produced with a high concentration of implanted hydrogen ($\sim 8 \times 10^{18}$ H/cm^3), the maximum HSD EPR signal is reached after annealing in the temperature range of 450–475°C (~ 20 min). In this case the HSD spectrum may be observed without illumination, although under illumination the intensity is increased several times. Electrical measurements (at 300 and 77 K) agree well with the HSD intensity for both the formation and decay stages found using isochronal annealing.

On further annealing of the same sample at 520–530°C (15 min), the intensity of the HSD EPR spectrum is decreased considerably. Under illumination, an intense NL51 spectrum appeared (Fig. 14a, b). In this case the intensities of the HSD and NL51 spectra are comparable and indicate concentrations of 8×10^{15} cm^{-3} or more. Note that only these two spectra are observed in samples prepared in this manner, and both spectra anneal out completely at 560–580°C.

Also note that the ($S = 1$) NL51 spectrum is observed only under illumination for samples with strongly different conditions of preparation (i.e., for different dose implantations and annealing temperatures). The intensity of the NL51 spectrum changes on annealing; it can exceed or be smaller than that of the HSD defect. Therefore, most probably the NL51 spectrum arises from an excited triplet state, while the ground state is the singlet $S = 0$ (Stallinga et al., 1994b).

The magnitude and anisotropy of the observed **g**- and **D**-tensors suggest that the nature of the zero-field splitting should be related to magnetic dipole-dipole interaction between the two unpaired electrons. We note that the magnitude of the zero-field splitting for NL51 is the least among known spin-triplet spectra in silicon (Lee et al., 1967, 1976). Using the point dipole approximation for the energy of a magnetic dipole-dipole interaction, it was estimated that the distance between two point dipoles should be about 9–10 Å along the $\langle 001 \rangle$ direction.

A careful search for ^{29}Si or other hf lines within 50 mT of the Zeeman lines was made for the NL51 (as well as for HSD) spectrum, but none was found. There is no hint of any structure in the Zeeman lines of the NL51 or HD defects that can be associated with ^{1}H or ^{29}Si hf interactions. The $\langle 001 \rangle$ symmetry of the **g**- and **D**-tensors, as well as the lack of any ^{29}Si hf interaction suggests that the nature of the defect wave function of the NL51 is antibonding. It is also important to note that the intensity of the NL51 spectrum is very sensitive to the sample temperature; it vanishes at temperatures above ~ 20 K. In view of these results, the model for the structure of the spin-triplet NL51 center, based on interacting, well-localized dipoles approximately 10 Å apart, is not very likely. More plausible is an extended, one-core defect in which the value of ~ 10 Å extracted from the zero-field splitting is associated with a helium-like effective-mass wave function for an excited triplet state (Stallinga *et al.*, 1994b).

Note that the possibility of NL51 and B3 arising from the same structure cannot be excluded. In this case it would be logical to assign the NL51 to an excited, neutral charge state of the di-interstitial (Si-Si)°; hydrogen's role would be to passivate side defects from the center. An additional argument for associating NL51 with the B3 center is that the anomalous ratio of line intensities for the equivalent positions of the B3 (~ 0.26 in contrast to normal ratio of 0.5) (Brower, 1976) is also observed for the NL51 spectrum (see Fig. 14a). However, hydrogen can be involved in the Si-Si complex, changing its electronic structure and passivating nearby dangling bonds. IR absorption spectroscopy studies show that there is an Si-H stretch mode at 2222 cm^{-1} with tetrahedral symmetry (Nielsen *et al.*, 1995). The 2222 cm^{-1} mode disappears on thermal annealing at $\sim 550°$C. Other tetrahedral symmetry defects stable up to 550°C were not detected by IR spectroscopy. The similarity of the properties of the 2222 cm^{-1} mode and the NL51 center allow us to propose that they arise from the same defect. In Nielsen *et al.* (1995), the 2222 cm^{-1} mode was identified as a hydrogenated vacancy (VH$_4$) complex; we cannot rule out the possibility that the core of the defect is a self-interstitial complex (like B3), while the hydrogen atoms belong to an NL51 shell.

In any case, it has been concluded that the NL51 center is a (helium-like) double donor in neutral charge state with a D_{2d} symmetry wave function with a singlet $S = 0$ ground state. The properties of the NL51 spectrum suggest that its neutral charge state should be a shallow donor.

Another important question that arises from these findings is the relationship between the NL51 and the (HSD) NL10-like spectra. Indeed, both NL51 and HSD are created in hydrogen-implanted layers of Fz silicon. In samples with either low or high hydrogen concentrations (differing by about two orders of magnitude), the intensities for both NL51 and HSD are proportional to the implantation dose, and both exhibit the same temperature-range stability. The resonant wave functions of the NL51 and (HSD) NL10-like centers have antibonding character, with very weak localization on their cores as for shallow donors. Note that after implantation and annealing, both samples have n-type conductivity ($\sim 10^{14}$ and $\sim 10^{16}$ e/cm^3). In view of these facts, it is believed that the nature of these centers should be very similar. The ($S = 1$) NL51 spectrum emerges on annealing, while the intensity of the ($S = 1/2$) NL10-like spectrum diminishes. The creation of the NL51 center cannot be caused by Fermi level shifts because both spectra are observed in samples with quite different concentrations of (HSD) NL10-like and NL51 centers. Hence it is logical to speculate that the emergence of the NL51 is caused by the HSD modification; it can result from the HSD decay or crowding.

Here it is important to note that the most probable candidate for the HSD core is a $\langle 001 \rangle$ Si di-interstitial complex (B3) created by high-energy collisions [i.e., neutron (Daly, 1971) or proton (Gorelkinskii et al., 1980) irradiation] to form defect clusters. Electron irradiation (1.5–3 MeV) of silicon (Watkins, 1994) does not produce Si interstitial-related complexes such as P6 (Lee and Corbett, 1976), A5 (Lee et al., 1972), or B3 (Brower, 1976). This fact suggests that self-interstitial complexes like B3 are distributed nonhomogeneously, mainly in defects clusters (Brower, 1976), after implantation and annealing. Earlier (Sec. IV.1.a) we argued that strong Si—H bonds also should promote the clustering (aggregation) of self-interstitial atoms during the annealing.

In view of the preceding, it is logical to propose that the weakly anisotropic ($S = 1/2$) SHD (NL10-like; Fig. 14b) spectrum results from the electron-exchange interaction of several ($S = 0$ or 1) helium-like double donors (NL51) that disrupt the spin-spin $S = 0$ or $S = 1$ correlation of the individual donor states. That is, we suggest that the NL10-like spectrum derives from multiple donors (i.e., it arises from aggregation of the NL51 double donors) (Gorelkinskii, 1992). It is important to note that in organic materials ($S = 1/2$), EPR centers do result from electron exchange by excited ($S = 1$) spin-triplet centers (Carrington and McLachlan, 1967). The local concentration of individual double shallow donors (NL51) may be too low to cause degeneracy of their levels in the forbidden gap but can nonetheless be sufficient for the destruction of the spin-spin correlation ($S = 0$ or $S = 1$). The separation of hydrogen-like shallow donors should be ~ 50 Å (assuming $E_C - 0.05$ eV distance between individual centers); i.e., the local concentration should be $\sim 10^{17}$ cm^{-3}. Therefore, an isotropic (for X-band) or slightly anisotropic (for K-band), poorly resolved (NL10-like) EPR spectrum arising from the HSD defect can be described as a defect with $S = 1/2$ (Fig. 14b).

During annealing, the local concentration of the HSD centers is decreased, and isolated HSD centers appear. This is the (excited, $S = 1$) NL51 spectrum that is observed simultaneously with the (HSD) NL10-like spectrum. It is important to note that the spin-lattice relaxation time of the NL51 is much longer than that of the NL10-like defect (it is in agreement with the proposed model), so optimal conditions for the observation differ strongly even at low-frequency (12 Hz) field modulation. In Fig. 14 the operating conditions are more optimal for NL51 detection; the real intensity NL10-like spectrum is several times larger.

From this perspective, the intensity of IR electronic transitions should depend strongly on the local density of helium-like shallow donors in the cluster, and deep transitions should be suppressed with increasing donor density. This suggests that the excited-state ($S = 1$) NL51 spectrum arises from the simple structure of a hydrogen-associated shallow double donor in its neutral charge state, while the NL10-like EPR spectrum arises from regions with a dense distribution of NL51 centers (Gorelkinskii, 1992). In this case, however, it is proposed that the positive charge state of HDD (NL51) would yield to the AA1 (like to NL8) spectrum; i.e., the **g**-tensor of th NL51 center in the positive charge state should be changed from D_{2d} to C_{2v} symmetry. In this model it is also understood that there should be a high sensitivity of the (HSD) NL10-like ($S = 1/2$) EPR spectrum to illumination independently of concentration and treatment of the sample.

Although these considerations are in agreement with all experimental data, it is clear that more theoretical and experimental studies will be required to obtain an exact model and structural differences for these three HSD-related EPR centers (AA1, NL51, and NL10-like). In particular, uniaxial stress experiments and ENDOR measurements (as well as IR studies of electronic transitions of the HSD center) may give unambiguous answers to the key questions.

4. EPR EVIDENCE OF HYDROGEN-ENHANCED DIFFUSION OF AL IN SILICON

Hydrogen-enhanced impurity migration in silicon has been discovered recently (Newman and Jones, 1994; Murray, 1991). As a first example, the presence of hydrogen atoms leads to an enhancement of the diffusivity of interstitial oxygen. This enhancement was observed at $T \approx 300°C$ and was detected using measurements of the O_i reorientation rate, the rates of O_i loss from solution, and the rate of thermal donor formation. Theoretical models (Newman and Jones, 1994; Estreicher, 1990) have been proposed to explain this phenomenon.

More recently (Gorelkinskii et al., 1997b) the observation of hydrogen-enhanced migration of aluminum in silicon has been reported at low-temperature (<200 K). The conclusion regarding long-range aluminum migration derived from the detection of a new EPR spectrum (labeled Si-AA15) that is clearly identified as a defect including an Al-Al pair. The migration of Al atoms, which occurred in these experiments at $T \leq 200$ K, is neither recombination-enhanced (Troxell et al., 1979) nor radiation-induced migration and is thus apparently a hydrogen-catalyzed process.

a. Experimental Procedure

Samples from Fz-grown silicon doped with 5×10^{16} aluminum/cm^3 were used in these experiments. Hydrogen was incorporated either by proton implantation or by thermal annealing in the presence of hydrogen species. Implantation was carried out at a sample temperature of ~ 80 K to a total dose of $(1-5) \times 10^{15}$ H/cm^2, which corresponds to a concentration of $(1-5) \times 10^{17}$ H/cm^3. Alternatively, the samples were heated in the presence of water vapor; the "as grown" samples were sealed in a quartz ampule containing some distilled water and were heated to 1200–1250°C for 1/2 hr and were then allowed either to cool naturally to room temperature or were quenched in water. After etching to remove all surface damage, these samples, along with control samples (without intentional hydrogen incorporation), were irradiated at 80 K with 30-MeV protons. Because the proton range is much larger than the sample thickness, hydrogen implantation in the bulk of the sample was completely avoided in this case. After bombardment, the samples were installed in the cavity of a 37-GHz EPR spectrometer without warmup. The EPR measurements were performed at 77 K in the absorption mode.

b. Experimental Results and Discussion

Proton bombardment of the samples at 80 K produces the known EPR spectra of divacancies and of Si-B2 B2 (Lütgemeier and Schnitzke, 1967; Gorelkinskii and Nevinnyi, 1991). New spectra labeled Si-AA15 and Si-AA16 (Fig. 15) emerge on thermal annealing at 180–200 K (Fig. 16) both in the hydrogen-implanted samples and in the samples heat treated in the presence of water vapor (Gorelkinskii et al., 1997; Abdullin et al., 1997). AA15 and AA16 were not observed in the hydrogen-free (control) samples. The increased hydrogen concentration contributes to the growth of the intensity of the AA15 and AA16 spectra. These spectra are characterized by 11 groups (for AA15) and 6 groups (for AA16) of lines indicating that one

3 HYDROGEN AND HYDROGEN-RELATED DEFECTS IN CRYSTALLINE SILICON 71

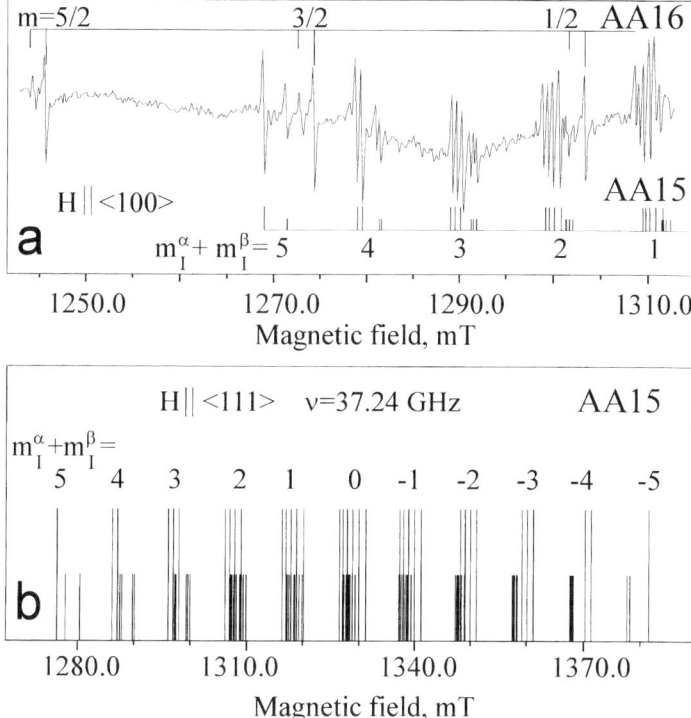

FIG. 15. (a) Low-field side of the Si-AA15 and Si-AA16 EPR spectra ($v = 37.28$ GHz) at 77 K, $\vec{H}\|[001]$. These spectra were observed after H implantation at 80 K of Fz-Si(Al) samples and annealing at 180 K for 10 min. (b) Computer fit of EPR line positions for the Si-AA15 spectrum, $\vec{H}\|\langle 111\rangle$, $v_0 = 37.24$ GHz. (Adapted from Abdullin et al., 1997.)

or two nuclei with spin $I_N = 5/2$ (respectively) are involved in the defect structure. These $I_N = 5/2$ nuclei in the silicon samples employed are 100% abundant and can only be ^{27}Al.

The AA15 spectrum can be described as arising from an anisotropic defect of C_{1h} symmetry with the spin Hamiltonian

$$H = \mu_B \vec{H} \vec{g} \vec{S} + \sum_{j=\alpha,\beta} \vec{S} \vec{A}_j \vec{I}_j \qquad (14)$$

with $S = 1/2$. The first term presents the electronic Zeeman interaction; the second term is the hyperfine interaction. The defect contains two Al nuclei (labeled α and β). The principal g and A values and axes (Fig. 17) were

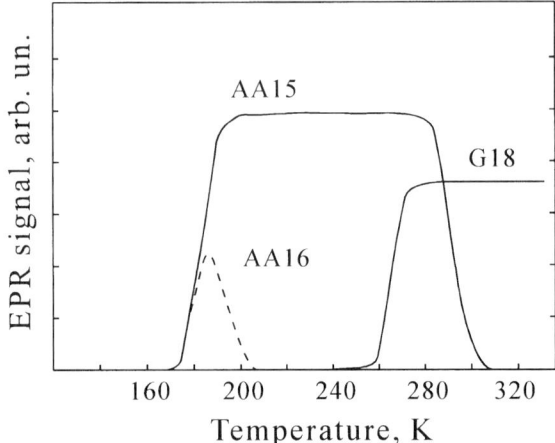

FIG. 16. Temperature region formation and stability of the AA15, AA16 spectra. An Fz-Si(Al) sample was implanted with protons at ~80 K. (Adapted from Gorelkinskii et al., 1997.)

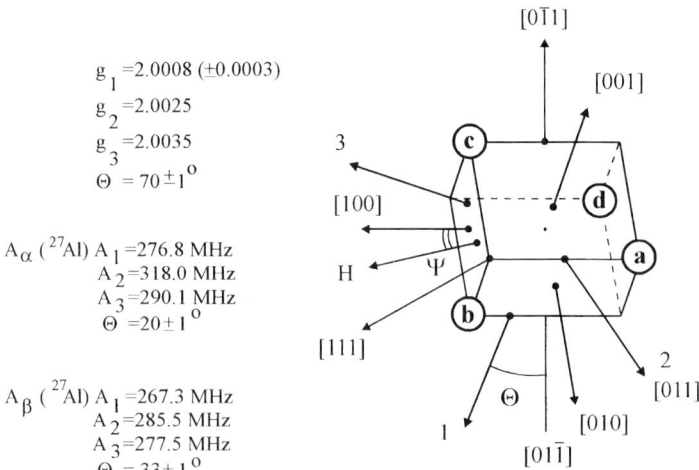

$g_1 = 2.0008\ (\pm 0.0003)$
$g_2 = 2.0025$
$g_3 = 2.0035$
$\Theta = 70 \pm 1^\circ$

$A_\alpha\ (^{27}\text{Al})\ A_1 = 276.8$ MHz
$A_2 = 318.0$ MHz
$A_3 = 290.1$ MHz
$\Theta = 20 \pm 1^\circ$

$A_\beta\ (^{27}\text{Al})\ A_1 = 267.3$ MHz
$A_2 = 285.5$ MHz
$A_3 = 277.5$ MHz
$\Theta = 33 \pm 1^\circ$

FIG. 17. Spin Hamiltonian constants and relevant principal axes for one of the twelve Si-AA15 defect orientations in the lattice. (Adapted from Abdullin et al., 1997.)

determined from the angular dependence of the experimental spectrum using second-order perturbation theory (Elkin and Watkins, 1968). There is a deviation of ~ 0.3 mT between calculated and experimental line positions when $H \| \langle 111 \rangle$. The discrepancy is likely to result from ignoring an anisotropic quadruple interaction of the Al nuclei. Although the AA15 spectrum is very intense, the signal-to-noise ratio of its individual lines was insufficient for the identification of forbidden transitions; the spectrum is split into 252 lines for an arbitrary magnetic field direction in the $\{011\}$ plane.

The second spectrum Si-AA16 ($S = 1/2$), has a nearly isotropic **g**-tensor with $\bar{g} \approx 2.0030$. Its ^{27}Al hyperfine interaction is nearly isotropic with weak trigonal distortion: $A_\| = 823$ MHz, $A_\perp = 836$ MHz (Abdullin et al., 1997). It is important to note that the width of the individual lines of the AA16 spectrum depends on the magnetic field orientation; this effect may be caused by weak hf interaction with protons.

The hyperfine interaction was analyzed in terms of a one-electron wave function for the unpaired electron (Watkins, 1975; Lee et al., 1976). Using the hyperfine parameters for $3s$ and $3p$ wave functions of ^{27}Al (Watkins and Corbett, 1964) ($|\psi_{3s}(0)|^2 = 20.4 \times 10^{24}$ cm^{-3} and $\langle r_{3p}^{-3} \rangle = 8.95 \times 10^{24}$ cm^{-3}), the observed hyperfine structure of the AA15 defect indicates that $\sim 18\%$ and $\sim 10\%$ of the unpaired spin wave function are located on the α and β aluminum atoms, respectively, and that it has $\sim 45\%$ $3s$ and $\sim 55\%$ $3p$ character. A resolved ^{29}Si hyperfine structure also can be seen whose intensity corresponds to ~ 4–6 neighboring sites; we allowed for the $\sim 5\%$ isotopic abundance. The same analysis for the AA16 center indicates that about 30% of the resonant wave function is localized on a single aluminum atom, with mostly $3s$-like character ($\sim 80\%$). Hyperfine interaction with the ^{27}Al nucleus in the case of an isolated aluminum interstitial (EPR spectrum G18; Watkins, 1964, 1994) is completely isotropic (100% $3s$).

The AA15 spectrum is dominant after annealing at 200 K, and its intensity does not change with annealing up to 300 K (see Fig. 16). After a 30-min anneal at room temperature, the AA15 spectrum disappears. We note that the Si-G18 spectrum (Al$_i$) (Watkins, 1994) is not observed in implanted or control samples immediately after 80 K irradiation, but that G18 appears in all samples annealed at ~ 260–280 K; there is no correlation of the AA15 and G18 spectral behavior. The second spectrum (AA16) disappears at 200–220 K. The narrow temperature region where the defect is observed by EPR may be due to mobile H in the BC position at these temperatures (see Sec. II.3.c); the AA16 is probably a precursor of the AA15 defect.

The AA15 defect formation that includes two aluminum atoms is strong evidence of low-temperature, long-range motion of Al atoms, at least over a distance of ~ 100 lattice constants. Aluminum interstitials usually migrate at ~ 500 K (Watkins, 1994). Recombination-enhanced migration may occur at room temperature (Troxell et al., 1979) but cannot explain our result because the characteristic time is too long at ~ 80 K. We suppose that aluminum atoms migrate by a hydrogen-enhanced mechanism; the AA15 spectrum is not observed in control samples and emerges only with increased hydrogen concentration. Consequently, the presence of a radiation defect and hydrogen in the samples are required to create Al-Al pairs (AA15 defect), and the diffusion takes place in the form of some radiation-induced interstitial $(Al + H)_i$ complex. This diffusion may occur in the process of irradiation at ~ 80 K or annealing at $T < 200$ K. Hydrogen's role may be to decrease the thermal barrier for the migration of Al atoms, similar to the influence of hydrogen on the oxygen migration (Newman and Jones, 1994).

The presence of an aluminum interstitial is necessary to create the $(Al_i + H)$ interstitial defect [by the reaction of substitution: $Al_s + Si_i \to Al_i + Si_s$ (Watkins, 1994)]. The Si-G18 spectrum (Al_i) has not been observed; consequently, it may be concluded that most of the Al_i are involved in $(Al_i + H)$ defects. These defects have to be mobile at temperature of 200 K or lower to ensure the formation of Al-Al pairs (AA15 complexes). The AA16 center is a suitable candidate for the $(Al_i + H)$ defect because the angular dependence of its linewidth may result from weak hyperfine interaction with protons. For the AA16 center, about 30% of the wave function is associated with an Al atom. This is comparable with the case of an Al_i tetrahedral interstitial ($\sim 38\%$) (Watkins and Corbett, 1964). The atomic orbital has $20\% p$ character. This fact implies that the Al atom is slightly displaced from the T_d site by some distortion.

In contrast to the case of known Al_i-Al_s pairs (G19, G20), where the hyperfine parameters for Al_i and Al_s atoms differ considerably, both Al atoms incorporated in the AA15 defect have similar hyperfine parameters. Hence both Al atoms are nearly equivalent and lie in interstitial positions that would be equivalent in a perfect lattice. An additional slight distortion lowers the defect symmetry down to C_{1h} and results in nonequivalency of the atom positions. The distortion may be due to the Jahn-Teller effect or to the presence of an additional defect close to one of the Al atoms; such a defect may be a hydrogen atom. It is known that hydrogen can be involved in the molecular structure of a defect and indeed can radically alter the defect properties, while the presence of hydrogen remains undetected by simple EPR methods (Martynov et al., 1995a; see also Sec. III.4).

As a result of the interaction between a mobile Al_i atom and a substitutional Al_s atom, a $\langle 110 \rangle$ or $\langle 100 \rangle$ split interstitial distorted to C_{1h} may be formed; the $\langle 110 \rangle$ split interstitial seems to be preferable. Indeed, the paramagnetic electron in case of the $\langle 100 \rangle$ split interstitial defect must be localized on two perpendicular orbitals with $3p$ character, as for the case of carbon interstitials (Watkins and Brower, 1976). The resonant wave function of the AA15 defect has mainly $3s$ character. Moreover, the axes of α and β orbitals are parallel. There are other conceivable models for the AA15 center, i.e., the vacancy model with two nearest substitutional Al atoms or two Al_i atoms located close to the two nearest hexagonal interstitial sites (Abdullin *et al.*, 1997). However, the probability of low-temperature formation of a complex involving two or more radiation defects is very small.

In summary, EPR evidence has been obtained that the presence of hydrogen can strongly enhance the migration rate of Al atoms in silicon at 200 K or lower. The unambiguous evidence appears from the observation of Al-Al pairs in Al-doped silicon after hydrogen implantation or irradiation at 80 K. Future investigations replacing hydrogen with deuterium and applying uniaxial stress can give important data on exact models of the EPR centers.

V. Summary and Conclusions

Methods of electron paramagnetic resonance have yielded much direct information about the molecular and electronic structures of hydrogen and hydrogen-related complexes in silicon. Advancements of recent EPR studies have been described briefly in this chapter.

Other EPR studies of atomic reorientation of the H in BC configuration unambiguously confirmed previous theoretical data that $H^+(BC)$ really exists and Si atoms relax outward from hydrogen. It has been shown experimentally that the $H^0(BC)$ [in contrast to $H^+(BC)$] cannot be reoriented between equivalent sites and jumps directly to a tetrahedral position with increasing temperature.

Several hydrogen-impurity complexes in silicon, such as PtH_2 and $(SH)^0$, with resolved hyperfine interaction from proton(s) have been discovered and exact models extracted. A new approach for incorporation of hydrogen into crystal at high temperature has been used successfully in these works and may stimulate future EPR experiments.

By means of ENDOR, hydrogen was found in the structure of oxygen thermal shallow donor. Thus the oxygen-related NL10 center has been identified as singly passivated by a hydrogen thermal double donor. These experiments directly demonstrate that hydrogen can play a vital role in the formation of extended defects such as thermal donors.

The EPR spectra and electrical properties of the hydrogen-associated shallow donor (HSD) and oxygen thermal donor are very similar and mysterious. Recent EPR measurements have shown unambiguously that the molecular structures of these defects are different. The presence of oxygen in silicon suppresses the HSD formation rate. It is suggested that (excited $S = 1$) the NL51 spectrum arises from a helium-like double donor in the neutral charge state and is the simplest structure of HSD. The NL10-like ($S = 1/2$) spectrum arises from the same centers (NL51) concentrated in clusters. Because of wave function overlapping, the states $S = 1$ and/or $S = 0$ are disordered, and a poorly resolved ($S = 1/2$) NL10-like spectrum is observed.

EPR evidence has been obtained showing that the presence of hydrogen can strongly enhance the migration rate of Al atoms in silicon at temperatures below 200 K. The unambiguous evidence appears from the observation of Al-Al pairs in Al-doped silicon after hydrogen implantation or irradiation of a hydrogen-containing sample at 80 K.

In the last several years the concerted effort by both theory and experiment has resulted in much progress in understanding of the behavior and properties of interstitial hydrogen in silicon (and other semiconductors). However, only by means of EPR has the bond-centered $H^0(BC)$ configuration in silicon been identified. It appears that additional successful EPR investigations of atomic hydrogen in its different positions in the lattice can be expected on perfect silicon crystals. To this general remark, an exception must be made for the high concentration of radiation defects introduced into silicon during hydrogen implantation. For example, a hydrogen species can be introduced above room temperature with subsequent liberation of atomic hydrogen by means of electron (or X-ray) irradiation at low temperature. In this case, the EPR observation of the $H^0(T_d)$ as well as other configurations of atomic hydrogen is expected. Thus it is not known whether the ground states of the $H^-(T_d)$ and $H^-(BC)$ are $S = 1$ or $S = 0$, although at low temperatures (helium region) in perfect n-type crystals we can hope to obtain unambiguous answers. Using a variety of n- and p-type silicon samples, we hope to discover the paramagnetic states of different forms of the hydrogen molecule that are (1) predicted theoretically and (2) observed by other methods. An EPR investigation of the simplest hydrogen intrinsic-defect complexes, such as H-self-interstitial, vacancy-hydrogen, and hydrogen-impurity complexes (with oxygen, carbon, transition metals, doped impurity, etc.) also would be important for the development of basic and applied semiconductor physics.

ACKNOWLEDGMENTS

I would like to thank Kh. A. Abdullin and N. N. Nevinnyi for useful discussions and contributions to this work. Kh. A. Abdullin read the manuscript and made valuable suggestions.

References

Abdullin, Kh. A., Mukashev, B. N., and Gorelkinskii, Yu. V. (1997). *Appl. Phys. Lett.*, **71**(12), 1703.
Ammerlaan, C. A. J., Gregorkiewicz, T., and Gorelkinskii, Yu. V. (1992). (Unpublished).
Bekman, H. H. P. Th., Gregorkiewicz, T., and Ammerlaan, C. A. J. (1988). *Phys. Rev. Lett.*, **61**, 227.
Bekman, H. H. P. Th., Gregorkiewicz, T., and Ammerlaan, C. A. J. (1989). *Phys. Rev. B*, **39**, 1648.
Benton, J. L., Kimmerling, L. C., and Stavola, M. (1983). *Physica*, **116B**, 271.
Bergman, K., Stavola, M., Pearton, J. S., and Lopata, J. (1988). *Phys. Rev. B*, **37**, 2770.
Binns, M. J., Londos, C. A., McQuaid, S. A., Newman, R. C., Semaltianos, N. G., and Tucker, J. H. (1996). *J. Mater. Sci. Mater. Electron.*, **7**, 347.
Blöchl, P. E., Van de Walle, C. G., and Pantelides, S. T. (1990). *Phys. Rev. Lett.*, **64**, 1401.
Boucher, D. E., and Deleo, G. G. (1994). *Phys. Rev. B*, **50**, 5247.
Bratus, V. Ya., Ischenko, S. S., Okulov, S. M., Vorona, I. P., von Bardeleben, H. J., and Schoisswohl, M. (1994). *Phys. Rev. B*, **50**, 1.
Bratus, V. Ya., Ishchenko, S. S., Okulov, S. M., Vorono, I. P., and von Bardeleben, H. J. (1995). *Mater. Sci. Forum*, **196–201**, 529.
Brower, K. L. (1976). *Phys. Rev. B*, **14**, 872.
Brower, K. L. (1983). *J. Appl. Phys.*, **43**, 1111.
Caplan, P. J., Poindexter, E. H., Deal, B. E., and Rasowr, R. R. (1979). *J. Appl. Phys.*, **50**, 5847.
Carrington, A., and McLachlan, A. D. (1967). *Introduction to Magnetic Resonance with Applications to Chemistry and Chemical Physics*. New York: Harper & Row.
Chang, K. J., and Chadi, D. J. (1989a). *Phys. Rev. Lett.*, **62**, 937.
Chang, K. J., and Chadi, D. J. (1989b). *Phys. Rev. B*, **40**, 11644.
Cox, S. F. J., and Symons, M. C. R. (1986). *Chem. Phys. Lett.*, **126**, 516.
Daly, D. F. (1969). *Appl. Phys. Lett.*, **18**, 267.
Daly, D. F. (1971). *J. Appl. Phys.*, **42**, 864.
Deák, P., Synder, L. C., Lindström, J. L., Corbett, J. W., Pearton, S. J., and Tavendale, A. J. (1988). *Phys. Lett. A*, **126**, 427.
Denteneer, P. J. H., Van de Walle, C. G., and Pantelides, S. T. (1990). *Phys. Rev. B*, **41**, 3885.
Elkin, E. L., and Watkins, G. D. (1968). *Phys. Rev.*, **174**, 881.
Estle, T. L., Estreicher, S. K., and Marynich, D. S. (1987). *Phys. Rev. Lett.*, **58**, 1547.
Estreicher, S. K. (1987). *Phys. Rev. B*, **36**, 9122.
Estreicher, S. (1990). *Phys. Rev. B*, **141**, 9886.
Gale, R., Feigel, F. J., Magee, C. W., and Young, D. R. (1983). *J. Appl. Phys.*, **54**, 6938.
Gerasimenko, N. N., Roll'e, M., Cheng, L. J., Lee, Y. H., Corelli, J. C., and Corbett, J. W. (1978). *Phys. Stat. Sol. B*, **90**, 689.
Gorelkinskii, Yu. V. (1992). (Unpublished).
Gorelkinskii, Yu. V., and Nevinnyi, N. H. (1981a). *Pis'ma Zn. Tekh. Fiz.*, **7**, 1044.
Gorelkinskii, Yu. V., and Nevinnyi, N. H. (1981b). In *Materials of the International Working Meeting on Ion Implantation in Semiconductors and Other Materials*, M. Setvak, ed. Prague: Technical University of Prague, p. 41.
Gorelkinskii, Yu. V., and Nevinnyi, N. N. (1983a). *Nucl. Inst. Meth.*, **209/210**, 677.
Gorelkinskii, Yu. V., and Nevinnyi, N. N. (1983b). *Radiation Effects*, **71**, 1.
Gorelkinskii, Yu. V., and Nevinnyi, N. N. (1987). *Sov. Tech. Phys. Lett.*, **13**, 45.
Gorelkinskii, Yu. V., and Nevinnyi, N. N. (1991). *Physica B*, **170**, 155.
Gorelkinskii, Yu. V., and Nevinnyi, N. N. (1996). *Mater. Sci. Engrg.*, **B36**, 133.
Gorelkinskii, Yu. V., Sigle, V. O., and Takibaev, Zh. S. (1974). *Phys. Stat. Sol. (a)*, **22**, K55.
Gorelkinskii, Yu. V., Nevinnyi, N. N., and Botvin, V. A. (1976). *Fiz. Tech. Poluprov.*, **10**, 2256.

Gorelkinskii, Yu. V., Nevinnyi, N. N., and Botvin, V. A. (1980). *Radiation Effects*, **49**, 161.
Gorelkinskii, Yu. V., Nevinnyi, N. N., and Ajazbaev, S. S. (1987). *Phys. Lett. A*, **125**, 354.
Gorelkinskii, Yu. V., Nevinnyi, N. N., and Lyuts, E. A. (1994). *Semiconductors*, **28**, 23.
Gorelkinskii, Yu. V., Mukashev, B. N., and Abdullin, Kh. A. (1997). *Mater. Sci. Forum*, **258–263**, 1773.
Gorelkinskii, Yu. V., Nevinnyi, N. N., and Abdullin, Kh. A. (1998). *J. Appl. Phys.*, **84**, 4847.
Gregorkiewicz, T., Bekman, H. H. P. Th., and Ammerlaan, C. A. J. (1988). *Phys. Rev. B*, **38**, 3998.
Gregorkiewicz, T., van Wezep, D. A., Bekman, H. H. P. Th., and Ammerlaan, C. A. J. (1987). *Phys. Rev. B*, **35**, 3810.
Hartung, J., and Weber, J. (1993). *Phys. Rev. B*, **48**, 14161.
Hartung, J., and Weber, J. (1995). *J. Appl. Phys.*, **77**(1), 118.
Hofmann, D. M., Meyer, B. K., Christmann, P., Wimbauer, T., Stadler, W., Nikolov, A., Scharmann, A., and Hofstätter, A. (1995). *Mater. Sci. Forum*, **196–201**, 1673.
Höhne, M., Juda, U., Martynov, Yu. V., Gregorkiewicz, T., Ammerlaan, C. A. J., and Vlasenko, L. S. (1994a). *Phys. Rev. B*, **49**, 13423.
Höhne, M., Juda, U., Martynov, Yu. V., Gregorkiewicz, T., Ammerlaan, C. A. J., and Vlasenko, L. S. (1994b). *Mater. Sci. Forum*, **143–147**, 1659.
Holm, B., Nielsen, K. B., and Nielsen, B. B. (1991). *Phys. Rev. Lett.*, **66**, 2360.
Irmscher, K., Klose, H., and Maass, K. (1984). *J. Phys. C*, **17**, 6317.
Johnson, N. M., and Herring, C. (1994). *Mater. Sci. Forum*, **143–147**, 867.
Johnson, N. M., Herring, C., and Chadi, D. J. (1986). *Phys. Rev. Lett.*, **56**, 769.
Johnson, N. M., Herring, C., and Van de Walle, C. G. (1994). *Phys. Rev. Lett.*, **73**, 130.
Johnson, N. M., Doland, C., Ponce, F., Walker, J., and Anderson, G. (1991). *Physica B*, **170**, 3.
Jones, R., ed. (1996). *Early Stages of Oxygen Precipitation in Silicon*. NATO series. Boston: Kluwer Academic Publishers.
Kadono, R., Matsushita, A., Macrae, R. M., Nishiyama, K., and Nagamine, K. (1994). *Phys. Rev. Lett.*, **73**, 2724.
Kaplyanskii, A. A. (1964). *Opt. Spektrosk. (USSR)*, **16**, 329.
Kiefl, R. F., Celio, M., Estle, T. L., Kreitzman, S. R., Luke, G. M., Riseman, T. M., and Ansaldo, E. J. (1988). *Phys. Rev. Lett.*, **60**, 224.
Kiefl, R. F., and Estle, T. L. (1991). In *Hydrogen in Semiconductors*, J. I. Pankove and N. M. Jonson, eds. San Diego: Academic Press.
Kimmerling, L. C. (1986). *Mater. Res. Soc. Symp. Proc.*, **59**, 83.
Kleinhenz, R. L., Lee, Y. H., Singh, V. A., Mooney, P. M., Jaworowskyi, A., Roth, L. M., Corelli, J. C., and Corbett, J. W. (1979). *Inst. Phys. Conf. Ser.*, **46**, 200.
Kreitzman, S. R., Hitti, B., Lichti, R. L., Estle, T. L., and Chow, K. H. (1995). *Phys. Rev. B*, **51**, 13117.
Kuten, S. A., Rapoport, V. I., Mudry, A. V., Gelfand, P. B., Pushkarchuk, A. L., and Ulyashin, A. G. (1988). *Hyp. Int.*, **39**, 379.
Lee, Y.-H., Kim, Y. M., and Corbett, J. W. (1972). *Radiation Effects*, **15**, 77.
Lee, Y.-H., and Corbett, J. W. (1973). *Phys. Rev. B*, **8**, 2810.
Lee, Y.-H., and Corbett, J. W. (1974). *Phys. Rev. B*, **9**, 4351.
Lee, Y.-H., and Corbett, J. W. (1976). *Phys. Rev. B*, **13**, 2653.
Lee, Y.-H., Gerasimenko, N. N., and Corbett, J. W. (1976). *Phys. Rev. B*, **14**, 4506.
Lightowlers, E. C. (1995). *Mater. Sci. Forum*, **196–201**, 817.
Lütgemeier, H., and Schnitzke, K. (1967). *Phys. Lett. A*, **25**, 267.
Lyding, J. W., Hess, K., and Kizilyalli, I. C. (1996). *Appl. Phys. Lett.*, **68**, 2526.
Markevich, V. P., Suezawa, M., Sumino, K., and Murin, L. I. (1994). *J. Appl. Phys.*, **76**, 7347.
Martynov, Yu. V., Gregorkiewicz, T., and Ammerlaan, C. A. J. (1995a). *Phys. Rev. Lett.*, **74**, 2030.

Martynov, Yu. V., Gregorkiewicz, T., and Ammerlaan, C. A. J. (1995b). *Mater. Sci. Forum*, **196–201**, 849.
McQuaid, S. A., Newman, R. C., Tucker, J. H., Lightowlers, E. C., Kubiak, R. A. A., and Goulding, M. (1991). *Appl. Phys. Lett.*, **58**, 2933.
McQuaid, S. A., Binns, M. J., Londos, C. A., Tucker, J. H., Brown, A. R., and Newman, R. C. (1995). *J. Appl. Phys.*, **77**, 1427.
Meilwes, N., Spaeth, J.-M., Emtsev, V. V., Oganesyan, G. A., Götz, W., and Pensl, G. (1994). *Mater. Sci. Forum*, **143–147**, 141.
Meng, X. T. (1991). *Physica B*, **170**, 249.
Michel, J., Niklas, J. R., and Spaeth, J.-M. (1986). *Mater. Res. Soc. Symp. Proc.*, **59**, 111.
Michel, J., Niklas, J. R., and Spaeth, J.-M. (1989). *Phys. Rev. B*, **40**, 1732.
Mukashev, B. N., Tamendarov, M. F., and Tokmoldin, S. Zh. (1989). *Mater. Sci. Forum*, **38–41**, 1039.
Muller, S. H., Sprenger, M., Sieverts, E. G., and Ammerlaan, C. A. J. (1978). *Solid State Commun.*, **25**, 987.
Murray, R. (1991). *Physica B*, **170**, 115.
Myers, S. M., Baskes, M. I., Birnbaum, H. K., Corbett, J. W., DeLeo, G. G., Estreicher, S. K., Haller, E. E., Jena, P., Johnson, N. M., Kirchheim, R., Pearton, S. J., and Stavola, M. J. (1992). *Rev. Mod. Phys.*, **64**, 559.
Newman, R. C. (1985). *J. Phys. C: Solid State Phys.*, **18**, L967.
Newman, R. C. (1996). *Mater. Sci. Engrg.*, **B36**, 1.
Newman, R. C., and Jones, R. (1994). In *Semiconductors and Semimetals*, **42**, F. Shimura, ed. San Diego: Academic Press, p. 289.
Newman, R. C., Tucker, J. H., Semaltianos, N. G., Lightowlers, E. C., Gregorkiewicz, T., Zevenbergen, I. S., and Ammerlaan, C. A. J. (1996). *Phys. Rev. B*, **54**, R6803.
Nickel, N. H., Jackson, W. B., Wu, I. W., Tsai, C. C., and Chiang, A. (1995). *Phys. Rev. B*, **52**, 7791.
Nielsen, B. B., Nielsen, K. B., and Byberg, J. R. (1994). *Mater. Sci. Forum*, **143–147**, 909.
Nielsen, B. B., Hoffmann, L., Budde, M., Jones, R., Goss, J., and Öberg, S. (1995). *Mater. Sci. Forum*, **196–201**, 933.
Nielsen, B. B., Johannesen, P., Stallinga, P., Nielsen, K. B., and Byberg, J. R. (1997). *Phys. Rev. Lett.*, **79**, 1507.
Ohmura, Y., Zohta, Y., and Kanazawa, M. (1972). *Solid State Commun.*, **11**, 263.
Ohmura, Y., Zohta, Y., and Kanazawa, M. (1973). *Phys. Stat. Sol. (a)*, **15**, 93.
Pankove, J. I., and Johnson, N. M., eds. (1991). *Hydrogen in Semiconductors*. San Diego: Academic Press.
Pankove, J. I., Carlson, D. E., Berkeyheiser, J. E., and Wance, R. O. (1983). *Phys. Rev. Lett.*, **51**, 2224.
Paschedag, N. (1996). PhD thesis, University of Zurich, Switzerland.
Paschedag, N., Suter, H., Maric, D., and Meier, P. (1993) *Phys. Rev. Lett.*, **70**, 154.
Patterson, B. D. (1988). *Rev. Mod. Phys.*, **60**, 69.
Pearton, S. J., Corbett, J. W., and Shi, T. S. (1987). *Appl. Phys. A*, **43**, 153.
Pearton, S. J., Corbett, J. W., and Stavola, M. (1992). *Hydrogen in Crystalline Semiconductors*. Berlin: Springer-Verlag.
Schwuttke, G. H., Brack, K., Gorey, E. F., Kahan, A., and Lowe, L. F. (1970). *Radiation Effects*, **6**, 103.
Sah, C. T., Sun, J. Y. C., and Tzou, J. J. T. (1983). *Appl. Phys. Lett.*, **43**, 204.
Seager, C. H., and Anderson, R. A. (1988). *Appl. Phys. Lett.*, **53**, 1181.
Seager, C. H., and Anderson, R. A. (1990). *Solid State Commun.*, **76**, 285.
Sieverts, E. G. (1983). *Phys. Stat. Sol. (b)*, **120**, 11.
Stallinga, P. (1994). Ph.D. thesis, University of Amsterdam, The Netherlands.

Stallinga, P., Gregorkiewicz, T., Ammerlaan, C. A. J., and Gorelkinskii, Yu. V. (1993). *Phys. Rev. Lett.*, **71**, 117.
Stallinga, P., Gregorkiewicz, T., Ammerlaan, C. A. J., and Gorelkinskii, Yu. V. (1994a). *Mater. Sci. Forum*, **143–147**, 853.
Stallinga, P., Gregorkiewicz, T., Ammerlaan, C. A. J., and Gorelkinskii, Yu. V. (1994b). *Solid State Commun.*, **90**, 401.
Stavola, M., and Lee, K. M. (1986). *Mater. Res. Soc. Symp. Proc.*, **59**, 95.
Stavola, M., Zheng, J.-F., Cheng, Y. M., Abernathy, C. R., and Pearton, S. J. (1995). *Mater. Sci. Forum*, **196–201**, 809.
Stein, H. J. (1975). *J. Electron. Mater.*, **4**, 159.
Stein, M. J. (1979). *Phys. Rev. Lett.*, **43**, 1030.
Tokmoldin, S. Zh., Mukashev, B. N., Tamendarov, M. F., and Chasnikova, S. S. (1990). In *Defect Control in Semiconductors*, K. Sumino, ed. New York: Elsevier Science Publishers, p. 425.
Torres, V. J. B., Öberg, S., and Jones, R. (1996). In *Shallow Lewels Centers in Semiconductors*, C. A. J. Ammerlaan and B. Pajot, eds. London: World Scientific, p. 501.
Trombetta, J. M., Watkins, G. D., Hage, J., and Wagner, P. (1997). *J. Appl. Phys.*, **81**, 1109.
Troxell, J. R., Chatterjee, A. P., Watkins, G. D., and Kimerling, L. C. (1979). *Phys. Rev. B*, **19**, 5336.
Uftring, S. J., Stavola, M. J., Williams, P. M., and Watkins, G. D. (1995). *Phys. Rev. B*, **51**, 9612.
Van de Walle, C. G. (1991). *Physica B*, **170**, 21.
Van de Walle, C. G. (1994). *Phys. Rev. B*, **49**, 4579.
Van de Walle, C. G., and Blöchl, P. E. (1993). *Phys. Rev. B*, **47**, 4244.
Van de Walle, C. G., and Street, R. A. (1994). *Phys. Rev. B*, **49**, 14766.
Van de Walle, C. G., and Street, R. A. (1995). *Phys. Rev. B*, **51**, 10615.
Van de Walle, C. G., and Nickel, N. H. (1995). *Phys. Rev. B*, **51**, 2636.
Van de Walle, P. J. H., Bar-Yam, Y., and Pantelides, S. T. (1988). *Phys. Rev. Lett.*, **60**, 2761.
Van de Walle, C. G., Denteneer, P. J. H., Bar-Yam, Y., and Pantelides, S. T. (1989). *Phys. Rev. B*, **39**, 10791.
Van Wieringen, A., and Warmoltz, N. (1956). *Physica*, **22**, 849.
Velorisoa, I. A., Stavola, M., Kozuch, D. M., Peale, R. E., and Watkins, G. D. (1991). *Appl. Phys. Lett.*, **59**, 2121.
Wagner, P., Gottschalk, H., Trombetta, J., and Watkins, G. D. (1987). *J. Appl. Phys.*, **61**, 346.
Wang, Z.-Y., and Lin, L.-Y. (1984). In *Neutron Transmutation Doping of Semiconductors Materials*, R. D. Larrabee, ed. New York: Plenum Press, p. 311.
Watkins, G. D. (1964). In *Radiation Damage in Semiconductors*, P. Baruch, ed. Paris: Dunod, pp. 97–113.
Watkins, G. D., and Corbett, J. W. (1965). *Phys. Rev.*, **138**, A343.
Watkins, G. D. (1975). *Phys. Rev. B*, **12**, 5824.
Watkins, G. D. (1994). *Mater. Sci. Forum*, **143–147**, 9.
Watkins, G. D. (1996). In *Early Stages of Oxygen Precipitation in Silicon*, R. Jones, ed. NATO series. Boston: Kluwer Academic Publishers.
Watkins, G. D., and Brower, K. L. (1976). *Phys. Rev. Lett.*, **36**, 1329.
Watkins, G. D., and Corbett, J. W. (1964). *Phys. Rev.*, **134**, A1359.
Williams, P. M., Watkins, G. D., Uftring, S., and Stavola, M. (1994). *Mater. Sci. Forum*, **143–147**, 891.
Zevenbergen, I. S., Gregorkiewicz, T., and Ammerlaan, C. A. J. (1995a). *Phys. Rev. B*, **51**, 16746.
Zevenbergen, I. S., Gregorkiewicz, T., and Ammerlaan, C. A. J. (1995b). *Mater. Sci. Forum*, **196–201**, 855.

Zevenbergen, I. S., Martynov, Yu. V., Rasmussen, F. B., Gregorkiewicz, T., and Ammerlaan, C. A. J. (1996). *Mater. Sci. Engrg.*, **B36**, 138.
Zheng, J.-F., and Stavola, M. (1996). *Phys. Rev. Lett.*, **76**, 1154.
Zhou, Y., Luchsinger, R., and Meier, P. F. (1995). *Mater. Sci. Forum*, **196–201**, 885.

CHAPTER 4

Hydrogen in Polycrystalline Silicon

*Norbert H. Nickel**

HAHN-MEITNER-INSTITUT BERLIN
DEPARTMENT OF APPLIED PHYSICS
BERLIN, GERMANY

I. INTRODUCTION 83
II. EXPERIMENTAL TECHNIQUES 85
   1. Sample Preparation and Characterization 85
   2. Hydrogen Passivation. 86
   3. Characterization 86
III. HYDROGEN DIFFUSION 87
   1. Hydrogen Diffusion from a Plasma Source 87
   2. Hydrogen Diffusion from a Silicon Layer 98
   3. Hydrogen Density of States 102
IV. HYDROGEN PASSIVATION OF GRAIN-BOUNDARY DEFECTS 110
V. METASTABILITY 124
   1. Light-Induced Defect Generation 125
   2. Metastable Changes in the Electrical Conductivity 132
VI. HYDROGEN-INDUCED DEFECTS DURING PLASMA EXPOSURE 143
   1. Generation of Acceptor-Like Defects 144
   2. Platelets 152
VII. SUMMARY AND FUTURE DIRECTIONS 159
   REFERENCES 161

I. Introduction

In contrast to single-crystal silicon (c-Si), polycrystalline silicon (poly-Si) is an inhomogeneous material in which identical crystallites are separated by grain boundaries. These two-dimensional boundaries accommodate high concentrations of strained Si—Si bonds and Si dangling bonds. In the past, polycrystalline silicon has been investigated intensively with the aim to obtain device-grade material suitable for photovoltaic applications (Shi and Green, 1994) and large-area liquid-crystal displays (Hawkins, 1986). A great

*This work was performed at the Xerox Palo Alto Research Center in Palo Alto, California.

effort has been dedicated to the passivation of grain-boundary defects in order to improve electronic and optical properties of poly-Si. Anneals of poly-Si in atomic hydrogen resulted in an improvement in the electrical characteristics of bulk (Seager and Ginley, 1979) and thin-film poly-Si (Kamins and Marcoux, 1980; Johnson et al., 1982b). The microscopic mechanism of hydrogen passivation was investigated by electron paramagnetic resonance (EPR) measurements. Polycrystalline silicon films were found to contain large concentrations of paramagnetic defects with a g value of 2.0055, which is characteristic for silicon dangling bonds. Introducing hydrogen or deuterium into poly-Si samples at elevated temperatures resulted in a significant decrease of the EPR resonance (Johnson et al., 1982a). Optical absorption measurements on fine-grain poly-Si films indicated that the Si dangling-bond defect is located 0.65 ± 0.15 eV below the conduction-band minimum in the grain boundary. In addition, from photothermal deflection spectroscopy measurements, Jackson et al. (1983) found evidence for exponential tailing of the band edges. It is widely believed that exponential band tails, observed in amorphous and polycrystalline silicon, are due to variations in the Si—Si bond angle. Hydrogen passivation experiments showed that the exponential slope of the subgap absorption varied from 120 meV for unhydrogenated poly-Si films to 85 meV for completely passivated samples (Jackson et al., 1983). This experiment indicated that hydrogen not only passivates Si dangling bonds but also breaks and passivates strained Si—Si bonds. The ability of hydrogen to passivate deep and shallow defects also was observed in current deep-level transient spectroscopy (DLTS) measurements performed on poly-Si thin-film transistors (Ayres, 1993).

In recent years, almost all research related to hydrogen in poly-Si focused on the optimization of the hydrogenation conditions. The influence of substrate temperature and hydrogenation time on the electrical and optical properties of poly-Si was explored using a variety of hydrogen sources and techniques such as electron-cyclotron resonance (ECR) plasma sources, radiofrequency (rf) glow-discharge systems, ion-beam systems, proton implantation, and electrochemical techniques. A detailed description of hydrogenation techniques can be found elsewhere (Seager, 1991).

An enhancement of the hydrogenation efficiency in poly-Si was observed after postultrasound treatment. An ultrasound treatment below 100°C caused a pronounced decrease in the sheet resistance by up to two orders of magnitude (Ostapenko et al., 1996). This is due to the release of hydrogen atoms from a reservoir. The excess hydrogen is *not* passivating preexisting Si dangling bonds and hence can be used to improve the electrical properties of poly-Si (Nickel et al., 1993).

4 HYDROGEN IN POLYCRYSTALLINE SILICON

In this chapter I review properties of hydrogen in poly-Si films that were discovered recently. The chapter is organized as follows: Section II briefly describes sample preparation, hydrogen passivation, and the experimental techniques used to characterize the specimens. The diffusion properties of hydrogen in poly-Si are discussed in Section III. In Section IV the kinetics of defect passivation are presented. As a consequence of the presence of hydrogen in poly-Si, metastable defect behavior is observed, and this is discussed in Section V. The plasma-induced generation of stable defects such as platelets and acceptor states is presented in Section VI. Finally, Section VII summarizes the main results of this chapter and gives a brief outlook to future directions.

II. Experimental Techniques

1. SAMPLE PREPARATION AND CHARACTERIZATION

Experiments described in this chapter were performed on a variety of poly-Si samples prepared by low-pressure chemical vapor deposition (LPCVD), solid-state crystallization, and laser crystallization. All poly-Si films were deposited on quartz and on single-crystal silicon wafers with a thin thermal oxide, 10 nm thick. LPCVD polycrystalline silicon films were deposited at 625°C to a thickness of 0.55 μm. Cross-sectional transmission electron microscopy (TEM) showed that this material is composed of columnar grains with a somewhat smaller grain size near the substrate as compared with the bulk. The average grain diameter is 15 nm, and the length of the grains is limited only by the film thickness. The second set of poly-Si samples was prepared by solid-state crystallization (SSC) of LPCVD *amorphous silicon* (*a*-Si) at 600°C. The sample thickness varied between 0.3 and 0.55 μm. According to TEM micrographs, grain boundaries in SSC poly-Si are randomly distributed, and the average grain size is 120 nm. A third set of poly-Si samples was fabricated by laser crystallization of 0.1-μm thick LPCVD amorphous silicon. Average grain sizes ranging from 80 nm to 1.5 μm were achieved using laser energy densities between 400 and 600 mJ/cm^2. TEM micrographs of samples with an average grain size of more than about 0.1 μm show that the grains extend from the substrate to the surface. Hence grain boundaries are predominantly aligned normal to the substrate. Moreover, the concentration of intragrain defects is significantly smaller than in SSC poly-Si.

Solid state crystallized poly-Si samples were doped by multiple phosphorous or boron implantations. The dopants were activated in a 30-min furnace anneal at 900°C, while doping of laser crystallized poly-Si was achieved by crystallizing doped LPCVD *a*-Si films.

For electrical measurements, coplanar or four-probe contacts consisting of either titanium-gold double layers, n^+ or p^+ amorphous silicon layers, or phosphorous or boron implants were used. In addition, the amorphous silicon contact layers and the P or B implants were covered with a protective chrome layer. Boron-doped contacts were used for p-type material, while phosphorous-doped contact layers were used for intrinsic and n-type samples.

2. HYDROGEN PASSIVATION

In the subsequent sections of this chapter, the term *hydrogenation* will be used interchangeably with *deuteration* because no significant differences in properties such as defect passivation, diffusion, and defect generation have been found for the two isotopes. Deuterium was used preferably for secondary-ion mass spectrometry because it is a readily identifiable isotope that duplicates H chemistry. Hydrogenation of poly-Si samples was accomplished through the following steps: Prior to plasma passivation, the samples were given a metaloxide semiconductor-grade detergent cleaning, and the native oxide was removed with dilute HF. Hydrogen or deuterium was introduced into poly-Si by exposing the samples to monatomic H or D generated in an optically isolated remote plasma system (Johnson et al., 1991b) at elevated temperatures. The microwave power was held to 70 W at 2.1 GHz. The gas pressure was 2 torr, and the D flow rate was 9 SCCM.

For the diffusion experiments described in Section III, layer hydrogenations were performed. First, the native oxide of the poly-Si samples was removed. Then a 0.1-μm-thick deuterated amorphous silicon (a-Si:H:D) layer was deposited at 150°C by diluting silane with 20% deuterium in the gas phase. The deuterated or source layer was sealed with a 0.5-μm-thick a-Si:H capping layer to minimize deuterium loss at the sample surface. These structures were annealed at different temperatures, causing hydrogen and deuterium to migrate into the unpassivated poly-Si films.

3. CHARACTERIZATION

The average grain size of the poly-Si samples was determined from cross-sectional electron transmission microscopy micrographs. Information on the defect density was obtained from electron paramagnetic resonance (EPR) measurements. Conductivity measurements were performed using four-probe and coplanar contacts. These measurements were performed in an oil-free vacuum or in a nitrogen atmosphere, since the coplanar dark conductivity is sensitive to the presence of surface adsorbates (Tanielian, 1982). Deuterium and hydrogen concentration profiles were measured by SIMS using a Cs^+ ion beam. A standard c-Si sample implanted with a deuterium dose of $1 \times 10^{14}\,cm^{-2}$ was used to calibrate the deuterium concentration. The absolute concentration values are accurate to within a factor of 2, but the relative precision when comparing two profiles is much better. The depth scales were obtained by measuring the depth of the sputtered craters using a mechanical stylus.

III. Hydrogen Diffusion

In general, the activation energy for diffusion of an impurity along grain boundaries is lower than for diffusion within a crystal grain (Gupta, 1988; Ballutaud et al., 1986). Therefore, until some years ago, it was widely believed that grain boundaries act as diffusion pipes for hydrogen. However, recently Jackson et al. (1992) demonstrated that H migration is extremely sensitive to the presence of trap states. In fact, comparing H diffusion in c-Si with diffusion in disordered silicon (poly-Si, a-Si:H) showed that grain boundaries act as efficient sinks for H, effectively reducing the diffusivity. This observation indicates that hydrogen diffusion in poly-Si is a very complex mechanism.

In this section I review hydrogen transport in undoped solid-state crystallized (SSC) and LPCVD-grown polycrystalline silicon. Hydrogen diffusion was studied as a function of time, temperature, and H concentration. From the experimental results, information on the hydrogen density of states is obtained.

1. HYDROGEN DIFFUSION FROM A PLASMA SOURCE

a. *Solid-State Crystallized Polycrystalline Silicon*

Two very important factors for hydrogen passivation experiments are the substrate temperature and the hydrogenation time. Typical deuterium concentration profiles measured in SSC poly-Si are shown in Fig. 1 for isochronal plasma hydrogenations at the indicated temperatures T_H. With increasing substrate temperature, deuterium diffuses deeper into the bulk. This is accompanied by a decrease in the D concentration at the poly-Si surface from $1 \times 10^{21}\,cm^{-3}$ at 250°C to $2.6 \times 10^{19}\,cm^{-3}$ at 450°C. The depth profiles were analyzed by fits of the data to the convolution of a comple-

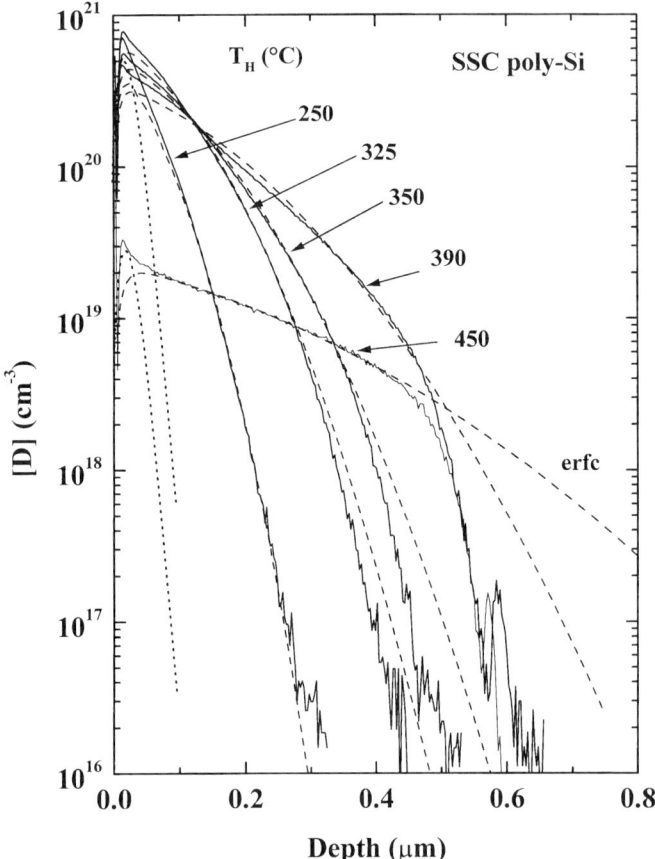

FIG. 1. Deuterium concentration depth profiles in solid-state crystallized poly-Si. The poly-Si films were exposed to monatomic deuterium for 1 hour at the indicated temperatures. Data are shown by the solid lines. The dashed lines depict a least-squares fit to the convolution of an *erfc* with the SIMS depth resolution function (Eq. 1). The dotted lines indicate a deuterium accumulation in the near-surface region of the poly-Si samples.

mentary error function (*erfc*) and the resolution function of the SIMS analysis (Santos and Jackson, 1992):

$$C(x, t) = C_0 \left[erfc\left(\frac{x}{2\sqrt{D_{eff}t}}\right) + r(x, t) \right] \quad (1)$$

where D_{eff} is the effective diffusion coefficient, C_0 is the deuterium surface concentration, x is the distance from the surface, t is the diffusion time, and

$r(x, t)$ is the resolution function of the SIMS analysis. For H and/or D diffusion from a plasma source, $r(x, t)$ is given by

$$r(x, t) = \exp\left(-\frac{x}{x_r}\right)$$
$$\times \left\{\exp\left(\frac{D_{\text{eff}}t}{x_r^2}\right)\left[\text{erfc}\left(\frac{\sqrt{D_{\text{eff}}t}}{x_r} - \frac{x}{2\sqrt{D_{\text{eff}}t}}\right) - \text{erfc}\left(\frac{\sqrt{D_{\text{eff}}t}}{x_r}\right)\right] - 1\right\}$$
(2)

with a SIMS depth resolution of $x_r \approx 8$ nm. At high D concentrations and for a depth greater than 0.1 μm, the fits represented by dashed lines are in good agreement with the data. In this region, H and D transport can be described by a single diffusion coefficient. Since hydrogen diffusion in disordered materials like poly-Si is strongly affected by a distribution of H traps, the analysis of the deuterium depth profiles provides an effective diffusion coefficient D_{eff}. At D concentrations below $\sim 4 \times 10^{18}$ cm^{-3}, the deuterium depth profiles exhibit a kink that becomes more pronounced with increasing hydrogenation temperature. Moreover, the depth profiles decay exponentially with depth, indicating that in this region the H migration process is dominated by the capture and release of H from deep hydrogen traps. Deviations between fit and data in the near-surface region are due to the formation of hydrogen-stabilized platelets. Platelet generation in poly-Si is discussed in detail in Section VI.2.

Figure 2 shows the time dependence of deuterium diffusion in SSC poly-Si at 350 and 450°C. The curves in Fig. 2a are shifted vertically for clarity. Independent of hydrogenation time and temperature, all deuterium depth profiles deviate from the fits to Eq. (1) (dashed lines) at a D concentration of $4-8 \times 10^{18}$ cm^{-3}. At lower concentrations, data are fitted to an exponential decay with a slope of $x_0 = 24.3$ nm that is independent of exposure time and temperature. With increasing passivation time, deuterium diffuses deeper into the film, resulting in an increase in the bulk D concentration. Hence, for a constant value of x_0, the gap between data and fit increases. Similar results have been reported for H diffusion in hydrogenated amorphous silicon (Jackson *et al.*, 1992; Branz *et al.*, 1993; Kemp and Branz, 1993).

The effective diffusion coefficient for the deuterium profiles shown in Figs. 1 and 2 were obtained from least-squares fits of the high-concentration portion of the diffusion profiles. In Fig. 3, D_{eff} is plotted versus $1/T$ (open triangles) and exhibits an activation energy of $E_A = 0.63$ eV. This diffusion energy is similar to the activation energy observed for diffusion from a plasma source into hydrogenated amorphous silicon (Santos and Jackson,

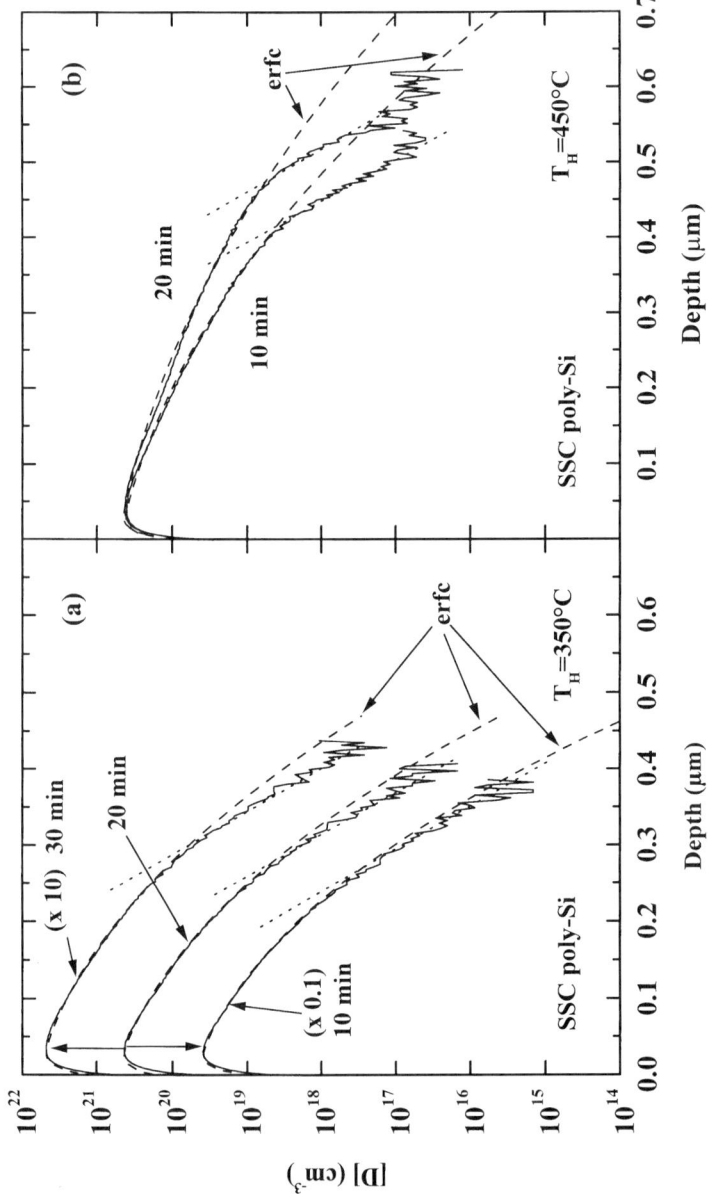

FIG. 2. Time dependence of deuterium depth profiles in SSC poly-Si. Hydrogenations were performed at (a) 350°C (*note*: offset for clarity) and (b) 450°C. Dashed lines represent a least-squares fit to Eq. (1), and the dotted lines depict the exponential decays with a characteristic length of $x_0 = 24.3$ nm.

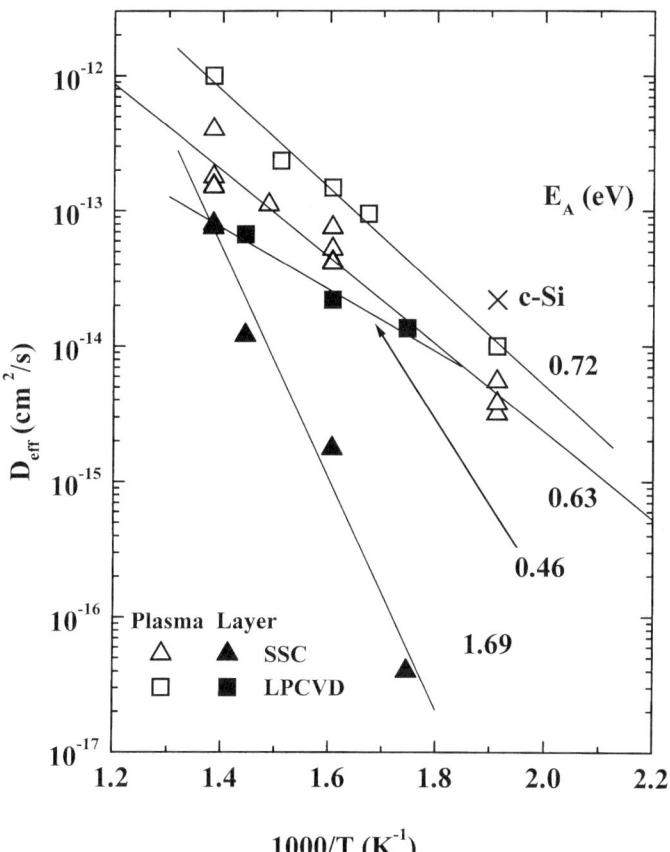

FIG. 3. Effective diffusion coefficient D_{eff} as a function of the reciprocal temperature for diffusion from a plasma (open symbols) and a solid source (solid symbols). Data shown by triangles were obtained from solid-state crystallized poly-Si. The cross represents the diffusion coefficient of c-Si, which was obtained from the high-concentration part of the deuterium depth profile shown in Fig. 9. The squares represent data from LPCVD-grown poly-Si. The activation energies E_A were obtained from the slopes of the lines.

1992). The time dependence of the effective diffusion coefficient is shown by the open triangles in Fig. 4. The effective diffusion coefficient exhibits a power-law dependency on deuteration time that can be described by (Street et al., 1987; Shinar et al., 1991)

$$D_{\text{eff}} \approx D_0(\omega t)^{-(1-\beta)} \qquad (3)$$

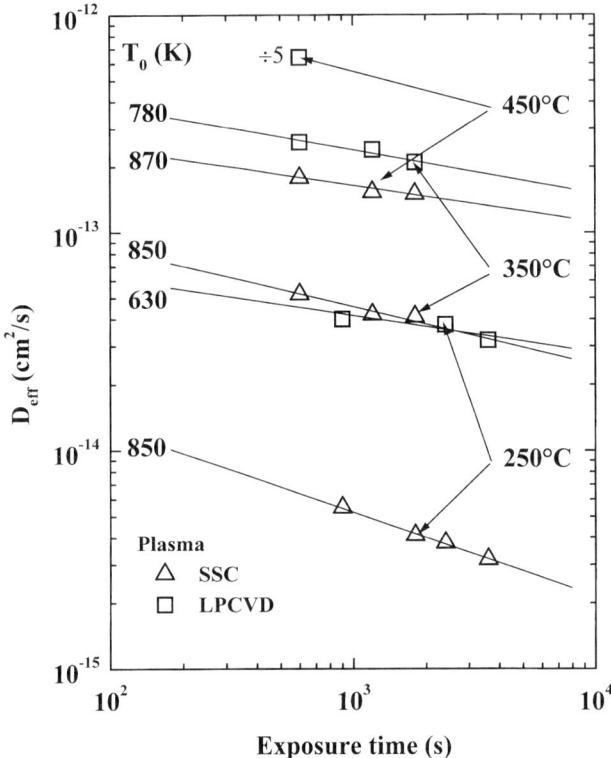

FIG. 4. Time dependence of the effective diffusion coefficient for various indicated hydrogenation temperatures. The triangles and squares represent data from solid-state crystallized and LPCVD-grown poly-Si, respectively. The lines were obtained from a least-squares fit of the data to Eq. (3).

where D_0 is a constant, ω is an attempt frequency, and $\beta = T/T_0$ is the dispersion parameter. T is the hydrogenation temperature, and T_0 is the characteristic disorder temperature. Fits of the data to Eq. (3) are shown by the lines in Fig. 4 with characteristic temperatures T_0 that vary between 630 and 870 K. This is within the range of 600–1100 K reported for H diffusion from a plasma in a-Si:H (Jackson and Tsai, 1992; Shinar et al., 1989). The observed decrease in D_{eff} with increasing exposure time is indicative of dispersive H diffusion.

The data shown in Figs. 1 and 2 indicate a significant difference in the diffusion mechanism for high and low D concentrations. Hence, for understanding microscopic diffusion processes, it is important to determine the concentration dependence of the diffusion. In the experiments described earlier, a change in the deuterium surface concentration always occurred in conjunction with a temperature change. In order to separate the two effects, the deuterium concentration at a fixed temperature was attenuated using two different methods: (1) A poly-Si film was mounted upside-down on the sample holder and a gap of 1 mm between sample holder and specimen was provided by two ceramic pillars. The concentration of monatomic D is attenuated due to recombination of atomic D on the surface of the sample holder. A control sample with the surface facing the downstream plasma was mounted similarly. (2) A 10-nm-thick silicon oxide layer was deposited on top of the poly-Si film. In this case, monatomic D fails to penetrate the oxide, thereby reducing the concentration of active deuterium in the surface region of the poly-Si film (Nickel *et al.*, 1995). Figures 5 and 6 illustrate the

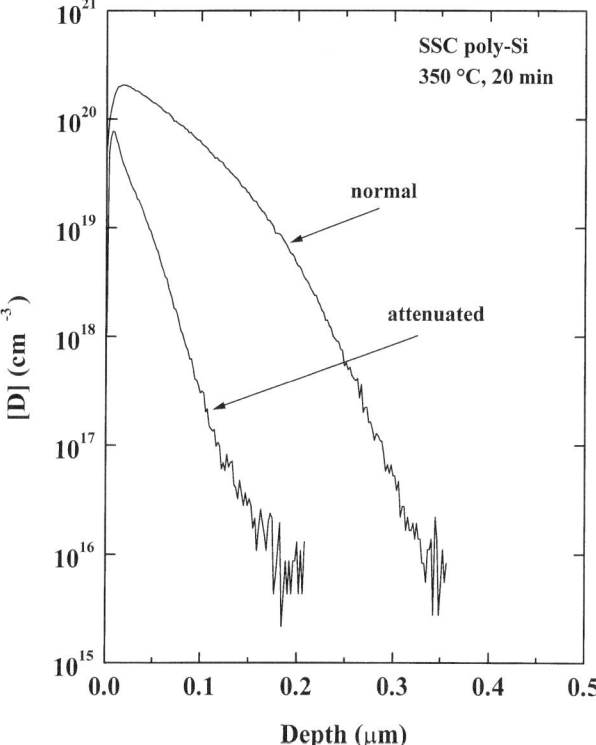

FIG. 5. Deuterium depth profiles in solid-state crystallized poly-Si after a "normal" and an "attenuated" plasma deuteration. Both samples were exposed simultaneously to monatomic D at 350°C for 20 min. The curve labeled "attenuated" was obtained from a sample that was mounted upside-down on the sample holder. A gap of 1 mm between sample holder and specimen was provided by two ceramic pillars. The diffusion profile labeled "normal" was taken on the control sample.

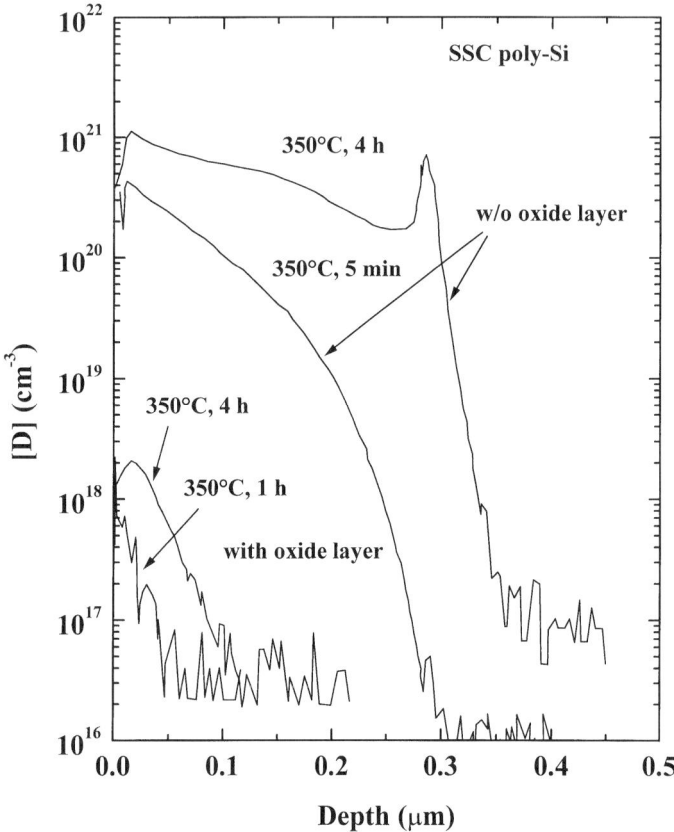

FIG. 6. Comparison of deuterium depth profiles in solid-state crystallized poly-Si with and without a 10-nm-thick SiO_2 capping layer. The specimens were deuterated at 350°C for the indicated times.

effect of an attenuated deuterium surface concentration. In Fig. 5, the "attenuated" depth profile represents the effect of an attenuated flux of monatomic D into the sample by mounting it upside-down on ceramic pillars. The D depth profile exhibits a simple exponential decay with a decay length of ~19.5 nm. For comparison, the diffusion profile of the unattenuated control sample labeled "normal" is shown. At high D concentrations ($>10^{18}$ cm^{-3}), the deuterium concentration decays according to a complementary error function (*erfc*) with depth.

Greater attenuation was achieved by depositing a 10-nm-thick SiO_2 layer on top of the poly-Si film. The presence of the oxide layer resulted in a decrease in the D concentration by almost 3 orders of magnitude. Again, at these low deuterium concentrations, the D profile decays exponentially with depth with a decay length of ~ 22 nm. Since the decay length exceeds the SIMS depth resolution by about a factor of 3, this observation indicates that D diffusion at low concentrations is dominated by deep deuterium or hydrogen traps. In addition, the data shown in Figs. 5 and 6 are consistent with data shown in Figs. 1 and 2, where trap-dominated diffusion was observed for D concentrations less than $4-8 \times 10^{18}$ cm^{-3}. The pronounced decrease in the D surface concentration is accompanied by a significant decrease in D_{eff}; e.g., for the data shown in Fig. 5, a decrease in D_{eff} from 6.2×10^{-14} cm^2/sec for the control sample to 6.7×10^{-15} cm^2/sec for the attenuated specimen was found. Hence the concentration dependence of the diffusion process provides an important insight into the microscopic diffusion process.

b. LPCVD-Grown Polycrystalline Silicon

In this subsection I compare hydrogen diffusion in poly-Si with randomly oriented grain boundaries with H diffusion in LPCVD-grown poly-Si. The temperature dependence of the deuterium diffusion is shown in Fig. 7. The LPCVD-grown samples and the SSC poly-Si specimens were exposed to monatomic D simultaneously. Hence, comparing the data of Fig. 1 and Fig. 7 clearly shows that deuterium diffusion in LPCVD-grown poly-Si is enhanced. This is confirmed by the temperature dependence of the diffusion coefficient (see open squares in Fig. 3). The effective diffusion coefficient is thermally activated with $E_A = 0.72$ eV, which is comparable with the result obtained for SSC poly-Si. As observed for SSC poly-Si, the deuterium surface concentration in LPCVD-grown material decreases from $C_0 = 4.2 \times 10^{20}$ to 2×10^{19} cm^{-3} as the passivation temperature increases from 250 to 450°C. In addition, at 450°C, the deuterium concentration is constant with depth. Least-squares fits of the data to Eq. (1) are shown by the dashed curves in Fig. 7. At high D concentrations, fits and data agree well. Deviations from the fits occur at concentrations below 4×10^{19} cm^{-3}. However, in contrast to SSC poly-Si, this is caused by the influence of the substrate on the diffusion profile at a depth of 0.55 μm rather than by deep hydrogen traps. The lack of a significant deviation between data and fit at low D concentrations may be due to either a low trap concentration ($<10^{18}$ cm^{-3}) or a shallow energetic position of the traps with respect to the H migration level.

The depth of H traps in LPCVD-grown poly-Si can be investigated simply by decreasing the deuteration time. Deuterium depth profiles mea-

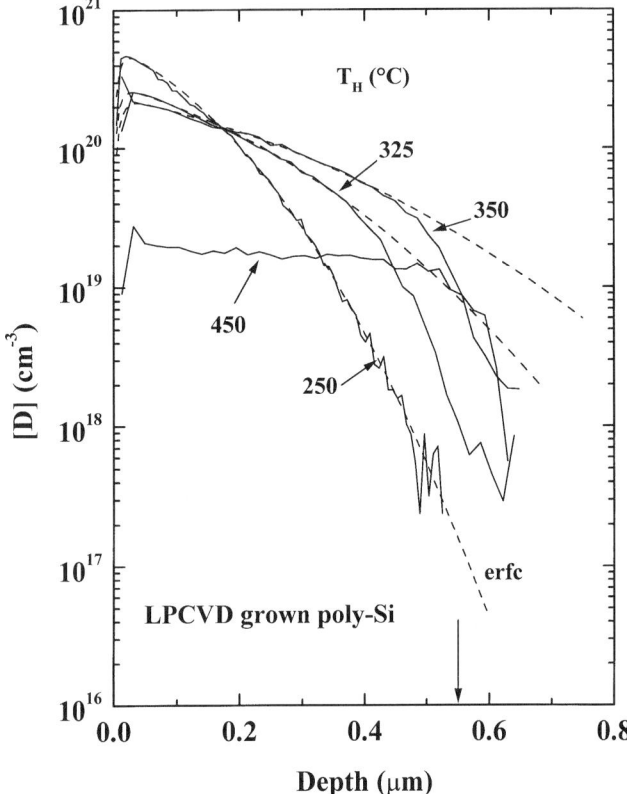

FIG. 7. Deuterium concentration depth profiles in LPCVD-grown poly-Si (solid lines). The poly-Si films were exposed to monatomic deuterium for 1 hour at the indicated temperatures. Data are shown by the solid lines. The dashed lines depict a least-squares fit to the convolution of an *erfc* with the SIMS depth-resolution function (Eq. 1). The poly-Si–substrate interface is located at a depth of 0.55 μm.

sured after an exposure to monatomic D for 5 min at 200 and 300°C are plotted in Fig. 8. The depth profiles are similar to those measured in SSC poly-Si (see Fig. 1). A high deuterium concentration in the near-surface region is observed that is even more pronounced than in SSC poly-Si (dotted curves in Fig. 1). Moreover, the data in Fig. 8 reveal a kink that shifts from a depth of 0.15 μm to 0.22 μm as the passivation temperature is increased from 200 to 300°C. This observation is similar to data obtained in c-Si, where the high deuterium concentration in the near-surface region was attributed to the generation of hydrogen-stabilized platelets during the

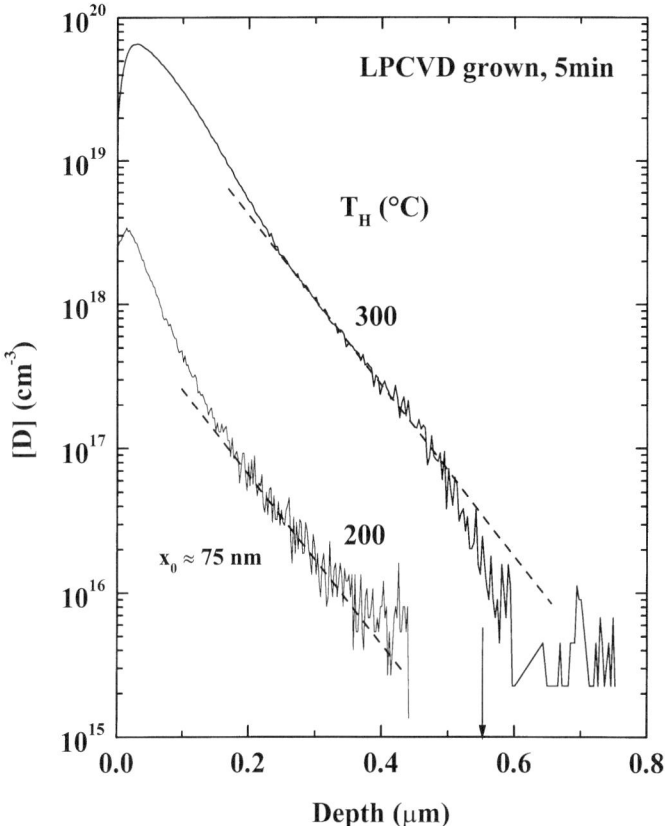

FIG. 8. Deuterium depth profiles measured in LPCVD-grown poly-Si after an exposure to monatomic D at 200 and 300°C for 5 min. The dashed lines represent a least-squares fit to an exponential decay with a slope of $x_0 = 75$ nm. The samples thickness of 0.55 μm is indicated by the arrow.

exposure to monatomic H (Johnson et al., 1987). The similarities of the deuterium diffusion profiles measured in LPCVD-grown poly-Si, SSC poly-Si, and c-Si suggest that hydrogen passivation of these materials also leads to the generation of platelets and/or hydrogenated stacking faults. This will be discussed in detail in Section VI.2. The dashed lines in Fig. 8 represent least-squares fits to an exponential decay with a slope of $x_0 = 75$ nm that is independent of temperature and time. Data and fit are in good agreement for a depth greater than 0.2 μm. This result suggests that the low deuterium concentration region represents D trapping into deep traps. The deuterium penetration of the entire film thickness of 0.55 μm at hydrogenation condition of 300°C for 5 min demonstrates again that H diffusion in LPCVD-grown poly-Si is enhanced compared with SSC poly-Si.

Enhanced diffusion in LPCVD-grown poly-Si becomes apparent when comparing the time dependence of the diffusivity. In Fig. 4 the effective diffusion coefficient for LPCVD-grown poly-Si is shown by the open squares. The time dependence is quite similar to that of SSC poly-Si. However, for all temperatures D_{eff} in LPCVD-grown poly-Si (open squares) is enhanced by about one order of magnitude compared with SSC poly-Si (open triangles). This comparison of diffusion in LPCVD-grown poly-Si and SSC poly-Si indicates that the crystalline structure of the material has a significant impact on the diffusion process.

The influence of the microscopic structure of the host material was investigated by comparing the diffusion process in LPCVD-grown and SSC poly-Si with diffusion in single-crystal silicon. SSC poly-Si, LPCVD-grown poly-Si, and single-crystal silicon samples were simultaneously exposed to monatomic deuterium at 250°C for 30 min. The deuterium depth profiles are plotted in Fig. 9. The data clearly show that deuterium diffusion in polycrystalline silicon is strongly reduced compared with the diffusion in c-Si. Again, it is observed that the diffusivity among the poly-Si samples is enhanced in LPCVD-grown material. It is important to note that the deuterium depth profile measured in c-Si exhibits a fast and a slow diffusion component. The fast diffusion component was fitted to an *erfc*, indicated by the dashed curve in Fig. 9. The fit yielded a deuterium surface concentration of $C_0 = 1.2 \times 10^{18}$ cm^{-3} and a diffusion coefficient of $D = 7.2 \times 10^{-11}$ cm^2/sec. It is interesting to note that the diffusion coefficient is 2–3 orders of magnitude larger than the effective diffusion coefficients obtained from plasma diffusion experiments in poly-Si (see Fig. 3). This illustrates that grain boundaries act as efficient traps for H and D, effectively reducing the diffusivity, as was suggested previously by Jackson *et al.* (1992). The enhanced deuterium concentration in the near-surface layer associated with the slow diffusion component is indicative of platelets (Johnson *et al.*, 1987). Of particular interest is the fact that the high concentration profiles of all the specimens are remarkably similar and independent of the macroscopic structure of the host material. The effective diffusion in the high-concentration limit does not appear to depend on the structure of the crystalline silicon host.

2. Hydrogen Diffusion from a Silicon Layer

In the preceding subsection I showed that it is important to understand low-concentration hydrogen diffusion. Diffusion from a deuterated and/or

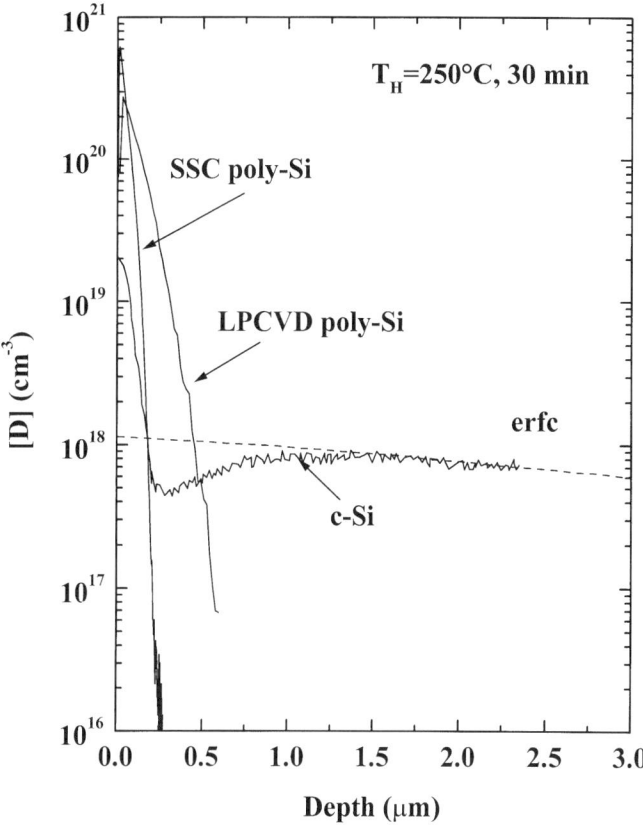

FIG. 9. Deuterium concentration depth profiles measured in SSC poly-Si, LPCVD-grown poly-Si, and c-Si. The specimens were deuterated simultaneously at 250°C for 30 min. The dashed curve represents a least-squares fit of the fast diffusion component in c-Si to a complementary error function.

hydrogenated layer into unhydrogenated polycrystalline silicon is an alternative method that allows us to study low-concentration diffusion processes. In addition, this method could reveal possible surface-limited processes such as different barriers for the incorporation of H from a plasma and a solid source.

Deuterium and hydrogen concentration profiles measured before (dashed curve) and after a layer diffusion experiment (solid curve) performed by annealing a-Si/a-Si:D:H/poly-Si multilayer structures at 420°C for 1 hour are shown in Figs. 10 and 11 for LPCVD-grown and SSC poly-Si,

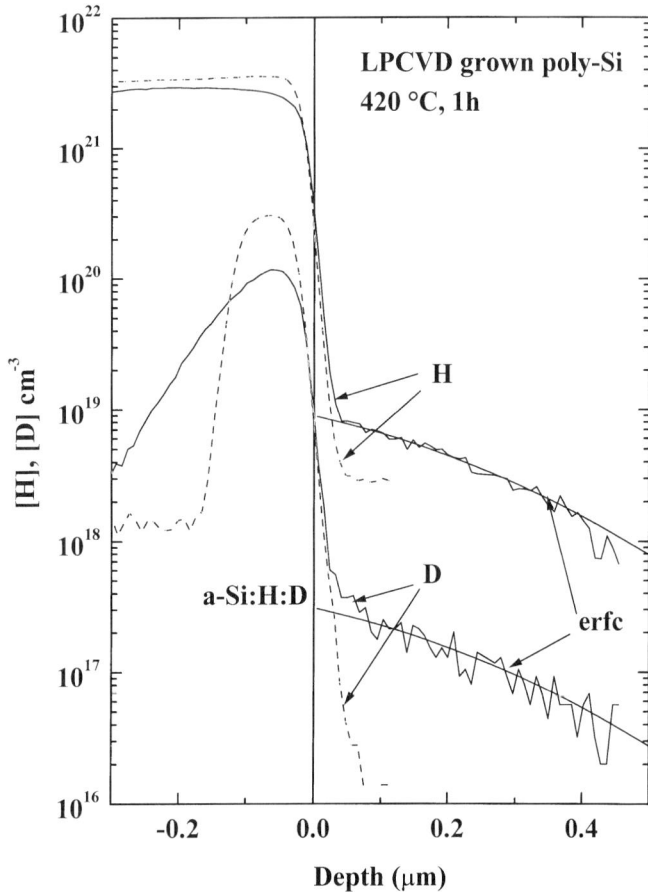

FIG. 10. Hydrogen and deuterium diffusion from a deuterated amorphous silicon (a-Si:D:H) layer into unhydrogenated LPCVD-grown poly-Si. H and D concentration profiles of the poly-Si/a-Si:D:H/a-Si:H multilayer structure prior to the diffusion experiment are indicated by dashed curves. Layer diffusion was performed by annealing the multilayer structure at 420°C for 1 hour. After the anneal, deuterium and hydrogen depth profiles (solid lines) decay according to an *erfc*. A depth of 0 μm denotes the interface between poly-Si and the a-Si:D:H layer.

respectively. The interface between poly-Si and the deuterated a-Si:H layer is marked by a solid line at a depth of 0. The exponential decay at the interface is determined by the SIMS depth resolution, which is limited by the intermixing of material during ion bombardment (Wilson et al., 1989). From the data, a SIMS depth resolution of $x_r = 8$ nm was obtained. The

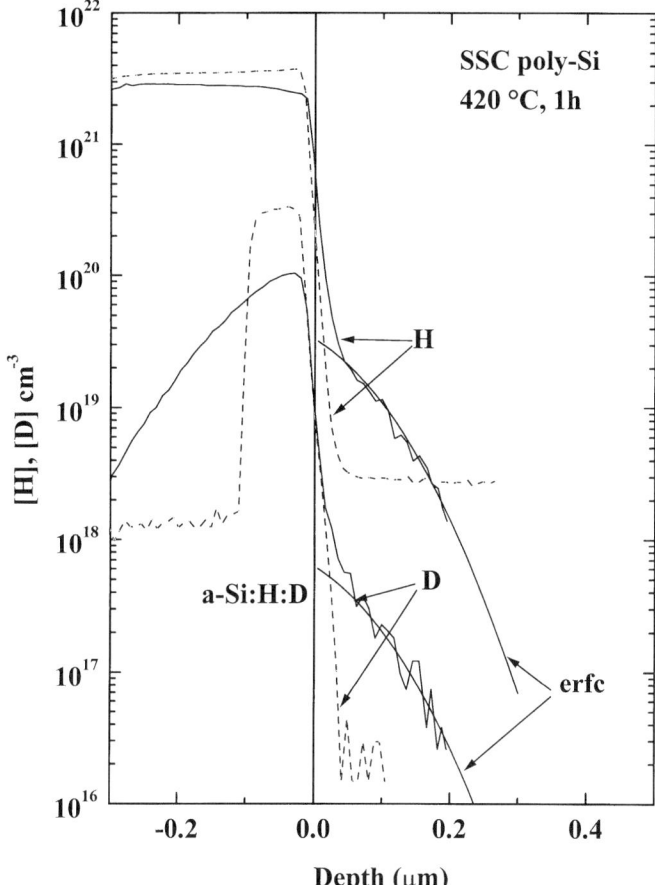

FIG. 11. Layer diffusion in unhydrogenated solid-state crystallized poly-Si. The specimen was annealed at 420°C for 1 hour.

deuterium and hydrogen surface concentrations and the effective diffusion coefficient were obtained by fitting the H and D depth profiles to Eq. (1). Since the hydrogen source is a solid layer, the resolution function $r(x, t)$ becomes (Santos and Jackson, 1992)

$$r(x, t) = \exp\left(\frac{D_{\text{eff}} t}{x_r^2} - \frac{x}{x_r}\right) erfc\left(\frac{\sqrt{D_{\text{eff}} t}}{x_r} - \frac{x}{2\sqrt{D_{\text{eff}} t}}\right) \quad (4)$$

For layer diffusion in LPCVD-grown poly-Si (Fig. 10), deuterium and hydrogen surface concentrations of 3.5×10^{17} and $\sim 10^{19}\,\text{cm}^{-3}$ were obtained, respectively. In SSC poly-Si, C_0 is a factor of 1.8 and 3 larger for H and D, respectively. In contrast to plasma hydrogenations, diffusion becomes appreciable above 300°C. Although the hydrogen and deuterium concentrations are rather low compared with diffusion from a plasma source, all diffusion profiles decay with depth according to an *erfc*. This is an important result indicating that the total H plus D concentration never exceeds the number of deep H traps, as in the case of plasma hydrogenation. In addition, the data in Figs. 10 and 11 show that layer diffusion is enhanced in LPCVD-grown poly-Si compared with SSC poly-Si, similar to the case of plasma hydrogenation.

The temperature dependence of the effective diffusion coefficient is shown by the solid squares and triangles in Fig. 3 for LPCVD-grown and SSC poly-Si, respectively. At low temperatures, the effective diffusion coefficient obtained from LPCVD-grown poly-Si is approximately 3 orders of magnitude higher than D_{eff} obtained from SSC poly-Si samples. At high temperatures (450°C), the effective diffusion coefficient is similar for both materials. The effective diffusion coefficient is thermally activated with $E_A = 1.69$ and 0.46 eV for SSC and LPCVD-grown poly-Si, respectively. The value for SSC poly-Si is in good agreement with results reported for layer diffusion in a-Si:H (Santos and Jackson, 1992; Carlson and Magee, 1978). The rather low activation energy found for LPCVD-grown poly-Si is surprising. Moreover, the absolute values of the diffusion coefficient are only a factor of 3 to 6 smaller compared with diffusion from a plasma source. This suggests that the H-trap concentration in LPCVD-grown material is either low or the energetic distance of the traps to the migration states is small.

3. HYDROGEN DENSITY OF STATES

Over the past years, a great deal of research on hydrogen bonding and transport has been done in polycrystalline and amorphous silicon. It is generally accepted that hydrogen motion is dominated by hydrogen capture and release by silicon-hydrogen bonds (Street *et al.*, 1987). Thus models for the hydrogen diffusion process based on a hydrogen density of states distribution consisting of energetically shallow states and deep traps were developed for hydrogenated amorphous silicon (Santos and Jackson, 1992). For polycrystalline silicon with small grains, hydrogen diffusion can be approximated by models developed for H diffusion in a-Si:H. In this subsection I discuss the data presented earlier in terms of a hydrogen density of states model. The data are analyzed to derive parameters such as the energies of shallow and deep H traps and trap concentrations.

a. Hydrogen Traps

At moderate hydrogenation temperatures, silicon self-diffusion is negligible because the diffusion coefficient at 450°C amounts to approximately 4×10^{-30} cm^2/sec (Jackson and Tsai, 1992). Hence the overall structure of the silicon lattice does not change. Transmission electron microscopy micrographs taken before and after a plasma exposure of poly-Si do not exhibit structural differences due to the incorporation of hydrogen. Thus hydrogen diffusion can be regarded as a trapping and releasing process of hydrogen atoms at various lattice sites.

The density of hydrogen traps can be regarded as more or less fixed for temperatures below 420°C, short hydrogenation times, and small changes in hydrogen concentration. On the other hand, for temperatures well above 420°C, long hydrogenation times, and significant changes in the total hydrogen concentration, the density of hydrogen-trapping sites can change. Under these conditions, the concentration of H-trapping sites depends on the hydrogen concentration (Jackson and Tsai, 1992). Equivalently, this can be described as a change in hydrogen solubility, as discussed by Beyer in Chap. 5 for hydrogen diffusion and solubility in amorphous silicon and amorphous silicon alloys.

Hydrogen introduced into poly-Si diffuses primarily as a monatomic species by a series of thermally activated random hops between lattice sites. These lattice sites are separated by an energy barrier, and hydrogen migration occurs by surmounting the barrier at a saddle point. This concept has been verified for hydrogen diffusion in single-crystal silicon, where the saddle point lies 0.2–0.5 eV above the Si—Si bond-center site that was identified as the global minimum for hydrogen diffusion (Van de Walle *et al.*, 1989). The lower value for the saddle point was determined from first-principles calculations (Van de Walle *et al.*, 1989), whereas the higher value was obtained from high-temperature H diffusion experiments (Van Wieringen and Warmholtz, 1956). An important constituent of poly-Si is disorder, which gives rise to a distribution of barriers between hydrogen transport sites and a distribution of hydrogen-trapping sites. The experimental data indicate the presence of a band of shallow H traps and a band of deep H traps separated in energy. Details of the hydrogen-trapping levels can be determined from the temperature and concentration dependence of the diffusion data. Assuming a simple two-level hydrogen density of states distribution, the effective diffusion coefficient is given by

$$D_{\text{eff}} = D_{\text{micro}} \frac{N_f}{N_t} \exp\left(-\frac{E_A}{kT}\right) \quad (5)$$

where D_{micro} is the microscopic diffusion prefactor, N_f is the density of free states, N_t is the density of trapped states, and E_A is given by

$$E_A = \begin{cases} E_M - E_{tr} & \text{when } C_H < N_{tr} \\ E_M - E_{BC} & \text{when } N_{tr} + N_{BC} > C_H > N_{tr} \end{cases} \quad (6)$$

E_M denotes the saddle point for hydrogen migration, E_{tr} is the energy of deep hydrogen traps, E_{BC} is the energy of the Si—Si bond center (BC) site, C_H is the hydrogen concentration, N_{tr} is the density of deep hydrogen traps, and N_{BC} is the concentration of bond-center sites. Equation (5) is valid for discrete levels and for a broadened H density of states that exhibits rather pronounced peaks at various band energies. Hence the energetic positions of the shallow and deep trap levels can be determined from the activation energies of the hydrogen diffusion. As shown in Fig. 3, the activation energies for high deuterium concentrations are $E_A = 0.72$ and $0.63\,\text{eV}$ for LPCVD-grown and SSC poly-Si, respectively. In the described two-band model, this energy corresponds to the shallow trap levels at E_{BC}. Hence, for LPCVD-grown and SSC poly-Si, the bond-center sites are located on the order of 0.4–0.6 eV below the migration saddle point E_M. This activation energy is quite similar to the value of 0.48 eV observed for hydrogen diffusion in crystalline silicon (Van Wieringen and Warmholtz, 1956; Seager *et al.*, 1987). D_{eff} determined from the high deuterium concentration in the near-surface region of single crystal silicon (Fig. 9) amounts to $2.2 \times 10^{-14}\,\text{cm}^2/\text{sec}$, which is quite similar to the effective diffusion coefficient from a plasma for LPCVD-grown poly-Si. Because the activation energy roughly represents the difference between the migration saddle point and the Si—Si bond-center position, the diffusion appears to be independent of the disorder or structure of the silicon lattice at high D concentrations. Typically, the concentration of bond-center sites vastly exceeds the number of deep hydrogen traps. Thus, at high H concentrations, the low-energy sites are occupied, and the average trap configurations available to H atoms are independent of the lattice structure. Since low diffusivity and high hydrogen concentration in the near-surface region of c-Si are clearly due to the presence of platelets, the diffusion profiles shown in Fig. 9 suggest that H-stabilized platelets may be prevalent and control diffusion in polycrystalline silicon as well.

For low deuterium concentrations, the diffusivity in SSC poly-Si is thermally activated with an activation energy of approximately 1.7 eV. According to Eq. (5), we can conclude that the deep hydrogen traps are located 1.7 eV below E_M. For low hydrogen concentrations, the diffusion is trap dominated, and the H concentration exhibits an exponential depth dependence of the form (Pearton et al., 1987; Herring and Johnson, 1991)

$$C_H(x) = N_{tr} \exp\left(-\frac{x}{x_0}\right) \quad (7)$$

where x_0 is the slope of the exponential decay, and x is the depth measured from the point where the free hydrogen concentration is equal to the H trap density. From the analysis of the deuterium depth profiles in Figs. 1 and 2, we can estimate the trap concentration. At high D concentrations ($C_D > N_{tr}$), the depth profiles decay according to a complementary error function, whereas at low D concentrations ($C_D \leq N_{tr}$), diffusion is controlled by deep traps, and hence the depth profile decays exponentially with depth. The concentration of deep H traps can be estimated from the kink in the diffusion profiles, e.g., the concentration at which the *erfc* and the exponential decay cross ($C_D \cong N_{tr}$). Thus for SSC poly-Si we can conclude that the trap concentration is $N_{tr} = 4-8 \times 10^{18}\,cm^{-3}$. LPCVD-grown poly-Si revealed an activation energy of about 0.4–0.5 eV (full squares in Fig. 3). Apparently, the traps are either significantly shallower than in SSC poly-Si, LPCVD-grown poly-Si has fewer H traps, or the trap states are inaccessible to migrating hydrogen.

The spatial concentration profiles can be analyzed further. The exponential slope x_0 represents the mean free path between trapping events and is approximately given by (Jackson and Tsai, 1992)

$$x_0 \approx \frac{1}{\sqrt{4\pi r_k N_{tr}}} \sqrt{1 + \frac{v}{\kappa N_{tr}}} \quad (8)$$

where r_k is the capture radius, κ is the capture coefficient, and v is the hydrogen release rate. In the case of low-concentration diffusion, the hydrogen release rate is small compared with κN_{tr}, and x_0 is independent of temperature. Hence, knowing the H trap concentration, the capture radius can be determined from Eq. (8). From the observed value of $x_0 \approx 24\,nm$, the capture radius computes to $r_k \approx 0.034\,nm$ for solid-state recrystallized poly-Si. A similar value for r_k was reported for H traps in amorphous silicon (Jackson and Tsai, 1992). For LPCVD-grown poly-Si, the mean free path is about a factor of 3 larger (75 nm). In this case, however, the trap concentration is difficult to determine because a clear demarcation between *erfc* and exponential decay could not be observed. The absence of a deep trap signature in LPCVD-grown poly-Si is consistent with the low activation energy for hydrogen diffusion from a solid source (full squares in Fig. 3).

b. Hydrogen Chemical Potential

Shallow and deep hydrogen traps are separated in energy by about 1.0–1.2 eV. In equilibrium, the hydrogen chemical potential μ_H separates the filled from the empty states. The position of the hydrogen chemical potential can be estimated by the following procedure. The effective diffusion coefficient is given by

$$D_{eff} = D_{micro} \exp\left\{-\frac{[E_M - \mu_H(T)]}{kT}\right\} \quad (9)$$

where T is the temperature, and the microscopic diffusion prefactor is determined by

$$D_{micro} = \tfrac{1}{6} v a^2 \quad (10)$$

For an attempt frequency of $v \approx 10^{12}$ sec^{-1} and a hopping distance of $a \approx 3$ Å, the microscopic diffusion prefactor computes to $D_{micro} \approx 10^{-3}$ cm^2/sec. Assuming this prefactor, the position of the hydrogen chemical potential can be calculated as a function of the temperature using the relation

$$E_M - \mu_H(T) = -kT \ln\left(\frac{D_{eff}}{D_{micro}}\right) \quad (11)$$

Since the results depend logarithmically on the microscopic diffusion prefactor, even an order of magnitude error will change the energy of the hydrogen chemical potential by less than 100 meV. In Fig. 12, $E_M - \mu_H(T)$ is plotted as a function of the passivation temperature for plasma and solid-state diffusion experiments in SSC and LPCVD-grown poly-Si. With increasing passivation temperature, the hydrogen chemical potential sinks deeper in energy with respect to the migration saddle point E_M with a rate of approximately 10^{-3} eV per K. Extrapolating the data to $T = 0$ yields the activation energies presented in Fig. 3. On the other hand, data obtained from layer diffusion experiments in solid-state crystallized (SSC) poly-Si do not show a temperature dependence of μ_H (solid triangles). When the hydrogen chemical potential resides in a large density of states, it cannot change rapidly with temperature. The data are therefore consistent with the chemical potential being pinned by H traps for the SSC material for layer diffusion. In order to exhibit the observed shift with temperature, the concentration of H-trapping sites must be small at μ_H. Therefore, the

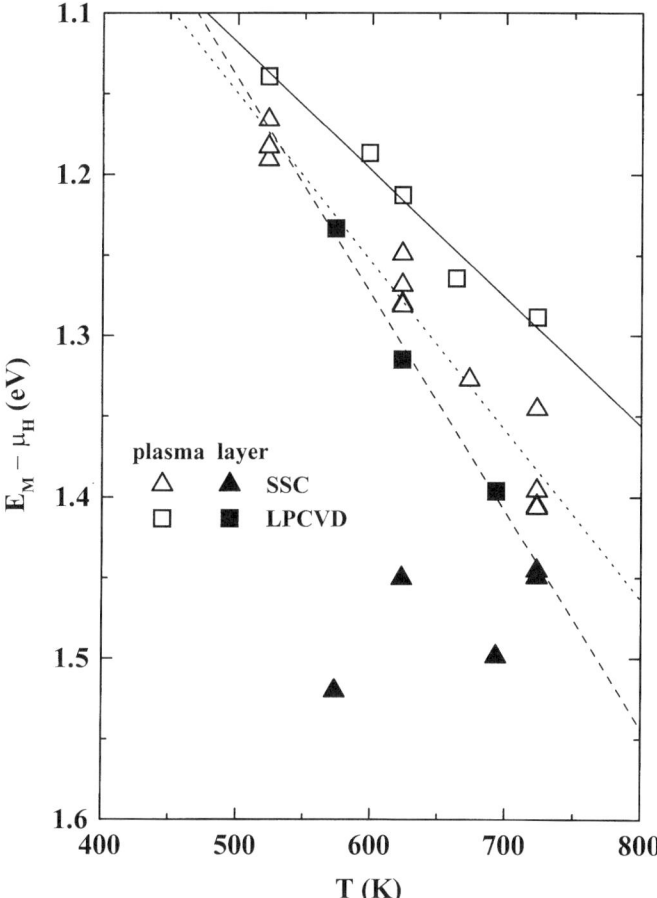

FIG. 12. $E_M - \mu_H$ as a function of the hydrogenation temperature. The open and solid triangles represent SSC plasma and layer diffusion, respectively, and the open and solid squares denote LPCVD plasma and layer diffusion.

hydrogen density of states must consist of two regions of high density of states separated by a deep minimum where the chemical potential normally resides.

The temperature dependence of μ_H may arise because there are more occupied states above μ_H than empty states below. The chemical potential usually moves away from energies of high concentrations of traps toward low density of states. A second explanation for the temperature dependence of the hydrogen chemical potential is related to the hydrogen quasi-chemical potential of the hydrogen plasma μ_P. The H chemical potential in the poly-Si films is determined by μ_P, which decreases with increasing passivation temperature. Thus, assuming that hydrogen in the semiconductor is in approximate equilibrium with H in the plasma, a decrease in μ_P causes an increase in $E_M - \mu_H$. The quasi-chemical potential of the monatomic H gas is given by (Kittel, 1969)

$$\mu_P = kT \ln(c_H V_Q) \quad (12)$$

where c_H is the concentration of atomic hydrogen in the gas, k is Boltzmann's constant, T is the gas temperature, and V_Q is the quantum volume. At room temperature, the quantum volume for hydrogen is approximately $V_Q = 10^{-24}$ cm^3. The concentration of monatomic H in an optically isolated remote plasma was measured by *in situ* EPR and determined to be about 5×10^{15} cm^{-3} (Johnson et al., 1991b). Since we do not know the temperature dependence of c_H, we can only determine the rate of change of μ_P with temperature, which computes to -0.0016 eV/K, and this agrees well with the experimentally determined value of -0.0013 eV/K. Thus the observed decrease in the hydrogen chemical potential with increasing passivation temperature is likely to arise from a decrease in the plasma chemical potential.

The main result of the preceding paragraphs is the connection between hydrogen chemical potential and temperature. This knowledge can be used to correct the data shown in Fig. 12 to calculate the position of μ_H with respect to the migration saddle point E_M at a *constant* temperature. Figure 13 shows the dependency of $E_M - \mu_H$ at 350°C as a function of the hydrogen surface concentration. Data of both sets of samples, SSC and LPCVD-grown poly-Si, show a similar behavior. At low H concentrations ($< 10^{19}$ cm^{-3}), the H chemical potential is pinned at ~ 1.3 and ~ 1.47 eV for LPCVD and SSC poly-Si, respectively. For SSC poly-Si, $E_M - \mu_H$ (350°C) decreases roughly by 0.2 eV when the H concentration exceeds the mid-10^{18} cm^{-3} level, whereas only a small change is observed for LPCVD-grown poly-Si (~ 0.1 eV). It appears that H diffusion in LPCVD-grown material is dominated by shallow states even at low H concentrations. A clear difference in the low H concentration diffusion properties can only be seen in SSC poly-Si.

The main results of this section are summarized in Fig. 14. Deep hydrogen-trapping sites are only present in large concentrations in solid-state crystallized poly-Si. They are located 1.5–1.7 eV below the migration saddle point E_M. The hydrogen chemical potential resides at 1.2–1.3 eV below E_M, and it moves about 0.2 eV in response to changes in the H concentration in SSC material. The trap density is within a factor of 3 to 4 of the spin density ($N_S \approx 2 \times 10^{18}$ cm^{-3}; see Sec. IV) in this material. Hence

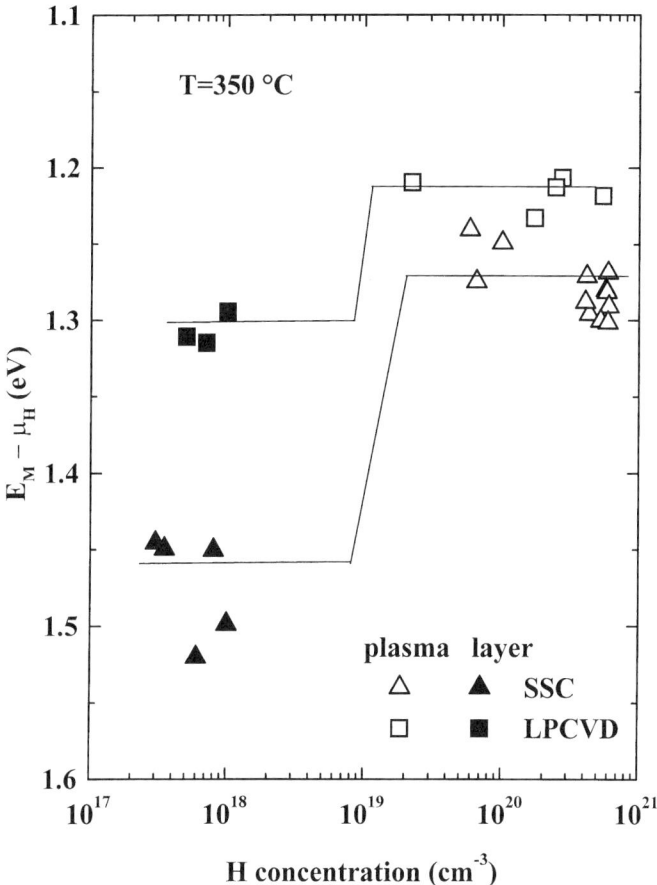

FIG. 13. $E_M - \mu_H(T)$ at 350°C using the temperature dependence observed in Fig. 12 as a function of the hydrogen surface concentration. Open squares, solid squares, open triangles, and solid triangles correspond to plasma LPCVD, layer LPCVD, plasma SSC, and layer SSC diffusion experiments, respectively.

one might think that the deep traps are silicon dangling-bond defects. While Si dangling bonds are energetically favorable for H trapping and passivation undoubtedly occurs, the H concentration necessary to passivate Si dangling bonds exceeds the spin density by almost two orders of magnitude (Nickel et al., 1993). Thus there must be other trapping sites in equal or higher concentration. In addition, these traps are not confined to the grain-boundary regions. When comparing the mean free path x_0 with the grain size of SSC poly-Si, it turns out that x_0 is small compared with the average grain size of 120 nm. If deep H traps were located predominantly at grain

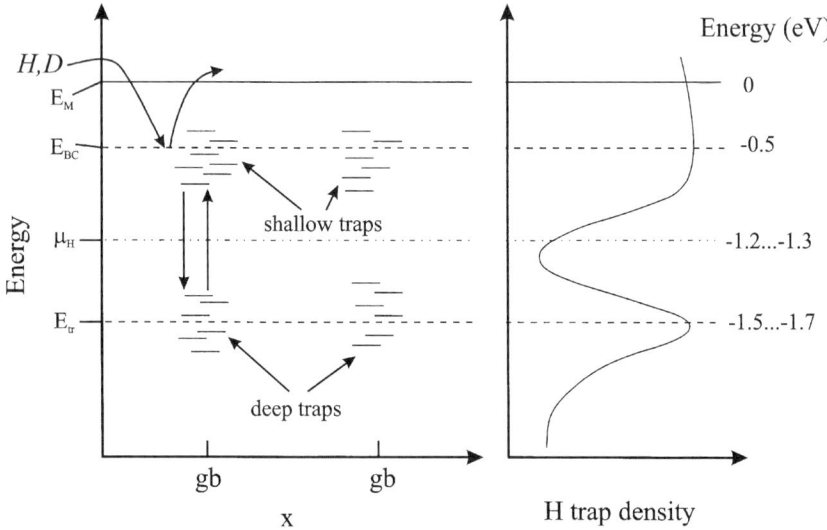

FIG. 14. Hydrogen density of states in poly-Si. The left part of the figure shows the energy versus distance. E_M, E_{BC}, and E_{tr} denote the energies of the migration saddle point, the Si—Si bond-center (BC) site, and the deep trap sites, respectively. The occupation of the various levels is determined by the hydrogen chemical potential μ_H. The H density of states distribution is illustrated on the right side. The energies for the various levels determined from the data are indicated with respect to the migration saddle point ($E_M = 0\,\text{eV}$).

boundaries, the mean free path should be on the order of the average grain size. It is conceivable that the high concentration of deep H traps is related to the presence of platelets that control trapping and release of H from these clusters. The possibility that platelets serve as traps for H is supported by the large deuterium concentration in the near-surface layer that is due to the presence of H-stabilized platelets in c-Si. On the other hand, a mean free path of 75 nm found for LPCVD-grown poly-Si suggests that if H traps are associated with grain boundaries, they are in low concentration, or hydrogen is guided away from grain boundaries.

IV. Hydrogen Passivation of Grain-Boundary Defects

In the preceding section I showed that hydrogen effectively can be incorporated into polycrystalline silicon simply by exposing the specimens to monatomic H at elevated temperatures. Hydrogen diffusion is governed by trapping and release of H from shallow and deep traps. In poly-Si, a prominent deep hydrogen trap predominantly located at grain boundaries is the silicon dangling bond. These defect states are located 1.9–2.5 eV below the hydrogen transport levels (Zellama et al., 1981; Street et al., 1988; Allan et al., 1982) and have been detected and identified by electron paramagnetic resonance (EPR) (Johnson et al., 1982a). A typical EPR spectrum is plotted in Fig. 15. At room temperature, the dangling-bond resonance occurs at a g value of 2.0055 and exhibits a line width of $\Delta H_{PP} = 6.1$ G. The spin concentration is obtained by integrating the spectrum in Fig. 15 twice. Unhydrogenated laser-crystallized, solid-state crystallized, and LPCVD-grown poly-Si reveal spin densities of 6.6×10^{18}, 2.2×10^{18}, and 1.8×10^{18} cm^{-3}, respectively (Fig. 16). The laser-crystallized specimens were

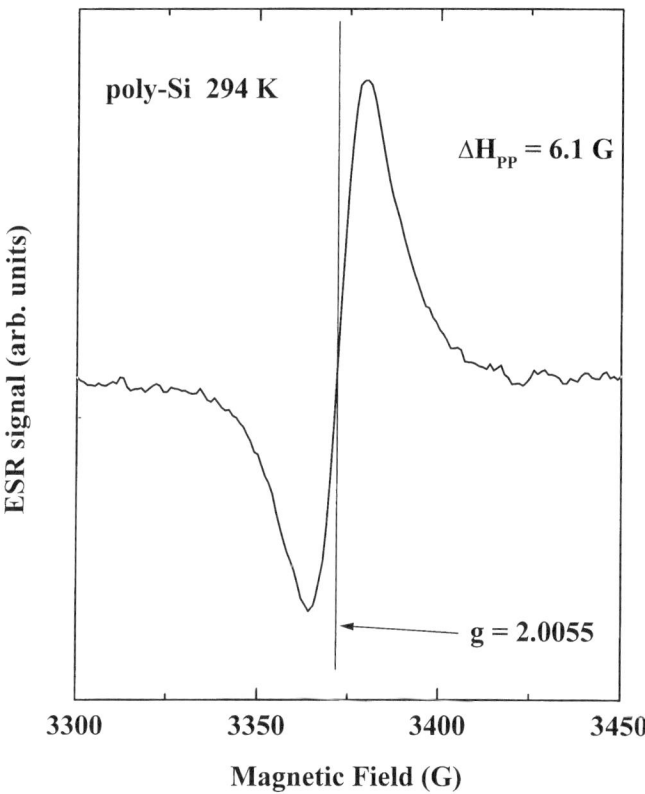

FIG. 15. Electron-spin resonance signal of silicon dangling-bond defects in poly-Si at 294 K. The resonance shows a g value of 2.0055 and a line width of $\Delta H_{PP} = 6.1$ G.

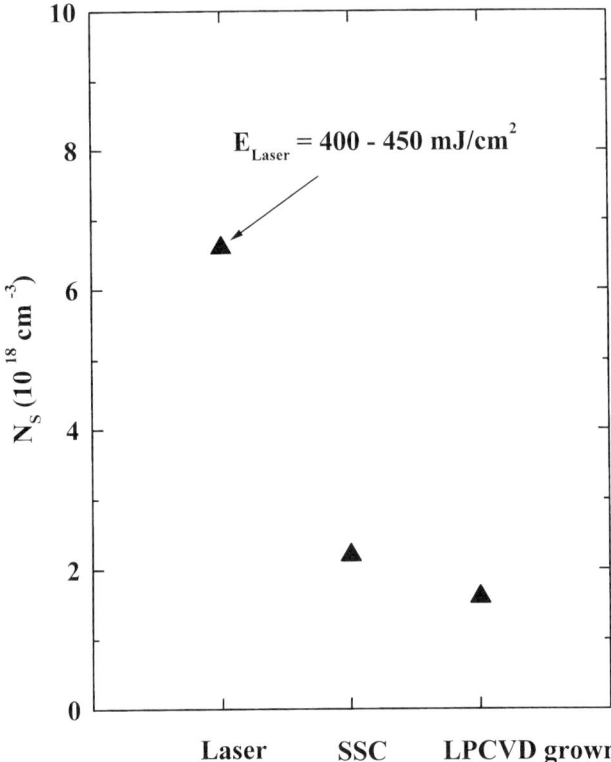

FIG. 16. Spin density N_S of laser-crystallized, solid-state crystallized, and LPCVD-grown polycrystalline silicon in the unhydrogenated state. The laser-crystallized samples were crystallized with laser energy densities from 400–450 mJ/cm².

crystallized using a laser fluence of 400–450 mJ/cm², which is known to produce material with an average grain size of 0.08–0.1 μm (Nickel et al., 1997b).

Hydrogen effectively passivates silicon dangling bonds, thereby improving the electrical properties of poly-Si films and devices, for example, by increasing carrier mobilities and reducing leakage currents and threshold voltages in thin-film transistors (Kamins and Marcoux, 1980; Thompson, 1991). Previously, reductions in the defect density by between a factor of 3 (Pandya and Kahn, 1987) and 10 (Jousse et al., 1989) have been reported. However, little is known on the defect passivation kinetics.

In this section I present the dependence of defect passivation on hydrogenation time and temperature in undoped polycrystalline silicon. For this purpose, poly-Si samples were hydrogenated through a sequence of 1-hour exposures to monatomic hydrogen at substrate temperatures ranging from 250–450°C. The poly-Si specimens were characterized by measuring the spin density with EPR prior to the hydrogenation schedule and after each exposure to monatomic hydrogen. The time dependence of the spin density is plotted in Fig. 17. At $t_H = 0$, the poly-Si sample exhibits a spin density of $N_S = 1.8 \times 10^{18}$ cm^{-3}. The corresponding EPR spectrum is depicted by the dashed curve in the inset. An exposure to monatomic hydrogen reduces only the intensity of the EPR resonance. Line width and g value remain unchanged, indicating that hydrogenation effectively passivates Si dangling-bond defects (solid curve in the inset of Fig. 17). Eventually, after five consecutive hydrogenations, the spin density saturates at $N_S = 9.6$

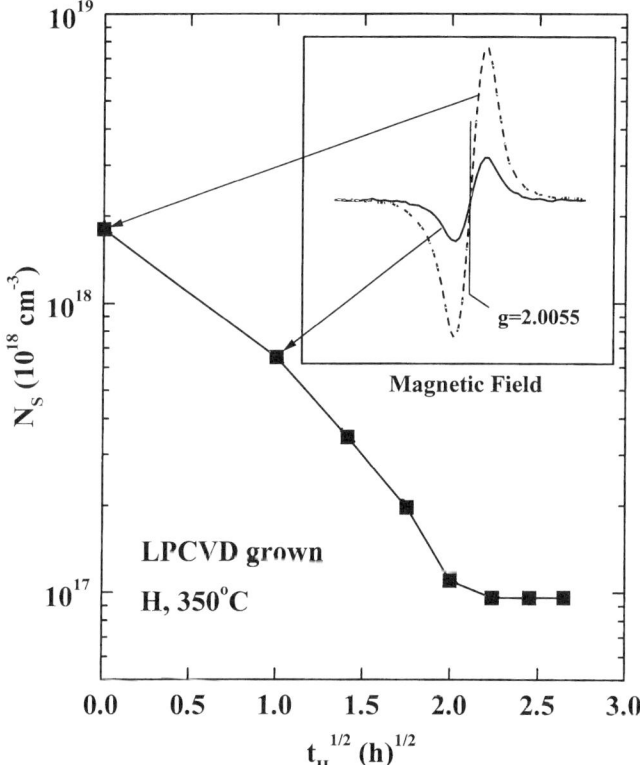

FIG. 17. Spin density N_S versus square root of the hydrogenation time t_H for LPCVD-grown poly-Si. Hydrogenation was performed at 350°C. The inset shows the EPR resonance of the specimen before (dashed curve) and after an exposure to monatomic H for 1 hour.

$\times 10^{16}$ cm^{-3}. The number of passivated silicon dangling bonds, $N_S^{max} - N_S$, is plotted in Fig. 18 as a function of the square root of the hydrogenation time for various hydrogenation temperatures. Initially, the concentration of passivated silicon dangling bonds increases proportional to the square root of the passivation time. At a given passivation time, the number of passivated spins increases with increasing hydrogenation temperature. This is consistent with the temperature dependence of the deuterium depth profiles shown in Fig. 7. Hence the data in Fig. 18 are in agreement with

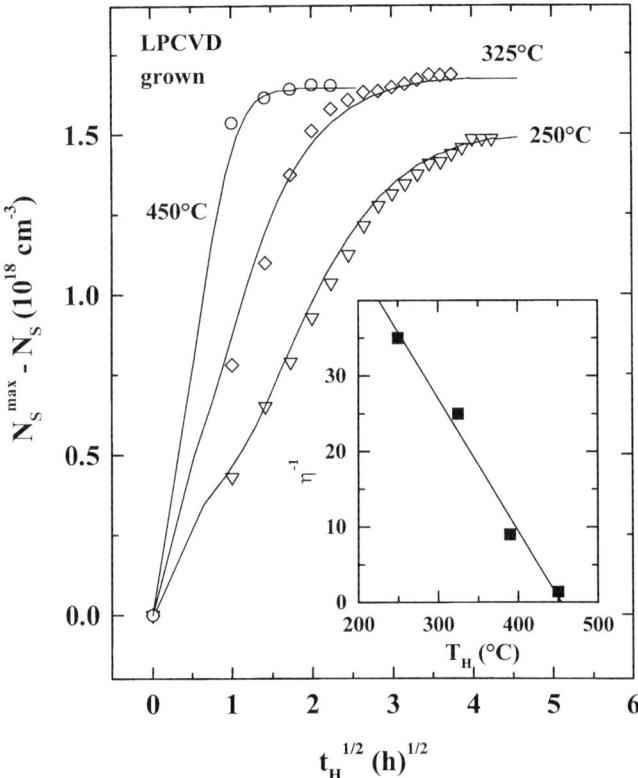

FIG. 18. Number of H-passivated silicon dangling bonds, $N_S^{max} - N_S$ in LPCVD-grown poly-Si as a function of the square root of the passivation time t_H for the indicated temperatures. The inset depicts the dependency of the inverse of the passivation efficiency η^{-1} on the hydrogenation temperature T_H. Data are depicted by the points, while the lines represent least-squares fits to Eq. (13). Details are described in the text. (From Nickel et al., 1993; reprinted with permission from "Hydrogen Passivation of Grain Boundary Defects in Polycrystalline Silicon Thin Films," Applied Physics Letters, **62**, 3285–3287, 1993. Copyright 1993 American Institute of Physics.)

the idea that a spin passivation front advanced from the sample surface by hydrogen diffusion. Similar observations have been reported for posthydrogenation experiments performed on amorphous silicon (Sol et al., 1980). At long hydrogenation times, the spin density, and thus $N_S^{max} - N_S$, deviates from a linear behavior and eventually saturates. A similar dependency of $N_S^{max} - N_S$ is observed for SSC poly-Si (Fig. 19).

The saturation values of the spin density were determined through the following procedure. The specimens were exposed to monatomic H at a given temperature for 1 hour, and subsequently, the spin density was measured. This process was repeated until the spin density remained

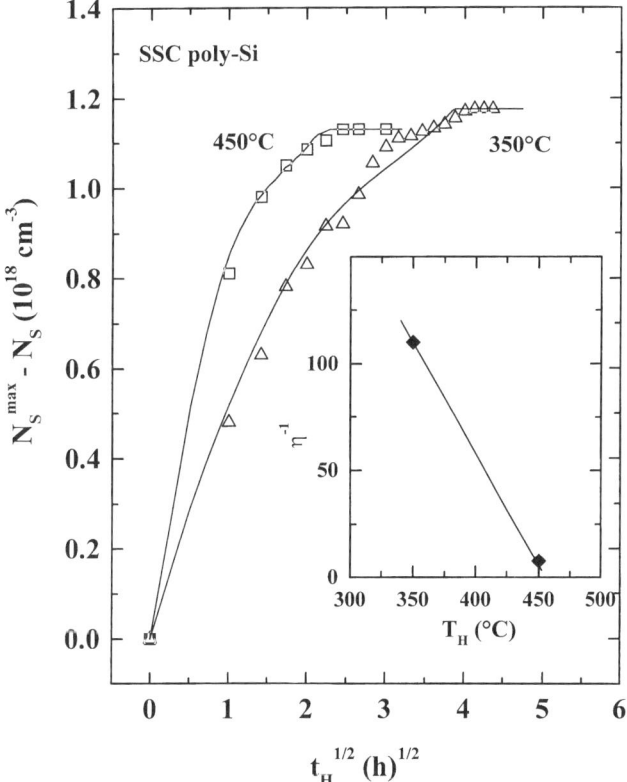

FIG. 19. Number of H-passivated silicon dangling bonds $N_S^{max} - N_S$ in solid-state crystallized poly-Si as a function of the square-root of the passivation time t_H for the indicated temperatures. The inset depicts the dependency of the inverse of the passivation efficiency η^{-1} on the hydrogenation temperature T_H. Data are displayed by the points, while the lines represent least-squares fits to Eq. (13). Details are described in the text.

unchanged for three consecutive hydrogenations. In Fig. 20 the saturation values of the spin density N_S^{sat} are shown by the full triangles and diamonds for LPCVD-grown and SSC poly-Si, respectively. The saturated spin density reveals a pronounced temperature dependence. LPCVD-growth poly-Si reveals the lowest spin density of $N_S^{sat} = 9.6 \times 10^{16}\,\text{cm}^{-3}$ at a passivation temperature of 350°C. Although Fig. 20 contains only two data points for SSC poly-Si, the temperature dependence of N_S^{sat} should be similar qualita-

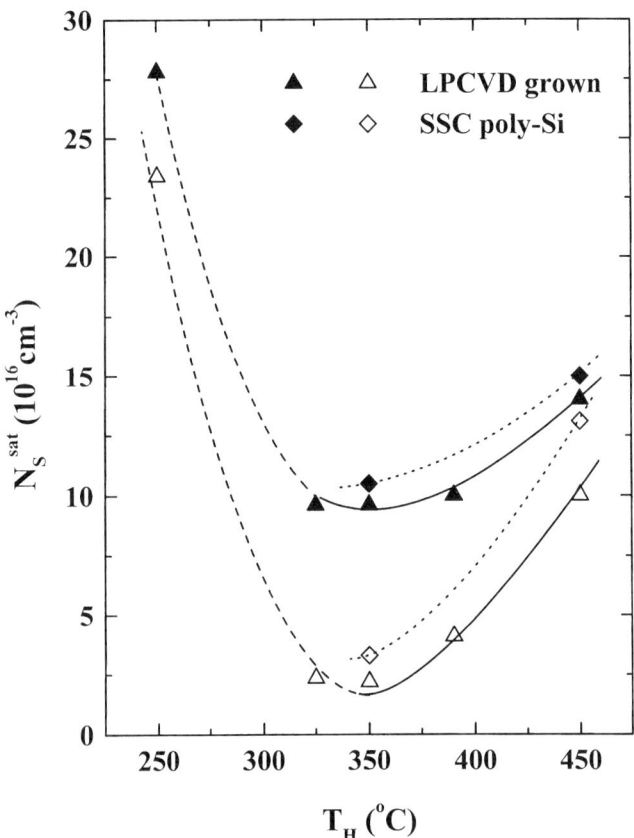

FIG. 20. Saturation value of the spin density N_S^{sat} as a function of the passivation temperature T_H. The solid triangles and diamonds represent LPCVD-grown and SSC poly-Si, respectively. The open symbols were obtained after an additional vacuum anneal at 160°C for 15.5 hours. (From Nickel et al., 1993; reprinted with permission from "Hydrogen Passivation of Grain Boundary Defects in Polycrystalline Silicon Thin Films," Applied Physics Letters, 62, 3285–3287, 1993. Copyright 1993 American Institute of Physics.)

tively to LPCVD-grown samples; e.g., the lowest spin density should occur at around $T_H = 350°C$. Thus defect passivation from a hydrogen plasma is most efficient at 350°C.

With increasing hydrogenation temperature, N_S^{sat} increases by about 50% independent of the macroscopic structure of the specimens. Most likely this is due to a decrease in the total hydrogen concentration in the samples by up to 1 order of magnitude (see Figs. 1 and 7 for SSC and LPCVD-grown poly-Si, respectively). On the other hand, the striking increase in the saturated spin density at lower temperatures cannot be explained by a decrease in the total H concentration since, according to Figs. 1 and 7, the hydrogen concentration increases with decreasing hydrogenation temperature. It is conceivable that N_S^{sat} at 250°C is not the saturation value of the spin density. The spin density could decrease further on a much longer time scale approaching the residual spin density obtained at $T_H = 325°C$. A second mechanism contributing to the high saturation spin density could be the formation of an equilibrium between defect passivation and defect generation. Defect generation occurs when a single hydrogen atom breaks a strained Si—Si bond forming a Si—H bond and a silicon dangling bond. The passivation of these bonds could be temperature dependent, causing their concentration to increase with decreasing passivation temperature.

Initially, the passivation of silicon dangling bonds shows a square-root time dependence, indicating that defect passivation is diffusion limited. This is depicted schematically in Fig. 21. The solid and dotted curves indicate the intermediate states of the hydrogen depth distribution for a sequence of hydrogenations. The hydrogen passivation front advances from the surface and eliminates silicon dangling-bond defects with an efficiency $\eta(T)$ until an equilibrium between defect passivation and defect generation is reached. The dashed curve depicts the depth dependence of the spin concentration $N(x)$, which was determined by etching off the poly-Si sample step by step in a CF_4 plasma. Since plasma etching is known to introduce surface and subsurface damage (Northrop and Oehrlein, 1986) that could result in an increase in the spin density, plasma etching was compared with chemical wet etching, which should not produce a damage layer in the surface region of the sample. A 0.3-μm-thick layer was removed by a plasma etch and a chemical wet etch, respectively. Subsequently, the spin density of both specimens was measured. The same values were obtained in both cases, indicating that plasma damage of the film surface did not result in a significant increase in the spin density. The solid and dashed lines in Fig. 22 represent $N(x)$ for LPCVD-grown and SSC poly-Si, respectively. Both classes of samples reveal an increase in $N(x)$ toward the interface and toward the surface. However, the most striking increase in $N(x)$ by about a factor of 6 is observed close to the interface in LPCVD-grown poly-Si.

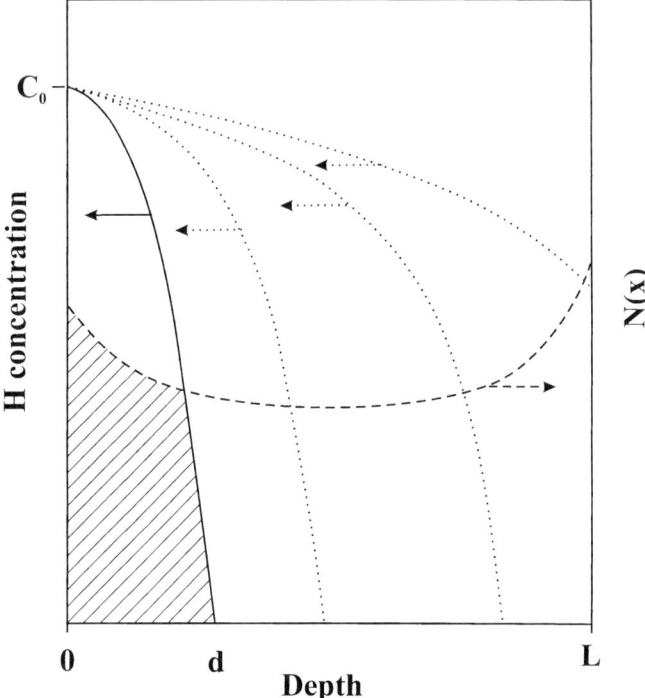

FIG. 21. Illustration of the diffusion-limited spin passivation process. The dashed line indicates the depth dependence of the spin density $N(x)$ (see Fig. 22). The solid and dotted lines show the evolution of hydrogen concentration depth profiles with increasing passivation time. An exposure to monatomic hydrogen for the time t passivates grain-boundary defects to a depth d (hatched area).

With the assumption that hydrogen passivates dangling bonds with an efficiency η, the time and temperature dependence of the density of passivated Si dangling bonds is given by

$$N_S^{max} - N_S(T, t) = \frac{1}{L}\left\{\int_0^L N(x)\,dx - \int_{d_L(t)}^L \left[N(x) - \eta(T)C_0\, erfc\left(\frac{x}{2\sqrt{D_{eff}t}}\right)\right]dx - \int_0^{d_L(t)} N_S^{sat}\,dx\right\}$$

(13)

where T is the passivation temperature, t is the hydrogenation time, and L is the sample thickness. For the hydrogen surface concentration C_0 and the

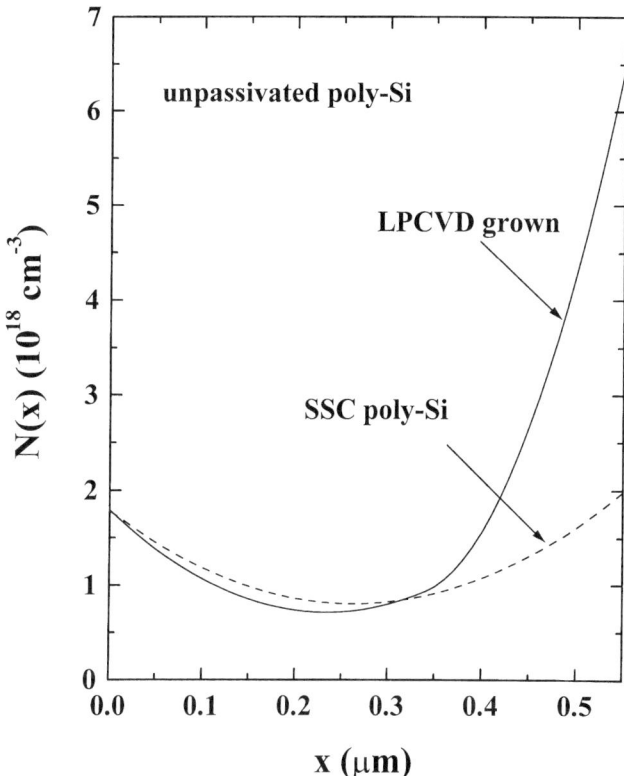

FIG. 22. Depth dependence of the spin density in unpassivated LPCVD-grown (solid line) and SSC poly-Si (dashed line). $x = 0$ and $x = 0.55$ mark the sample surface and the interface to the substrate, respectively. (From Nickel et al., 1993; reprinted with permission from "Hydrogen Passivation of Grain Boundary Defects in Polycrystalline Silicon Thin Films," Applied Physics Letters, **62**, 3285–3287, 1993. Copyright 1993 American Institute of Physics.)

effective diffusion coefficient D_{eff}, values obtained by fitting the depth profiles in Figs. 1 and 7 to Eq. (1) were used (see Sec. III.1). The depth of complete defect passivation d_L is determined by a solution of

$$N(d_L) = \eta(T)C_0 \, erfc\left(\frac{d_L}{2\sqrt{D_{\text{eff}}t}}\right) \qquad (14)$$

Hence the only free parameter in Eq. (13) is the hydrogen passivation efficiency $\eta(T)$. Using Eqs. (13) and (14), the time and temperature dependence of the number of passivated silicon dangling bonds was calculated. The fits are represented by the solid lines in Figs. 18 and 19 and are in good agreement with the data. The inverse of passivation efficiency is plotted in the insets of Figs. 18 and 19 as a function of the hydrogenation temperature. For both SSC and LPCVD-grown poly-Si, the passivation efficiency increases with increasing hydrogenation temperature. For LPCVD-grown poly-Si, the lowest passivation efficiency of $\eta = 0.03$ was obtained at $T_H = 250°C$. With increasing passivation temperature, the passivation efficiency increases to 0.7 at 450°C. A similar tendency is observed for SSC poly-Si; however, the η values are more than an order of magnitude smaller; at 350°C, a passivation efficiency of 0.009 was obtained.

Hydrogen passivation of LPCVD-grown and SSC poly-Si yields total hydrogen concentrations that exceed the defect concentration by more than 2 orders of magnitude; e.g., LPCVD-grown poly-Si contains about 1.76 $\times 10^{18}$-cm^{-3} preexisting Si dangling bonds, and at a hydrogenation temperature of $T_H = 250°C$, the total hydrogen concentration amounts to approximately 5×10^{20} cm^{-3}. Since hydrogenation passivates Si dangling bonds, most hydrogen atoms must be accommodated in locations where dangling bonds exist neither before nor after H passivation. From these data we can draw a very important conclusion: Hydrogen atoms must enter the specimens in pairs, since a single H atom would either passivate an additional Si dangling bond or generate a new defect by breaking a strained or weak Si—Si bond, leading to the formation of a Si—H and a Si dangling bond. This is characteristic for a system with a negative correlation energy. Explicit possibilities for diatomic H complexes include H_2^* (Chang and Chadi, 1989) and hydrogen-stabilized platelets (Johnson et al., 1987; Nickel et al., 1996). The H_2^* complex consists of one hydrogen atom in a silicon bond-center site forming a Si—H. The second hydrogen atom is accommodated in the antibonding interstitial site, forming a Si—H with the remaining Si atom (Chang and Chadi, 1989). Platelets, on the other hand, are generally believed to be composed of aggregated H_2^* complexes (Jackson and Zhang, 1991). The generation and the properties of platelets in polycrystalline silicon are discussed in Section VI.2.

Another possibility is the incorporation of hydrogen as interstitial H_2 molecules, which would neither generate nor eliminate dangling-bond defects. However, nuclear magnetic resonance (NMR) measurements have not detected large amounts of interstitial molecular H_2 in single-crystal silicon after posthydrogenation treatments (Boyce et al., 1992). Hence, with the excess hydrogen predominantly accommodated in small H_{2n}^* clusters, the passivation efficiency η measures the ratio of hydrogen passivating preexisting Si dangling bonds to hydrogen residing in platelet-like clusters.

The temperature dependence of η is consistent with the temperature dependence of the size and concentration distribution of platelets. Hydrogen stabilized platelets are preferably generated at low hydrogenation temperatures (Johnson et al., 1992). In addition, in this temperature range, the mobility of H is low. Thus a large concentration of hydrogen is necessary to passivate dangling-bond defects, since most H atoms are accommodated in clusters rather than at grain-boundary defects. On the other hand, with increasing passivation temperature, the probability of generating platelets decreases which might be in part due to enhanced H diffusion. In summary, for LPCVD-grown poly-Si passivated at 250°C, 34 hydrogen atoms are required to passivate one silicon dangling-bond defect. For SSC poly-Si hydrogenated at 350°C, the passivation of one silicon dangling bond requires roughly 110 H atoms. With further increasing passivation temperature, these values decrease to ~ 1 and ~ 8 for LPCVD-grown and SSC poly-Si, respectively, indicating that the defect-passivation efficiency increases with increasing temperature.

Plasma hydrogenation of polycrystalline silicon is commonly performed at high temperatures to take advantage of the enhanced diffusivity of hydrogen. Under these conditions, a steady state is established between the rate of defect creation and defect passivation. After complete passivation, the specimen is cooled down to room temperature, and hence the established equilibrium is frozen in. Since the hydrogenations are performed at relatively high temperatures, it should be possible to establish a new equilibrium at a lower temperature, at which the number of passivated silicon dangling bonds increases while the amount of H atoms accommodated in platelets decreases. After poly-Si films were hydrogenated until the spin density reached its saturation value, the specimens were subjected to a low-temperature vacuum anneal ($T = 160°C$) for 15.5 hours, and subsequently, the spin density was measured again. Independent of the hydrogenation temperature, the low-temperature anneal resulted in a significant decrease in the spin density (open diamonds and triangles in Fig. 20). The lowest spin density of 2.2×10^{16} cm^{-3} was achieved in LPCVD-grown poly-Si after hydrogen passivation at 350°C for 7 hours, followed by a vacuum anneal at 160°C for 15.5 hours. To the best of my knowledge, this is the lowest residual defect density ever reported for posthydrogenated poly-Si. It is interesting to note that the functional dependency of N_S^{sat} is maintained after the low-temperature vacuum anneal.

The time dependency of the decay of the normalized spin density N_S^0 at various annealing temperatures is shown in Fig. 23. The spin density exhibits a stretched exponential decay. In order to establish a new equilibrium with a lower defect density, poly-Si specimens have to be annealed for

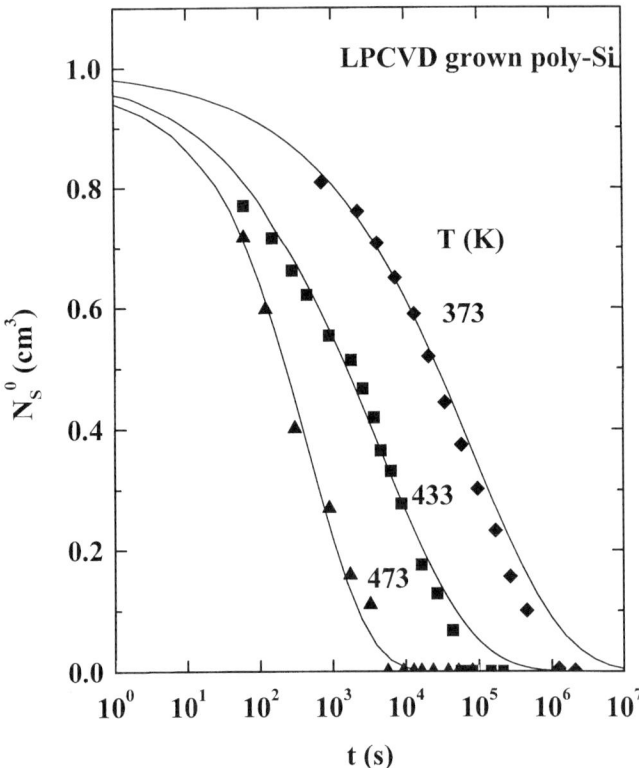

FIG. 23. Normalized spin density N_S^0 versus annealing time at the indicated temperatures. The decays exhibit a stretched exponential behavior.

2×10^6 and 10^4 sec at 373 and 473 K, respectively. However, the minimum residual spin density N_S^{min} varies with the annealing temperature (see Table I). The lowest value for N_S^{min} was obtained at $T_{anneal} = 433$ K. The time constant of the transients was obtained by fitting the data to a stretched exponential decay. The temperature dependence of the time constant τ is plotted in Fig. 24 and reveals an activation energy of $E_A = 0.95 \pm 0.02$ eV. This energy is consistent with the idea that H_{2n}^* clusters are the source or reservoir of H atoms. A similar time and temperature dependence of the spin density was observed for SSC poly-Si. The implications are discussed in detail in the next section.

The defect elimination and generation processes are summarized by the following reactions:

TABLE I

TEMPERATURE DEPENDENCE OF THE MINIMUM
SPIN DENSITY N_S^{min}

T_{anneal} (K)	N_S^{min} (cm^{-3})
373	3.9×10^{16}
433	2.2×10^{16}
473	3.3×10^{16}

$$(\text{Si—Si})_{\text{strained}} + \text{Si}_{\text{DB}} + (\text{Si—H}) + \text{H} \Leftrightarrow (\text{Si—Si})_{\text{strained}} + 2(\text{Si—H}) \quad (15)$$

$$(\text{Si—Si})_{\text{strained}} + \text{H} \Leftrightarrow (\text{Si—H}) + \text{Si}_{\text{DB}} \quad (16)$$

$$\text{Si—H} \Leftrightarrow \text{Si}_{\text{DB}} + \text{H} \quad (17)$$

where Si_{DB} denotes a silicon dangling-bond defect. Dangling-bond passivation arises from double or zero occupancy of Si—Si bonds (reaction 15) or hydrogen capture from isolated silicon dangling bonds (reverse reaction 17). Dangling-bond generation occurs when a single H atom is introduced into a strained or weak Si—Si bond (reaction 16) and when a hydrogen atom is desorbed from a passivated Si dangling bond (reaction 17). The occupancy of the density of states for hydrogen in silicon is determined by the position of the chemical potential μ_H (see also Sec. III.3). Hydrogen in silicon has a negative correlation energy (Zafar and Schiff, 1989). Hence the single occupied state resides above μ_H, and the double occupied state resides below the chemical potential. When the hydrogen plasma is turned off, a significant amount of hydrogen remains in the higher-energy singly occupied state. Thus the concentration of unoccupied isolated dangling bonds is enhanced. The low-temperature vacuum anneal causes hydrogen atoms to form diatomic complexes, thereby eliminating singly occupied states. Hence a new thermal equilibrium is established with a lower concentration of silicon dangling bonds.

In this section I presented the time and temperature dependence of the hydrogen-induced defect passivation in poly-Si. The lowest spin density was obtained at a hydrogenation temperature of $T_H = 350°C$. With a single model based on hydrogen diffusion we can account for the kinetics of the passivation process and the temperature dependence of the passivation efficiency. A further decrease in the spin density was achieved by subjecting the specimens to a low-temperature vacuum anneal. The lowest residual spin density of $N_S^{min} = 2.2 \times 10^{16}$ cm^{-3} was obtained after a vacuum anneal at 160°C for 15.5 hours. This low spin density is the essential precondition to observe light-induced metastability.

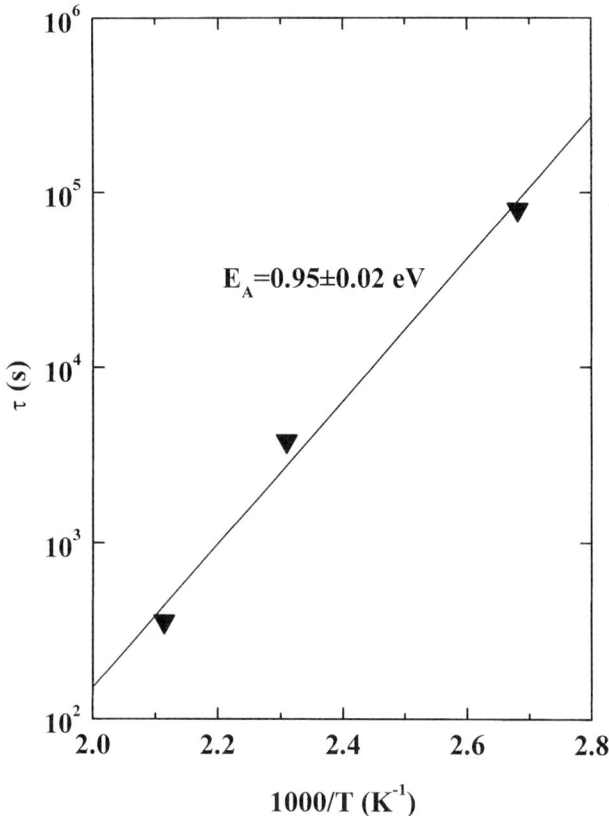

FIG. 24. Temperature dependence of the relaxation time constant τ obtained by fitting the data in Fig. 23 to a stretched exponential decay.

V. Metastability

Metastability is a well-known phenomenon in hydrogenated amorphous silicon. It manifests itself in a variety of effects ranging from light-induced reversible structural changes to charge-induced degradation of devices such as solar cells (Pfleiderer *et al.*, 1984) and thin-film transistors (TFTs) (Powell *et al.*, 1991; Nickel *et al.*, 1991). In *a*-Si:H, metastability was first reported by Staebler and Wronski (1980). They found that the electrical dark conductivity σ_D of amorphous silicon films decreased by several orders of magnitude after prolonged illumination with white light. This is accompanied by an increase in the spin concentration (Dersch *et al.*, 1981). Hence metastability is the major drawback of a-Si:H for photovoltaic applications. The initial values of σ_D and the spin density could be restored simply by annealing the sample at temperatures around 180°C. An increase in the defect density also was observed for charge injection in P-N junctions (Pfleiderer et al., 1984; Den Boer et al., 1984) and for charge accumulation in the active channel of TFTs (Jackson, 1990; Nickel et al., 1991; Powell et al., 1991). In amorphous semiconductors these phenomena have been documented extensively and are considered to arise from the unique properties of the amorphous phase.

In this section I show that metastable defects also can form in hydrogen-passivated polycrystalline silicon. The presence of light-induced defect generation in poly-Si is demonstrated first, and then I demonstrate a cooling rate-dependent metastable change in the direct-current (dc) dark conductivity of hydrogen-passivated polycrystalline silicon. While the first effect is similar to light-induced defect generation in a-Si:H, the latter is due to the formation and dissociation of a hydrogen complex at the grain boundaries. However, in both phenomena, the participation of hydrogen is evident from the experimental results.

1. LIGHT-INDUCED DEFECT GENERATION

Previously, there was no evidence of any deleterious aspects due to the presence of hydrogen in poly-Si. This is mainly related to the fact that the residual spin concentration, obtained after complete hydrogenation of poly-Si, remains very high compared with device-grade amorphous silicon. In undoped hydrogenated amorphous silicon, the spin density amounts to $N_S \approx 5 \times 10^{15}\,\text{cm}^{-3}$, which is about a factor of 20 lower than the spin density measured in poly-Si after 7 consecutive hydrogenations at 350°C ($N_S = 9.6 \times 10^{16}\,\text{cm}^{-3}$). This vast difference is generally attributed to the different means of hydrogen incorporation. In the case of a-Si:H, hydrogen is incorporated during the growth process, whereas in the case of poly-Si, a posthydrogenation step is required to introduce hydrogen. An enhanced spin concentration also was reported when unhydrogenated amorphous silicon was plasma hydrogenated (Sol et al., 1980). Apparently, the defect sites that can be accessed by hydrogen are different for growth and posthydrogenation. Therefore, a light-induced increase of the dangling-bond concentration is difficult to observe. However, hydrogenation followed by a low-temperature vacuum anneal results in a further decrease in the spin concentration for all passivation temperatures (open symbols in Fig. 20). The lowest residual spin density of $N_S^{\min} = 2.2 \times 10^{16}\,\text{cm}^{-3}$ was obtained after an anneal at 160°C for 15.5 hours, thus fulfilling the essential precondition to observe light-induced defect generation in hydrogen-passivated poly-Si.

The minimum residual spin concentration of $N_S^{min} = 2.2 \times 10^{16}\,\text{cm}^{-3}$ obtained according to the passivation schedule just described remained constant even for anneals up to 60 hours. The samples were then mounted on a copper block in a chamber that provided convective cooling or heating. A thermal conductive paste was used to minimize temperature gradients between sample and copper block. For illumination, water-filtered light from a xenon arc lamp was used.

In the following I present experimental data taken on LPCVD-grown poly-Si. Similar data were obtained on SSC poly-Si, indicating that the underlying microscopic mechanisms are independent of the macroscopic structure of the poly-Si samples. EPR absorption spectra of undoped poly-Si and a-Si:H are displayed in Fig. 25. As discussed in Section IV, the indicated g value of 2.0055 is characteristic of a singly occupied silicon dangling bond. The highest spin concentrations were obtained for unhydrogenated specimens (dotted curves). It is interesting to note that the spin density of unhydrogenated amorphous silicon ($N_S = 2 \times 10^{19}\,\text{cm}^{-3}$) exceeds the N_S of unpassivated poly-Si by 1 order of magnitude. This spin-density difference is related to the different microscopic structure of the samples, indicating that the greater degree of order present in poly-Si results in a smaller number of dangling bonds. On the other hand, device-grade a-Si:H deposited by rf glow discharge contains between 8 and 10 at.% hydrogen and, consequently, exhibits spin concentrations as low as $N_S \approx 5 \times 10^{15}\,\text{cm}^{-3}$. Note that the corresponding EPR resonance-labeled state A in Fig. 25b is multiplied by a factor of 10. In order to obtain device-grade poly-Si, the samples have to be posthydrogenated at elevated temperatures. Moreover, to lower the spin density further to $2.2 \times 10^{16}\,\text{cm}^{-3}$, the specimens have to be annealed at around 160°C for 15.5 hours (state A in Fig. 25a).

Polycrystalline silicon and amorphous silicon samples in state A were illuminated with white light at a temperature of 303 K for 15.5 hours. This resulted in an increase of the EPR resonance signal of both samples (state B in Fig. 25), indicating the generation of additional silicon dangling bonds. In poly-Si and a-Si:H, intense illumination generated 1.5×10^{16} and $4 \times 10^{16}\,\text{cm}^{-3}$ additional defects, respectively. In posthydrogenated poly-Si, the average volume density of newly generated defects is smaller than in amorphous silicon, since defect generation is confined to the grain-boundary regions. As in a-Si:H, the newly generated dangling bonds in poly-Si are metastable, and an anneal at 160°C completely restores state A. There is, however, a significant difference in the annealing properties. While in a-Si:H a 20-min anneal at 160°C is sufficient to restore state A, poly-Si requires a 15-hour anneal at the same temperature to completely restore the initial

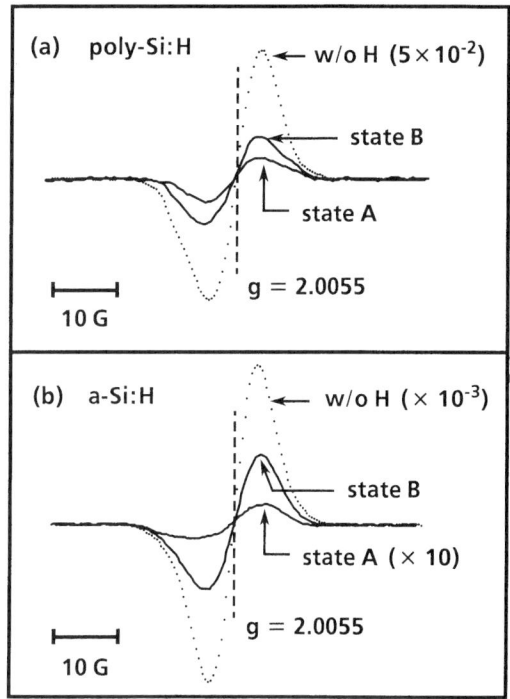

FIG. 25. EPR absorption spectra of (a) undoped poly-Si and (b) undoped amorphous silicon at 300 K. The resonance exhibits a g value of 2.0055 that is characteristic for Si dangling-bond defects. State A is ascribed to the annealed sample, while state B refers to the degraded specimen. Light-induced degradation was achieved by exposing the samples to white light with $P = 7$ W cm^{-2} at 303 K for 15.5 hours. For comparison, EPR absorption spectra of unhydrogenated poly-Si (a) and amorphous silicon (b) are shown by the dotted curves. Note that some of the EPR spectra are multiplied by scaling factors.

state A. Most likely this is due to the fact that light-induced defect creation is confined to the grain-boundary regions, where the degree of disorder is small compared with amorphous semiconductors. The experimental data presented in Fig. 25 are the first evidence of light-induced metastable defect generation in H-passivated poly-Si. Moreover, the data clearly demonstrate that the degree of disorder commonly present at grain boundaries plus the presence of hydrogen is sufficient for metastability. From the data in Fig. 25, it also can be concluded that the vast degree of disorder of amorphous semiconductors is *not* essential for metastability.

In Fig. 26 the temperature dependence of the light-induced defect generation in poly-Si is compared with that of a-Si:H. The concentration of light-induced dangling bonds ΔN_S in poly-Si shows the same temperature dependence as in amorphous silicon. In both materials, ΔN_S is thermally activated with an energy of $E_A = 50$ meV. This result supports the idea that in poly-Si the source of metastable silicon dangling bonds is strained or weak Si—Si bonds, as has been suggested for a-Si:H (Stutzmann, 1989).

The importance of hydrogen in the light-induced defect-generation process becomes evident when defect-creation and defect-annealing processes are repeated several times on the same poly-Si specimen. The data of such a cycling experiment are plotted in Fig. 27. Starting at cycle 0, the open square represents the saturated spin density N_S^{sat} obtained after 7 consecutive exposures to monatomic H at 350°C (see Sec. IV). As already shown in the preceding section, an additional low-temperature anneal at $T_A = 160$°C for 15.5 hours lowers the spin density to its minimum residual value of $N_S^{min} = 2.2 \times 10^{16}$ cm^{-3} (open triangle at cycle 1). Then the poly-Si samples were degraded and annealed several times. Light soaking was performed with white light at a substrate temperature of 80°C for 15.5 hours, and defect annealing was achieved at 160°C for 15.5 hours. Intense illumination caused the spin density to increase to $N_S = 4.4 \times 10^{16}$ cm^{-3} during the first two

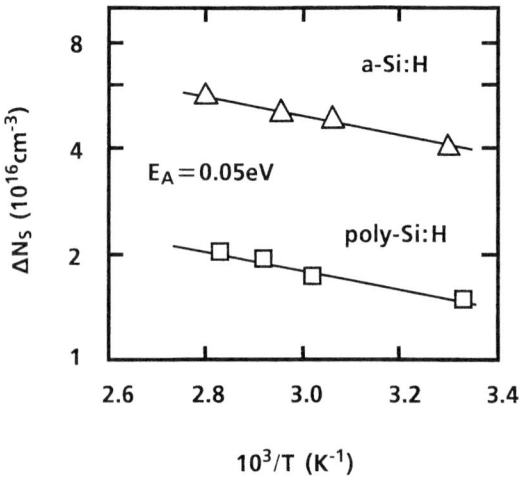

FIG. 26. Temperature dependence of the newly created metastable silicon dangling bonds ΔN_S in polycrystalline (squares) and amorphous silicon (triangles). The specimens were illuminated with white light ($P = 7$ W cm^{-2}) for 15.5 hours. Both materials reveal activated behavior with an activation energy of $E_A = 50$ meV.

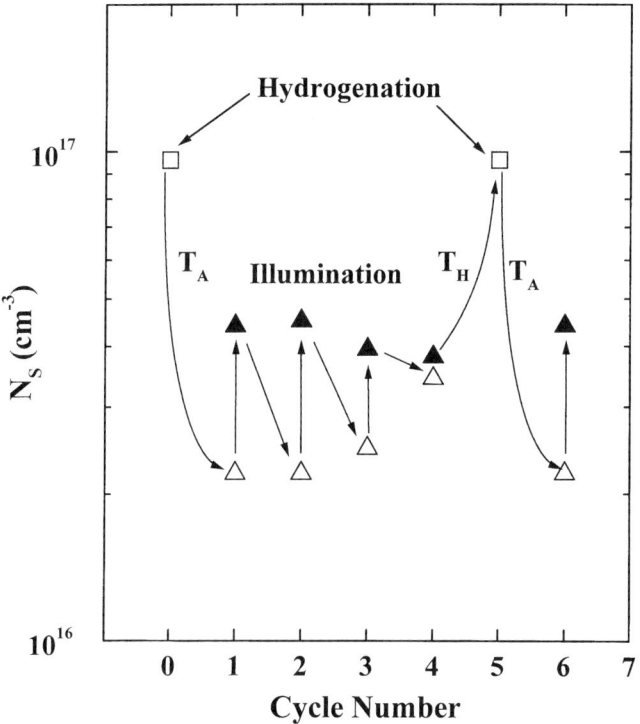

FIG. 27. Spin density N_S of hydrogen-passivated poly-Si for several defect creation and annealing cycles. The open squares denote the saturated spin concentration N_S^{sat} after complete H passivation in a remote hydrogen plasma at 350°C. T_A denotes the low-temperature anneal at 160°C for 15.5 hours. The open and full triangles represent the annealed state A and the degraded state B after light soaking with white light at $T = 80°C$ and $P = 7\,\text{W cm}^{-2}$ for 15.5 hours, respectively. T_H indicates a reexposure to monatomic H at 350°C for 1 hour.

cycles, and annealing restored the initial state A only after the first cycle (cycle number 2). In the following degradation and annealing cycles, the number of light-induced defects decreases with increasing cycle number. This defect density decrease is accompanied by an increase in the defect density in the annealed state A (cycles 2–4), and eventually, after four degradation and annealing cycles, the spin concentrations in states A and B are almost identical. At this time, the poly-Si sample was reexposed to monatomic hydrogen at 350°C for 1 hour (cycle 5). Surprisingly, this caused the spin concentration to increase to the same saturation value N_S^{sat} obtained after complete hydrogen passivation (cycle 0). Annealing at $T_A = 160°C$

again reduced the spin concentration to its minimum residual value N_S^{min} = 2.2 × 10^{16} cm^{-3} known as state A. Moreover, the light-induced defect generation and the annealing properties were fully recovered (cycle 6) with the same values as in cycle 1. Since the recovery of the effect is achieved by reexposing the sample to monatomic H, my results clearly demonstrate that hydrogen is actively involved in the light-induced defect generation.

Analyzing the data of the cycling experiment provides a number of important implications for light-induced defect generation in poly-Si *and* amorphous silicon. The EPR data shown in Fig. 25 and the temperature dependence of the net change in the spin concentration (Fig. 26) suggest that the underlying microscopic mechanism for light-induced defect generation is similar in poly-Si and *a*-Si:H. Previously, it was suggested that light-induced defect creation in *a*-Si:H could be due to the activation and passivation of impurities (Redfield and Bube, 1990). However, since the reexposure of poly-Si to monatomic H does not cause the impurity concentration to change, it is unlikely that impurities other than H are solely responsible. Although the interaction of hydrogen with impurities cannot be excluded, the connection between hydrogen passivation and metastability shows that hydrogen is directly involved in the formation and annealing of defects.

Based on the data just presented, it is difficult to specify the precise microscopic mechanism responsible for light-induced metastability. However, one possibility consistent with results for hydrogen in c-Si and *a*-Si:H is the following: As discussed in Section IV, hydrogenation of poly-Si establishes an equilibrium in which the hydrogen concentration exceeds the number of Si dangling bonds by more than 2 orders of magnitude. Most of the excess hydrogen is accommodated in locations that do not require the existence of dangling bonds either before or after hydrogenation, such as strained Si—Si bonds forming H_{2n}^* complexes (Chang and Chadi, 1989). Hydrogen passivation establishes a steady state between the rate of defect passivation and defect generation. A low-temperature vacuum anneal establishes a new quasi-equilibrium with a lower spin concentration by releasing H atoms from small hydrogen clusters (cycle 1 in Fig. 27). Illumination of the poly-Si film lowers the formation energy of dangling bonds, therefore establishing a new steady state with an enhanced defect density. Repeated illumination and annealing cycles cause hydrogen pairs to reversibly dissociate and re-form and the spin concentration to increase and decrease accordingly. However, over time, a fraction of the hydrogen atoms participating in the annealing and degradation process finds more stable configurations, where they can no longer be released by illumination or annealing. Due to the lack of mobile, unpaired hydrogen, the spin density of the annealed state A increases (open triangles in Fig. 27), and the number of additional generated metastable defects (solid triangles) decreases. It is important to note that the increase in the spin density in state A occurs only if the specimens are illuminated between the anneals. Simply annealing the sample even up to 60 hours does not produce an increase in the minimum spin density. Therefore, light soaking is essential and necessary for the cycling experiment. Rehydrogenation at 350°C increases the concentration of weakly bound hydrogen in paired and unpaired configurations, thus causing the spin density to increase to its saturation value. Then the entire process starts over again. Possible H complexes for strongly bound hydrogen include large H-stabilized platelets (Johnson et al., 1987; Nickel et al., 1996) and molecular interstitial H_2. In the case of H-stabilized platelets, the hydrogen binding energy depends strongly on the size of the platelets. Small platelets have a low H binding energy, and thus these H complexes are also likely candidates for the H reservoir in the low-temperature vacuum anneal. Large H clusters tend to grow at the expense of small platelets. Therefore, with each degradation and annealing cycle, more H atoms become strongly bound through attachment to larger platelets (Jackson et al., 1993), resulting in an increase in the number of dangling bonds that remain incompletely hydrogenated. In addition, these unpassivated defects reduce the light-induced carrier density due to enhanced defect recombination.

The experiments presented in this subsection clearly show that defect metastability exists in hydrogen-passivated poly-Si. Compared with amorphous silicon, the effect is small and should not be of any concern for device applications except possibly solar cell applications, which require a long minority carrier lifetime to ensure carrier collection over the long absorption depth in poly-Si. The key to observing this effect is a low spin concentration in the annealed samples. The rejuvenation of light-induced metastability observed after rehydrogenation clearly establishes that hydrogen directly participates in the underlying microscopic mechanisms. Based on the similarities of light-induced defect generation in poly-Si and a-Si:H, I propose that only H atoms, strained Si—Si bonds, and silicon dangling bonds are necessary to observe the effect. Moreover, my results suggest that hydrogen is directly involved in light-induced defect generation in amorphous silicon as well. This has some important implications for light-induced defect creation in amorphous silicon. Previously, there was no direct evidence of the participation of hydrogen in this effect, but most models require hydrogen to stabilize newly generated defects (Dersch et al., 1981; Santos and Jackson, 1991). A key experiment providing indirect evidence for the involvement of H was reported by Santos et al. They found that hydrogen diffusion in a-Si:H is enhanced during illumination (Santos et al., 1991). On the other hand, some models developed for amorphous silicon do not require the presence of hydrogen. In light of the preceding experiments, this, however, is hard to believe. Considering the structure of poly-Si, strained or weak Si—Si bonds are predominantly located at grain boundaries. When a strained or weak Si—Si bond at a grain boundary ruptures, two silicon dangling bonds are created in close proximity. In order to stabilize both dangling bonds, large relaxations of the surrounding silicon network and especially the nearest-neighbor Si atoms are required. This would involve at least two to four silicon lattice planes in each direction. In poly-Si it is very unlikely that such a relaxation could be accommodated within the grain boundaries alone. Even if the lattice could serve this requirement, it would not result in the formation of two isolated silicon dangling bonds because they would be passivated by hydrogen atoms. Consequently, it is difficult to reconcile metastable defect-generation models without hydrogen. The interaction of hydrogen is necessary at least to stabilize light-induced defects.

2. METASTABLE CHANGES IN THE ELECTRICAL CONDUCTIVITY

In polycrystalline silicon, metastable defect creation is a small effect and should not be of great concern for device applications. However, the presence of hydrogen in poly-Si gives rise to a second metastable effect that manifests itself by changes in the electrical dark conductivity of more than 8 orders of magnitude. This hitherto unexpected phenomenon is only observed in hydrogen-passivated LPCVD-grown and SSC poly-Si and thus clearly is due to the presence of hydrogen in poly-Si.

The temperature dependence of the dc electrical dark conductivity σ_D of undoped poly-Si prior to and after hydrogenation is plotted in Fig. 28. The open circles represent the unhydrogenated state and show two regimes. At temperatures above 250 K, the dark conductivity shows activated behavior with an activation energy of $E_A = 0.55$ eV. Below 250 K, a $T^{-1/4}$ behavior is observed, which is indicative of variable range hopping (Mott, 1968). It is very important to note that the temperature dependence of the dark conductivity of unhydrogenated poly-Si is independent of any thermal treatment prior to the measurement; e.g., annealing at high temperatures followed by a rapid thermal quench does not influence the temperature dependence of σ_D. Hydrogen passivation of silicon dangling bonds causes the activation energy to decrease to $E_A = 0.45$ eV. Moreover, activated behavior is observed in the entire temperature range. At temperatures above 150 K this activation energy decrease is accompanied by an increase in the dark conductivity by up to a factor of 35 and is, most likely, due to a decrease in the grain-boundary potential barriers caused by hydrogenation of dangling bonds.

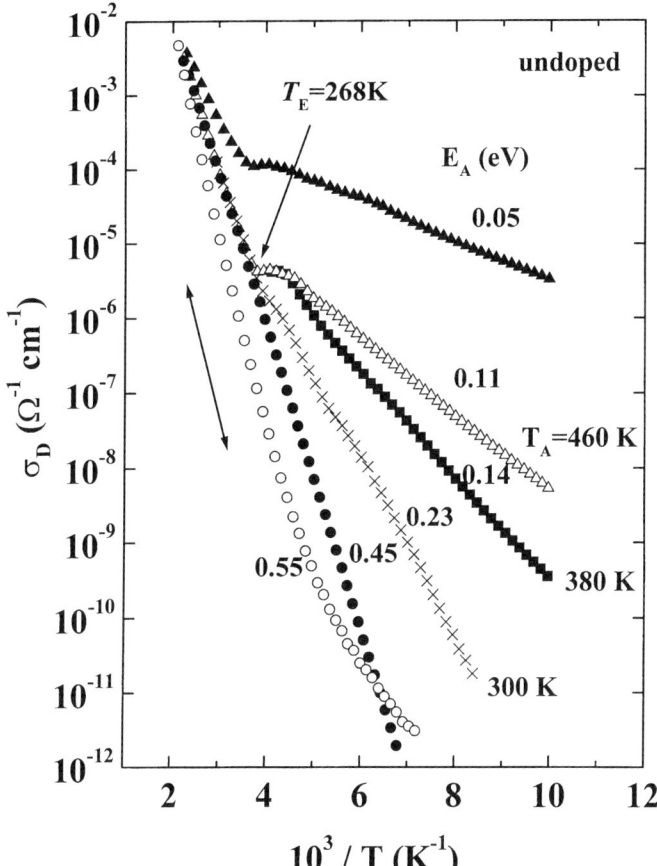

FIG. 28. Arrhenius plots of the dc electrical dark conductivity σ_D of undoped poly-Si prior to (open circles) and after hydrogen passivation. The solid circles represent the temperature dependence of σ_D in the relaxed state obtained after slowly cooling the specimen. The data represented by crosses, squares, and open triangles were obtained after the poly-Si film was annealed at the indicated temperatures T_A and subsequently thermally quenched with a cooling rate of 1.5 K/sec. The solid triangles depict σ_D after a rapid thermal quench from 460 K with a cooling rate of 3 K/sec. The data were taken while the poly-Si sample was heated at a rate of 0.1 K/sec.

Hydrogenated polycrystalline silicon is very sensitive to thermal treatment prior to the conductivity measurement. In order to obtain the data represented by the full circles in Fig. 28, the specimen was annealed at 460 K, subsequently slowly cooled to 77 K, and then the data were taken while the poly-Si sample was heated at a rate of 0.1 K/sec. Deviations from this measurement procedure result in striking changes in the electrical dark conductivity. The influence of the annealing temperature followed by rapid thermal quenching was determined through the following procedure: The sample was annealed at the indicated temperature T_A for 10 min and subsequently thermally quenched to 77 K at a quenching rate of 1.5 K/sec. Then the electrical dark conductivity was measured while the sample was heated at a rate of 0.1 K/sec. The data obtained after annealing at 300, 380, and 460 K are represented by the crosses, squares, and open triangles in Fig. 28, respectively. With increasing annealing temperature, the dark conductivity measured below room temperature increases significantly. All curves exhibit activated behavior according to

$$\sigma_D = \sigma_0 \exp\left(-\frac{E_A}{kT}\right) \qquad (18)$$

The activation energy of the low-temperature branch decreases from 0.45 to 0.11 eV with increasing annealing temperature. Simultaneously, the prefactor σ_0 decreases, which is indicative of a Meyer-Neldel behavior (Meyer and Neldel, 1937). At a measurement temperature of 268 K, all curves merge together, and the dark conductivity becomes independent of the thermal treatment.

The enhancement of the dark conductivity also should depend on the quenching rate. A hydrogen-passivated poly-Si film was annealed at 460 K for 10 min and subsequently cooled to 77 K by immersing it in liquid nitrogen. This resulted in an increase in the cooling rate to 3 K/sec and produced the most striking enhancement of σ_D by more than 8 orders of magnitude (solid triangles in Fig. 28). An enhancement of the dark conductivity even above 268 K was observed. This is due to the high heating rate at which the data were taken, which does not allow for complete relaxation of the quenched-in state during the measurement time. Therefore, after rapid thermal quenching from 460 K at a quenching rate of $dT_Q/dt = 1.5$ K/sec, the temperature dependence of the dark conductivity was measured as a function of the heating rate (Fig. 29). With a decreasing heating rate from $dT_H/dt = 0.1$–0.02 K/sec, a conductivity decay starting at about 210 K is observed, and eventually σ_D passes through a minimum at $T = 256$ K. The magnitude of the minimum strongly depends on the heating rate. Thus a lower heating rate provides more time for the quenched-in state to relax. This result is consistent with measurements of the relaxation kinetics discussed below.

Similar thermal quenching experiments were performed on phosphorus- and boron-doped poly-Si films. Arrhenius plots of the dark conductivity of phosphorus-doped poly-Si before and after H passivation and after rapid

4 HYDROGEN IN POLYCRYSTALLINE SILICON 135

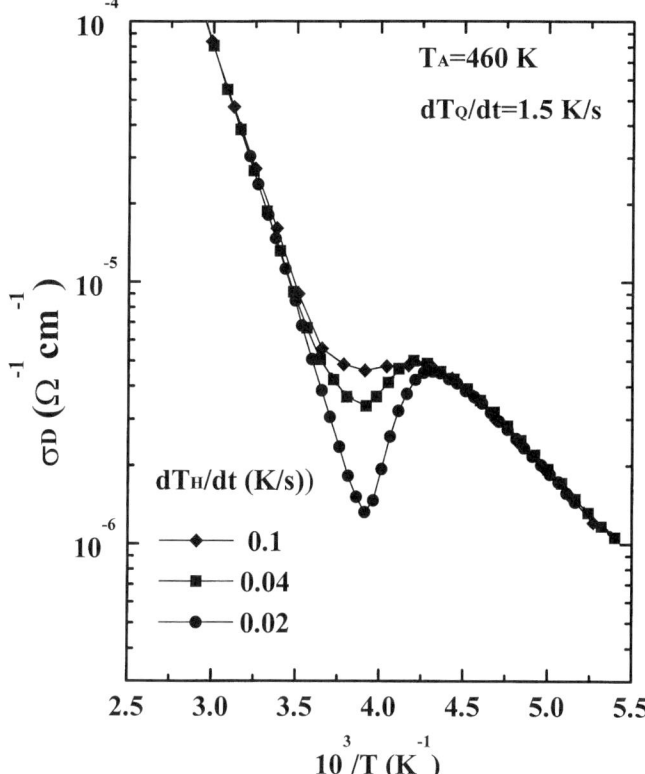

FIG. 29. Temperature dependence of the electrical dark conductivity σ_D measured with various heating rates dT_H/dt. Prior to each measurement, the hydrogen-passivated poly-Si sample was annealed at 460 K for 10 min and then thermally quenched at a quenching rate of $dT_Q/dt = 1.5$ K/sec.

thermal cooling are shown in Fig. 30. The curve depicted by the open circles was obtained on the unhydrogenated sample, and the full circles represent the slow cooled or relaxed state of the hydrogen-passivated specimen. Thermal treatment prior to the conductivity measurement causes similar changes in the temperature dependence of σ_D as in undoped poly-Si film (see Fig. 28). The magnitude of the quenched-in dark conductivity is somewhat smaller, which most likely is due to a smaller grain-boundary volume, since the average grain size in the phosphorus-doped samples is 150 nm. Again, no conductivity enhancement was observed in unhydrogenated samples.

With increasing phosphorus concentration, the quenching-induced conductivity enhancement decreases and eventually disappears at a phosphorus

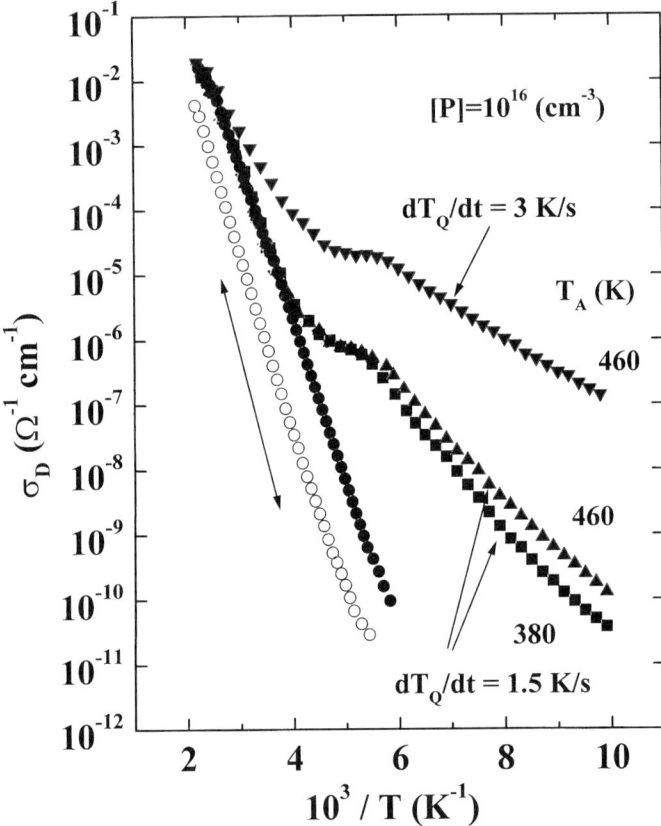

FIG. 30. Temperature dependence of the electrical dark conductivity measured on phosphorus-doped poly-Si. The open and full circles represent the unhydrogenated sample and the slow-cooled state, respectively. The curves depicted by squares and up triangles were obtained after annealing the sample at $T_A = 380$ and 460 K, respectively, and cooling them with a cooling rate of $dT_Q/dt = 1.5$ K/sec. The data shown by the down triangles were obtained after an anneal at 460 K and a rapid thermal quench with a cooling rate of $dT_Q/dt = 3$ K/sec. The poly-Si sample was doped with a phosphorus concentration of 10^{16} cm^{-3}.

concentration of 10^{18} cm^{-3}. In this set of experiments, the average grain size remained constant, and only the phosphorus concentration changed. Hence, the magnitude of quenching-induced enhancement of σ_D should be similar in all samples. However, the increase in conductivity due to phosphorus doping screens the quenching-induced enhancement of σ_D at high phosphorus concentrations.

On the other hand, rapid thermal quenching of boron-doped polycrystalline silicon exhibits a new feature. The temperature dependence of σ_D of a boron-doped sample with a doping concentration of 5×10^{16} cm^{-3} is plotted in Fig. 31. The dark conductivity shown by the full circles was measured after the hydrogenated poly-Si film was slowly cooled to 100 K. An enhancement of the conductivity was achieved only with a quenching rate of 3 K/sec. At 100 K, an increase in σ_D of approximately a factor of 30

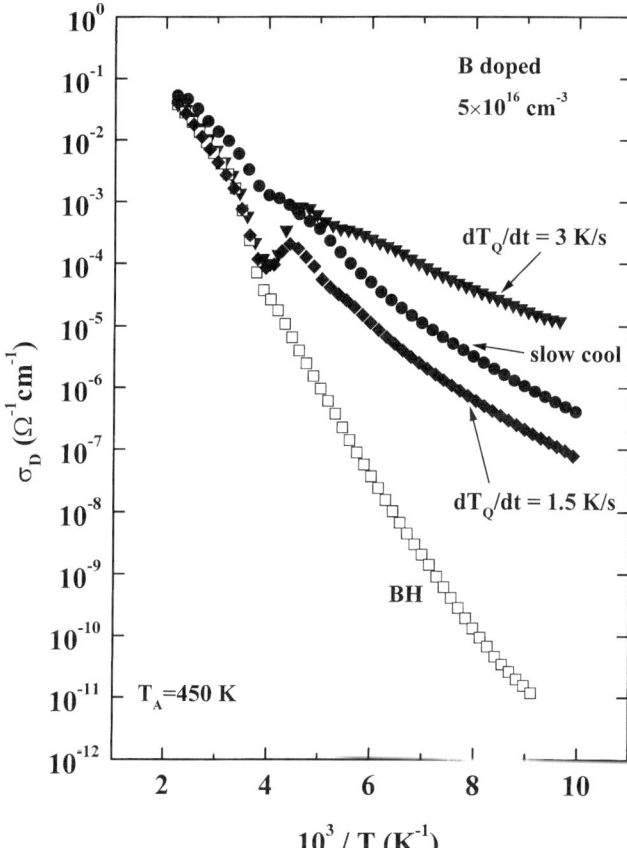

FIG. 31. Temperature dependence of the electrical dark conductivity measured on hydrogenated boron-doped poly-Si. The circles represent the slow-cooled state. Data shown by the diamonds and triangles were obtained after thermal quenching from 450 K at cooling rates of $dT_Q/dt = 1.5$ and 3 K/sec, respectively. The open squares show the influence of dopant passivation. The poly-Si samples were doped with a boron concentration of 5×10^{16} cm^{-3}. Experimental details are described in the text.

is observed. This increase is small compared with undoped and phosphorus-doped poly-Si. On the other hand, a surprising result is obtained when the same sample is thermally quenched with a smaller cooling rate of $dT_Q/dt = 1.5$ K/sec. Compared with the slow-cooled state, the dark conductivity is reduced (diamonds in Fig. 31). Moreover, independent of the quenching rate used to produce the metastable state, the dark conductivity decays in the temperature range between 230 and 268 K. This behavior indicates that the curve obtained after a slow cool does not represent the relaxed state with respect to the quenching-enhanced conductivity. The ground state with respect to the quenched-in state was obtained through the following procedure: First, a boron-doped poly-Si sample was thermally quenched from 450–100 K at a cooling rate of 3 K/sec, and subsequently, the temperature dependence of σ_D was measured up to 260 K. At this temperature, the decay of the dark conductivity was measured as a function of time until σ_D reached a stable value. Then the sample was slowly cooled to 100 K, and the temperature dependence of σ_D was measured again. The data are shown by the open squares in Fig. 31. The decrease in σ_D and the increase in the activation energy of the slow-cooled state are due to the passivation of boron atoms. At 450 K, hydrogen is released from a reservoir and migrates through the specimen. At the time when the poly-Si film is rapidly cooled, a fraction of the migrating hydrogen atoms is captured by boron atoms, forming BH complexes that are electrically inactive (Pankove, 1991). The acceptor passivation leads to a decrease in the dark conductivity and an increase in the activation energy (open squares in Fig. 31). This important result shows that dopants are *not* participating in the quenching-induced metastable conductivity enhancement.

The changes in the electrical dark conductivity of both undoped and doped hydrogenated polycrystalline silicon are completely reversible. An anneal at elevated temperatures followed by a slow cool completely restores the temperature dependence of the relaxed state shown by the full circles in Figs. 28, 30, and 31. From the data presented earlier, an important conclusion can be drawn regarding the microscopic origin of the quenching metastability. Since quenching-induced conductivity enhancements occur only in hydrogen-passivated poly-Si films, it is evident that hydrogen participates in the phenomenon. Moreover, the changes in the dark conductivity are due to the formation and dissociation of an electrically active hydrogen complex.

Hall-effect measurements performed on boron-doped poly-Si in the slow-cooled state, after thermal quenching and after acceptor passivation, provide further insight into the microscopic mechanism of the quenching-enhanced conductivity. In the slow-cooled state and in the relaxed state, labeled BH in Fig. 31, the majority carriers are holes. However, after rapid thermal quenching, the majority carriers are electrons (diamonds and triangles in

Fig. 31). Evidently, the electrically active hydrogen complex is a donor overcompensating the electrically active acceptor states.

Further information on the microscopic mechanism responsible for the quenching metastability can be obtained from the relaxation kinetics. A poly-Si sample was rapidly quenched from 460 K by immersing the sample in liquid nitrogen. Then the specimen was heated to a temperature T at which the time dependence of the conductivity decay was recorded. Normalized dark-conductivity decays of undoped poly-Si are shown by the points in Fig. 32. Similar conductivity decays were obtained from phosphorus- and boron-doped poly-Si samples. The lines in Fig. 32 represent least-squares fits to a stretched exponential decay

$$\frac{\sigma_D(\infty) - \sigma_D(t)}{\sigma_D(\infty) - \sigma_D(0)} = \frac{\Delta\sigma_D(t)}{\Delta\sigma_D(0)} = \exp\left[-\left(\frac{t}{\tau}\right)^\beta\right] \quad (19)$$

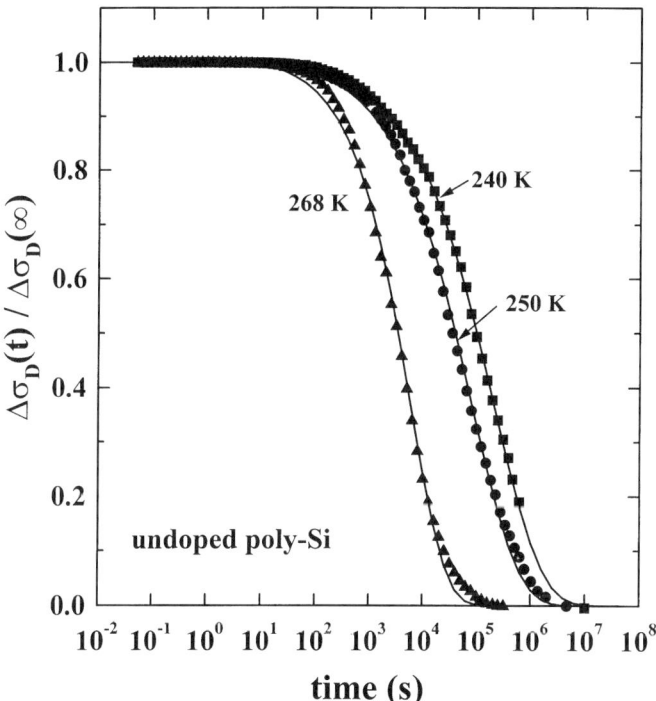

FIG. 32. Normalized conductivity transients measured on undoped hydrogen-passivated poly-Si. The data are represented by the points, and the lines are a least-squares fit to a stretched exponential decay (Eq. 19).

and are in good agreement with the data. Fitting parameters are the dispersion parameter β, which varies between 0.49 and 0.66, and the relaxation-time constant τ. The temperature dependence of the relaxation-time constant is plotted in Fig. 33 for undoped (circles) and boron-doped poly-Si films (triangles). For both samples, τ exhibits activated behavior with $E_\tau \approx 0.74\,\text{eV}$ and an exponential prefactor of $\tau_0 \approx 5 \times 10^{-11}$ and $8 \times 10^{-11}\,\text{sec}$ for boron and undoped poly-Si, respectively. This indicates that the microscopic mechanism responsible for quenching-induced conductivity enhancements is a first-order process and can be described by a two-level system. In this model, the energetically higher level corresponds to the quenched-in metastable state, while the energetically lower level is identified with the relaxed or slow-cooled state. This is sketched on the left-hand side of Fig. 34.

During an anneal at T_A, weakly bound hydrogen from the reservoir is released and migrates through the poly-Si lattice. The reservoir includes configurations such as the H_2^* complex (Chang and Chadi, 1989) and

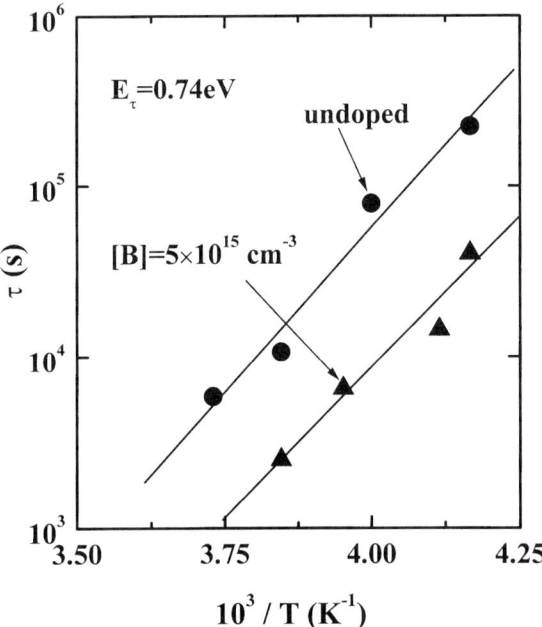

FIG. 33. Temperature dependence of the relaxation time constant τ for undoped (circles) and boron-doped poly-Si (triangles). τ is thermally activated with an activation energy of $E_\tau \approx 0.74\,\text{eV}$.

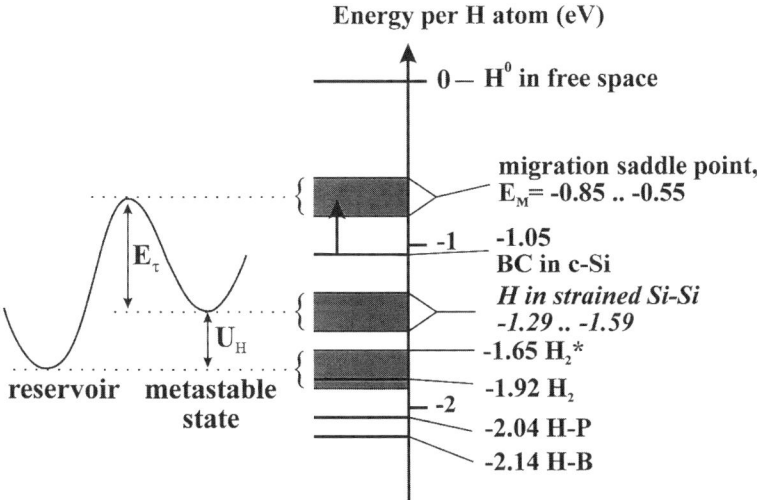

FIG. 34. Calculated energies for various hydrogen configurations in silicon. The zero of energy corresponds to a free H atom, and the energy scale shows the energy per H atom to form the various H complexes. The energy levels of a two-level model are adjusted at the saddle point for H migration. $E_\tau = 0.74$ eV is the activation energy of the relaxation time constant τ, and $U_H \approx 0.35$ eV is the formation energy of the electrically active hydrogen donor complex.

hydrogen-induced platelets (see Sec. VI). Rapid thermal cooling forces the poly-Si samples into a nonequilibrium state that causes the electrical dark conductivity to increase by more than 8 orders of magnitude (Fig. 28). This requires the generation of approximately 10^{15} cm^{-3} hydrogen donor complexes. The formation energy of the H donors was estimated from Arrhenius plots of the quenching-induced change in the electrical dark conductivity as a function of $1/T_A$. Since the slope of the low-temperature branches of the enhanced conductivity decreases with increasing annealing temperature T_A, only a rough estimate of the formation energy of $U_H \approx 0.35$ eV was obtained. Hence the total barrier height between H reservoir and metastable state is given by $E_{total} = E_\tau + U_H \approx 1.09$ eV.

According to the time and temperature dependence of the defect formation and dissociation, the metastable hydrogen complex is a donor, and it has to be low enough in energy to dissociate below room temperature. A likely candidate is thermal oxygen donors that can form during plasma hydrogenation (Michel and Kimerling, 1994). However, hydrogen can passivate oxygen thermal donors (Hoelzlein et al., 1986; Pearton et al., 1986), and these complexes are stable at temperatures well above room temperature. Also, annihilation of aggregated thermal donors cannot explain my observations because this process occurs at temperatures above 500°C and is thermally activated with 3.4 eV (Michel and Kimerling, 1994). Thus the only hydrogen complex that is low enough in energy to disassociate below room temperature is in a Si—Si H bond center site. This H configuration has been identified previously as a donor state in single-crystal silicon (Gorelkinskii and Nevinnyi, 1987; Holm et al., 1991). However, in c-Si this H complex is unstable at temperatures above 100 K, and dissociation is thermally activated with only 0.29 eV (Holm et al., 1991). On the other hand, the nature of polycrystalline silicon provides a simple mechanism that allows for hydrogen in a Si—Si bond center [H(BC)] to be stable up to room temperature. At grain boundaries, the Si—Si bond length deviates from its intragrain value of 2.35 Å. In fact, for polycrystalline germanium, it has been reported that bond distortions are asymmetric, with stretched bonds outnumbering compressed bonds. Moreover, the presence of tensile stress causes bond-length distortions of up to 15%, or 0.35 Å (Tarnow et al., 1990). This has direct impact on the energy required to insert a hydrogen atom into the bond-center site. According to first-principles calculations performed for tensile stress and bond-angle distortions, the formation energy of the H(BC) complex decreases at a rate of 0.4 eV per 0.1 Å bond distortions (Van de Walle and Nickel, 1995). Therefore, grain-boundary disorder can account for the stabilization of bond-centered hydrogen complexes above 100 K. Detailed information on the first-principles calculations is presented in Chap. 6.

The simple two-level system can be adjusted to a relative energy scale where the zero of energy corresponds to a free H atom and the scale shows the energy per H atom to form the various complexes. The maximum of the two-level model corresponds to the migration saddle point E_M for hydrogen. In single-crystal silicon, E_M has been determined experimentally and theoretically. From first-principles calculations (Van de Walle et al., 1989) and high-temperature diffusion experiments (Van Wieringen and Warmholtz, 1956), the migration saddle point was determined to occur 0.2–0.5 eV above the ground state of the hydrogen interstitial, respectively. Thus the maximum of the two-level system is located at $E_M = -0.55$ to -0.85 eV with respect to a free hydrogen atom (see Fig. 34). Since the electrically active hydrogen complex dissociates over a barrier of $E_\tau = 0.74$ eV, the metastable state is located at an energy of -1.29 to -1.59 eV, which is 0.24–0.54 eV *below* the interstitial site for hydrogen in c-Si. According to first-principles calculations, this energy gain translates into a bond-length distortion of 0.06–0.14 Å (Van de Walle and Nickel, 1995). This result is consistent with bond-length distortions reported for grain boundaries in polycrystalline germanium (Tarnow et al., 1990). Subtracting the formation energy U_H of the metastable state places the hydrogen reservoir or relaxed state to energies of -1.65 to $-1.92\,\mathrm{eV}$. This energy range includes interstitial molecular hydrogen, H_2^*, and hydrogen-stabilized platelets, H_{2n}^*.

The total barrier height of the two-level system and the energetic position of the reservoir are consistent with results obtained from the time and temperature dependence of the spin passivation caused by a low-temperature vacuum anneal (see Sec. IV). A low-temperature anneal establishes a new equilibrium with a lower spin concentration by activating weakly bound H from a reservoir. The characteristic time of the spin passivation reveals activated behavior with an activation energy of $E_A = 0.95 \pm 0.02\,\mathrm{eV}$. Since the spin passivation is due to release of hydrogen from a reservoir and capture of H by silicon dangling bonds, the migration saddle point is identical to that indicated in the two-level model in Fig. 34. From this experiment the energetic position of the reservoir is estimated to be $E_M - E_A = -1.5 \pm 0.02$ to $-1.8 \pm 0.02\,\mathrm{eV}$, which is in good agreement with the energy range obtained from the quenching experiments. Thus the energies for the reservoir are consistent with H_2^* complexes and H-stabilized platelets, and the metastable state is consistent with the identification of bond-center hydrogen as the H-induced donor complex responsible for the quenching metastability.

In this section I have demonstrated that the presence of hydrogen in poly-Si gives rise to metastability. Light-induced defect generation was observed in completely hydrogen-passivated samples. I propose that only hydrogen, silicon dangling bonds, and strained Si—Si bonds are necessary to observe this effect. The rejuvenation of the light-induced metastability simply by reexposing the sample to monatomic hydrogen establishes the participation of H atoms in the defect-generation process. The cooling rate–dependent conductivity enhancement is clearly due to the formation of a hydrogen donor complex. I propose bond-center hydrogen as the active hydrogen donor that is stable up to room temperature due to disorder in the grain-boundary regions. Moreover, BC(H) as the active donor complex is completely consistent with experimental and theoretical studies of hydrogen in silicon.

VI. Hydrogen-Induced Defects During Plasma Exposure

The metastable phenomena presented in preceding sections occur in hydrogen-passivated poly-Si specimens. Commonly, hydrogen is incorporated into poly-Si by exposing the samples to a hydrogen plasma at elevated temperatures, and it is assumed that the main effect of hydrogen is the passivation of preexisting dangling bonds. In this section I show that in addition to grain-boundary passivation, a simple posthydrogenation step can cause the formation of acceptor-like defects and hydrogen-stabilized platelets. Acceptor-like defects are generated on a time scale of 10^4 sec at 350°C and produce conductivity-type conversion, whereas hydrogen-stabilized platelets nucleate and grow at lower temperatures ($<300°C$) and are predominantly oriented along $\{111\}$ crystallographic planes. For all experiments, the influence of plasma radiation is negligible because an optically isolated remote hydrogen system (Johnson et al., 1991b) was used.

1. GENERATION OF ACCEPTOR-LIKE DEFECTS

Commonly, information on hydrogen diffusion is obtained after the diffusion experiment has been performed; e.g., a specimen is exposed to a hydrogen plasma, and then the H depth distribution is measured with SIMS. These experiments provide limited information on microscopic diffusion processes. A deeper insight in the diffusion process can be obtained from *in situ* conductivity measurements during hydrogen passivation at elevated temperatures.

The specimens investigated here had a thickness of 0.1 μm and were fabricated on quartz substrates. Prior to the *in situ* conductivity measurements, the native oxide was removed, and subsequently, an 8-nm-thick oxide layer was deposited in an optically isolated remote plasma system. Then the poly-Si films were exposed to monatomic H in the same system, and the *in situ* conductivity σ_{plasma} was measured as a function of exposure time and temperature. Typical *in situ* conductivity curves are shown in Fig. 35 for various indicated temperatures. The *in situ* conductivity exhibits a steep increase immediately after the plasma is ignited. With increasing hydrogenation time, σ_{plasma} reaches a maximum and eventually decays on a very long time scale. Because these measurements were performed in the presence of a remote H plasma, the observed changes also could be due to plasma heating or the generation of fixed charge in the oxide layer. This, however, is not the case. The influence of plasma heating on the *in situ* conductivity was evaluated carefully by exposing poly-Si films to other plasma species such as oxygen. In the temperature range where the experiments were performed, plasma heating caused an increase of the *in situ* conductivity of less than 5% with respect to the initial electrical dark conductivity and hence cannot account for the observed increase of σ_{plasma}. On the other hand, fixed charge in the top oxide easily could cause the observed changes of σ_{plasma}. In order to rule out the influence of fixed charge on σ_{plasma}, the oxide layer of a poly-Si sample was removed after an exposure for 8 hours

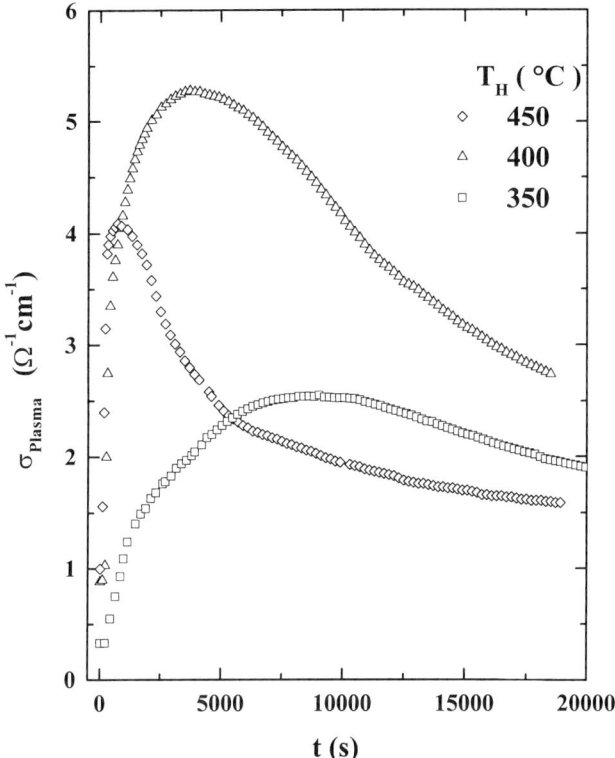

FIG. 35. *In situ* conductivity σ_{plasma} as a function of the exposure time t. Nominally undoped poly-Si was exposed to monatomic H at the indicated temperatures. The samples had an average grain size of 1.5 μm and were 0.1 μm thick.

to monatomic H, and the electrical dark conductivity was remeasured. Removing the oxide layer resulted in a conductivity change by only 9%, which is small compared with the observed conductivity enhancement during hydrogen exposure. Therefore, the observed changes of σ_{plasma} are due to neither plasma heating nor the presence of fixed charge in the oxide layer.

The time for the *in situ* conductivity to reach the maximum decreases with increasing hydrogenation temperature, which is consistent with the temperature dependence of the effective diffusion coefficient (Fig. 3). On the other hand, the absolute values of the *in situ* conductivity data plotted in Fig. 35 cannot be compared directly because they depend on the electrical dark conductivity σ_{plasma} ($t = 0$), which for undoped samples is thermally activated at 0.52 eV (see Fig. 37). In order to account for the temperature dependence of the dark conductivity, the data in Fig. 35 were normalized to unity at $t = 0$. The normalized *in situ* conductivity σ_{norm} curves are shown in Fig. 36. The largest increase of σ_{norm} by a factor of 7.6 is observed at 350°C. This corresponds to approximately $10^{17}\,cm^{-3}$ additional carriers. With increasing exposure temperature, the maximum value of σ_{norm} decreases. Moreover, the conductivity maximum is inversely proportional to the hydrogenation temperature and the grain size of the polycrystalline silicon films. This is consistent with the observation that the hydrogen concentration decreases with increasing substrate temperature (see Sec. III). An exposure to an attenuated hydrogen plasma revealed that the plasma-induced increase in the *in situ* conductivity is proportional to the concentration of monatomic H, while the time to reach the maximum of σ_{norm} is inversely proportional to the concentration of monatomic H. Also, *in situ*

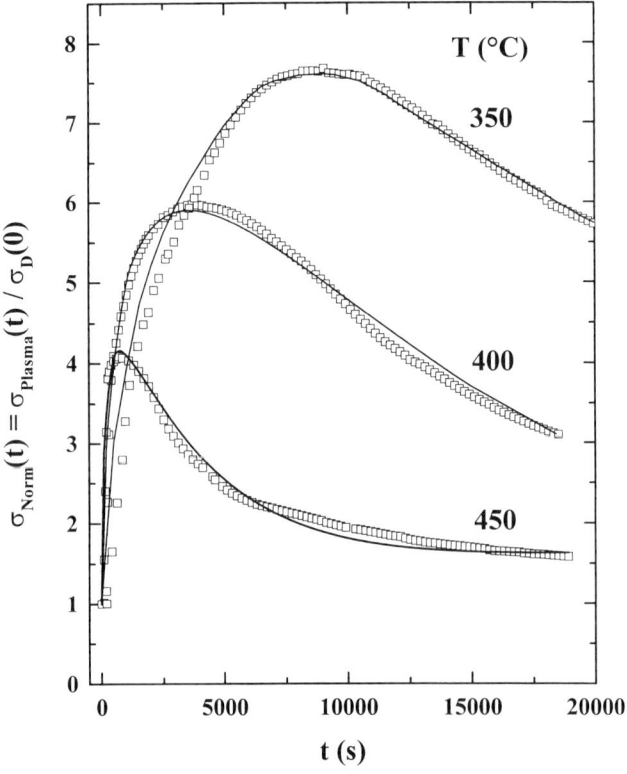

FIG. 36. *In situ* conductivity curves of Fig. 35 normalized to unity at $t = 0$. The data are shown by the open symbols, and the lines represent the calculated *in situ* conductivity.

conductivity transients, as observed in Fig. 36, were obtained only when the specimens were exposed to monatomic hydrogen. From these results it can be concluded that the observed conductivity enhancement is due to migrating H atoms in the poly-Si host. Moreover, since the poly-Si films are undoped and the Fermi energy resides approximately at midgap, hydrogen introduced from the plasma diffuses in the positive charge state, H^+, donating electrons.

When the hydrogen plasma is turned off, the *in situ* conductivity decreases to a residual value somewhat larger than σ_{plasma} ($t = 0$). This is due to defect passivation at the grain boundaries that lowers the potential barriers and hence causes an increase of the electrical dark conductivity. According to the time and temperature dependence of the defect passivation (see Sec. IV), the defect concentration reaches a minimum during the first 2000 sec at 350°C. At higher temperatures, this time decreases. Moreover, hydrogen passivation of grain-boundary defects enhances the conductivity by only ~15% of the total conductivity increase observed in Fig. 36. Hence this effect cannot be the cause of the overall enhancement of *in situ* conductivity.

At long hydrogenation times, the *in situ* conductivity decreases exponentially with time constants of 2.8×10^4 and 2×10^3 sec at hydrogenation temperatures of 350 and 475°C, respectively. A change in the hydrogen concentration cannot account for the decay, since the H^+ concentration and hence the additional concentration of electrons are constant with time once the *in situ* conductivity reaches the maximum.

Important information on the microscopic origin of the conductivity decay was obtained from temperature-dependent dark-conductivity and Hall-effect measurements. Figure 37 shows the temperature dependence of σ_D before (squares) and after hydrogenation at 400°C for 6×10^3 (triangles) and 3.5×10^4 sec (diamonds), respectively. The results obtained from Hall-effect measurements are depicted by the open symbols, whereas the solid symbols were obtained with coplanar contacts. In the temperature range where both techniques could be applied, the data are in good agreement. The unhydrogenated sample exhibits a thermally activated dark conductivity with an activation energy of $E_A = 0.52\,eV$, and the majority carriers are electrons (squares in Fig. 37). This is consistent with the data shown in Section V.2. Hall-effect measurements revealed an electron mobility of $\mu_e \approx 1.9\,cm^2/V \cdot sec$ at 400 K. The triangles in Fig. 37 were measured after an exposure to monatomic H for 6000 sec at 400°C. The dark conductivity reveals a noticeable increase in the entire temperature range that is accompanied by a decrease in the activation energy to $E_A = 0.26\,eV$. Surprisingly, however, the majority carriers have changed from electrons to *holes*. With increasing hydrogenation time, this effect becomes even more pronounced. Data shown by the diamonds in Fig. 37 were taken after the

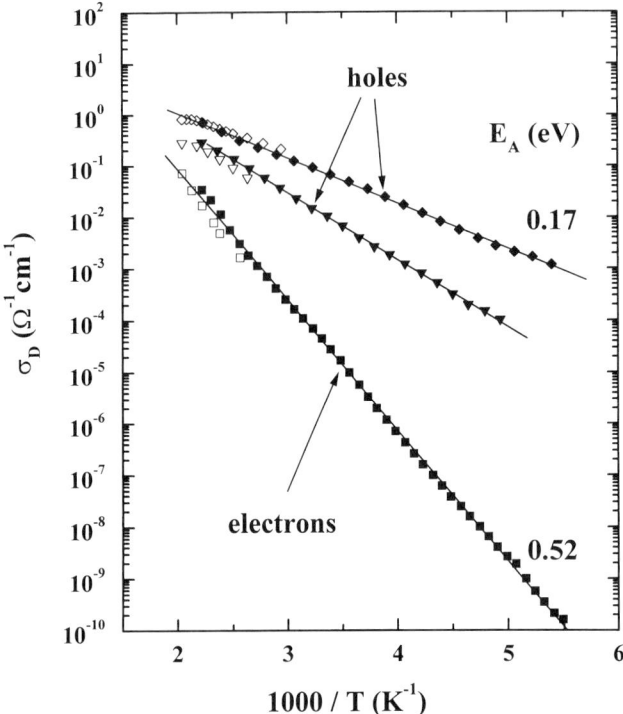

FIG. 37. Temperature dependence of the electrical dark conductivity of undoped poly-Si. The solid symbols represent data obtained with coplanar contacts, and the open symbols depict data from Hall-effect measurements. The lines are least-squares fits to the data. The squares depict the dark conductivity before hydrogenation, and the majority carriers are electrons. The triangles and diamonds represent σ_D after hydrogenation for 6×10^3 and 3.5×10^4 sec at 400°C, respectively. The majority carriers are holes, indicating conductivity-type conversion of the poly-Si sample.

in situ conductivity reached steady state ($t = 3.5 \times 10^4$ sec). The activation energy decreased to $E_A = 0.17$ eV, and at 400 K, a hole mobility of $\mu_h = 12.4$ cm^2/V · sec was measured. The activation energy and the hole mobility reveal a time dependence because hydrogenation and hence the conductivity enhancement and the acceptor generation advance from the sample surface and are diffusion limited. These results demonstrate that prolonged hydrogenation at elevated temperatures causes the generation of acceptor-like states that lead to conductivity-type conversion. Extrapolating the data in Fig. 36 shows that the steady-state value of the *in situ* conductivity is temperature dependent. Thus the concentration of newly generated acceptor states is determined by the hydrogen concentration and not limited by the number of available sites.

The data shown in Figs. 35 to 37 suggest that changes in the *in situ* conductivity are governed by three processes: (1) passivation of grain-boundary defects, (2) monatomic H from the gas phase diffusing in the positive charge state H$^+$ and contributing electrons, and (3) the generation of compensating acceptor states. The first two mechanisms contribute to an increase in the *in situ* conductivity, whereas mechanism 3 reduces σ_{plasma}. Since grain-boundary passivation occurs on a very short time scale of 500–2000 sec at 450 and 350°C, respectively, and causes conductivity enhancements of less than 15%, the *in situ* conductivity is mainly determined by the diffusion of monatomic H in the positive charge state and by the generation of acceptor states. These mechanisms contribute to the *in situ* conductivity as follows:

$$\Delta\sigma_{\text{plasma}}(t) = \sigma_{\text{plasma}}(t) - \sigma_D(t) \cong q\mu_e \left[\frac{1}{L} \int_0^L C_H(x,t)\,dx - \frac{\mu_h}{\mu_e} N_{\text{acc}}(t) \right] \quad (20)$$

where $\sigma_D(t)$ is the dark conductivity, q is the electric charge, L is the sample thickness, and μ_e and μ_h are the electron and hole mobility, respectively. The hydrogen concentration is given by Eq. (1), and the acceptor concentration increases according to

$$N_{\text{acc}}(t) = N_0 \left[1 - \exp\left(-\frac{t}{\tau}\right) \right] \quad (21)$$

with a time constant τ and a saturated value N_0. The free parameters are the concentration of positively charged H atoms, which is smaller than the total H concentration in the samples, the effective diffusion coefficient of H$^+$ N_0, and the time constant for acceptor generation.

The specimens are covered with an 8-nm-thick oxide layer that attenuates the hydrogen flux into the underlying poly-Si layer (Nickel *et al.*, 1995). Therefore, the hydrogen surface concentration C_0 is time dependent. Since the hydrogen diffusion length is much larger than the thickness of the oxide layer, $C_0(t)$ can be approximated by a linear increase until the saturated surface concentration is reached. In Fig. 36 the calculated *in situ* conductivity normalized to unity at $t = 0$ is overplotted on the data (lines). The fits are in good agreement with the data, indicating that the *in situ* conductivity is mainly determined by the hydrogen-induced conductivity enhancement due to the reaction H$^0 \rightarrow$ H$^+$ + e$^-$ and by the generation of acceptor states. Since the hydrogen concentration saturates by the time the *in situ* conductivity reaches the maximum, the exponential decay can be analyzed in terms of acceptor generation.

The temperature dependence of the fitting parameters provides further insight into the microscopic processes. The temperature dependence of the effective diffusion coefficient is plotted in Fig. 38. D_{eff} is thermally activated with an activation energy of $E_A = 0.63\,\text{eV}$, which is in good agreement with results obtained from SIMS measurements (Fig. 3). The values of the effective diffusion coefficient obtained from the fits are about a factor of 10 smaller than the experimentally obtained data. Most likely this is due to the presence of a thin oxide layer that causes a significant decrease in H concentration, which is accompanied by a decrease in the effective diffusion coefficient (Nickel et al., 1995). This result supports the idea that the conductivity enhancement is due to hydrogen migrating in the positive charge state.

The time constant for acceptor generation is thermally activated at $E_A = 1.62\,\text{eV}$ (squares in Fig. 39). This, however, is not enough information

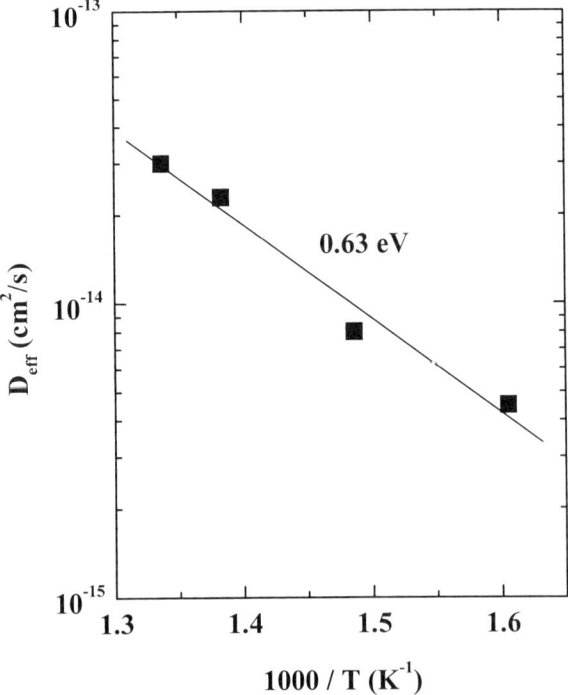

FIG. 38. Temperature dependence of the effective diffusion coefficient D_{eff} obtained from the calculated in situ conductivity. D_{eff} is thermally activated with $E_A = 0.63\,\text{eV}$.

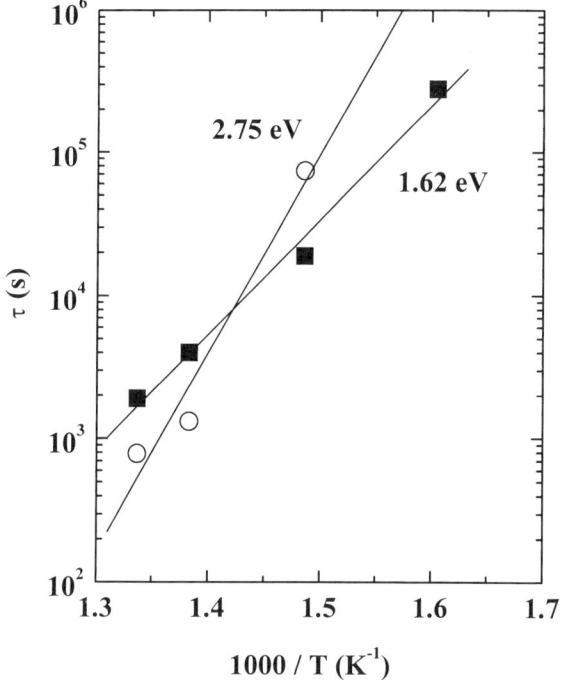

FIG. 39. Temperature dependence of the time constant for acceptor generation τ (squares) and acceptor annealing (circles). The activation energies were obtained from least-squares fits to the data.

to identify the acceptor-like state or the microscopic process that leads to acceptor generation. A first attempt has been made by Van de Walle and Neugebauer (1995), who investigated complexes involving one and two H atoms and a Si self-interstitial using first-principles calculations (see also Chap. 6). A Si self-interstitial with one or two H atoms introduces donor and acceptor states in the silicon bandgap. Both states are located in the lower half of the bandgap. Since the poly-Si films used for this study are nominally undoped and the Fermi energy is located in the upper half of the gap ($E_C - E_F = 0.52\,\text{eV}$), hydrogenated silicon self-interstitials would be in the negative charge state and act as acceptors. This is consistent with the data in Figs. 35 to 37. However, further investigations are necessary to identify the microscopic processes involved in this phenomenon.

The newly generated acceptor-like states are thermally stable up to temperatures of 350°C. The annealing kinetics were measured using the following procedure: The plasma was terminated and the poly-Si sample was subjected to isochronal anneals at elevated temperatures ($>350°C$). Then the plasma was reignited, and the *in situ* conductivity was measured again. The difference between the *in situ* conductivity under steady-state conditions and σ_{plasma} after the anneal is proportional to the number of annealed acceptor states. The dissociation of Si—H bonds does not influence this experiment, since the dissociation rate is negligible below 500°C (Zellama et al., 1981). The temperature dependence of the time constant for acceptor annealing exhibits activated behavior with an activation energy of $E_A = 2.75\,eV$ (circles in Fig. 39), which is consistent with the high thermal stability of these defects.

This set of experiments has shown a new property of hydrogen in silicon. Prolonged exposure of poly-Si to monatomic H at temperatures above 300°C produces conductivity-type conversion by generating acceptor-like defects. This is clearly attributed to the interaction of H with the silicon lattice, since the number of acceptors is limited only by the hydrogen concentration and not by the number of available sites. Acceptor generation is thermally activated with $E_A = 1.62\,eV$. The acceptor-like states are stable up to 350°C and anneal over a barrier of 2.75 eV. Moreover, the *in situ* conductivity experiments have shown that hydrogen in undoped poly-Si migrates in the positive charge state H^+.

2. PLATELETS

Hydrogen passivation in the low-temperature range ($<300°C$) also produces defects. The enhanced deuterium concentration in the near-surface region of poly-Si (Fig. 1) suggests the presence of hydrogen-stabilized platelets. In the case of single-crystal silicon, this correlation is well established (Johnson et al., 1987, 1992). Previously, some general features of the platelets and their formation conditions have been determined. The defects appear only in *n*-type silicon within 1000 Å of the sample surface, are planar in shape with a diameter of up to 1000 Å, and are predominantly oriented along {111} crystallographic planes. Platelet formation occurs predominantly at low hydrogenation temperatures ($<300°C$) as a consequence of enhanced nucleation at low temperatures. Raman spectroscopy provided direct evidence that platelets arise from the coordination formation of Si—H bonds. Specimens containing platelets revealed a broad peak centered at $2140\,cm^{-1}$ that shifts to $1570\,cm^{-1}$ in deuterated specimens (Johnson et al., 1987, 1991a). The vibrational frequency and the isotope shift are characteristic of Si—H bonds, suggesting a higher-order Si—H complex (Lucovsky et al., 1979). Moreover, platelets introduce electrically active gap states that have been detected by photoluminescence spectroscopy and deep-level transient spectroscopy (DLTS). In photoluminescence spectroscopy measurements, several radiative transitions were observed, with the main peak located at 0.98 eV. On the other hand, DLTS detects two hydrogen-induced gap states at approximately 0.06 and 0.51 eV below the conduction band edge (Johnson et al., 1987).

In this subsection I show that hydrogen passivation of poly-Si films results in the formation of platelets despite the presence of high concentrations of H traps. The experiments were performed on nominally undoped and phosphorus-doped laser-crystallized polycrystalline silicon films. In order to generate high concentrations of platelets, the samples were passivated according to a two-step hydrogenation schedule. The first passivation was performed at 150°C for 20 min, followed by an exposure to monatomic H at 275°C for 60 min. This procedure is known to generate high concentrations of platelets with a length of up to 100 nm. The nucleation rate is temperature dependent and decreases with increasing hydrogenation temperature. Platelets nucleated at low temperatures continue to grow during the second exposure step (Johnson et al., 1992).

In Section III.1 deuterium concentration depth profiles measured on undoped poly-Si indicated the presence of platelets. Since platelets, generated with an optically isolated remote plasma, occur only in n-type samples (Nickel et al., 1999), deuterium depth profiles obtained on undoped and phosphorus-doped poly-Si samples are compared in Fig. 40. The undoped specimen was exposed to monatomic D for 60 min at 200°C, and the D depth profile is shown by the solid curve. The dashed line represents a least-squares fit to Eq. (1). In the near-surface region the fit deviates from the data revealing a deuterium accumulation that, most likely, is due to the presence of deep H traps. For a depth greater than 0.1 μm, data and fit are in good agreement. The phosphorus-doped samples were exposed to monatomic D at 150°C for 5 and 30 min, respectively. Both depth profiles reveal a peak in the deuterium concentration within a depth of 300 Å that is similar to data reported for n-type c-Si (see Fig. 9 and Johnson et al., 1987).

The similarities of the deuterium depth profiles measured in poly-Si to data obtained from single-crystal silicon strongly suggest that hydrogenation of poly-Si also leads to the generation of hydrogen-stabilized platelets. On the other hand, it is also conceivable that the enhanced D surface concentration found in poly-Si arises from hydrogen accommodated in preexisting defects such as stacking faults and grain boundaries. Hence cross-sectional TEM micrographs were taken on undoped and phosphorus-doped poly-Si films. Cross-sectional TEM micrographs taken before and after hydrogenation are shown in Fig. 41. Prior to the hydrogenations, the only defects visible are grain boundaries (Fig. 41a). Undoped and phosphorus-doped poly-Si are shown in Fig. 41b and c, respectively, after a

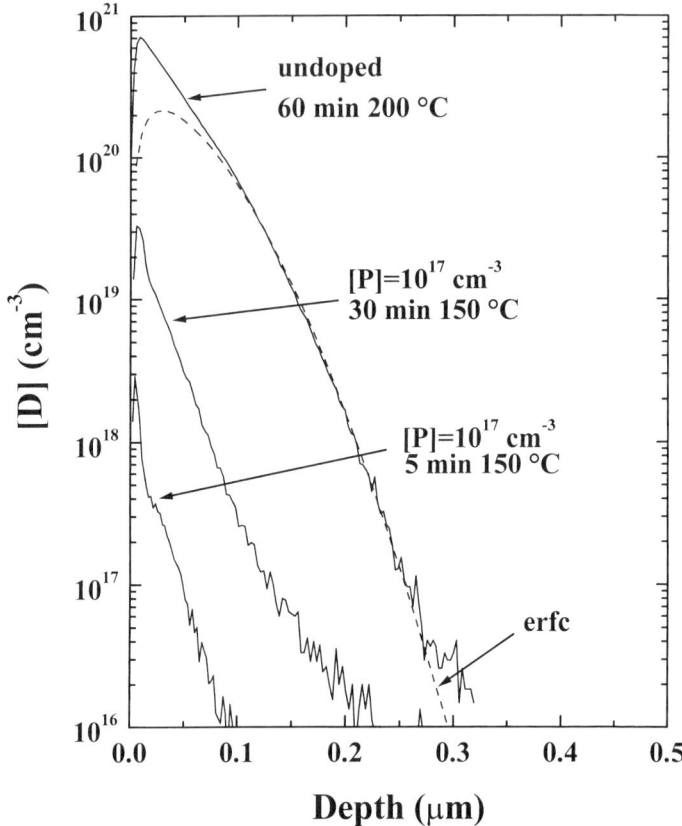

FIG. 40. Deuterium concentration depth profiles in undoped and phosphorus-doped poly-Si. The undoped specimen was passivated at 200°C for 60 min. The phosphorus-doped specimens were hydrogenated at 150°C for 5 and 30 min, respectively. The dashed curve depicts a least-squares fit to Eq. (1). Note the enhanced deuterium concentration in the near-surface region of all samples. (From Nickel et al., 1996; reprinted from "Hydrogen-Induced Platelets in Disordered Silicon," Solid State Communications, 99, 427–431. Copyright 1996 by Elsevier Science, Ltd.)

two-step hydrogenation. In both specimens, micro defects identical to hydrogen-stabilized platelets are observed. The micrographs show the poly-Si cross section close to a $\langle 110 \rangle$ zone axis. Since platelets are predominantly oriented along $\langle 111 \rangle$ planes, they appear as linear defects in two of the four planes. From Fig. 41 a distinct difference in platelet concentration can be seen. In undoped poly-Si, the platelet concentration amounts to approximately 5×10^{15} cm^{-3}, whereas phosphorus-doped

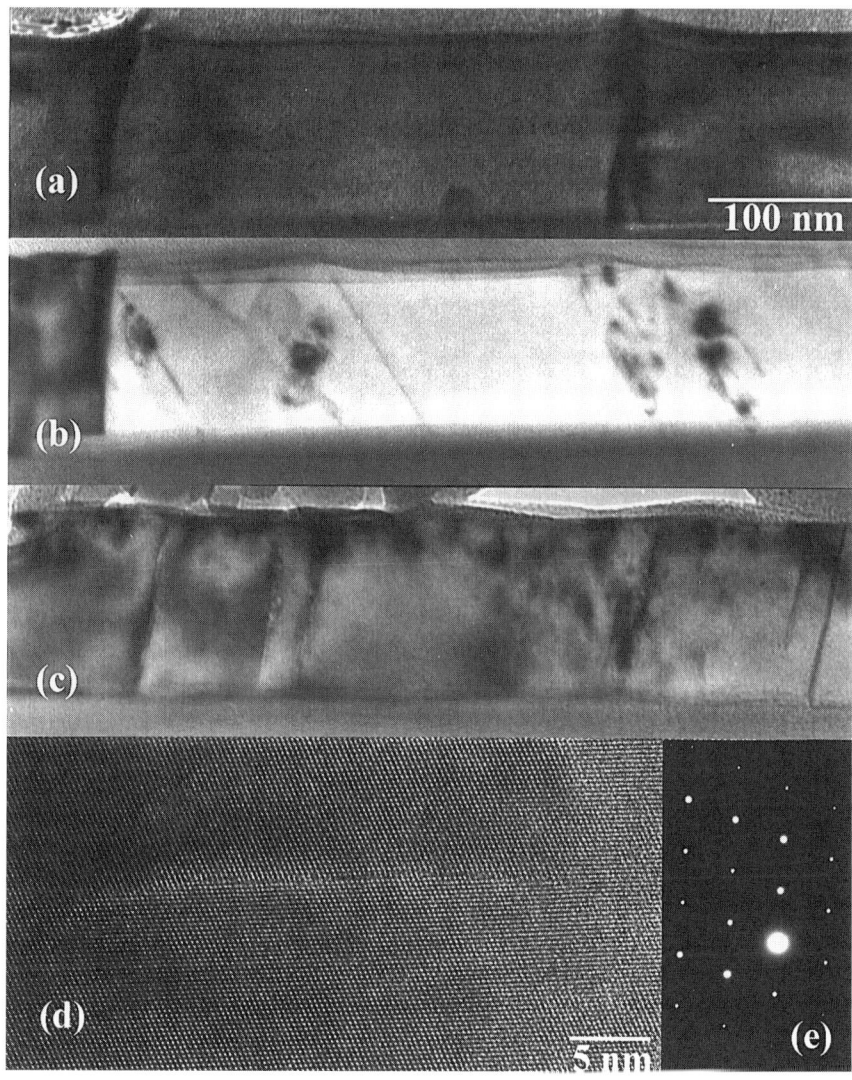

FIG. 41. Cross-sectional TEM micrographs viewed in a $\langle 110 \rangle$ projection. (a) Bright-field image of unhydrogenated poly-Si. (b) Undoped poly-Si exposed to monatomic H at 150 and 275°C for 20 and 60 min, respectively. (c) Poly-Si doped with a P concentration of 1×10^{20} cm^{-3} hydrogenated simultaneously with the specimen shown in (b). (d) High-resolution lattice image of a hydrogen-stabilized platelet in P-doped poly-Si in $\langle 110 \rangle$ projection. (e) Electron diffraction pattern of the grain shown in (d). (From Nickel et al., 1996; reprinted from "Hydrogen-Induced Platelets in Disordered Silicon," Solid State Communications, **99**, 427–431. Copyright 1996 by Elsevier Science, Ltd.)

poly-Si contained about 1.5×10^{16} cm^{-3} platelets. This is consistent with previous results showing that the platelet concentration increases monotonically with the phosphorus concentration (Johnson et al., 1992; Nickel et al., 1999). A high-resolution lattice image of a platelet in a {111} crystallographic plane is shown in Fig. 41d, and the corresponding electron diffraction pattern is shown in Fig. 41e. A Burgers circuit analysis of the platelet shown in Fig. 41d did not indicate a displacement of the lattice, eliminating dislocations as the origin of the platelets. Moreover, contrast typical of stacking faults (Ponce et al., 1981) was not observed, eliminating the possibility that the observed defects are interstitial or vacancy loops.

Considering the high concentration of hydrogen traps (see Sec. III), the generation of H-stabilized platelets in poly-Si is a surprising result indicating that the H concentration in the near-surface region of the specimens plays an important role. Platelet formation induces strain in the host lattice that can be observed in TEM micrographs (Fig. 41). Hence one would expect that grain boundaries are preferential sites for platelet nucleation. This, however, is not the case. Figure 42b shows a high-resolution cross-sectional TEM micrograph of a grain boundary. The periodicity of the silicon lattice appears for one grain only due to a different orientation of the other grain. The high-resolution images taken at grain boundaries showed neither evidence for nucleation nor for the penetration of grain boundaries by platelets. In fact, no platelets were found within 40 nm of the grain boundaries. The average platelet concentration in undoped poly-Si is plotted in Fig. 43 as a function of the distance from the grain boundary. For this purpose, platelet concentrations were averaged over a large number of grains. In Fig. 43, the grain boundary is located at $d = 0$, and the center of the grain is situated at $d = 600$ nm. With increasing distance from the grain boundary into the grain, the platelet concentration increases monotomically from 0 at the grain boundary to approximately 5×10^{15} cm^{-3} in the center of the grain. Similar results were obtained on phosphorus-doped poly-Si samples, clearly demonstrating that the presence of lattice strain does not promote the generation of platelets.

Since platelets arise from the coordination formation of Si—H bonds (Johnson et al., 1987), the amount of hydrogen accommodated in these micro defects can be estimated from platelet size and concentration. The average platelet diameter amounts to ~ 80 nm, accommodating approximately 6×10^4 Si—H bonds. In undoped poly-Si the platelet concentration amounts to $\sim 5 \times 10^{15}$ cm^{-3}. Thus the concentration of H accommodated in platelet structures computes to roughly 3×10^{20} cm^{-3}, which is consistent within a factor of 3 with the H concentration in the near-surface layer obtained from SIMS measurements. This important result clearly shows that H accumulated in the near-surface region of the specimens is accommodated in platelets.

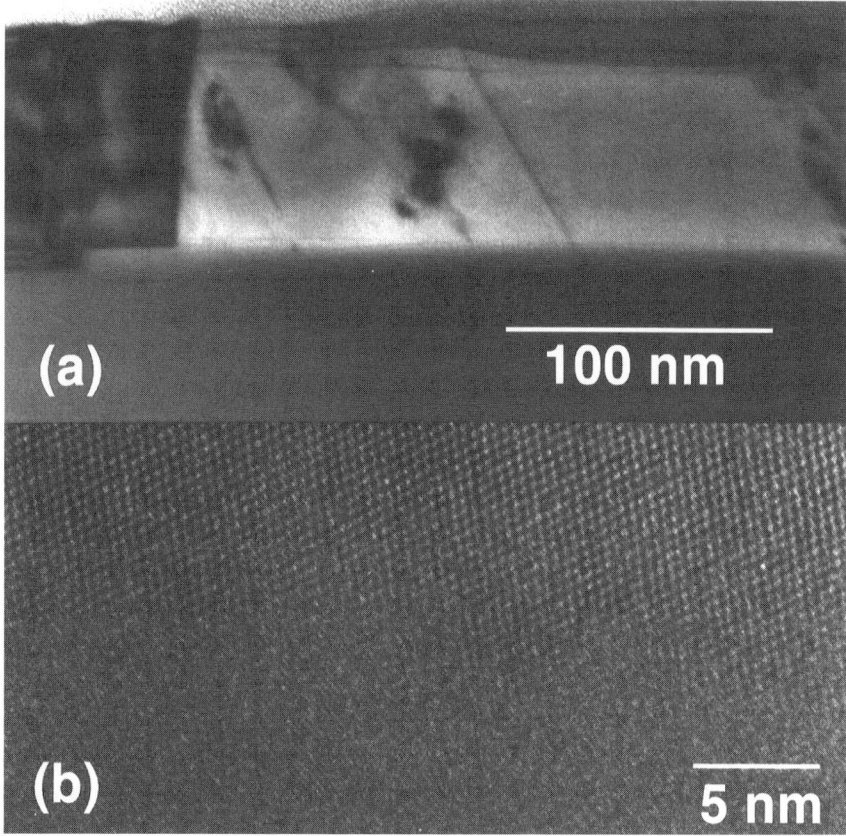

FIG. 42. Cross-sectional TEM micrographs of grain boundaries in low (a) and high (b) resolution. The image in (b) is rotated by 90°. The upper half of (b) shows the lattice periodicity of a grain that does not appear in the lower half due to a different orientation of the grain. (From Nickel et al., 1996; reprinted from "Hydrogen-Induced Platelets in Disordered Silicon," Solid State Communications, **99**, 427–431. Copyright 1996 by Elsevier Science, Ltd.)

Platelet nucleation studies performed on c-Si clearly show that platelet formation occurs only in n-type specimens with a Fermi-level position of $E_C - E_F \leq 0.45$ eV (Nickel et al., 1999). The observation of platelets in phosphorus-doped poly-Si is consistent with these results. On the other hand, TEM micrographs and H concentration depth profiles obtained from nominally undoped poly-Si clearly establish the presence of platelets in this material. This raises an interesting question. The temperature dependence of the dark electrical conductivity, measured on undoped laser-crystallized

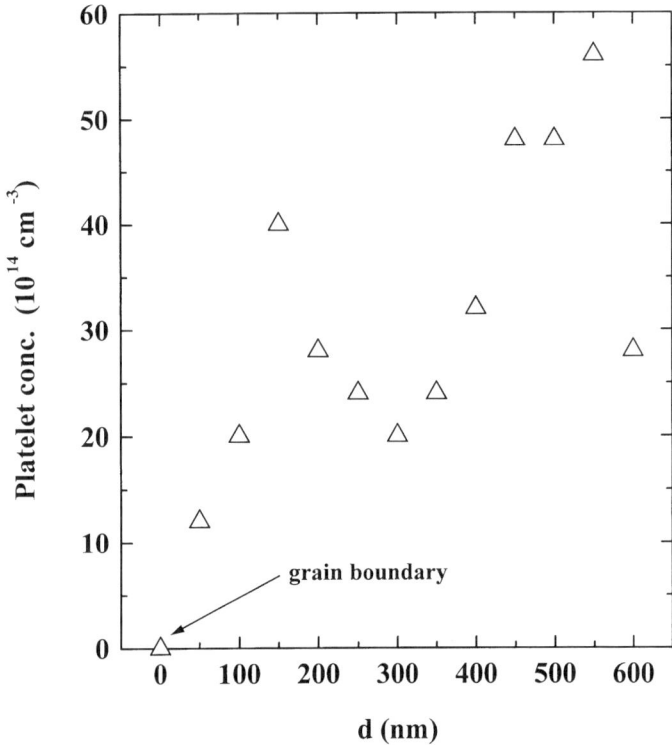

FIG. 43. Platelet concentration in undoped polycrystalline silicon versus distance from the grain boundary. $d = 0$ indicates the grain boundary. (From Nickel et al., 1996; reprinted from "Hydrogen-Induced Platelets in Disordered Silicon," Solid State Communications, **99**, 427-431. Copyright 1996 by Elsevier Science, Ltd.)

poly-Si prior to hydrogenation, revealed activated behavior with an activation energy of 0.46 eV. This suggests that either the amorphous silicon was contaminated or some degree of contamination occurred during laser crystallization.

Strained Si—Si bonds predominantly located at grain boundaries introduce localized states in the bandgap. These states form exponential band tails commonly observed in disordered semiconductors such as amorphous (Fritzsche, 1988) and polycrystalline silicon (Jackson et al., 1983). Since the generation of platelets is accompanied by the generation of lattice strain, one would expect that platelet nucleation is favorable at grain boundaries. According to the results presented earlier, platelets do not nucleate or penetrate grain boundaries. Moreover, I did *not* observe that two platelets cross each other. There are several mechanisms that can prevent platelet formation at and close to grain boundaries: (1) grain boundaries contain high concentrations of deep hydrogen traps such as Si dangling bonds, (2) hydrogenation of poly-Si reduces strain by decreasing the concentration of weak Si—Si bonds; this could effectively prevent the generation of extended platelets, and (3) in n-type poly-Si, a depletion layer is formed at the grain boundaries due to fixed charge. Hence at grain boundaries the Fermi energy resides deeper in the bandgap compared with the center of the grain. In order for platelets to nucleate, the Fermi-level position has to be in the upper half of the bandgap ($E_C - E_F \leq 0.45$ eV) (Nickel et al., 1999). Thus platelet nucleation does not occur close to grain boundaries.

In this section I have presented experimental evidence for the nucleation and growth of hydrogen-stabilized platelets in polycrystalline silicon. As in c-Si, platelets are predominantly oriented along {111} crystallographic planes. In undoped and phosphorus-doped poly-Si, platelet concentrations of approximately 5×10^{15} and 1.5×10^{16} cm^{-3} were achieved by exposing the specimens to monatomic hydrogen at 150 and 275°C for 20 and 60 min, respectively. The amount of H accommodated in platelets is in good agreement with the H concentration in the near-surface layer measured by SIMS. Although platelets induce strain in the host lattice, preexisting lattice strain does not promote platelet generation. The results presented here are consistent with recent data obtained on single-crystal silicon (Johnson et al., 1987).

VII. Summary and Future Directions

In this chapter I have shown that the properties of hydrogen in polycrystalline silicon extend well beyond the simple passivation of silicon dangling bonds. Although hydrogen is readily introduced into poly-Si, the diffusion process is highly complex and depends on experimental parameters as well as the microstructure of the poly-Si samples. In general, hydrogen diffusion in LPCVD-grown poly-Si is enhanced compared with SSC specimens. At high hydrogen concentrations, both materials exhibited dispersive hydrogen diffusion, which is consistent with multiple trapping within a distribution of barriers between transport sites. The data are consistent with a simple two-level model used to describe H diffusion in amorphous silicon. Information on the H density of states was obtained applying this model. The hydrogen chemical potential was found to reside about 1.2 eV below the transport sites. Moreover, in response to an increase in the H concentration in SSC poly-Si from 10^{17} to mid-10^{21} cm^{-3}, the hydrogen chemical potential increases abruptly by about 0.2 eV. In this material, shallow and deep

H traps are located at 0.5 and 1.5–1.7 eV below the transport states, respectively. However, at low hydrogen concentrations, the H chemical potential depends markedly on the method used to prepare the samples, and clear evidence for deep traps was found only in SSC poly-Si.

The most prominent deep hydrogen trap predominantly located at grain boundaries is the silicon dangling bond. Hydrogen introduced into poly-Si from a plasma source effectively passivates dangling bonds. This results in an improvement in the electrical properties of poly-Si films and devices. The lowest spin density was obtained at a hydrogenation temperature of 350°C. Defect passivation is diffusion limited. Model calculations of defect passivation suggest that the amount of H required to passivate defects exceeds the defect concentration by a factor that is larger than unity and which depends on the passivation temperature as well as on the macroscopic structure of the poly-Si samples.

The presence of hydrogen gives rise to metastability. Hydrogen is directly involved in the light-induced defect creation and in the cooling rate–dependent change in the electrical dark conductivity. In the first effect, the participation of H manifests itself by the ability to rejuvenate the light-induced defect generation by reexposure to monatomic H, whereas the latter effect is clearly due to the generation and dissociation of a H donor complex. I suggest that this complex is bond-centered H, which is completely consistent with experimental and theoretical studies of hydrogen in silicon.

In addition to metastability and defect passivation, a very important property of hydrogen is the generation of defects during plasma exposure. Two different kinds of defects are generated. At high passivation temperature ($>300°C$), prolonged exposure of poly-Si to monatomic H causes the formation of acceptor states. This results in conductivity-type conversion. The concentration of the newly generated acceptor states is limited by the H concentration in the specimens. In addition, model calculations show that H introduced into undoped poly-Si diffuses in the positive charge state. Acceptor generation is thermally activated with $E_A = 1.62$ eV. The acceptor states are thermally stable up to 350°C. The time constant for annealing is thermally activated with $E_A = 2.75$ eV. On the other hand, at hydrogenation temperatures below 300°C, the exposure of poly-Si to monatomic H causes the formation of hydrogen-stabilized platelets. As in c-Si, these defects appear within a depth of 0.1 μm below the sample surface and are oriented along $\{111\}$ crystallographic planes. Platelet concentrations of up to 1.5×10^{16} cm^{-3} were observed in phosphorus-doped poly-Si. The amount of H accommodated in platelets is in good agreement with the H concentration in the near-surface layer measured by SIMS. Preexisting lattice strain does not promote platelet growth and nucleation because platelets were not observed in the vicinity of grain boundaries.

4 HYDROGEN IN POLYCRYSTALLINE SILICON

Although a surprising richness of phenomena directly related to the interactions of hydrogen and poly-Si have been discovered, a lot of questions are still open. Most of these features are doubtlessly related to the grain boundaries. The work presented in this chapter by no means provides an exhausting picture of the properties of H in poly-Si. A lot of questions are still open, and additional experimental and theoretical work is required to develop a complete microscopic picture of the interactions of hydrogen with the grain boundaries.

ACKNOWLEDGMENTS

I am grateful to G. B. Anderson, D. K. Biegelsen, J. B. Boyce, C. Herring, W. B. Jackson, N. M. Johnson, R. I. Johnson, R. A. Street, Chris G. Van de Walle, and J. Walker for an interesting and successful collaboration. I would like to thank W. B. Jackson for a critical reading of the manuscript. This work was supported in part by NREL and by the Alexander von Humboldt Foundation, Federal Republic of Germany.

REFERENCES

Allan, D. C., Joannopoulos, J. D., and Pollard, W. B. (1982). *Phys. Rev. B*, **25**, 1065.
Ayres, J. R. (1993). *J. Appl. Phys.*, **74**, 1787.
Ballutaud, D., Aucouturier, M., and Babonneau, F. (1986). *Appl. Phys. Lett.*, **49**, 1620.
Boyce, J. B., Johnson, N. M., Ready, S. E., and Walker, J. (1992). *Phys. Rev. B*, **46**, 4308.
Branz, H. M., Asher, S. E., and Nelson, B. P. (1993). *Phys. Rev. B*, **47**, 7061.
Carlson, D. E., and Magee, C. W. (1978). *Appl. Phys. Lett.*, **33**, 81.
Chang, K. J., and Chadi, D. J. (1989). *Phys. Rev. Lett.*, **62**, 937.
Den Boer, W., Geerts, M. J., Ondris, M., and Wentinck, H. M. (1984). *J. Non-Cryst. Solids*, **66**, 363.
Dersch, H., Stuke, J., and Beichler, J. (1981). *Appl. Phys. Lett.*, **38**, 456.
Fritzsche, H. (ed.). (1988). *Amorphous Silicon and Related Materials*. Singapore: World Scientific.
Gorelkinskii, Y. V., and Nevinnyi, N. N. (1987). *Sov. Technol. Phys. Lett.*, **13**, 45.
Gupta, D. (1988). In *Diffusion Phenomena in Thin Films and Microelectronic Materials*, D. Gupta and P. S. Ho, eds. Park Ridge, NJ: Noyes, chap. 1.
Hawkins, W. G. (1986). *IEEE Trans. Electron. Devices*, **33**, 477.
Herring, C., and Johnson, N. M. (1991). In *Hydrogen in Silicon Semiconductors and Semimetals*, **54**, J. I. Pankove and N. M. Johnson, eds. San Diego: Academic Press, chap. 10.
Hoelzlein, K.-H., Pensl, G., Schulz, M., and Johnson, N. M. (1986). In *Oxygen, Carbon, Hydrogen, and Nitrogen in Crystalline Silicon*, J. C. Mikkelsen, S. J. Pearton, J. W. Corbett, and S. J. Pennycook, eds. Pittsburgh: MRS, pp. 481–486.

Holm, B., Nielsen, K., and Nielsen, B. (1991). *Phys. Rev. Lett.*, **66**, 2360.
Jackson, W. B. (1990). *Phys. Rev. B*, **41**, 1059.
Jackson, W. B., and Tsai, C. C. (1992). *Phys. Rev. B*, **45**, 6564.
Jackson, W. B., and Zhang, S. B. (1991). *Physica*, **B170**, 197.
Jackson, W. B., Johnson, N. M., and Biegelsen, D. K. (1983). *Appl. Phys. Lett.*, **43**, 195.
Jackson, W. B., Johnson, N. M., Tsai, C. C., Wu, I.-W., Chiang, A., and Smith, D. (1992). *Appl. Phys. Lett.*, **61**, 1670.
Jackson, W. B., Santos, P. V., and Tsai, C. C. (1993). *Phys. Rev. B*, **47**, 9993.
Johnson, N. M., Biegelsen, D. K., and Moyer, M. D. (1982a). *Appl. Phys. Lett.*, **40**, 882.
Johnson, N. M., Biegelsen, D. K., and Moyer, M. D. (1982b). In *Grain Boundaries in Semiconductors*, C. H. Seager, G. E. Pike, and H. J. Leamy, eds. New York: Elsevier.
Johnson, N. M., Ponce, F. A., Street, R. A., and Nemanich, R. J. (1987). *Phys. Rev. B*, **35**, 4166.
Johnson, N. M., Doland, C., Ponce, F. A., Walker, J., and Anderson, G. B. (1991a). *Physica*, **B170**, 3.
Johnson, N. M., Walker, J., and Stevens, K. S. (1991b). *Appl. Phys. Lett.*, **69**, 2631.
Johnson, N. M., Herring, C., Doland, C., Walker, J., Anderson, G. B., and Ponce, F. (1992). *Mater. Sci. Forum*, **83–87**, 33–38.
Jousse, D., Delage, S. L., and Iyer, S. S. (1989). *Philos. Mag. B*, **63**, 443.
Kamins, T. I., and Marcoux, P. J. (1980). *IEEE Electron. Device Lett.*, **EDL-1**, 159.
Kemp, M., and Branz, H. M. (1993). *Phys. Rev. B*, **47**, 7067.
Kittel, C. (1969). *Thermal Physics*. New York: Wiley.
Lucovsky, G., Nemanich, R. J., and Knights, J. C. (1979). *Phys. Rev. B*, **19**, 2064.
Meyer, W., and Neldel, H. (1937). *Z. Technol. Phys.*, **18**, 588.
Michel, J., and Kimerling, L. C. (1994). In *Oxygen in Silicon*, **42**, F. Shimura, ed. San Diego: Academic Press, pp. 251–289.
Mott, N. F. (1968). *J. Non-Cryst. Solids*, **1**, 1.
Nickel, N. H., Fuhs, W., and Mell, H. (1991). *J. Non-Cryst. Solids*, **137&138**, 1221.
Nickel, N. H., Johnson, N. M., and Jackson, W. B. (1993). *Appl. Phys. Lett.*, **62**, 3285.
Nickel, N. H., Jackson, W. B., Wu, I.-W., Tsai, C. C., and Chiang, A. (1995). *Phys. Rev. B*, **52**, 7791.
Nickel, N. H., Anderson, G. B., and Walker, J. (1996). *Solid State Commun.*, **99**, 427–431.
Nickel, N. H., Anderson, G. B., Johnson, N. M., and Walker, J. (1999). In press.
Nickel, N. H., Anderson, G. B., and Johnson, R. I. (1997b). *Phys. Rev. B*, **56**, 12065.
Northrop, G. A., and Oehrlein, G. S. (1986). In *Defects in Semiconductors*, **10–12**, H. J. v. Bardeleben, ed. Paris: Materials Science Forum, p. 1253.
Ostapenko, S., Jastrzebski, L., and Lagowski, J. (1996). *Appl. Phys. Lett.*, **68**, 2873.
Pandya, R., and Kahn, B. A. (1987). *J. Appl. Phys.*, **62**, 3244.
Pankove, J. I. (1991). In *Hydrogen in Semiconductors*, **34**, J. I. Pankove and N. M. Johnson, eds. San Diego: Academic Press, pp. 91–112.
Pearton, S. J., Chantre, A. M., Kimerling, L. C., Cummings, K. D., and Dautremont-Smith, W. C. (1986). In *Oxygen, Carbon, Hydrogen, and Nitrogen in Crystalline Silicon*, J. C. Mikkelsen, S. J. Pearton, J. W. Corbett, and S. J. Pennycook, eds. Pittsburgh: MRS, pp. 475–480.
Pearton, S. J., Corbett, J. W., and Shi, T. S. (1987). *Appl. Phys. A*, **43**, 153.
Pfleiderer, H., Kusian, W., and Krühler, W. (1984). *Solid State Commun.*, **49**, 493.
Ponce, F. A., Yamashita, T., Bube, R. H., and Sinclair, R. (1981). In *Defects in Semiconductors*, J. Narayan and T. Y. Tan, eds. Amsterdam: North-Holland, p. 503.
Powell, M. J., Deane, S. C., French, I. D., Hughes, J. R., and Milne, W. I. (1991). *Philos. Mag. B*, **63**, 325.

Redfield, D., and Bube, R. H. (1990). In *Twenty-First IEEE Photovoltaic Specialists Conference*, 1673. New York: IEEE, p. 1506.
Santos, P. V., and Jackson, W. B. (1991). *Phys. Rev. B*, **44**, 10937.
Santos, P. V., and Jackson, W. B. (1992). *Phys. Rev. B*, **46**, 4595.
Santos, P. V., Johnson, N. M., and Street, R. A. (1991). *Phys. Rev. Lett.*, **67**, 2686.
Seager, C. H. (1991). In *Hydrogen in Semiconductors*, **34**, J. I. Pankove and N. M. Johnson, eds. San Diego: Academic Press, p. 17.
Seager, C. H., and Ginley, D. S. (1979). *Appl. Phys. Lett.*, **34**, 337.
Seager, C. H., Anderson, R. A., and Panitz, J. K. G. (1987). *J. Mater. Res.*, **2**, 96.
Shi, Z., and Green, M. A. (1994). *Solid State Phenomena*, **37–38**, 459.
Shinar, J., Shinar, R., Mitra, S., and Kim, J.-Y. (1989). *Phys. Rev. Lett.*, **62**, 2001.
Shinar, J., Shinar, R., Wu, X. L., Mitra, S., and Girvan, R. F. (1991). *Phys. Rev. B*, **43**, 1631.
Sol, N., Kaplan, D., Dieumegard, D., and Dubreuil, D. (1980). *J. Non-Cryst. Solids*, **35&36**, 291.
Staebler, D. L., and Wronski, C. R. (1980). *J. Appl. Phys.*, **51**, 3262.
Street, R. A., Hack, M., and Jackson, W. B. (1988). *Phys. Rev. B*, **37**, 4209.
Street, R. A., Tsai, C. C., Kakalios, J., and Jackson, W. B. (1987). *Philos. Mag. B*, **56**, 305.
Stutzmann, M. (1989). *Philos. Mag. B*, **60**, 531.
Tanielian, M. (1982). *Philos. Mag. B*, **45**, 435.
Tarnow, E., Dallot, P., Bristowe, P. D., Joannopoulos, J. D., Francis, G. P., and Payne, M. C. (1990). *Phys. Rev. B*, **42**, 3644.
Thompson, M. J. (1991). *J. Non-Cryst. Solids*, **137&138**, 1209.
Van de Walle, C. G., and Neugebauer, J. (1995). *Phys. Rev. B*, **52**, R14320.
Van de Walle, C. G., and Nickel, N. H. (1995). *Phys. Rev. B*, **51**, 2636–2639.
Van de Walle, C. G., Denteneer, P. J., Bar-Yam, Y., and Pantelides, S. T. (1989). *Phys. Rev. B*, **39**, 10791.
Van Wieringen, A., and Warmholtz, N. (1956). *Physics*, **22**, 849.
Wilson, R. G., Stevie, F. A., and Magee, G. W. (1989). *Secondary Ion Mass Spectrometry: A Practical Handbook for Depth Profiling and Bulk Impurity Analysis*. New York: Wiley.
Zafar, S., and Schiff, E. A. (1989). *Phys. Rev. B*, **40**, 5235.
Zellama, K., Germain, P., Squelard, S., Bourbon, B., Fontenille, J., and Danielou, R. (1981). *Phys. Rev. B*, **23**, 6648.

CHAPTER 5

Hydrogen Phenomena in Hydrogenated Amorphous Silicon

Wolfhard Beyer

INSTITUT FÜR SCHICHT- UND IONENTECHNIK
FORSCHUNGSZENTRUM JÜLICH GMBH
JÜLICH, GERMANY

I. INTRODUCTION 165
II. MATERIAL CHARACTERIZATION BY HYDROGEN EFFUSION AND INFRARED
   ABSORPTION 166
   1. Measurement Techniques and Material Preparation 166
   2. Hydrogen Effusion Data. 167
   3. Infrared Absorption Data 174
III. EXPERIMENTAL HYDROGEN DIFFUSION AND SOLUBILITY DATA 182
   1. Hydrogen Diffusion Data 182
   2. Hydrogen Solubility Data 192
IV. HYDROGEN DIFFUSION AND EFFUSION EFFECTS 199
   1. Hydrogen Diffusion Processes 199
   2. Hydrogen Density of States Distribution and Hydrogen Chemical Potential 206
   3. Temperature Shift of Hydrogen Chemical Potential and Meyer-Neldel
      Rule of Hydrogen Diffusion 212
   4. Time Dependence of Hydrogen Diffusion Coefficient 217
   5. Deviations from Error-Function Diffusion Profiles 220
   6. Plasma In-Diffusion Versus Layer Diffusion 225
   7. Relation Between SIMS Diffusion Data and Hydrogen Effusion Data 226
   8. Interrelation of IR Absorption Spectra and Effusion Transients 229
V. HYDROGEN SOLUBILITY EFFECTS 230
   1. Solubility in Compact Material 231
   2. Hydrogen-Related Void Formation 234
VI. CONCLUSIONS 235
   REFERENCES 236

I. Introduction

One major reason for the continuous interest in the role of hydrogen in amorphous silicon, germanium, and related alloys during the past two decades is hydrogen's property to saturate dangling bonds of the host material and thus to reduce the concentration of defect states. This property was recognized in the early seventies by various groups (Beyer and Stuke, 1974; Le Comber et al., 1974; Lewis et al., 1974). Since then, hydrogenated amorphous silicon (a-Si:H) and a-Si–based alloys have found a wide range of application in thin-film transistors, thin-film solar cells, and other electronic devices. Much work has been done to characterize the state of hydrogen in these materials and to study processes of hydrogen motion and hydrogen release. Besides nuclear magnetic resonance (NMR) measurements (see, e.g., Boyce and Ready, 1991), some of the most successful characterization methods were infrared (IR) absorption, hydrogen effusion, and hydrogen diffusion measurements by secondary ion mass spectroscopy (SIMS), which were first applied by Lewis et al. (1974), Triska et al. (1975), and Carlson and Magee (1978), respectively. In this chapter, these latter three methods are applied to various series of hydrogenated amorphous silicon films with greatly different hydrogen concentrations. Results of these investigations provide new insights into various hydrogen phenomena in a-Si:H, such as the hydrogen effusion, hydrogen-related IR absorption, hydrogen-related microstructure, hydrogen diffusion, and hydrogen solubility. We confine the discussion to undoped a-Si:H.

II. Material Characterization by Hydrogen Effusion and Infrared Absorption

1. MEASUREMENT TECHNIQUES AND MATERIAL PREPARATION

Our study of various hydrogen phenomena in a-Si:H is based on several series of a:Si:H thin films of thicknesses ranging between 0.4 and 3 μm deposited on crystalline silicon or quartz substrates. Two series of radiofrequency (rf; 13.56 MHz) plasma-deposited A-Si:H samples (rf1 and rf2) can be considered as standard glow-discharge samples. Using capacitively coupled glow-discharge reactors of somewhat different geometry and different deposition parameters (see Table I), electronic-grade a-Si:H was obtained in both cases at the substrate temperatures $T_s = 200$–$250°C$. As a process gas, undiluted silane (SiH_4) or deuterosilane (SiD_4) was used. In the second reactor, we also employed silane diluted 10:1 with hydrogen (rf2, H-dil.). Another series of samples (ECR a-Si:H) was prepared by electron cyclotron resonance (ECR) plasma deposition using a remote plasma arrangement. We employed a flow of the argon carrier gas of 30 sccm and of (undiluted) silane of 15 sccm, a pressure of 10^{-2} mbar, and a microwave

TABLE I
Deposition Parameters of Standard rf a-Si:H Samples

	rf1	rf2
Electrode area (cm^2)	28	50
Electrode spacing (cm)	4	2.5
Gas pressure (mbar)	0.5	0.2–0.3
rf power (W)	10	10–20
SiH$_4$ flow rate (sccm)	2	5–10

(2.45 GHz) power of 400 W. For comparison, we also discuss results of plasma-hydrogenated (deuterated) a-Si:H films (Beyer and Zastrow, 1996), of implantation-hydrogenated (IH) amorphous and crystalline Si samples that were prepared by hydrogen implantation of rf1 a-Si:H films, and of Si(100) wafers, respectively (Beyer and Zastrow, 1998a, 1998b), as well as of microcrystalline silicon (μc-Si:H) films (Beyer et al., 1997a).

Infrared (IR) absorption and hydrogen effusion measurements provide rapid characterization for the bonding configurations and the thermal stability of hydrogen, respectively, in hydrogenated silicon materials. We performed the IR absorption measurements using a Fourier transform spectrometer. The measured transmission spectrum was normalized to the transmission spectrum of the crystalline silicon substrate, and the absorption coefficient was determined by applying Lambert-Beer's law. Hydrogen effusion was measured as described by Beyer et al. (1991) using a heating rate of 20 K/min. These measurements give an absolute value for the hydrogen (deuterium) density $N_H(N_D)$. The hydrogen concentration c_H was determined from the hydrogen density assuming a constant silicon density of $N_{Si} = 5 \times 10^{22}$ cm^{-3}; i.e., no correction was made accounting for an influence of hydrogen concentration on film density. According to literature data for the density of a-Si:H films versus preparation conditions (Tatsumi and Ohsaki, 1989), the error is expected to be small up to hydrogen concentrations of about 10 at.%. At higher concentrations, however, we may underestimate the hydrogen concentration by as much as 20 at.%.

2. Hydrogen Effusion Data

a. Substrate Temperature and Concentration Dependence of Hydrogen Effusion Transients

Typical effusion transients for a series of rf1 a-Si:D films prepared at various substrate temperatures T_s are shown in Fig. 1. Plotted is the

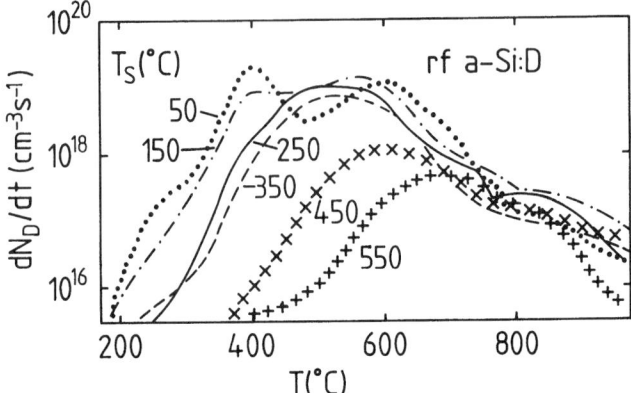

FIG. 1. Deuterium effusion rate dN_D/dt versus temperature for rf plasma-grown a-Si:D films of 0.4 μm thickness deposited at various substrate temperatures T_s. (From Beyer, 1996.)

deuterium effusion rate dN_D/dt as a function of temperature for films with a thickness of about 0.4 μm. Samples deposited at $T_s = 50$ and 150°C show two effusion maxima (near $T = 400$ and $T = 550$–600°C), whereas for higher substrate temperature a single effusion peak is present that shifts from about 500 to almost 700°C when the substrate temperature is raised from 250 to 550°C. Similar two-peak and single-peak hydrogen effusion transients for different deposition conditions of plasma-grown a-Si:H films have been reported earlier (Brodsky et al., 1977b; Fritzsche et al., 1979; Beyer and Wagner, 1981). Literature data suggest that there is no significant difference between the two isotopes hydrogen and deuterium; a-Si:H and a-Si:D films deposited by rf glow discharge under similar conditions have approximately equal hydrogen and deuterium concentrations, respectively, and show similar effusion transients for H and D (Beyer et al., 1988). The results of the present investigation support this. In particular, we find, as shown in Fig. 2, that the substrate temperature dependence of the hydrogen and deuterium concentration in rf1 a-Si:H and a-Si:D, respectively, is almost identical. In both cases, the hydrogen (deuterium) concentration drops from about 20 at.% ($T_s = 150°C$) to 1–2 at.% for $T_s = 500°C$ material. For higher substrate temperatures, only deuterium effusion from a-Si:D gave signals above the background of the effusion apparatus.

At first glance, the effusion results in Fig. 1 may be interpreted in terms of a broad distribution of deuterium (hydrogen) bonding states and suggest that with increasing substrate temperature the low-energy side of the distribution is removed. A detailed analysis of hydrogen effusion involving films prepared under identical conditions with different film thicknesses as

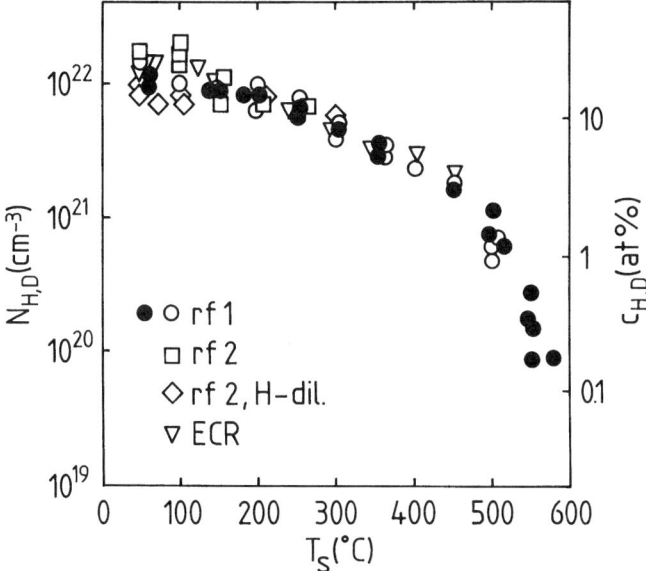

FIG. 2. Hydrogen and deuterium density $N_{H,D}$ and concentration $c_{H,D}$ for various series of a-Si:H and a-Si:D films (open symbols: a-Si:H; closed circles: a-Si:D) as a function of substrate temperature T_s.

well as layer structures of hydrogenated and deuterated material was performed by Beyer and coworkers (Beyer et al., 1982; Beyer and Wagner, 1983; Beyer, 1985, 1991). The results showed that a-Si:H films may undergo, depending on hydrogen concentration, microstructure, doping, etc., various structural changes during the effusion experiment and that different hydrogen (deuterium) release processes may be active as the temperature rises. Accordingly, the distribution of hydrogen-bonding states cannot be inferred straightforwardly from the effusion transients. For the films of Fig. 1, we attribute the step-like structure in deuterium effusion near 750°C to crystallization. The effusion peak near 400°C, termed the *low-temperature* (LT) *effusion peak*, observed for the present series of samples for $T_s \leq 150°C$ material has been associated with desorption of deuterium (hydrogen) bound to void surfaces followed by rapid out-diffusion of the deuterium (hydrogen) molecules through an interconnected void network (Beyer and Wagner, 1983; Beyer, 1985). This LT effusion peak is independent of film thickness. For effusion peaks at higher temperatures (HT peaks), diffusion by atomic deuterium (hydrogen) is found to limit the effusion process. Accordingly, the effusion peaks shift with increasing film thickness to higher temperatures. The shift of the HT effusion peaks for a fixed film thickness in Fig. 1 points to a variation in the diffusion coefficient. The appearance of a (diffusion-limited) HT peak following the (void-related) low-temperature effusion in rf plasma-deposited a-Si:H of $T_s = 50-150°C$ has been attributed to a collapse of interconnected voids on LT hydrogen release and to the subsequent formation of a compact material within which molecular hydrogen cannot diffuse. A decrease in the film thickness on LT hydrogen desorption has been observed (Beyer and Wagner, 1983) that may be related in part to collapse of the void structure.

While the effusion transients shown in Fig. 1 are typical for rf plasma-grown a-Si:H (a-Si:D) films prepared by decomposition of undiluted SiH_4 (SiD_4) gases, different effusion curves are obtained for samples prepared from hydrogen-diluted silane as a process gas, in particular when deposited at or near room temperature. The effusion transients of rf2 a-Si:H films deposited at $T_s = 50°C$ with and without hydrogen dilution are plotted in Fig. 3. It is seen that hydrogen dilution can cause an almost complete disappearance of the LT effusion peak, i.e., of the interconnected void structure. Note that the absolute hydrogen concentration is also changed; in

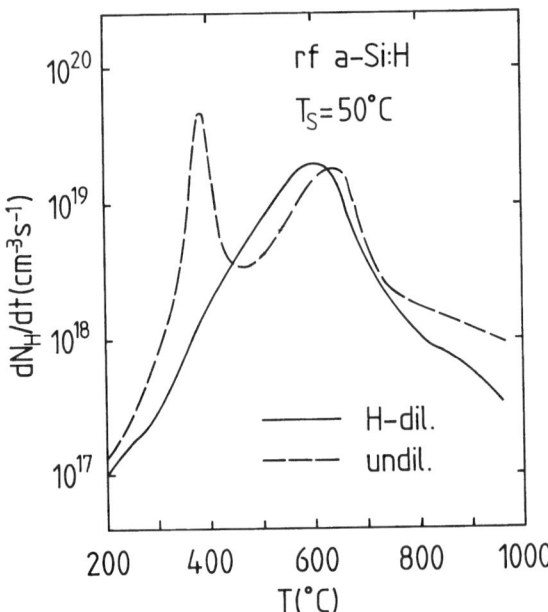

FIG. 3. Hydrogen effusion rate dN_H/dt versus temperature for a-Si:H samples deposited by rf plasma with and without hydrogen dilution at $T_s = 50°C$ (film thickness $d = 0.95$ and 1.4 μm, respectively).

the sample deposited with hydrogen dilution ($c_H \approx 17$ at.%), the concentration is about a factor of 1.3 smaller than in the reference sample deposited from undiluted silane. Our effusion results for a-Si:H prepared by the hydrogen dilution method agree closely with recent data by Zellama et al. (1996).

The presence of a void structure for (rf) a-Si:H of relatively high hydrogen concentration and its disappearance when the hydrogen concentration is reduced (by increasing the substrate temperature or by employing hydrogen-diluted silane) suggest that the void structure in rf plasma-grown a-Si:H films is hydrogen related. It was proposed that voids form in a-Si:H for hydrogen concentrations exceeding 15 to 20 at.% because the material loses its connectiveness due to extensive hydrogen incorporation at positions interrupting Si—Si bonds (Beyer and Wagner, 1983). Hydrogen (deuterium) implantation experiments support this concept. In Fig. 4a, deuterium effusion transients are shown for a-Si:H of low hydrogen concentration ($c_H \approx 0.3$ at.%) and in Fig. 4b for a-Si:H of high hydrogen concentration ($c_H \approx 14$ at.%), both implanted with various doses N_{imp} of deuterium. While up to a dose of 10^{17} cm^{-2} (equivalent to a maximum deuterium concentration $c_D \approx 13$ at.%) implanted deuterium effuses from a-Si:H of low c_H (Fig. 4a) in a HT peak near 600°C, higher doses result in an LT peak or in a shoulder near 400°C, i.e., in the formation of an interconnected hydrogen-related (deuterium-related) void structure. For the material of higher hydrogen concentration (Fig. 4b), the threshold for void formation is at implantation doses between 10^{16} and 10^{17} cm^{-2}, i.e., lower than for low c_H material.

Various deposition conditions of a-Si:H result in material with a pronounced microstructure. For rf plasma deposition, the growth of material with a columnar microstructure was reported by Biegelsen et al. (1979, 1980) when a strong argon dilution and/or a high rf power were employed. Our material deposited by ECR plasma (see Sec. II.1) also shows a columnar microstructure according to electron microscopy. In Fig. 5 the hydrogen effusion transients for such material grown at various substrate temperatures T_s are plotted. Film thickness was 1–2 μm. All samples show a predominant effusion peak at 400–500°C that is independent of film thickness, in agreement with an effusion process not limited by hydrogen diffusion in a compact material. The presence of an interconnected void network in this type of material also shows up in a bulk oxidation when stored under ambient conditions, similar to that reported for a-Si:H films with a columnar morphology by Masson et al. (1987) and for low T_s a-Ge:H films by Abo Ghazala et al. (1991). Figure 2 shows that the substrate temperature dependence of hydrogen concentration in ECR plasma-grown and rf plasma-grown a-Si:H agrees closely.

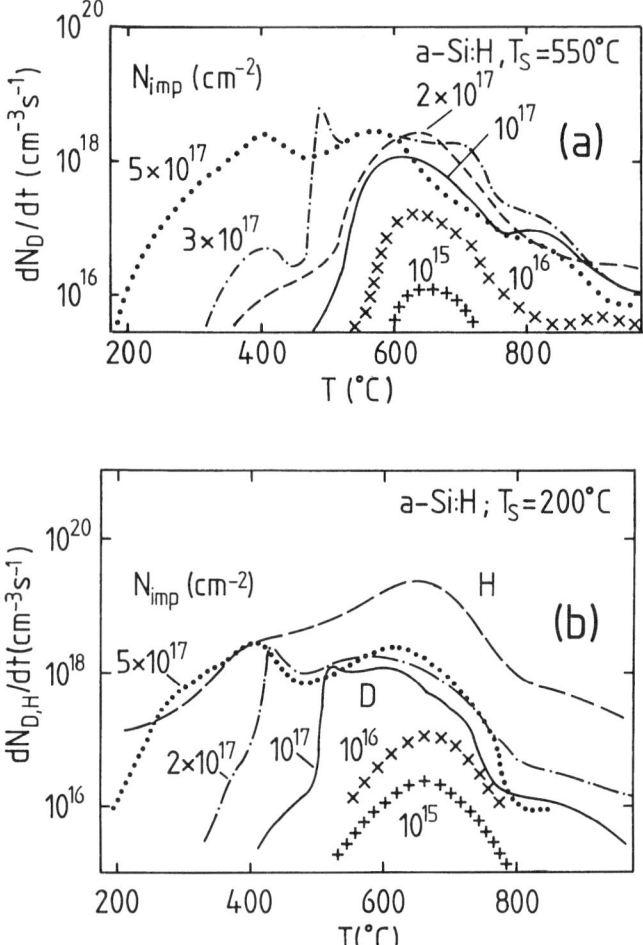

FIG. 4. Deuterium effusion rate dN_D/dt versus temperature for rf plasma-grown a-Si:H of (a) $T_s = 550°C$ and (b) $T_s = 200°C$ implanted with deuterium at the energy of 40 keV at various implantation doses N_{imp}. Hydrogen effusion rate dN_H/dt of the unimplanted $T_s = 200°C$ sample is shown for comparison. Film thickness was approximately 1.5 µm.

b. *Structural Characterization of a-Si:H Films by Hydrogen Effusion*

As suggested previously (Beyer, 1987; Finger et al., 1992), the fraction of the hydrogen density released in the LT effusion peak N_{LT}/N_H may be used as a measure for the microstructure of Si:H material. This definition, however, bears the problem that the LT effusion process often shows up

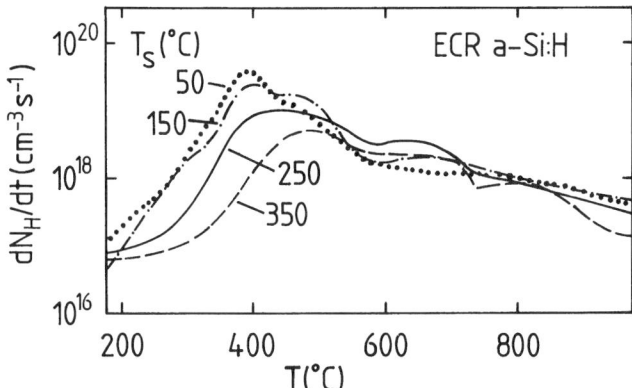

FIG. 5. Hydrogen effusion rate dN_H/dt versus temperature for ECR plasma-grown a-Si:H films deposited at various substrate temperatures T_s. Film thickness was 1–2 μm.

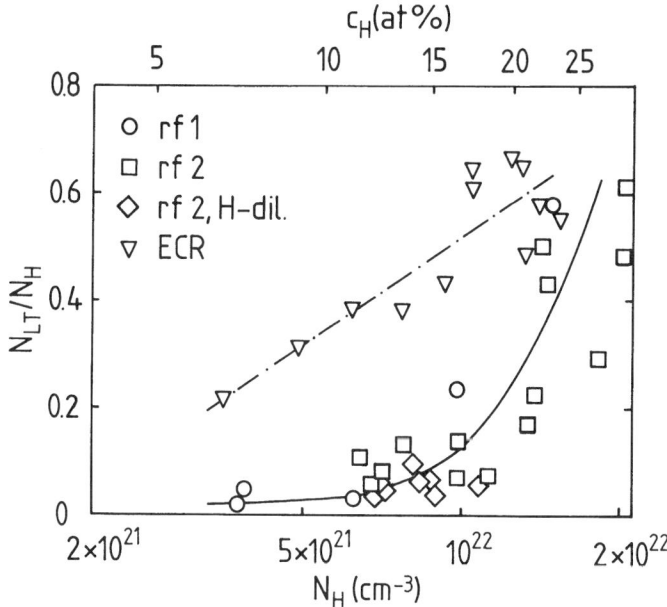

FIG. 6. Microstructure parameter N_{LT}/N_H for various series of a-Si:H films (film thickness $d \geq 1$ μm; substrate temperature $T_s \leq 300°C$) versus hydrogen density and concentration. Lines are guides to the eye.

as a shoulder or tail of the bulk diffusion-related effusion transient only. Since deconvolution of the effusion transients in LT and HT processes is not straightforward because of possible reconstruction effects (Beyer and Wagner, 1983), we use, for simplicity, for N_{LT} just the hydrogen density released up to a temperature of 450°C (confining to films of a film thickness $d \geq 1$ μm). Data on the microstructure parameter N_{LT}/N_H for the various series of a-Si:H films are plotted in Fig. 6 as a function of hydrogen density N_H and hydrogen concentration c_H. Significant microstructural effects with $N_{LT}/N_H > 0.2$ are observed for the ECR a-Si:H films at hydrogen concentrations as low as 7 at.%. In contrast, rf a-Si:H films show an almost negligible microstructure parameter N_{LT}/N_H up to a hydrogen concentration of about 10 at.% and a sizable microstructure parameter ($N_{LT}/N_H > 0.2$) only for concentrations exceeding about 15 at.%.

3. INFRARED ABSORPTION DATA

a. *Substrate Temperature and Concentration Dependence of Hydrogen-Related IR Vibrational Modes*

IR absorption measurements allow the study of hydrogen-bonding configurations and of the microstructure of as-deposited material. Typical IR absorption spectra for plasma-grown (rf2) a-Si:H films (deposited from undiluted silane) are shown in Fig. 7. Similar spectra have been reported frequently (Brodsky et al., 1977a; Knights et al., 1978; Freeman and Paul, 1978; Langford et al., 1992). A decrease in substrate temperature from 200 to 50°C (corresponding to an increase in hydrogen concentration from 12 at.% to 23 at.%) is seen to result in a general increase in the absorption peak height and in significant changes of the spectral distribution. In particular, a transition of the Si—H stretching absorption from a peak near 2000 cm^{-1} to another near 2100 cm^{-1} is observed, and bending modes of SiH$_2$ at 845 and 890 cm^{-1} show up. Only the Si—H wagging absorption appears to remain almost unchanged in its position near 640 cm^{-1}. A more detailed analysis shows, however, that all hydrogen-related IR absorption peaks are not constant in position but tend to shift with increasing hydrogen concentration to higher frequencies.

The hydrogen concentration dependence of various hydrogen-related vibrational frequencies in rf2 and ECR a-Si:H films is shown in Figs. 8a and b, respectively. The integrated absorption of the Si—H wagging mode, $I_{640} = \int \omega^{-1} \alpha(\omega) \, d\omega$ [with $\alpha(\omega)$ the absorption coefficient at the frequency ω], serves as a measure of the hydrogen concentration (Fang et al., 1980).

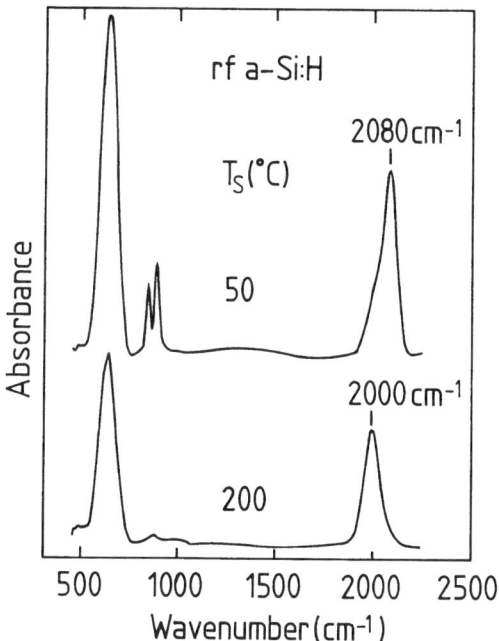

FIG. 7. IR absorption spectra of rf plasma-grown a-Si:H films deposited at $T_s = 50$ and 200°C.

The frequencies of the stretching peaks were determined by deconvolution of the stretching band into two Gaussian peaks. Also shown (full data points) is the mean stretching frequency ω_s. Over the hydrogen concentration range investigated, all absorption peaks are seen to shift by 10–30 cm^{-1} to higher frequencies. A much stronger variation of 70–90 cm^{-1} is observed for ω_s. A comparison of Figs. 8a and 8b shows differences, however, so that the hydrogen concentration cannot be the only parameter determining ω_s and the frequency of the absorption peaks. Shifts of IR absorption peaks in a-Si:H with changing hydrogen concentration have been reported by various authors (Cardona, 1983; Tsu and Lucovsky, 1987; Gracin et al., 1992; Maeda et al., 1995; Daey Ouwens and Schropp, 1995). Tsu and Lucovsky (1987) attributed the shift of the Si—H stretching absorption near 2000 cm^{-1} to a changed induction effect (see Lucovsky, 1979) of hydrogen atoms in backbonds to the Si—H oscillator. Tentatively, we also associate the hydrogen concentration dependence of the other vibrational frequencies with this effect.

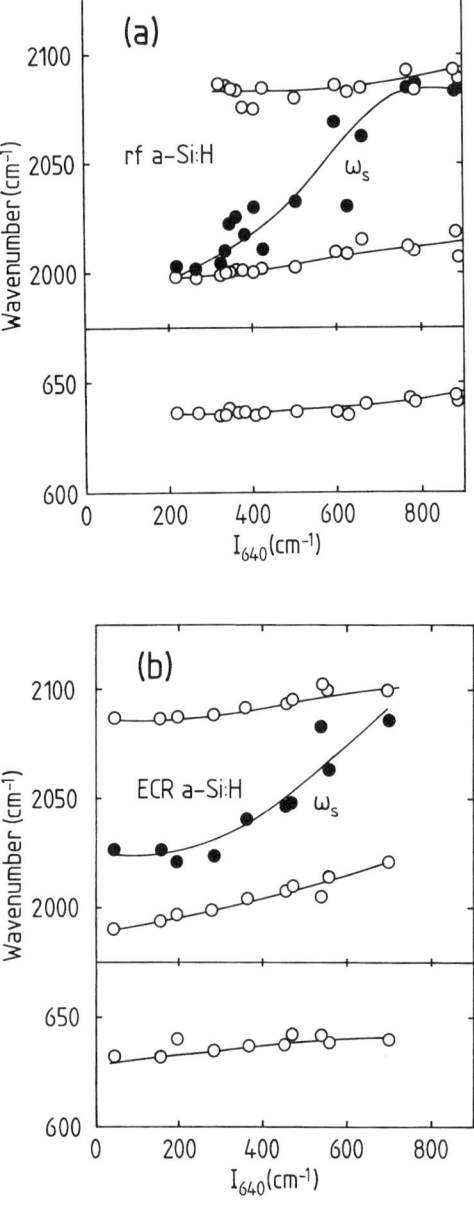

FIG. 8. Frequency of IR absorption peaks versus integrated IR absorption of the Si—H wagging mode. Also shown is the mean frequency of the Si—H stretching band ω_s: (a) rf plasma-grown, (b) ECR plasma-grown a-Si:H films. Lines are guides to the eye.

The results of Fig. 8 suggest that the shift of the individual vibrational frequencies with changing hydrogen content is an effect distinct from the transition of the dominant Si—H stretching absorption from a peak near 2000 cm^{-1} to another near 2100 cm^{-1} measured by ω_s. In agreement with Brodsky et al. (1977a), this change of ω_s is generally explained in terms of a transition from monohydride to dihydride. Brodsky's assignment was based on a detailed analysis of IR spectra of gaseous silanes in comparison with spectra of a-Si:H and a-Si:D films. This assignment has been questioned, however, by various authors (Freeman and Paul, 1978; Cardona, 1983; Wagner and Beyer, 1983). As pointed out by Wagner and Beyer (1983), both mono- and dihydride groups on crystalline Si surfaces have stretching vibrations near 2100 cm^{-1}, and it is difficult to see, accordingly, why in amorphous Si the frequency of silicon mono- and dihydride vibrations should differ by about 100 cm^{-1}. Recent work by Lee et al. (1996) also gives for amorphous Si surfaces monohydride stretch frequencies of 2082–2098 cm^{-1}. Various authors (Cardona, 1983; Wagner and Beyer, 1983; Gracin et al., 1992; Manfredotti et al., 1994) suggested to relate the frequency difference of the two Si—H stretching absorption peaks in a-Si:H not primarily to the frequency difference of Si—H stretching vibrations in SiH and SiH$_2$ bonding configurations but rather to a different structure of the material causing a different solid-state effect. This latter effect arises if a medium of high dielectric constant is in close vicinity to a vibrating dipole. Screening of the electrical field of the dipole by the image dipole tends to lower the effective force constant and thus the frequency of dipole-active (i.e., of Si—H stretching) vibrations (Wagner and Beyer, 1983). A decrease of the dielectric constant with rising hydrogen concentration (Langford et al., 1992) or an increase in the size of the cavity surrounding an Si—H oscillator would then cause an increase in the stretching frequencies (Cardona, 1983). In this interpretation, the 2000-cm^{-1} absorption peak is due to isolated SiH$_x$ groups (predominantly monohydride) embedded in rather small cavities in the a-Si solid, while the 2100 cm^{-1} peak originates from clustered SiH$_x$ groups (predominantly dihydride) in larger cavities. This assignment is supported by the correlation effects between hydrogen effusion transients and IR absorption spectra discussed in Sections II.3.b and IV.8. We note, however, that the relation between the Si—H stretching absorptions observed in amorphous and crystalline silicon is not clear so far. For c-Si:H, stretching absorptions of Si—H between 1800 and 2250 cm^{-1} have been reported for hydrogen-implanted material (Stein, 1975) and for other hydrogenation methods (Deák et al., 1991; Johnson et al., 1991). These absorption bands are found to depend strongly on sample preparation (e.g., hydrogen implantation conditions), annealing state, stress, and other parameters (Stein, 1979; Deák et al., 1991; Stavola and Pearton, 1991).

b. *Structural Characterization of a-Si:H Films by IR Absorption Measurements*

In agreement with the assignment of the two Si—H stretching absorption peaks to different structural environments, the use of the parameter $R^{IR} = I_{2100}/(I_{2000} + I_{2100})$, with I_{2000} and I_{2100} the integrated absorptions near $\omega = 2000$ and $\omega = 2100 \text{ cm}^{-1}$, respectively, was proposed for characterization of the microstructure in a-Si:H material (Mahan et al., 1987). It must be noted, however, that R^{IR} will provide reliable information on film microstructure only if the absorption strength values A defined as the proportionality factors between hydrogen density N_H and the integrated absorption I,

$$A = N_H/I \tag{1}$$

are constant (sample independent) for the two stretching modes, and if the void surfaces are covered by hydrogen. While a fairly stable coverage of void surfaces by hydrogen can be expected at temperatures significantly below the temperature of hydrogen surface desorption ($T \approx 400°C$ according to hydrogen effusion), the constancy of A remains a matter of discussion. A sample dependence of the absorption strength A_{str} of Si—H stretching modes was suggested by Shanks et al. (1980) and Oguz et al. (1980). Langford et al. (1992), on the other hand, recently concluded that the absorption strength values of Si—H vibrations in a-Si:H are essentially sample independent.

A way to explore a possible sample dependence of A (and to determine ratios of A values of the stretching and wagging absorptions) is a plot of I_{2000}/I_{640} as a function of I_{2100}/I_{640} (Beyer and Abo Ghazala, 1998). Assuming that any Si—H oscillator contributes to the wagging absorption band and to either of the two stretching bands, the relation

$$A_{640}I_{640} = A_{2000}I_{2000} + A_{2100}I_{2100} \tag{2}$$

is valid (Langford et al., 1992). This equation can be transformed to

$$\frac{I_{2000}}{I_{640}} = \alpha - \beta \frac{I_{2100}}{I_{640}} \tag{3}$$

with

$$\alpha = \frac{A_{640}}{A_{2000}} \tag{4}$$

and

$$\beta = \frac{A_{2100}}{A_{2000}} \quad (5)$$

Thus, for sample-independent absorption strength values, a plot of I_{2000}/I_{640} versus I_{2100}/I_{640} should give a straight line with the intercept α and the slope β yielding the absorption strength ratios. In Fig. 9, such a plot for various series of a-Si:H and c-Si:H samples is shown. While the data points scatter appreciably, a least-squares fit (dashed line) gives $\alpha = 0.185$ and $\beta = 0.97$, suggesting that the absorption strength values A_{2000} and A_{2100} are essentially equal. The same conclusion has been reached by Basrour et al. (1989), Gracin et al. (1992), and Daey Ouwens and Schropp (1995), whereas Amato et al. (1991) reported $\beta \approx 2.8$ and Langford et al. (1992) reported $\beta \approx 2.4$. With regard to a possible sample dependence of A, we note that the data points for I_{2100}/I_{640} ($I_{2000}/I_{640} \to 0$) in Fig. 9 scatter significantly stronger than the I_{2000}/I_{640} ($I_{2100}/I_{640} \to 0$) data points. This

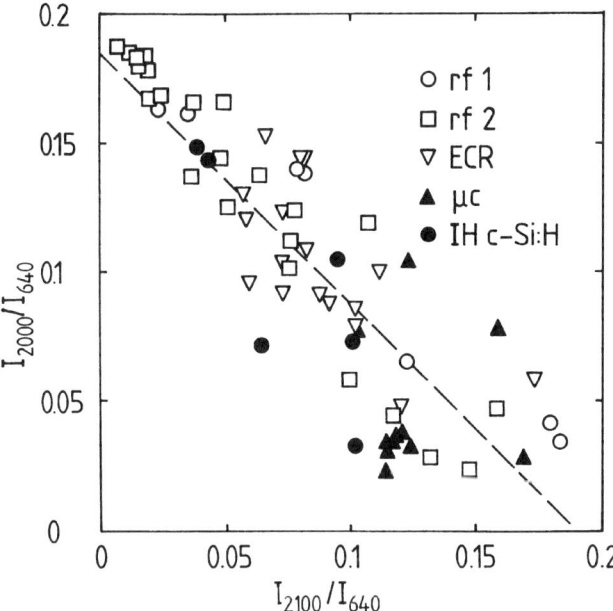

FIG. 9. Absorption strength of $2000\,\text{cm}^{-1}$ Si—H stretching mode, normalized to absorption strength of the $640\,\text{cm}^{-1}$ wagging mode, versus normalized absorption strength of $2100\,\text{cm}^{-1}$ stretching mode for various series of Si:H samples. Dashed line is the least squares fit. (From Beyer and Abo Ghazala, 1998.)

result suggests some sample dependence of A_{2100}, whereas A_{2000} appears to be well defined.

Well aware of the limited reliability of R^{IR} for structure characterization due to the sample dependence of A_{2100}, we use R^{IR} for structural characterization of our various series of a-Si:H films. Results are compiled in Fig. 10, showing R^{IR} plotted as a function of hydrogen density and concentration. At low hydrogen density, only the ECR samples reveal high R^{IR} values, in agreement with the high N_{LT}/N_H values observed (see Fig. 6). Most other samples show small R^{IR} values at low c_H and a rapid increase in R^{IR} only for hydrogen concentrations exceeding 10 to 15 at.%. The dotted line in Fig. 10 indicates R^{IR} versus N_H for the statistical incorporation of SiH and SiH$_2$ bonds in the amorphous network (Zastrow, 1997). To obtain this relation, it was assumed that the IR absorption peaks at 2000 and 2100 cm^{-1} are due to SiH and SiH$_2$ bonding configurations, respectively, and that the absorption strength for both stretching absorptions is equal. Only a few samples grown from hydrogen-diluted silane are found to agree with this curve, while all other R^{IR} values exceed the calculated data

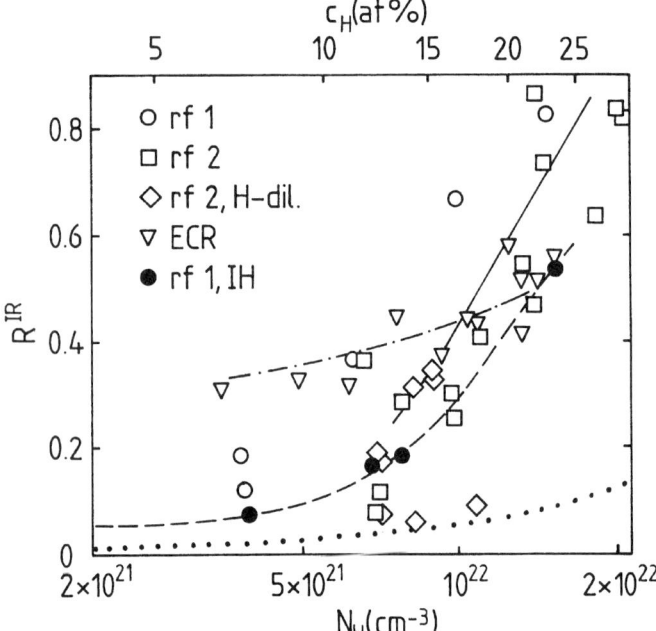

FIG. 10. Microstructure parameter R^{IR} for various series of a-Si:H films (film thickness $d \geq 1\ \mu m$, $T_s \leq 300°C$) versus hydrogen density and concentration. Dotted line calculated assuming statistical incorporation of SiH and SiH$_2$. Other lines are guides to the eye.

considerably. The presence of SiH_2 at concentrations exceeding the statistical incorporation of hydrogen supports the clustering of SiH_2 and thus the presence of hydrogen-related voids. The results suggest that clustering of SiH_2 occurs not only for deposition at low substrate temperatures using undiluted silane but also for hydrogen implantation into material deposited at high substrate temperatures. A reduction of SiH_2 clustering for deposition using hydrogen-diluted silane is in agreement with recent findings by Zellama et al. (1996).

The relation between the microstructure parameters R^{IR} from IR absorption and N_{LT}/N_H from hydrogen effusion is shown in Fig. 11. The two microstructure parameters are found to be correlated differently for ECR and rf plasma-grown a-Si:H. N_{LT}/N_H measuring interconnected voids and R^{IR} measuring the total hydrogen covered void surface, the results suggest a higher fraction of voids that are interconnected in the ECR plasma material than in rf plasma a-Si:H. The reason presumably is the columnar growth morphology of the ECR grown samples. Also shown in Fig. 11 are data points for microcrystalline Si films. These results show the presence of compact μc-Si:H material with regard to effusion at much higher R^{IR} values than observed for a-Si:H samples (Beyer et al., 1997a, 1997b).

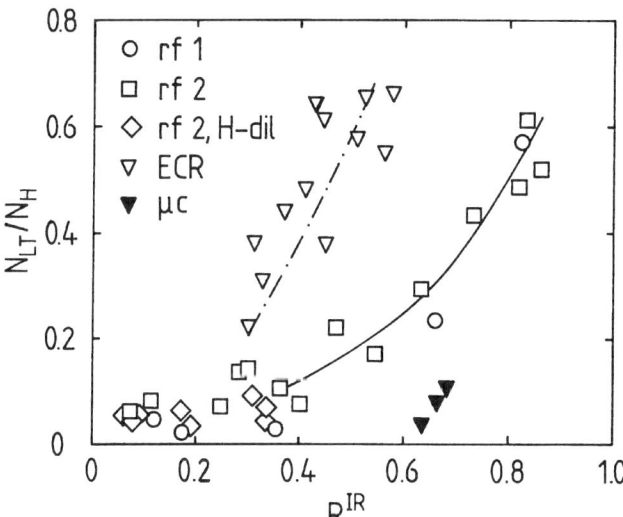

FIG. 11. Microstructure parameter N_{LT}/N_H from effusion versus microstructure parameter R^{IR} from IR absorption measurements for various series of a-Si:H samples (film thickness $d \geq 1\,\mu m$, $T_s \leq 300°C$). Lines are guides to the eye. Results for μc-Si:H films (μc) are shown for comparison.

III. Experimental Hydrogen Diffusion and Solubility Data

1. HYDROGEN DIFFUSION DATA

a. SIMS Depth Profiles of Hydrogen and Deuterium

For studying hydrogen diffusion in a-Si:H, various methods have been applied, such as hydrogen effusion (Beyer and Wagner, 1982; Chou et al., 1987), measurements of the annealing dependence of hydrogen-related IR absorption peaks (Zellama et al., 1981; Mahan et al., 1995) and of hydrogen out-diffusion profiles (Zellama et al., 1980, 1981; Reinelt et al., 1983; Tang et al., 1990a). The most widely applied experiment, however, is the study of deuterium and hydrogen interdiffusion by SIMS depth profiling (Carlson and Magee, 1978). In order to achieve steep initial deuterium and hydrogen depth profiles, mostly a-Si:D and a-Si:H:D layers in contact with a-Si:H were used (Carlson and Magee, 1978; Street et al., 1987; Beyer et al., 1988; Shinar et al., 1989; Branz et al., 1993a, 1993b). Other methods employed to introduce deuterium profiles in a-Si:H were ion implantation or plasma treatment (Beyer, 1996, 1997a). An advantage of the SIMS depth profiling method to measure hydrogen diffusion is the detection of deuterium or hydrogen motion without the necessity for deuterium (hydrogen) to leave the sample, as required for the other methods. Thus hydrogen diffusion can be measured down to rather low temperatures. In the high-temperature range, the different methods are found to yield comparable results (Beyer and Wagner, 1982; Reinelt et al., 1983; Beyer et al., 1988; Mahan et al., 1995). We confine the discussion in this section to compact (rf1) a-Si:H, i.e., to material with relatively small microstructure parameters $R^{IR} < 0.4$ and $N_{LT}/N_H < 0.10$.

For SIMS measurements, either oxygen (O_2^+) or cesium (Cs^+) sputtering beams were employed, as described by Zastrow et al. (1993, 1998). Typical SIMS depth profiles of hydrogen and deuterium in layered structures of (rf1) a-Si:H and a-Si:D prior to and after annealing are shown in Fig. 12. In Fig. 12a, the deuterium source layer concentration N_D equals approximately the hydrogen concentration N_H in the probed layer. Heat treatment results in a spreading of both hydrogen and deuterium distributions. The profiles can be fitted by complementary error functions throughout the sample, as expected from the solution of Fick's second law. Best fits are obtained when the deuterium (or hydrogen) concentration is normalized to the total hydrogen (H + D) concentration. Almost equal deuterium and hydrogen diffusion coefficients are obtained in agreement with hydrogen/deuterium effusion data (Beyer et al., 1988). In Fig. 12b, D-H interdiffusion is studied with N_H in the probed layer considerably smaller than N_D in the source layer. On annealing, deuterium is found to penetrate the hydrogenated layer

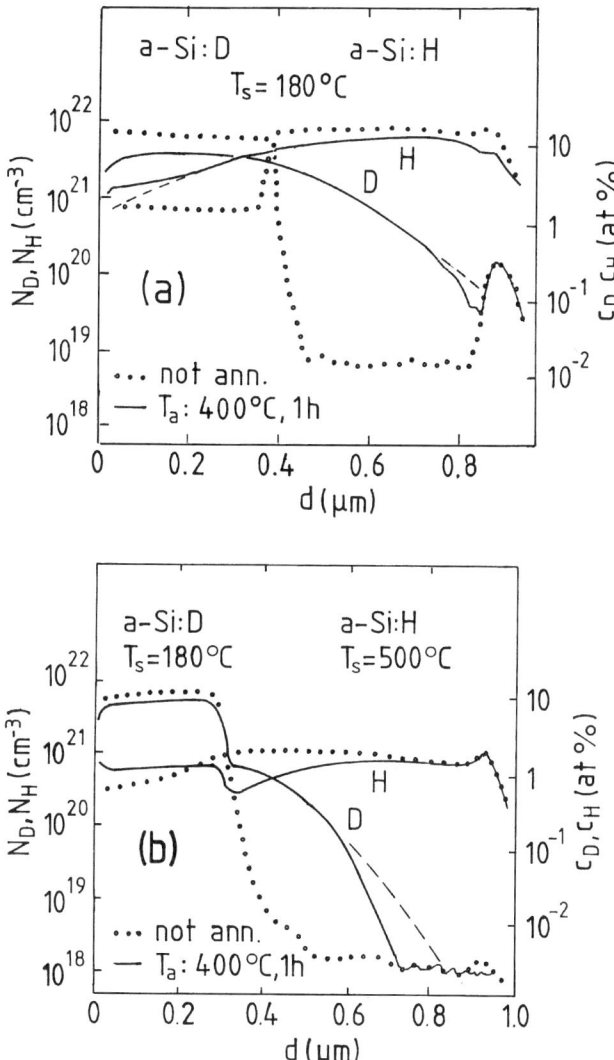

FIG. 12. SIMS depth profiles of deuterium and hydrogen in layered structures of a-Si:H and a-Si:D prior to and after heat treatment (dashed curves: error function fits). (a) $c_H \approx c_D$; (b) $c_H < c_D$. (From Beyer, 1997a.)

only up to the level of the (original) hydrogen concentration, resulting in a step in the deuterium concentration at the H/D interface. Within the hydrogenated layer, the deuterium profile follows an error function dependence (as indicated by the dashed line), with deviations at low deuterium concentration. After annealing, the hydrogen depth profile shows a step at the H/D interface too. Hydrogen is depleted in the a-Si:H layer near the H/D interface and accumulates in the a-Si:D layer. Yet the local total (H + D) hydrogen concentration remains approximately unchanged by heat treatment so that in the H-D interdiffusion process hydrogen and deuterium atoms are only exchanged. In the following, we shall denote diffusion structures with the source layer $(N_D + N_H)$ concentration equal to the concentration N_H in the probed layer as type A and structures with a source layer concentration exceeding the concentration in the probed layer as type B samples.

b. *Substrate Temperature Dependence of Hydrogen Diffusion Coefficient*

In Fig. 13, the diffusion coefficient of hydrogen in rf1 a-Si:H is plotted as a function of $1/T$ for various substrate temperatures. Figure 13a shows results for a-Si:D/a-Si:H layer samples of type A; Fig. 13b, of type B. In the latter case, the deuterium source concentration was $c_D \approx 10$ at.%. Also shown in Fig. 13b are results for in-diffusion from a deuterium plasma (Beyer and Zastrow, 1996). The diffusion data in Fig. 13 were obtained by fitting deuterium profiles in a-Si:H layers at various annealing stages by complementary error functions or superpositions of error functions. For a fixed annealing temperature, the time dependence of the diffusion coefficient (see Sec. IV.4) was measured. The data points in Fig. 13 (and for the diffusion coefficients throughout this chapter) refer to a diffusion length $L = 2(Dt)^{1/2} = 2 \times 10^{-5}$ cm. The results show that for D-H layer diffusion the diffusion coefficient always follows a straight line in the Arrhenius plot. A comparison with data by Carlson and Magee (1978) for $T_s = 315°C$ material (dotted line in Fig. 13a) yields an excellent agreement. While for type A samples the $1/T$ dependence of the diffusion coefficient varies significantly with substrate temperature and, accordingly, with temperature concentration, for type B samples (Fig. 13b) the substrate temperature dependence of the diffusion coefficient is much weaker. The diffusion coefficient for plasma in-diffusion in Fig. 13b is found to agree with layer diffusion at high temperatures but shows a kink at about 400°C and a considerably lower diffusion activation energy at lower temperatures. Almost no influence of the substrate temperature T_s and thus of the original hydrogen concentration on the diffusion coefficient for plasma in-diffusion is observed. In particular, no significant difference is seen between diffusion with an in-diffused deuterium concentration exceeding the original hydrogen concentration ($T_s = 500°C$ material up to a plasma treatment temperature of $T = 500°C$) and an

5 HYDROGEN PHENOMENA IN HYDROGENATED AMORPHOUS SILICON 185

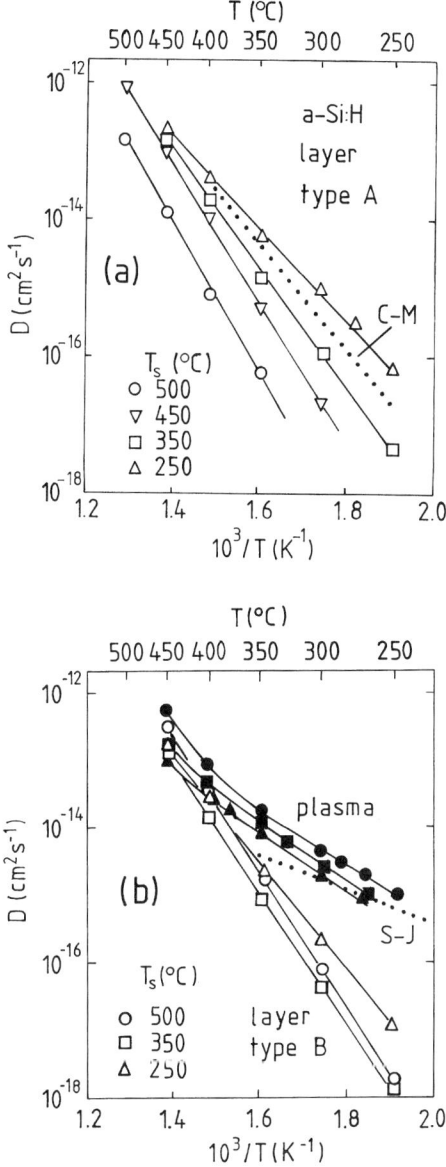

FIG. 13. Hydrogen diffusion coefficient D as a function of reciprocal temperature for a-Si:D/a-Si:H layers of (a) type A ($c_H \approx c_D$) and (b) type B ($c_H < c_D \approx 10$ at.%) (open symbols) and (b) D-plasma-treated a-Si:H (closed symbols) of various substrate temperatures T_s. Solid lines are guides to the eye (Beyer, 1997a). Dotted lines are literature data by Carlson and Magee (1978) (C–M) and by Santos and Jackson (1992b) (S–J).

in diffused deuterium concentration smaller than the original hydrogen concentration ($T_s = 250°C$ material at $T > 250°C$) (Beyer and Zastrow, 1996). Our results for plasma in-diffusion agree well with previously published data by Santos and Jackson (1992b) shown as a dotted line in Fig. 13b and with data by Widmer et al. (1983) for H_2 plasma treatment. The diffusion coefficient for plasma in-diffusion of $D \approx 10^{-14} \, \text{cm}^{-2} \, \text{sec}^{-1}$ estimated by Abeles et al. (1987a, 1987b) for the temperature of 160°C is also in fair agreement with our data considering the rather different diffusion lengths involved. Note that the accuracy of the diffusion data for plasma in-diffusion is less than for layer diffusion due to stronger deviations from error function profiles (see Sec. IV.5) and because measurement of the time dependence of the in-diffusion coefficient involves evaluation of data of a series of samples rather than of various annealing stages of a single sample.

Fitting our layer diffusion data by $D = D_0 \exp(-E_D/kT)$, with k as Boltzmann's constant, the substrate temperature (and hydrogen concentration) dependence shown in Fig. 14 is obtained. Also shown are data for samples with an implanted deuterium source concentration. For these latter samples, the diffusion coefficient was evaluated in the depth range not affected by implantation. The results show no significant difference between H/D layer and D-implanted diffusion structures. The diffusion energy E_D and the diffusion prefactor D_0 are found to increase with rising substrate temperature (and decreasing hydrogen concentration) for both type A (Fig. 14a) and type B (Fig. 14b) samples. Only for $T_s = 200-250°C$ (i.e., for high c_H) material, the theoretical diffusion prefactor

$$D_{00} = \frac{v_{ph} a^2}{6} \approx 10^{-3} \, \text{cm}^2 \, \text{sec}^{-1} \tag{6}$$

for a jump width of $a = 2.5 \, \text{Å}$ and a phonon frequency of $v_{ph} = 10^{13} \, \text{sec}^{-1}$ is observed.

Figure 15 shows a plot of the logarithm of the diffusion prefactor D_0 as a function of the diffusion energy E_D. The data points are found to follow almost a straight line, suggesting that for the hydrogen diffusion parameters D_0 and E_D in a-Si:H, a Meyer-Neldel (MN) rule (Meyer and Neldel, 1937) (or compensation law) is valid. Including data for deuterium plasma in-diffusion by Beyer and Zastrow (1996) and by Santos and Jackson (1992b) and for diffusion of implanted hydrogen in a-Si by Roth et al. (1993), the hydrogen diffusion prefactor is found to vary over almost 15 orders of magnitude at a variation of the diffusion energy by a factor of about 5. Fitting all data by

$$D_0 = D_0^{MN} \exp(E_D/kT^{MN}) \tag{7}$$

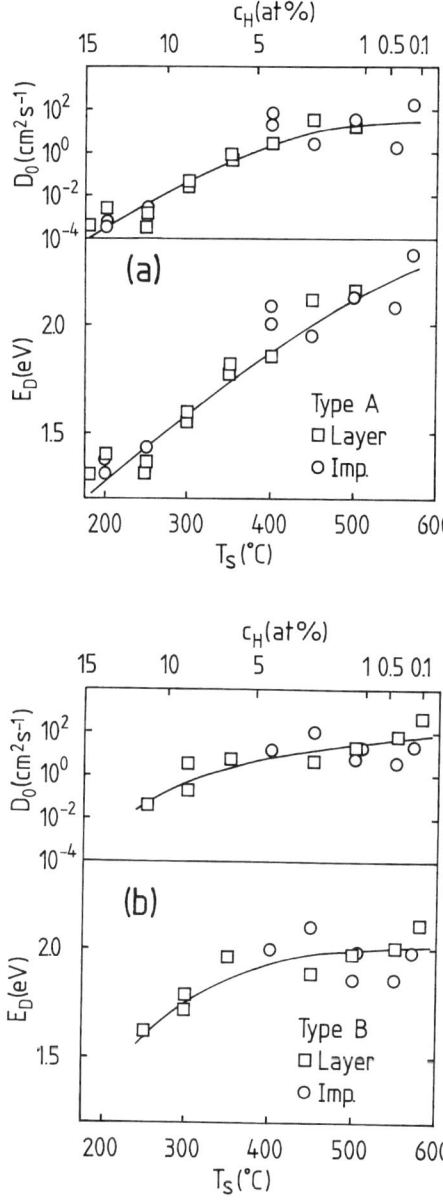

FIG. 14. Diffusion prefactor D_0 and diffusion energy E_D as a function of substrate temperature (and hydrogen concentration) for a-Si:D/a-Si:H layers (squares) and deuterium-implanted a-Si:H films (circles). (a) Type A ($c_H \approx c_D$); (b) type B ($c_H < c_D \approx 10$ at.%) samples. Lines are guides to the eye. (After Beyer, 1996.)

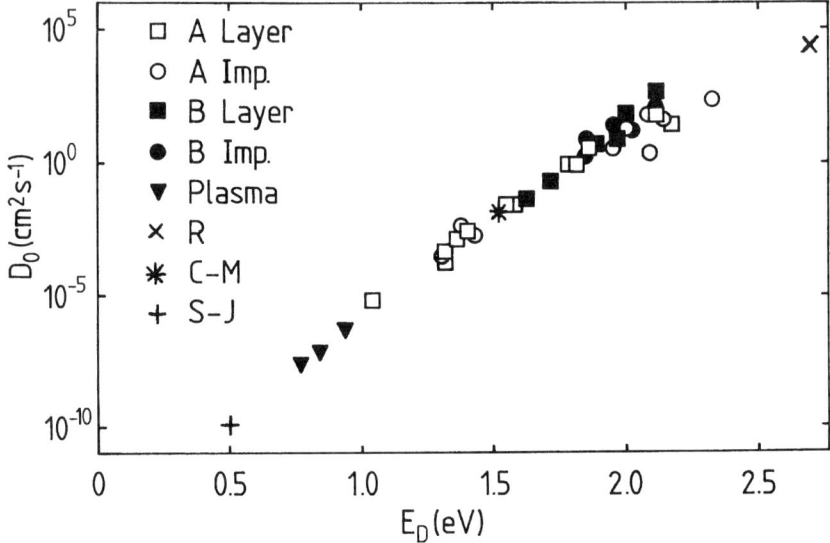

FIG. 15. Diffusion prefactor D_0 versus diffusion energy E_D for various H/D diffusion samples. (After Beyer, 1997a.) (R), (C–M), and (S–J) refer to literature data by Roth et al. (1993), Carlson and Magee (1978), and Santos and Jackson (1992b), respectively.

the Meyer-Neldel parameters $D_0^{MN} \approx 6 \times 10^{-13}$ cm^2 sec^{-1} and $T^{MN} \approx 760$ K result. These data are close to previously published values by Shinar et al. (1991) ($D_0^{MN} = 3.1 \times 10^{-14}$ cm^2 sec^{-1}, $T^{MN} = 730$ K). By confining to type A or type B diffusion samples, somewhat different values are obtained (Beyer, 1996). It is interesting to note that Meyer-Neldel dependences of diffusion prefactors and diffusion energies (with various Meyer-Neldel parameters) have been observed widely not only for hydrogen diffusion in various (disordered) metals and alloys but also, for example, for diffusion of carbon, oxygen, and phosphorus in metal base alloys and for boron in silicon (Kirchheim and Huang, 1987).

The substrate temperature (and hydrogen concentration) dependence of the H diffusion coefficient in rf plasma-grown a-Si:H at three different temperatures is plotted in Fig. 16. In Fig. 16a data of type A samples and in Fig. 16b of type B samples are shown. In agreement with the results of Fig. 13, the data show for type B samples a diffusion coefficient nearly independent of substrate temperature (and hydrogen concentration), whereas for type A samples the H diffusion coefficient decreases with increasing substrate temperatures and decreasing hydrogen concentration considerably. Note that for the type A samples this concentration dependence of the

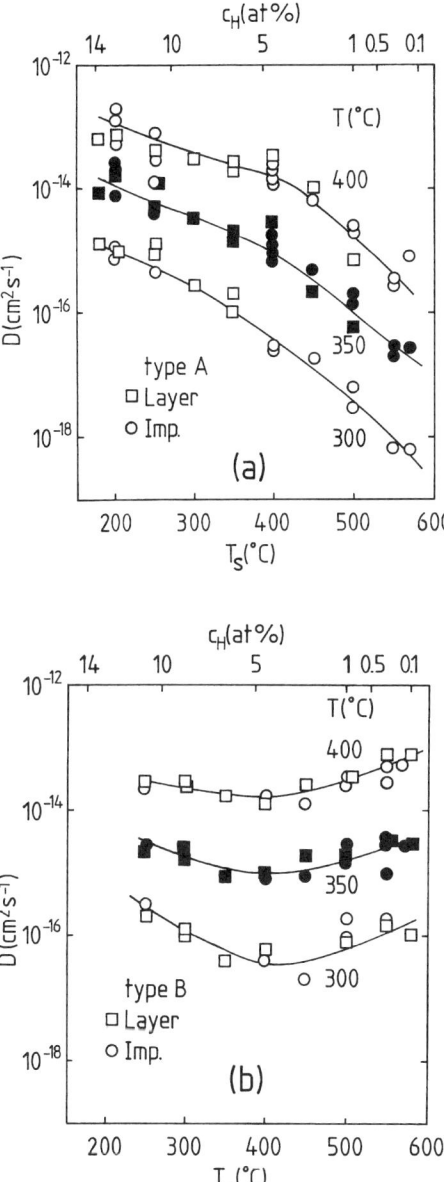

FIG. 16. Diffusion coefficient D at 300, 350, and 400°C for H/D layers (squares) and D-implanted a-Si:H films (circles) versus substrate temperature and hydrogen concentration of the probed layer. (a) Type A ($c_H \approx c_D$); (b) type B ($c_H < c_D \approx 10$ at.%) samples. Lines are guides to the eye.

diffusion coefficient agrees closely with data published by Carlson and Magee (1978) and by Shinar et al. (1992, 1993). Since the samples of type A and type B differ only in their deuterium source concentration, a dependence of the diffusion coefficient (in low c_H a-Si:H) on the diffusion source concentration must be concluded. This effect is confirmed by the results of Fig. 17, showing the deuterium diffusion coefficient at $T = 400°C$ in a-Si:H of low c_H ($T_s = 550°C$) as a function of the dose of implanted deuterium that served as a diffusion source. Up to a dose of about 10^{17} cm^{-2} (peak deuterium concentration $c_D \approx 13$ at.%), the diffusion coefficient in the probed a-Si:H layer is found to increase with increasing deuterium source concentration, in agreement with the results of Fig. 16. The plateau and slight decrease in the diffusion coefficient for implantation doses exceeding 10^{17} cm^{-2} are presumably related to the void formation in the implanted diffusion source as detected by effusion measurements (see Fig. 4a). Also shown in Fig. 17 is the level N_D^* of deuterium in-diffusion into the hydrogenated material as a function of implanted dose. It is seen that N_D^* has a similar dependence on implantation dose as the diffusion coefficient.

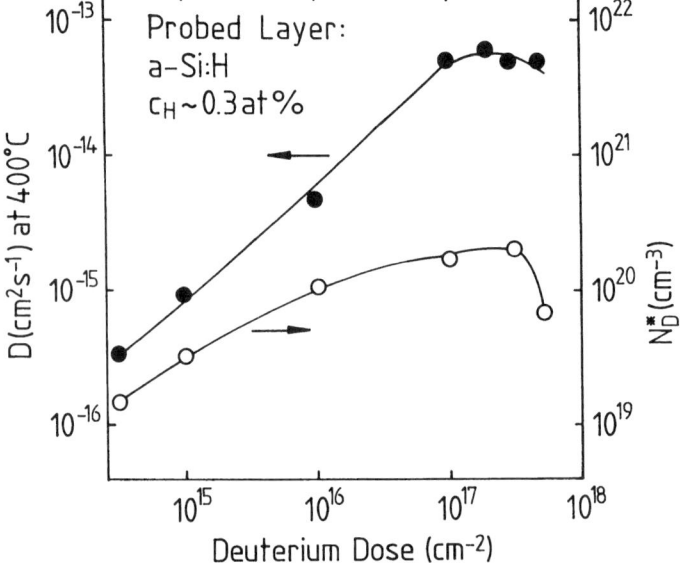

FIG. 17. Diffusion coefficient D at 400°C and level of deuterium in-diffusion N_D^* in a-Si:H ($T_s = 550°C$, $c_H \approx 0.3$ at.%) as a function of dose of implanted deuterium that served as a diffusion source (implantation energy was 40 keV). Lines are guides to the eye.

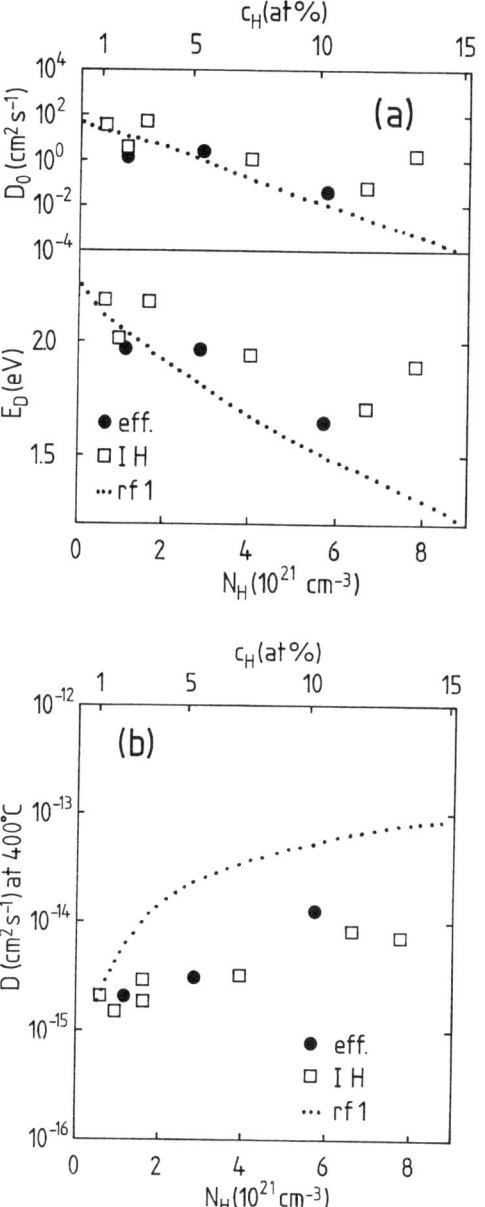

Fig. 18. (a) Diffusion prefactor D_0 and diffusion energy E_D and (b) diffusion coefficient D at 400°C versus hydrogen density and hydrogen concentration for hydrogen effused (eff.) and implantation-hydrogenated (IH) a-Si:H samples. Data of as-deposited rf1 a-Si:H samples (dotted line) are shown for comparison. (After Beyer and Zastrow, 1998b.)

c. *Influence of a Decrease in Hydrogen Concentration on Hydrogen Diffusion*

In order to study the effect of a decreased hydrogen concentration on hydrogen diffusion, we annealed hydrogen-rich (rf1, $T_s = 250°C$) a-Si:H films at various temperatures and time periods in order to effuse various amounts of hydrogen. Into these samples, deuterium source profiles of low concentration ($c_D < c_H$) were implanted, and hydrogen diffusion was studied (outside the implantation zone) as a function of the hydrogen concentration in the material. Results for the diffusion prefactor and the diffusion energy are shown in Fig. 18a and for the diffusion coefficient at $T = 400°C$ in Fig. 18b (eff.). It is seen that with decreasing hydrogen concentration the diffusion coefficient decreases, and the diffusion energy and prefactor both increase. Qualitatively, the results thus follow the same trends as obtained for rf samples deposited at various substrate temperatures (see Figs. 14a and 16a) and marked by the dotted lines in Fig. 18. It is interesting to note, however, that for a given hydrogen concentration, the diffusion coefficient (at 400°C) in as-grown (rf1) material is quite generally higher than in effused material.

d. *Influence of an Enhanced Hydrogen Concentration on Hydrogen Diffusion*

The hydrogen concentration in (low c_H) a-Si:H can be enhanced by hydrogen plasma treatment and hydrogen implantation. In order to study the influence of an enhanced hydrogen concentration on hydrogen (deuterium) diffusion, we prepared implantation-hydrogenated (IH) a-Si:H films by implanting hydrogen at various implantation doses into a-Si:H of low hydrogen concentration ($c_H \approx 0.3$ at.%). Aiming to achieve a fairly constant hydrogen concentration within the samples, a series of implantation energies was used. Hydrogen diffusion was studied by measuring the spreading of an implanted deuterium profile with $c_D < c_H$ (Beyer and Zastrow, 1998a, 1998b). The resulting hydrogen concentration dependence of D_0, E_D, and D (400°C) is also shown in Fig. 18 (IH). The data agree closely with the results of the effused samples.

2. HYDROGEN SOLUBILITY DATA

Anneal-stable steps in hydrogen (deuterium) concentration at contacts of hydrogen- (deuterium-) rich and poor material were noted by various authors (Street *et al.*, 1987; Street, 1991a; Beyer *et al.*, 1988). Beyer and coworkers (Beyer and Zastrow, 1993; Beyer and Wagner, 1994; Beyer, 1997a) associated the H concentration steps with steps in hydrogen solubility. As an example, Fig. 12b shows a deuterium concentration step after short-term annealing of 1 hour at $T = 400°C$. Figure 19a gives results for the same sample after long-term annealing of 30 hours at $T = 400°C$, demonstrating that the step remains present even after replacement of a large fraction of hydrogen in the a-Si:H layer by deuterium. Another example is shown in Fig. 19b. Here, the annealing behavior of a mesa-type deuterium concentration profile is studied, obtained by decreasing the substrate temperature of rf1 a-Si:D from 350°C to 200°C for a short period of time. The results show no spreading of the mesa structure after annealing at 400°C for 5 hours, despite a high diffusivity of deuterium. The high diffusivity is evidenced by the almost complete interdiffusion of deuterium with hydrogen from a ($T_s = 350°C$) surface a-Si:H layer. After annealing, a mesa-type structure also shows up in the hydrogen profile (Beyer and Wagner, 1994). These results demonstrate that H-D interdiffusion is essentially an H-D exchange at a fixed solubility and that in H/D layer structures diffusing hydrogen (deuterium) cannot break the majority of Si—Si bonds (Beyer and Wagner, 1994). Note, however, that this latter statement applies strictly to macroscopic diffusion only, i.e., to areas of several hundred angstroms in size. It cannot be excluded from the present data that on a local scale strong Si—Si bonds are broken by diffusing hydrogen if the same number of strong Si—Si bonds form nearby. If diffusing hydrogen (deuterium) would break strong Si—Si bonds on a macroscopic scale, however, the mesa structure should spread, and diffusion of hydrogen and deuterium should be largely independent of each other, opposite to the results of Fig. 19.

The dependence of hydrogen solubility in (rf1) a-Si:H on hydrogen concentration is studied in Fig. 20 (Beyer and Wagner, 1994; Beyer, 1997a). Plotted is the level N_D^* of deuterium in-diffusion into (rf1) a-Si:H films as a function of hydrogen density N_H and hydrogen concentration c_H. N_H was varied by changing the substrate temperature. The deuterium source concentration was always $c_D \approx 10$ at.%. It is seen that N_D^* follows the hydrogen density N_H within a factor of 2 down to a concentration of about 1 at.% and then becomes constant. Thus the hydrogen solubility in (rf1) a-Si:H films is defined by hydrogen concentration over a wide concentration range.

Both by hydrogen ion implantation and by hydrogen plasma treatment (the latter performed at temperatures below the deposition temperature), the hydrogen solubility in plasma-grown a-Si:H can be enhanced (Beyer and Zastrow, 1996). It can be decreased by hydrogen effusion. These effects are demonstrated in Figs. 21 to 23. Figure 21 shows depth profiles of implanted deuterium (40 kV implantation energy, 10^{17} cm^{-2} implantation dose) prior

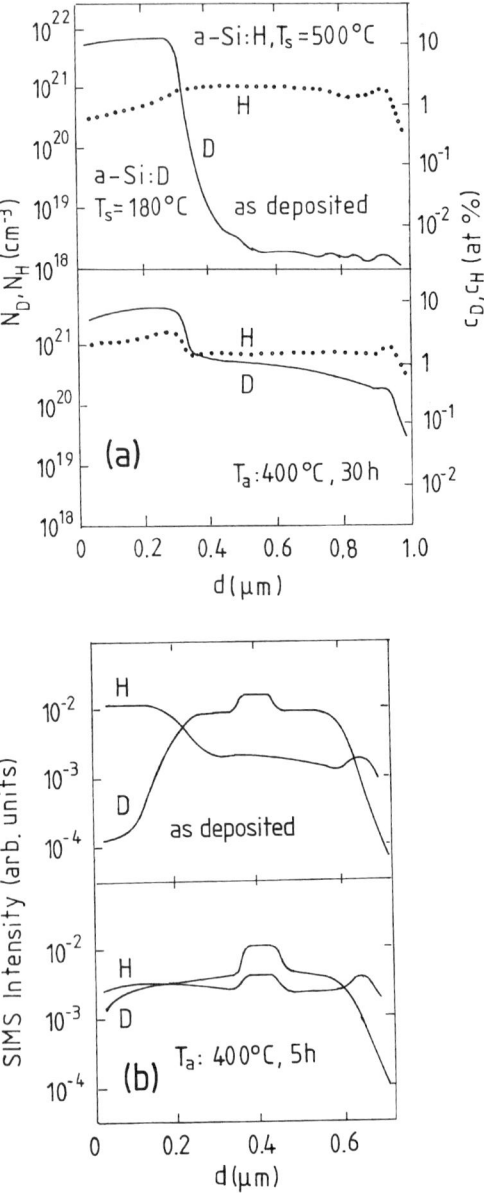

FIG. 19. SIMS depth profiles of hydrogen and deuterium for layered structures of a-Si:H and a-Si:D. (a) Type B sample, as deposited and after extended annealing; (b) type A sample with a-Si:D layer of mesa-type deuterium concentration profile, as deposited and after annealing. (From Beyer and Wagner, 1994.)

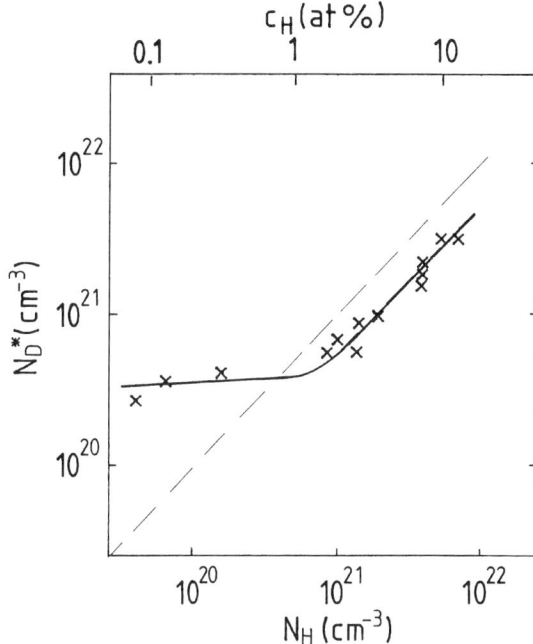

FIG. 20. Level of deuterium penetration N_D^* into a-Si:H, as a function of hydrogen density N_H (and hydrogen concentration c_H) for a-Si:D/a-Si:H layer structures with $c_D \approx 10$ at.%. Solid line serves as a guide to the eye; dashed line: $N_D^* = N_H$. (From Beyer, 1997a.)

to and after heat treatment for (rf1) a-Si:H deposited at $T_s = 500°$C. Similar to Fig. 12b, deuterium is seen to penetrate the unimplanted a-Si:H layer on the level of local hydrogen concentration. In the opposite direction (i.e., within the range of implantation defects), deuterium spreads at a higher level, indicating an enhanced hydrogen solubility in the presence of implantation defects. We estimate in this range the solubility to be about 5 at.%. Solubility effects are also visible in the hydrogen depth profiles; after annealing, hydrogen is seen to accumulate in the range of implanted deuterium, and it is depleted in the range where diffusing deuterium penetrates unimplanted a-Si:H.

Depth profiles of H and D in deuterium plasma-treated a-Si:H prior to and after heat treatment are shown in Fig. 22. The error function–like deuterium in-diffusion profile is seen to change to an s-like depth profile after heat treatment with deuterium penetrating the hydrogenated material on a slightly lower level than the original hydrogen concentration. This is

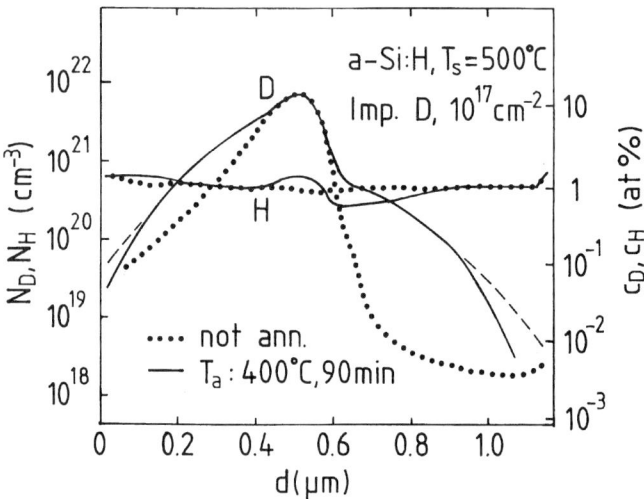

FIG. 21. SIMS depth profile of deuterium and hydrogen in a-Si:H ($T_s = 500°$C) implanted with 10^{17} cm^{-2} deuterium, prior to and after heat treatment (dashed line: error function fit). (From Beyer, 1996.)

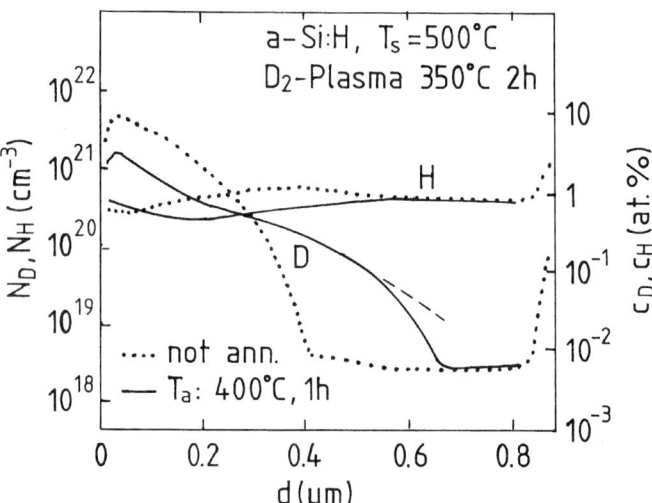

FIG. 22. SIMS depth profiles of deuterium and hydrogen in a-Si:H ($T_s = 500°$C) exposed for 2 hours to a D_2 plasma at $T = 350°$C. Dotted line: not annealed; solid line: after annealing at 400°C for 1 hour; dashed line: error function extrapolation. (From Beyer, 1997a.)

in agreement with a transformation of low c_H a-Si:H by plasma treatment to a material of higher c_H and higher H solubility.

Figure 23 demonstrates the effect of hydrogen out-diffusion on hydrogen solubility. The dashed line shows the hydrogen depth profile of the as-deposited sample ($T_s = 250°C$). After hydrogen effusion at 500°C for 5 hours, the dotted hydrogen profile resulted. Into this latter material, deuterium was implanted at a dose of $2 \times 10^{17}\,cm^{-2}$ (dotted deuterium line). After heat treatment (solid lines), deuterium is seen to penetrate the hydrogenated material up to the level of hydrogen concentration of the actual (hydrogen depleted) sample, not up to the original hydrogen concentration. This result demonstrates that by hydrogen effusion from (rf) plasma-grown a-Si:H the hydrogen solubility is being reduced.

Hydrogen solubility effects similar to those in a-Si:H are also present in hydrogen (deuterium) implanted crystalline silicon (c-Si). Figure 24 shows depth profiling results for (near-intrinsic) c-Si implanted with hydrogen at an energy of 30 keV and with deuterium at an energy of 150 keV. The dose was $10^{16}\,cm^{-2}$ in both cases. After heat treatment (400°C for 10 hours), deuterium is seen to accumulate in the range of the hydrogen implantation

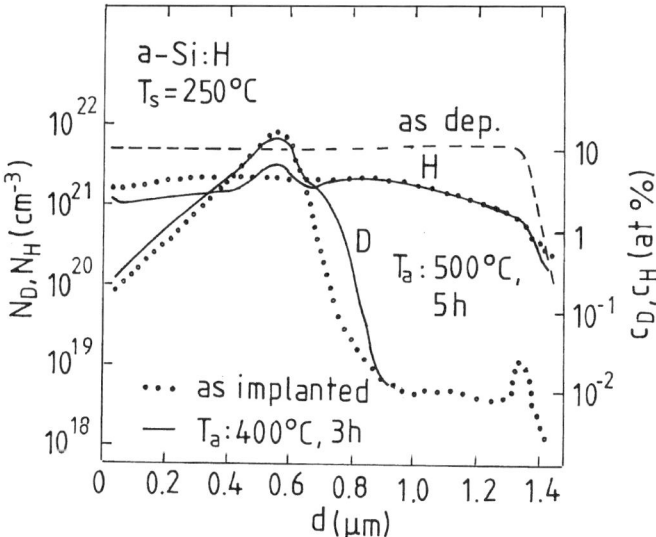

FIG. 23. SIMS depth profiles of hydrogen and implanted deuterium (energy: 40 keV; dose: $2 \times 10^{17}\,cm^{-2}$) for a-Si:H ($T_s = 250°C$) with hydrogen effused by heat treatment at $T_a = 500°C$ for 5 hours. Dashed line: hydrogen profile for as-deposited material; dotted lines: H and D profiles after effusing hydrogen and implanting deuterium; solid lines: after annealing at 400°C for 3 hours. (From Beyer, 1997a.)

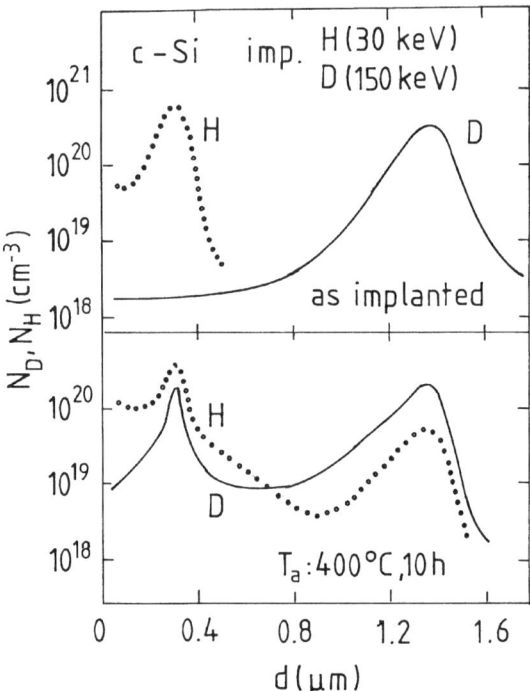

FIG. 24. SIMS depth profiles of hydrogen and deuterium in crystalline Si implanted by 30 keV H$^+$ and 150 keV D$^+$, as implanted and after annealing at 400°C for 10 hours. Implantation dose was 10^{16} cm^{-2}. (From Beyer, 1997a.)

peak, whereas hydrogen accumulates in the range of the deuterium implantation peak. Note, however, that the microstructure of a-Si:H and c-Si:H of $c_H > 1$ at.% seems to be different; recent studies of implantation-hydrogenated (IH) crystalline and amorphous silicon samples show a threshold for hydrogen-related void formation (see Sec. V.2) to be by a factor of about 10 lower in c-Si:H than in a-Si:H (Beyer and Zastrow, 1998a).

As a consequence of solubility effects that limit hydrogen incorporation in a-Si:H, a hydrogen concentration (and solubility) varying with depth will modify deuterium (hydrogen) diffusion profiles significantly. One example is shown in Fig. 19b, where a mesa-type deuterium concentration (and solubility) profile causes a mesa-type hydrogen diffusion profile. Another example is shown in Fig. 25. Here, a-Si:H of low hydrogen concentration was implanted by hydrogen of two implantation energies at the dose of 10^{17} cm^{-2} and by deuterium at the dose of 10^{16} cm^{-2}. On heat treatment, the deuterium profile is not found to spread following an error function, as

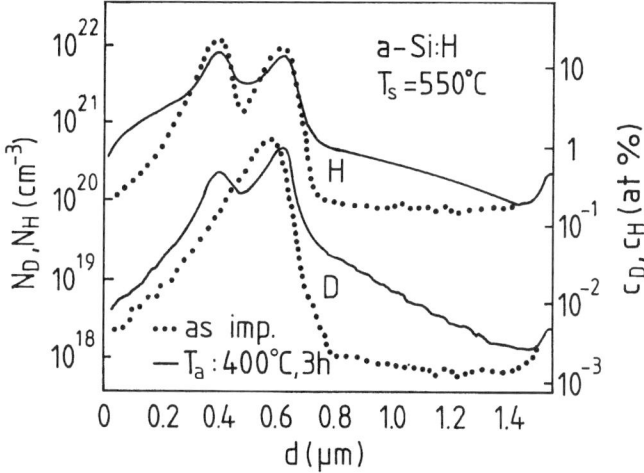

FIG. 25. SIMS depth profiles of deuterium and hydrogen in a-Si:H ($T_s = 550°C$) implanted by 40 and 70 keV H$^+$ (10^{17} cm^{-2}) and by 50 keV D$^+$ (10^{16} cm^{-2}) prior to and after annealing. (From Beyer, 1997a.)

one would expect for random diffusion in the silicon network. Instead, a double-peak depth profile shows up that resembles the hydrogen depth profile. According to these results, error function fitting of diffusion profiles of hydrogen or deuterium for a depth-dependent hydrogen solubility can lead to considerable errors for evaluation of the diffusion coefficient. We note, however, that by normalization to the total hydrogen (H + D) concentration, error function–type profiles often result. When evaluating such profiles to obtain the hydrogen diffusion coefficient, erroneous results still may be obtained for a strongly concentration-dependent hydrogen diffusion coefficient. In this case, variations of hydrogen concentration (and solubility) with depth may act as a diffusion barrier (Beyer and Zastrow, 1998a).

IV. Hydrogen Diffusion and Effusion Effects

1. HYDROGEN DIFFUSION PROCESSES

a. Hydrogen Trapping Model

Since hydrogen in a-Si:H is predominantly bound to silicon, the diffusion of hydrogen in (compact) a-Si:H is widely regarded as a trapping process, with hydrogen detrapping (hydrogen emission) being the rate-limiting step for hydrogen motion. Most commonly, Si dangling bonds (Carlson and Magee, 1978; Jackson et al., 1990) and/or weak Si—Si bonds (Street et al., 1987) are considered the traps, although other trapping sites have been taken into consideration from the beginning (Carlson and Magee, 1978). Recently, hydrogen clusters have been proposed as (shallow) hydrogen-trapping sites by Jackson and coworkers (Jackson et al., 1991b, 1992, 1993; Jackson and Tsai, 1992). Branz et al. (1993c) suggested that Si—H bonds act as traps. The direct exchange of plasma-induced interstitial hydrogen (deuterium) with bonded hydrogen (deuterium) as a possible hydrogen diffusion process has been proposed by Abeles et al. (1987a, 1987b).

Connected with the type of trapping sites, the concentration of hydrogen traps also was a matter of discussion. Street et al. (1988) proposed that only (preexisting) Si dangling bonds and weak Si—Si bonds act as traps. Jackson and Tsai (1992), based on posthydrogenation results, proposed that essentially all Si—Si bonds can be hydrogenated by diffusing hydrogen. The same was suggested by Branz et al. (1993b) discussing deuterium tracer diffusion and by Beyer (1985), who proposed the motion of pairs of hydrogen atoms in Si—H H—Si configurations. The hydrogen diffusion energy was associated directly with the depth of trapping levels in a series of recent publications (Jackson and Tsai, 1992; Santos and Jackson, 1992b; Mahan et al., 1995). Hydrogen-trapping levels at 0.5 eV (Santos and Jackson, 1992b), 1.3–1.5 eV (Jackson and Tsai, 1992; Santos and Jackson, 1992b; Mahan et al., 1995), and 1.9–2.5 eV (Jackson and Tsai, 1992; Mahan et al., 1995) have been identified. The time dependence of hydrogen diffusion was attributed to dispersive transport of hydrogen in a distribution of traps (Street et al., 1987; Kakalios et al., 1987). Kinks in the diffusion profiles of deuterium after deuterium plasma treatment of a-Si:H were attributed by Jackson and coworkers (Jackson et al., 1990; Jackson and Tsai, 1992) to a difference in deuterium diffusion when traps for hydrogen are empty or occupied.

The present results show, however, that many of these latter concepts are not well supported by experimental data. The smooth variation over 15 orders of magnitude of the diffusion prefactor D_0 as a function of the diffusion energy E_D (Meyer-Neldel rule; see Fig. 15) points against a direct association of the diffusion energy with depths of trapping levels. According to Jackson and Tsai (1992), trapping with fixed energy levels can explain a variation in the diffusion prefactor by up to 2 orders of magnitude only. Since the smooth variation of D_0 versus E_D includes both plasma in-diffusion and layer interdiffusion samples, a significant difference in the diffusion process between plasma in-diffusion and layer interdiffusion as proposed by Abeles et al. (1987a, 1987b) is unlikely. A significant change in the hydrogen diffusion coefficient for the transition from empty traps to filled traps as proposed by Jackson et al. (1990) and Santos and Jackson

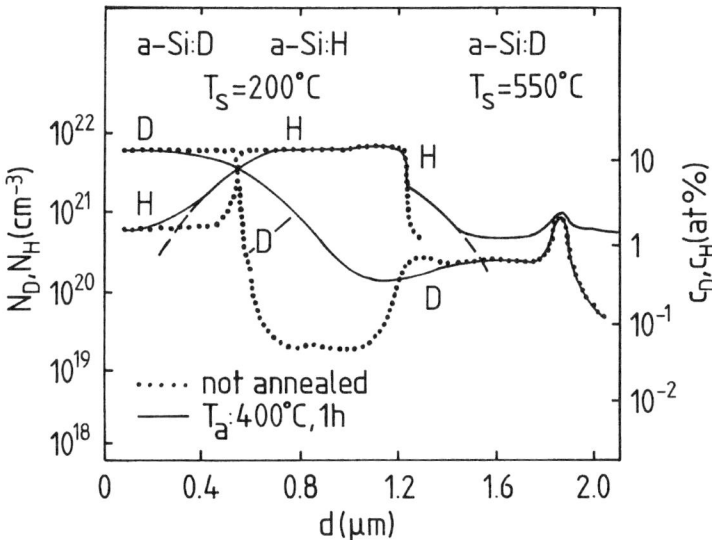

FIG. 26. SIMS depth profiles of deuterium and hydrogen in three-layer structure of a-Si:H and a-Si:D, prior to (dotted line) and after annealing at 400°C for 1 hour (solid line).

(1992a, 1992b) is also unlikely. Plasma postdeuteration experiments (see Fig. 13b) do not show a significant change in the diffusion coefficient if the level of deuterium penetration exceeds that of hydrogen present ($T_s = 500$°C material in Fig. 13b) or lies below ($T_s = 250$°C material in Fig. 13b) (see Beyer and Zastrow, 1996). The results of Fig. 26 demonstrate that there is also for layer diffusion no significant difference between the deuterium diffusion coefficient for H/D interdiffusion and the hydrogen diffusion coefficient for in-diffusion into low c_D a-Si:D if the source concentration is the same. In a three-layer structure, H-D interdiffusion in $T_s = 200$°C material and hydrogen in-diffusion into a-Si:D of low concentration ($T_s = 550$°C, $c_D \approx 0.8$ at.%) are studied in the same sample. Almost equal diffusion coefficients are observed for annealing at 400°C.

b. *Defect Motion Models*

Defect motion models assume that silicon-hydrogen bonds are highly stable and that hydrogen motion results from mobile defects triggering hydrogen bond-switching effects or hydrogen release into the interstitial state. Pantelides (1987) suggested that hydrogen diffusion proceeds through the motion of Si floating bonds. Data by Jackson et al. (1990), however, contradicted this assignment. In order to explain the observed high diffusion prefactor and high diffusion energy in a-Si of low hydrogen concentration ($c_\mathrm{H} \leq 2 \times 10^{19}\,\mathrm{cm}^{-3}$), Roth et al. (1993) proposed for this latter material hydrogen diffusion due to the migration of Si dangling bonds. Note, however, that Roth's model requires a fundamental change in the diffusion process as a function of hydrogen concentration, namely, from the motion of hydrogen (with a diffusion energy $E_\mathrm{D} \approx 1.5\,\mathrm{eV}$ and a diffusion prefactor of $D_0 \approx 10^{-2}\,\mathrm{cm^2\,sec^{-1}}$) at high hydrogen concentration to the motion of dangling bonds (with $E_\mathrm{D} \approx 2.7\,\mathrm{eV}$ and $D_0 \approx 10^4\,\mathrm{cm^2\,sec^{-1}}$) at low hydrogen concentration. Thus, instead of a gradual variation of D_0 versus E_D (Meyer-Neldel rule), as observed in Fig. 15, a step-like behavior would be expected. Moreover, one should see a kink in the Arrhenius plot of the diffusion coefficient at intermediate hydrogen concentration because the one or the other process will dominate hydrogen diffusion in the different temperature ranges. No such kinks are observed for H/D layer diffusion (see Fig. 13a), pointing against the validity of Roth's model.

c. *Equilibrium Energy Band Model*

In the trapping models discussed in Section IV.1.a, hydrogen motion is treated essentially as a nonequilibrium process by focusing on the detrapping process of a hydrogen atom from a particular bonding (trapping) site and/or by invoking the motion of free hydrogen that eventually gets trapped. An alternative model describes hydrogen diffusion in a-Si:H as an equilibrium process taking into account that hydrogen atoms in a-Si:H are generally diffusing in large numbers. Hydrogen is expected to be immobile when in bond with silicon and considered to be mobile only when activated to a transport state. If n_tr and N_H are mobile and total (mobile + immobile) hydrogen concentrations, respectively, and $n_\mathrm{tr}/N_\mathrm{H}$ is constant in space, the effective diffusion coefficient D defined by Fick's first law,

$$F = -D\,\mathrm{grad}\,N_\mathrm{H} \qquad (8)$$

(F is the flux of atoms) becomes

$$D = (n_\mathrm{tr}/N_\mathrm{H})D_\mathrm{H0} \qquad (9)$$

where D_H0 is the diffusion coefficient of the mobile species (Jackson and Tsai, 1992).

A likely reaction to describe hydrogen diffusion is the breaking of Si—H

5 HYDROGEN PHENOMENA IN HYDROGENATED AMORPHOUS SILICON

bonds to give silicon dangling bonds (Si—) and interstitial hydrogen atoms (H) (Street, 1991a)

$$\text{Si—H} \rightleftarrows \text{Si—} + \text{H} \tag{10}$$

The mass action law for this reaction gives for $N_{\text{Si—H}} \ll N_{\text{Si}}$:

$$n_{\text{tr}} \approx k_1 \frac{N_{\text{Si—H}} N_{\text{Si}}}{N_\text{D}} \tag{11}$$

where N_{Si}, $N_{\text{Si—H}}$, and N_D are the concentration of silicon atoms, bonded hydrogen, and Si dangling bonds, respectively, and k_1 is the reaction constant.

The Si dangling-bond concentration will be governed by another chemical reaction that describes the formation of Si—Si bonds from silicon dangling bonds

$$\text{Si—} + \text{Si—} \rightleftarrows \text{Si—Si} \tag{12}$$

For this reaction, the mass action law gives with the reaction constant k_2

$$N_\text{D} \approx k_2^{1/2} N_{\text{Si}} \tag{13}$$

Note that both reactions (Eqs. 10 and 12) may be affected by the constraints of chemical reactions proceeding in a solid. The solubility effects (see Secs. III.2 and V.1) show, in particular, that for diffusion in a-Si:H/a-Si:D layer structures the reactions in Eqs. (10) and (12) do not involve all silicon atoms but only those where hydrogen fits into the network.

An equilibrium energy-band model of hydrogen diffusion in a-Si:H was first proposed by Street *et al.* (1988). The participation of both Si—H bond rupture and Si—Si bond reconstruction in hydrogen diffusion and hydrogen release processes in a-Si:H was suggested earlier (Biegelsen *et al.*, 1979; McMillan and Peterson, 1979a; Beyer *et al.*, 1982; Beyer and Wagner, 1982). Street and coworkers (1988, 1991b) assumed that hydrogen atoms diffuse in transport states at or above the energy E_{tr} and that the hydrogen concentration in the transport states is determined by an equilibrium hydrogen chemical potential μ_H. This chemical potential is expected to lie (for $T = 0$) near the upper edge of the band of hydrogen states and will thus mark the filling level of a hydrogen density of states distribution. Hydrogen atoms at bonding sites of the energy $E < E_{tr}$ are considered to be localized. The hydrogen diffusion coefficient D is then given by (Street, 1991a, 1991b)

$$D = D_{H0} \exp[-(E_{tr} - \mu_H)/kT] \qquad (14)$$

where D_{H0} is the diffusion coefficient in the transport path. Many analogies to electronic transport in amorphous semiconductors exist. Both E_{tr} and μ_H may shift as a function of temperature. Assuming a linear temperature dependence,

$$E_{tr} - \mu_H(T) = (E_{tr} - \mu_H)_0 + \beta T \qquad (15)$$

the diffusion energy E_D and the diffusion prefactor D_0 become

$$E_D = (E_{tr} - \mu_H)_0 \qquad (16)$$

and

$$D_0 = D_{H0} \exp(-\beta/k) \qquad (17)$$

(Street, 1991b). In an equivalent thermodynamic picture, the diffusion free energy $\Delta G = (E_{tr} - \mu_H)$ may involve an entropy ΔS and enthalpy ΔH term:

$$\Delta G = \Delta H - T\Delta S \qquad (18)$$

so that

$$E_D = \Delta H \qquad (19)$$

and

$$D_0 = D_{H0} \exp(\Delta S/k) \qquad (20)$$

Thus the wide variation in the diffusion prefactors can be explained in the equilibrium energy-band model by a variation in the temperature shift of the hydrogen chemical potential with respect to the transport states or, in the thermodynamic description, by a varying entropy of reaction. As recently proposed by Beyer and Wagner (Beyer and Wagner, 1994; Beyer, 1996, 1997a) and discussed in Section IV.3, a variable linear temperature shift of the chemical potential also can give rise to the Meyer-Neldel dependence of the diffusion parameters, similar to the fact that a variable temperature shift of the Fermi level can cause the Meyer-Neldel rule for dark conductivity in a-Si:H (Beyer and Overhof, 1984). For a temperature-dependent hydrogen chemical potential (or diffusion free energy), the activation energy of hydrogen diffusion does not give direct information about trapping levels. Yet the position of the chemical potential relative to the transport path (or the diffusion free energy) can be obtained from the diffusion coefficient via Eq. (14), provided the diffusion prefactor D_{H0} is known. In Fig. 27 we present results for $(E_{tr} - \mu_H)$ in type A and type B samples. The diffusion coefficient D and $(E_{tr} - \mu_H)$ at $T = 400°C$ are plotted as a function of substrate temperature and hydrogen concentration. For D_{H0}, the theoretical diffusion prefactor $D_{00} = 10^{-3} \text{cm}^2 \text{sec}^{-1}$ (Eq. 6) was assumed. The results suggest that the chemical potential for type A samples decreases with decreasing hydrogen content by more than 0.3 eV. For type B samples [i.e., for a diffusion source layer of constant (high) concentration], μ_H in the probed a-Si:H layer lies independently of hydrogen concentration at about 1.4 eV below the transport states. It will be seen in Section IV.2 that this long-range interaction of source and probed layers is easily understood in the equilibrium energy-band model. The discussion in Sections IV.4 and IV.5 will show that various diffusion-related effects con-

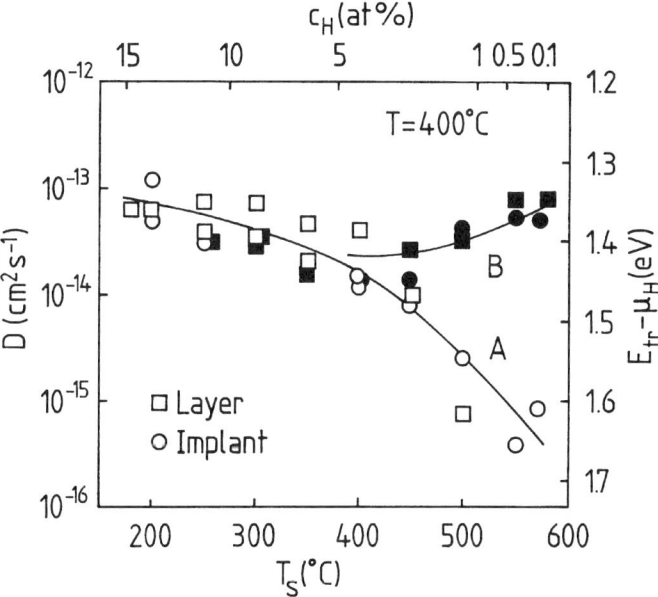

FIG. 27. Hydrogen diffusion coefficient D and position of the chemical potential $E_{tr} - \mu_H$ at $T = 400°C$ as a function of substrate temperature and hydrogen content for type A (open symbols) and type B (closed symbols) samples. Squares: a:Si:D/a-Si:H layers; circles: D-implanted a-Si:H films. (From Beyer, 1997a.)

sidered to be characteristic for trap-dominated hydrogen transport, such as the time dependence of the diffusion coefficient (Street et al., 1987), kinks in the diffusion profiles (Jackson et al., 1990), and short-time exponential wings (Branz et al., 1993a), are likely caused by mechanisms other than trap-limited diffusion in a compact material. Thus the equilibrium energy-band model seems to provide a better description of the present experimental diffusion data for compact a-Si:H than the description in terms of the kinetics of individual detrapping processes, as discussed in Section IV.1.a. We note, however, that the separation in compact and void-rich a-Si:H based mainly on hydrogen effusion measurements may not be justified on the atomic scale. Minor microstructural differences could then cause trapping or equilibrium features. We note, furthermore, that even if one deals with a microscopically compact material, hydrogen diffusion, like most transport phenomena, will involve both equilibrium and kinetic aspects and has both equilibrium and kinetic descriptions. The basic diffusion process in the equilibrium-band model (Eq. 10) involves the rupture of an Si—H bond and thus also may be viewed as a detrapping event. Accordingly, the basic diffusion steps are barely distinguishable in the two models, justifying, for example, that hydrogen diffusion also may be viewed in the equilibrium model as a process of jumps of hydrogen atoms from one site to another. A difference in the present trapping and equilibrium models is that trapping models commonly consider reactions Eq. (10) only, whereas the equilibrium model includes Eq. (12) as well. While our data suggest that macroscopic hydrogen diffusion in compact a-Si:H is rate limited by equilibrium effects, we stress that for a full understanding of hydrogen diffusion in a-Si:H, experimental and theoretical work describing the kinetics of the microscopic motion of hydrogen atoms or molecules is of great importance.

2. HYDROGEN DENSITY OF STATES DISTRIBUTION AND HYDROGEN CHEMICAL POTENTIAL

a. Hydrogen Density of States Distribution Inferred from Diffusion and Effusion Experiments

For the hydrogen transport states, bond-centered (BC) interstitial sites similar to those in single-crystalline Si (Van de Walle, 1991) were suggested by Street (1991a, 1991b). Various models have been proposed for the hydrogen density of states distribution below the transport states (Zafar and Schiff, 1989; Street, 1991a, 1991b, 1991c; Jackson and Tsai, 1992; Powell and Deane, 1996). Street assumed next to the transport states the presence of a tail of (empty) hydrogen-bonding states due to weak Si—Si bonds where hydrogen can insert. He related the slope of this tail to the slope of the electronic valence band tail (Street, 1991b). In some publications (Street et al., 1988; Kakalios and Jackson, 1988) the time dependence of hydrogen diffusion was related to the width of this tail (see Sec. IV.4). In order to determine the hydrogen density of states distribution of deeper states below the chemical potential, the analysis of hydrogen effusion experiments, in particular, of step-by-step hydrogen effusion experiments, has been proposed (Franz et al., 1998). In the step-by-step hydrogen effusion experiment (Jackson et al., 1994), the activation energy E_{eff} of the effusion rate dN_H/dt is determined for small effused hydrogen concentrations ΔN_{eff}. If E_{eff} can be related to the position of the chemical potential μ_H, the hydrogen density of states $g_H(E)$ is obtained by

$$g_H(E) = \frac{\Delta N_{eff}}{\Delta \mu_H} \qquad (21)$$

provided $g_H(E)$ and E_{tr} are independent of the occupation state. However, as discussed in Section II.2, different effusion processes such as diffusion-limited effusion or effusion via voids may contribute to the effusion transients, and samples may undergo structural changes as a function of temperature. Thus an analysis of hydrogen effusion transients in terms of a hydrogen density of states distribution is possible only if all processes limiting the effusion rate are carefully identified (Franz et al., 1998). Moreover, temperature shifts of the chemical potential will prevent identification of E_{eff} with $E_{tr} - \mu_H$. This problem is overcome, however, if (for the onset of effusion) a fixed effusion prefactor A_{eff} can be reasonably assumed in analogy to D_{H0} in Eq. (14); that is,

$$dN_H/dt = A_{eff} \exp[-(E_{tr} - \mu_H)]/kT \qquad (22)$$

for effusion limited by diffusion of atomic hydrogen. In this case, even single effusion transients will yield $(E_{tr} - \mu_H)$ versus N_{eff} and thus the hydrogen density of states distribution $g_H(E)$ with reasonable accuracy.

Hydrogen density of states distributions obtained by step-by-step effusion experiments (Franz et al., 1998) are qualitatively similar to density of states distributions obtained from analysis of hydrogen diffusion experiments (Beyer, 1996). In these latter experiments, the change in the diffusion coefficient on hydrogen effusion (see Fig. 18) was analyzed. Furthermore, by analysis of hydrogen in-diffusion from a (step-by-step) increased source concentration (see Fig. 17), the density of empty hydrogen states above the chemical potential also was obtained. The results of this analysis are shown

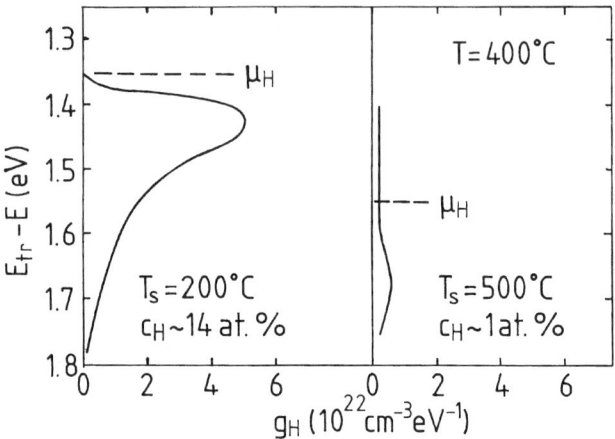

FIG. 28. Hydrogen density of states $g_H(E)$ versus energy E relative to the hydrogen transport level E_{tr} for a-Si:H of $T_s = 200$ and $500°C$. Hydrogen chemical potentials μ_H at $400°C$ are indicated. (From Beyer, 1997a.)

in Fig. 28. To determine $(E_{tr} - \mu_H)$ from the diffusion coefficient, for D_{H0} in Eq. (14) the theoretical diffusion prefactor $D_{00} \approx 10^{-3}\,\text{cm}^2\,\text{sec}^{-1}$ (see Eq. 6) was assumed. The results suggest for material of relatively high hydrogen concentration (low substrate temperature, $T_s = 200°C$) a maximum of $g_H(E)$ just below the chemical potential at $E_{tr} - E \approx 1.4–1.5\,\text{eV}$ and a tail of hydrogenated states at higher binding energy. Indications for a second density of states peak of higher binding energy were found by Franz et al. (1998). The corresponding effusion peak near $800°C$ tentatively attributed by Jackson et al. (1994) to the release of hydrogen bound to isolated dangling bonds also shows up in our effusion transients (see Fig. 1). Since in the effusion experiment our material crystallizes near $750°C$, we attribute this peak to the release of (mostly molecular) hydrogen trapped in the crystallized Si structure. We note that for a-Ge:H films, a similar crystallization-related dip in the effusion curves occurs near $500–550°C$ (Beyer et al., 1991). Also shown in Fig. 28 is the hydrogen density of states distribution for material of low hydrogen concentration (high substrate temperature, $T_s = 500°C$). The hydrogen chemical potential is lower, and hydrogen bonding states of low binding energy have disappeared. Furthermore, a nearly constant density of empty hydrogen states of $g_H = 0.5–5 \times 10^{21}\,\text{cm}^{-3}\,\text{eV}^{-1}$ (characteristic for a given sample) is inferred for energies above the chemical potential (Beyer, 1996).

The qualitatively similar concentration dependence of the hydrogen diffusion parameters independent of the method by which the hydrogen concentration is varied (see Fig. 18) suggests the presence of a rather unique hydrogen density of states distribution for plasma-grown (compact) a-Si:H. Thus the major difference between a-Si:H samples of different hydrogen concentrations would be the different degree of removal of Si—H states of low binding energy and, consequently, a different position of the hydrogen chemical potential. However, some quantitative differences between plasma-grown material, on the one hand, and hydrogen effused and implanted material, on the other, occur, showing up in particular in a different absolute value of the diffusion coefficient for a given temperature and given hydrogen concentration (see Fig. 18b). A different microstructure could be the origin. As will be discussed in Section IV.4, microstructural effects likely will affect the position of the chemical potential. It is also conceivable, however, that for differently prepared samples of the same hydrogen concentration, the hydrogen density of states distribution depends to some degree on the preparation conditions.

b. Source Concentration Dependence of the Hydrogen Chemical Potential

A source concentration dependence of hydrogen diffusion is observed in hydrogen-poor a-Si:H if such material is brought in contact with a-Si:D or a-Si:H source layers of higher hydrogen (plus deuterium) concentration (see Figs. 16, 17, and 27). One expects no step in the energy E_{tr} of the hydrogen transport states at the interface of the two materials because the energy of bond-centered hydrogen is unlikely strongly affected by different concentrations of Si—H bonds ($N_H \ll N_{Si}$). Owing to the gradient in the chemical potential, hydrogen will diffuse from the material of higher hydrogen concentration into the material of lower hydrogen concentration and will fill up the unoccupied hydrogen states in the latter material until the chemical potentials (relative to E_{tr}) become equal. Due to the low density of unoccupied weak Si—Si bond states in low c_H material (see Fig. 28) in-diffusion of only small amounts of hydrogen (deuterium) is sufficient to induce a shift of the chemical potential in the hydrogen-poor material to the same position as in the hydrogen-rich source layer. A common hydrogen chemical potential throughout samples with different hydrogen (deuterium) concentrations in the source and probed layers can explain the different concentration dependence observed for diffusion in type A and type B samples, in particular the rather concentration-independent diffusion coefficient of type B samples with equal source concentration (see Figs. 16b and 27). Likewise, a common hydrogen chemical potential can explain the strong influence of an implanted deuterium source concentration on hydrogen diffusion in the unimplanted part of a-Si:H samples of low hydrogen concentration (see Fig. 17).

c. *Energy Dispersion of Hydrogen Density of States and Mechanism of Hydrogen Exchange*

With regard to the relatively broad density of states distribution for bonded hydrogen in a-Si:H (Fig. 28), two problems require further attention, the nature of the energy dispersion of the hydrogen density of states and the mechanism of hydrogen exchange within the density of states distribution of bonded hydrogen necessary for establishing equilibrium. Considering that Si—H bonds are largely decoupled from distortions of the silicon network, Street *et al.* (1988) assumed a narrow energy distribution of Si—H bonds. A variation of the Si—H bond strength could be expected through Lucovsky's induction effect (Lucovsky, 1979; see also Sec. II.3) if at the backbonds of a given Si—H bonding site Si—Si bonds are replaced by Si—H bonds. Yet this mechanism would cause an increase in hydrogen binding energy with rising hydrogen concentration opposite to what is observed experimentally. A viable explanation of the energy dispersion is a different degree of the Si network distortion when hydrogen is incorporated at different bonding sites. For example, hydrogen bonding to a preexisting vacancy is expected to cause less network distortion than insertion of two hydrogen atoms into a strong Si—Si bond. This network distortion (i.e., Si-Si reconstruction) energy would be gained on hydrogen removal, thus decreasing the effective hydrogen emission energy. Small amounts of hydrogen causing little network distortion will be more tightly bound to silicon, and the hydrogen emission energy will be higher compared with the large amounts of hydrogen that can be incorporated under strong network distortion only. Recent calculations of hydrogen energies in crystalline and amorphous silicon (Chang and Chadi, 1989; Van de Walle, 1994; Van de Walle and Street, 1994, 1995) support this picture. Finally, it should be noted, that by interaction with electrons or holes, variations of Si—H binding and dissociation energies also can be expected (Carlson, 1986; Street *et al.*, 1987; Van de Walle and Street, 1995).

Jackson (1993) provided clear experimental evidence for exchange processes between hydrogen atoms of low binding (diffusion) energy penetrating a-Si:H from a hydrogen plasma and tightly bound hydrogen present in the material. The data of Fig. 29 demonstrate the presence of similar hydrogen-deuterium exchange processes also in a layer diffusion sample. A thin film of high c_H a-Si:H ($T_s = 200°C$) was deposited on top of low c_D a-Si:D ($T_s = 580°C$), and the diffusion profiles of hydrogen and deuterium of the as-deposited sample (not annealed) and after 1 hour heat treatment at 400°C were measured by SIMS. According to the deuterium (hydrogen) diffusion coefficient in $T_s = 580°C$ material of $D < 10^{-15}$ cm^2 sec^{-1} at 400°C (see Fig. 16a), a diffusion length for deuterium of $L = 2(Dt)^{1/2} < 0.04\,\mu$m is

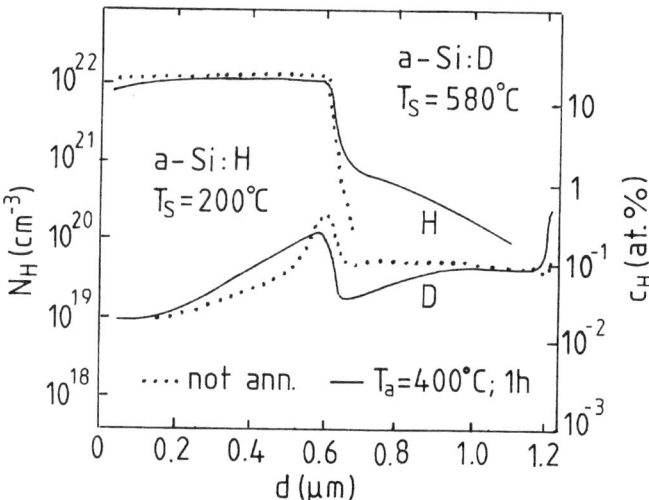

FIG. 29. SIMS depth profiles of hydrogen and deuterium prior to (dotted line) and after annealing at 400°C for 1 hour (solid line) for a-Si:H layer of high hydrogen concentration on top of a-Si:D layer of low deuterium concentration.

expected after heat treatment. The experimental data of Fig. 29 show, in contrast, in the a-Si:D layer a nearly equal diffusion length for (out-diffusing) deuterium and for (in-diffusing) hydrogen of $L \approx 0.3$ µm. Apparently, due to an H-D exchange process, deuterium atoms occupying originally deep states in the density of states distribution change to an energy level similar to the in-diffusing hydrogen atoms. A viable mechanism for the exchange process is the activation (in a bond-switching process) of deuterium from a deep bonding site to a bonding site in an empty weak Si—Si bond, i.e., into states located above the chemical potential. The empty deuterium bonding site is then free for in-diffusing hydrogen. In-diffusion of hydrogen into the vicinity of strongly bound deuterium, furthermore, may modify the local network distortion with the result that deuterium becomes weakly bound too. A direct exchange between free hydrogen or deuterium in the transport states with bonded hydrogen or deuterium is also conceivable, as proposed by Kemp and Branz (1995) and Branz et al. (1995). This process would not change significantly the diffusion coefficient if the free hydrogen is generated by thermal activation, not by direct insertion from the plasma. A direct exchange between bonded hydrogen and/or deuterium atoms, however, appears unlikely in compact material. Such a process is conceivable only for next-neighbor Si—H groups and would result in an additional hydrogen diffusion path independent of $(E_{tr} - \mu_H)$. No indications for such a diffusion process are observed. Note that the problem of hydrogen exchange within the distribution of bonded hydrogen has been addressed recently by Van de Walle and Street (1995).

3. TEMPERATURE SHIFT OF THE HYDROGEN CHEMICAL POTENTIAL AND MEYER-NELDEL RULE OF HYDROGEN DIFFUSION

As pointed out in Section IV.1.c, the wide variation in diffusion prefactors and diffusion energies in a-Si:H suggests the presence of a rather mobile hydrogen chemical potential with quite different temperature shifts of μ_H for material of low and high hydrogen concentration. In the energy-band model of hydrogen diffusion, the hydrogen chemical potential determines according to Eq. (14) the hydrogen diffusion coefficient. Thus, from measurements of the hydrogen diffusion coefficient, the temperature-dependent position of the chemical potential can be extracted, provided the diffusion prefactor D_{H0} is known. In Fig. 30, we plot the hydrogen chemical potential for (rf1) samples of Fig. 13 as a function of temperature. For D_{H0}, the theoretical diffusion prefactor $D_{00} = 10^{-3} \, \text{cm}^2 \, \text{sec}^{-1}$ (see Eq. 6) was assumed. The results show for layer diffusion samples of relatively high hydrogen concentration (low T_s) a nearly temperature-independent hydro-

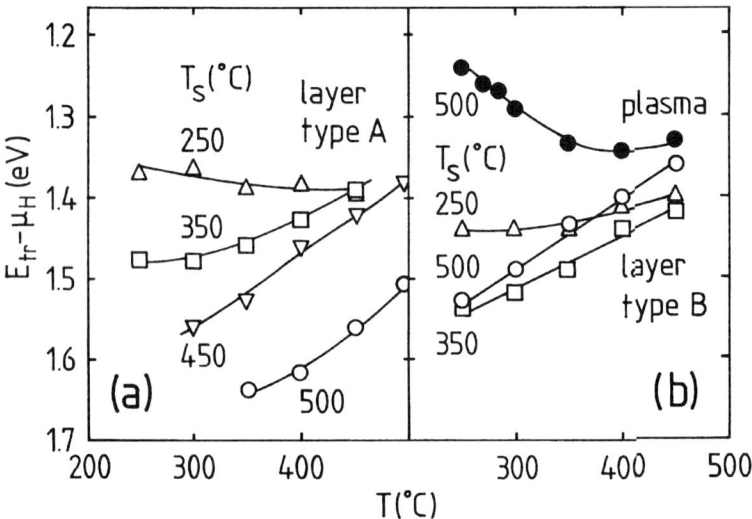

FIG. 30. Hydrogen chemical potential μ_H relative to hydrogen transport level E_{tr} as a function of temperature for samples of Fig. 13. (From Beyer, 1997a.)

gen chemical potential. A rather strong shift of the chemical potential is obtained for layer diffusion in material of low hydrogen concentration and for plasma in-diffusion. In the first case, the chemical potential shifts with rising temperature toward the transport energy E_{tr}; in the second case (in the low-temperature range), it shifts away from the transport energy. For plasma in-diffusion, the temperature dependence of the hydrogen chemical potential has been explained by a temperature dependence of the surface hydrogen chemical potential; as the temperature is increased, the surface hydrogen chemical potential is assumed to decrease due to hydrogen recombination and surface desorption (Beyer and Zastrow, 1996). For layer diffusion, the temperature dependence could be due to a statistical temperature shift of the chemical potential (Street, 1991b). For a low concentration of empty weak Si—Si bond states above the chemical potential and for a high hydrogen density of states below, the chemical potential would move toward the transport states with rising temperature. The shift for $T_s = 450°C$ material in Fig. 30a of $\beta \approx -9k$ (k is Boltzmann's constant) agrees well with Street's value of $\beta = -11k$ (Street, 1991b). However, to a first approximation, one would expect a stronger shift for a higher hydrogen concentration (lower T_s), opposite to what is observed experimentally. An alternative model introduced below to explain the Meyer-Neldel rule of diffusion prefactor and diffusion energy assumes that the temperature shift of $(E_{tr} - \mu_H)$ does not arise from a statistical temperature shift of the chemical potential but rather from temperature dependences of binding energies involved or of multiphonon excitations. Note, however, that all our information on absolute values and temperature shifts of $(E_{tr} - \mu_H)$ relies on the assumption of D_{H0}. Clearly, more work is needed for a better understanding of this subject.

It is also not clarified so far if the occurrence of Meyer-Neldel rules and of qualitatively similar concentration dependences of the diffusion parameters for hydrogen diffusion in amorphous metals (Kirchheim et al., 1982) and in a-Si:H indicate the presence of related hydrogen diffusion processes. Various models have been proposed to explain Meyer-Neldel (compensation) laws observed for electronic and hydrogen transport in amorphous and crystalline semiconductors (Wert and Zener, 1949; Beyer and Overhof, 1984; Kirchheim and Huang, 1987; Jackson, 1988; Yelon and Movaghar, 1990; Khait et al., 1990; Shinar et al., 1991; Yelon et al., 1992; Branz et al., 1994; Boisvert et al., 1995; Khait and Weil, 1995).

In the thermodynamic description of the diffusion coefficient in terms of a free energy of diffusion (Eq. 18), a Meyer-Neldel rule arises if the activation entropy is proportional to the activation enthalpy, that is,

$$\Delta H = a \Delta S + b \quad (23)$$

with constants a and b. This leads with Eqs. (19) and (20) directly to the Meyer-Meldel dependence (Eq. 7) with $T^{MN} = a$ and $D_0^{MN} = D_{H0} \exp(-b/kT^{MN})$. According to Eq. (18), the constant b equals the free energy ΔG at the Meyer-Neldel temperature, that is, $b = \Delta G^{MN}$, and thus

$$D_0^{MN} = D_{H0} \exp(-\Delta G^{MN}/kT^{MN}) \qquad (24)$$

Accordingly, the Meyer-Neldel dependence of diffusion prefactors and diffusion energies is equivalent to a temperature dependence of the free energy of diffusion,

$$\Delta G(T) = \Delta G^{MN} + \Delta S(T^{MN} - T) \qquad (25)$$

or, in the energy-band picture, equivalent to a temperature shift of the chemical potential relative to the transport states,

$$E_{tr} - \mu_H(T) = (E_{tr} - \mu_H)^{MN} - \beta(T^{MN} - T) \qquad (26)$$

Thus the observation of a Meyer-Neldel rule and of widely different temperature dependences of the parameters $(E_{tr} - \mu_H)$ and ΔG are directly connected. As mentioned earlier, such temperature shifts can arise in a specific density of states distribution due to occupation statistics, as has been proposed to explain the Meyer-Neldel dependence of dark conductivity in doped a-Si:H films (Beyer and Overhof, 1984). Yet the present results (see Fig. 30) suggest a stronger temperature shift of the hydrogen chemical potential for hydrogen-poor compared with hydrogen-rich materials, and it is presently not obvious how the respective hydrogen density of states distributions could account for such shifts.

An alternative model relates the temperature dependence of the parameters ΔG and $(E_{tr} - \mu_H)$ to temperature dependences of binding energies. In the equilibrium energy-band model, the energy required to raise hydrogen into the transport path involves two components, the actual splitting of Si—H bonds and the reconstruction of Si—Si bonds. Assuming (in first approximation) for the reaction constants of the reactions in Eqs. (10) and (12)

$$k_1 \approx \exp[-(E_{tr} - E_B)/kT] \qquad (27)$$

and

$$k_2 \approx \exp(E_{Si-Si}/kT) \qquad (28)$$

where E_B and E_{Si-Si} are the Si—H and Si—Si binding energies, respectively, we obtain with Eqs. (9), (11), (13), and (14) for $n_{tr} \ll N_{Si-H}$

$$E_{tr} - \mu_H = \Delta G = E_{tr} - E_B + \frac{E_{Si-Si}}{2} \quad (29)$$

Considering that a-Si:H is known to contain weak and strong Si—Si bonds, a plausible way to bring Eq. (29) into the form of Eqs. (25) and (26) is to assume that for hydrogen-poor a-Si:H (with a rather inflexible network at ambient temperature) the strength of the Si—Si bonds encountered in the diffusion path depends on temperature. While at high temperatures (i.e., near the Meyer-Neldel temperature T_{MN}) due to silicon network vibrations hydrogen diffusion would involve rather strong Si—Si bonds, at lower temperatures hydrogen diffusion could confine itself to paths with weaker Si—Si bonds resulting according to Eq. (29) in an increase in $(E_{tr} - \mu_H)$, as binding energies are negative. This effect could cause $(E_{tr} - \mu_H)$ to vary according to Eq. (26), with β depending on the flexibility of the network. For hydrogen-rich a-Si:H, when most hydrogen is presumably in Si—H H—Si bonds, the silicon network is expected to be highly flexible so that rather strong Si—Si bonds can form upon hydrogen release quite independently of temperature. Accordingly, $(E_{tr} - \mu_H)$ would attain the value for strong Si—Si bonds, see Eq. (29), over a wide temperature range. Other extrinsic effects such as hydrogen in-diffusion from a plasma also could cause deviations from Eq. (29) at low temperatures that disappear at high temperatures. Thus this latter model can explain (qualitatively) both the validity of a Meyer-Neldel rule and the presence of the different temperature shifts of the hydrogen chemical potential for a-Si:H materials of different hydrogen concentrations or when a hydrogen plasma is applied.

A drawback of this model, however, is the specific origin of the temperature shift of $(E_{tr} - \mu_H)$, which is attributed for layer diffusion to a temperature dependence of the average Si—Si binding energy encountered by diffusing hydrogen atoms. Yet Meyer-Neldel dependences in a-Si:H are also observed for processes where Si—Si reconstruction is unlikely involved. For example, a Meyer-Neldel rule has been reported for the prefactor and the activation energy of hydrogen surface desorption from void-rich a-Si:H when the doping level is varied (Khait et al., 1990). In this case, the free energy of desorption is associated with the rupture of two Si—H bonds and the formation of H_2. Thus more general effects than a dependence of the Si—Si binding energy on temperature and hydrogen concentration seem to cause the Meyer-Neldel rules of diffusion and desorption in a-Si:H. Indeed, various authors have given a more general explanation of the Meyer-Neldel (compensation) law, associating the proportionality of activation entropy and enthalpy with multiphonon excitations necessary for hydrogen atoms or electrons to surmount activation barriers (Yelon and Movaghar, 1990; Khait et al., 1990; Yelon et al., 1992). Accordingly, the different temperature shifts of the hydrogen chemical potential would arise from different multiphonon excitations. Khait et al. (1990) relates the Meyer-Neldel temperature T^{MN} to the energy difference between energy levels involved in electron transitions during the desorption event. It is presently not obvious, however, how the multiphonon transitions are changed by different hydrogen concentrations with the result of a stronger temperature dependence of ΔG for hydrogen-poor than for hydrogen-rich a-Si:H.

Whatever the reasons for the Meyer-Neldel dependence of hydrogen diffusion, the present experimental data suggest that the Meyer-Neldel prefactor (Eq. 24) and thus the diffusion free energy ΔG^{MN} [or the chemical potential position $(E_{tr} - \mu_H)^{MN}$] at T^{MN} are related to intrinsic or nearintrinsic material parameters. From the Meyer-Neldel parameters (see Sec. III.1.b) of a:Si:H, $T^{MN} \approx 760$ K and $D_0^{MN} \approx 6 \times 10^{-13}$ cm^2 sec^{-1}, one obtains with $D_{H0} \approx D_{00} = 10^{-3}$ cm^2 sec^{-1} (Eq. 6):

$$\Delta G^{MN} = (E_{tr} - \mu_H)^{MN} \approx 1.39 \text{ eV} \qquad (30)$$

Apparently, Eq. (30) and the MN dependence in Fig. 15 are characteristic for transport of hydrogen incorporated in Si—H bonding sites in compact Si:H material. Data for diffusion of hydrogen entirely incorporated in BC sites (in c-Si:H) reported by Van Wieringen and Warmoltz (1956) ($D_0 = 9.4 \times 10^{-3}$ cm^2 sec^{-1}, $E_D = 0.48$ eV) deviate considerably from the Meyer-Neldel dependence of Fig. 15, as do data for diffusion of hydrogen in a-Si:H with a microstructure ($E_D \approx 1$ eV, $D_0 \approx 3 \times 10^{-3}$ cm^2 sec^{-1}) inferred from results by Street and Tsai (1988). Data for diffusion of hydrogen in implantation-hydrogenated c-Si:H, on the other hand, agree well (Beyer and Zastrow, 1998b). For hydrogen diffusion in a-Ge:H, the value of $(E_{tr} - \mu_H)^{MN} \approx 1.28$ eV was reported (Beyer, 1996). Completely different values compared with a-Si:H are obtained for hydrogen diffusion in other materials. For example, for hydrogen diffusion in disordered palladium, $(E_{tr} - \mu_H)^{MN} \approx 0.19$ eV is obtained from experimental data by Kirchheim and Huang (1987). Thus the present data suggest that $(E_{tr} - \mu_H)^{MN} = \Delta G^{MN}$ is an important quantity characterizing a diffusion system. Indeed, we can use Eq. (30) to calculate intrinsic properties of a-Si:H. With the literature data $E_{tr} \approx -1.05$ eV and $E_B = E_{Si-H} \approx -3.55$ eV (referred to the Si dangling bond) by Van de Walle and Street (1995), Eqs. (29) and (30) yield $E_{Si-Si}/2 \approx -1.1$ eV. This value is in reasonable agreement with estimates for the formation energy of a silicon dangling bond in c-Si of $E \approx -1.38$ eV (Van de Walle and Street, 1994).

4. TIME DEPENDENCE OF HYDROGEN DIFFUSION COEFFICIENT

The time dependence of the hydrogen diffusion coefficient in a-Si:H was first reported by Street et al. (1987). For deuterium diffusion in a layered (boron-doped) a-Si:D/a-Si:H sample, these authors observed

$$D = D_0 t^{-\alpha} \tag{31}$$

where α is 0.2–0.25 for annealing time periods between ~ 10 min and nearly 200 hours. The authors explained the time dependence in analogy to dispersive electron transport by multiple trapping of hydrogen atoms in a distribution of (empty) trapping sites of various depths. After an extended period of time, a diffusing hydrogen atom would get trapped in the deepest trap available, and thus the H diffusion coefficient was expected to decrease as a function of time. This concept was elaborated by Kakalios et al. (1987), who found, over a limited temperature range, α to depend on temperature T, with $\alpha = 1 - T/T_0$. The quantity kT_0 was related to the characteristic energy of an exponential distribution of trapping sites (Kakalios and Jackson, 1988; Kakalios, 1991). However, it remained unclear whether the dispersion arises from the distribution of weak Si—Si bonds, the distribution of interstitial energies, or the distribution of Si—H bonds (Street, 1991a).

In Fig. 31 we plot for (type A) rf1 a-Si:H films the parameter α at a temperature of 400°C as a function of the substrate temperature and hydrogen content. It is seen that α attains rather high values ($\alpha \approx 0.5$) for low substrate temperature and high hydrogen concentration, while at substrate temperatures above 450°C, α is found to approach zero. A similar dependence of α on substrate temperature ($T_s < 250$°C) was reported by Tang et al. (1990a). A decrease in α with decreasing hydrogen concentration from $\alpha \approx 0.5$ to $\alpha \approx 0$ (Fig. 31) could be explained in the trapping model by a decrease in the characteristic energy kT_0 of the band of traps from about 0.1 eV to kT or by a change from one trapping level to another (with a smaller characteristic energy). However, this result contrasts to the fact that amorphous silicon of low hydrogen concentration is known to contain a high density of silicon dangling bonds and thus of hydrogen traps, so that rather a broad trap distribution and a sizable characteristic energy kT_0 are expected for this type of material. Moreover, as shown in Fig. 32, one finds α also to depend on the deuterium source concentration. Figure 32a shows for $T_s = 350$°C deposited (rf1) a-Si:H ($c_H \approx 6$ at.%) the parameter α as a function of the substrate temperature (and the deuterium concentration) of the deuterium source layer; in Fig. 32b, α of low c_H a-Si:H (rf1, $T_s = 550$°C) is plotted as a function of the implanted deuterium dose used as the diffusion source (see Fig. 17). In both cases, a rising deuterium source concentration

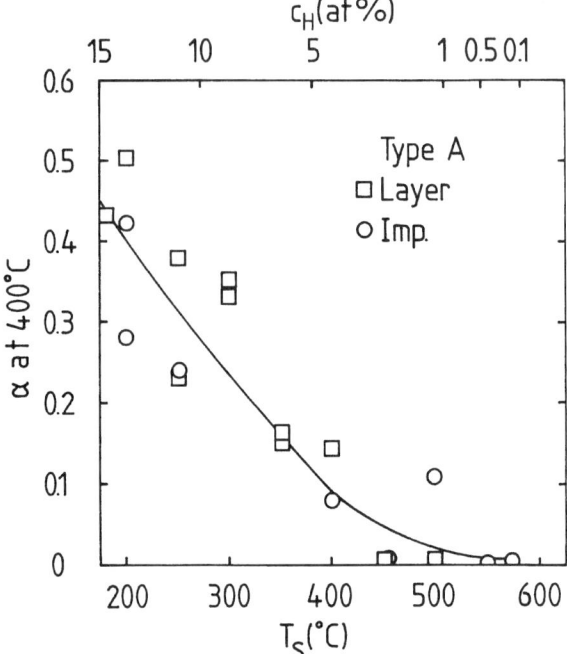

FIG. 31. Time-dependence parameter α at 400°C versus substrate temperature T_s (and hydrogen concentration) for type A (rf1) a-Si:D/a-Si:H layer structures and D-implanted a-Si:H films. (After Beyer and Wagner, 1994.)

results in an increase in α. In a trapping model, this result is difficult to explain, since it is difficult to see why the characteristic energy of the band of (empty) traps in the probed a-Si:H layer should increase with rising deuterium concentration in the source layer.

As noted previously (Beyer et al., 1989; Mitra et al., 1990), mechanisms other than dispersive hydrogen transport in a distribution of trapping sites also may result in a time dependence of the diffusion coefficient. Desorption of H_2 at void surfaces will inhibit long-range diffusion of hydrogen in a void-rich material. Furthermore, hydrogen out-diffusion into voids or to the film surface will decrease the hydrogen chemical potential, thus decreasing according to Eq. (14) the H diffusion coefficient. Both these effects would result in a time dependence of the hydrogen diffusion coefficient. Since we did not find, however, a significant dependence of α on film thickness (Beyer, 1997b), hydrogen out-diffusion from compact a-Si:H (to the film surface and film-substrate interfaces) seems not to play a dominant role. Finally, the

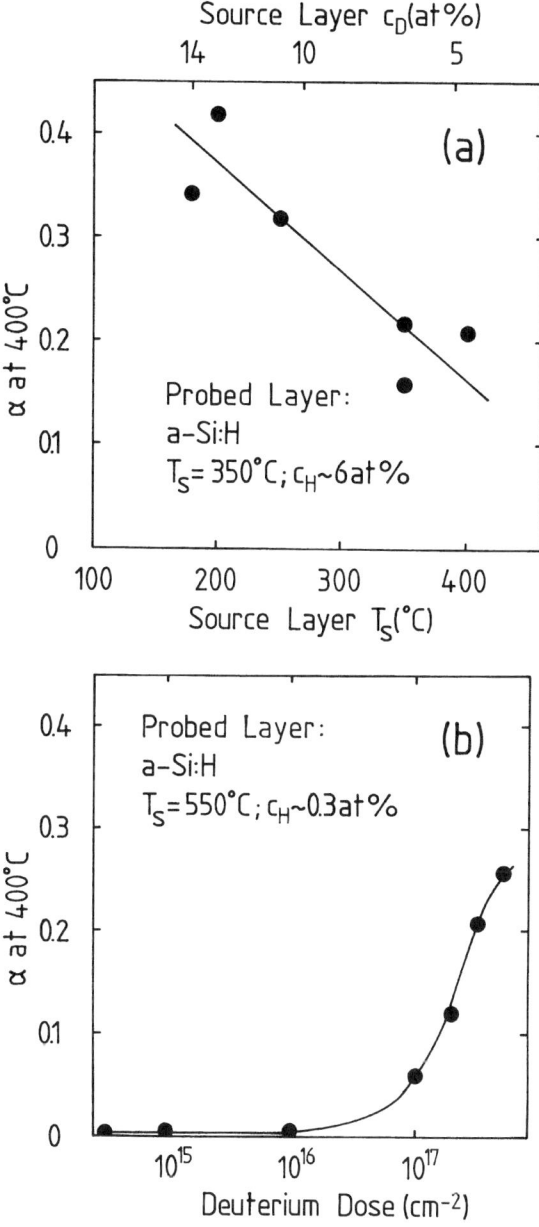

FIG. 32. Parameter α at 400°C as a function of source layer deuterium concentration (a) for (rf1) a-Si:H deposited at $T_s = 350°C$ and a-Si:D source layer deposited at various substrate temperatures T_s and (b) for (rf1) a-Si:H of $T_s = 550°C$ implanted with deuterium at various doses.

time dependence could be due to a general drop in the hydrogen chemical potential through a relaxation of the material. While a contribution of various factors to the time dependence of the H diffusion coefficient cannot be excluded, we note that the dependence of α on the hydrogen (deuterium) concentration in the source layer (Fig. 32) points to a time dependence of the hydrogen chemical potential as the major reason for the time dependence of the diffusion coefficient. According to Fig. 31, an increasing source layer concentration would then result in a stronger time dependence of the hydrogen (deuterium) chemical potential in the source layer and, since the chemical potentials in the source layer and in the diffusion layer adjust to each other, also in the (probed) diffusion layer. This interaction of source and diffusion layers via a common chemical potential thus can explain the results of Fig. 32. The major reason for the decrease in H chemical potential as a function of annealing time is presumably out-diffusion of hydrogen into voids, as suggested by the similarity between the concentration dependence of α (Fig. 31) and of the IR microstructure parameter (Fig. 10). Note that the (anomalous) negative values of α reported by Shinar *et al.* (1992) could be due to a collapse of voids in void-rich material with increasing annealing time, resulting in the reconstruction of compact material with an increased hydrogen diffusion length. The increase in α after annealing reported by Tang *et al.* (1990b, 1991) also may be due to changes in the film microstructure. More work, however, appears necessary to clarify this interesting subject.

5. Deviations from Error-Function Diffusion Profiles

From the solution of the diffusion equation, hydrogen and deuterium profiles following a complementary error function are expected for diffusion in H/D layer samples (Carlson and Magee, 1978). Such profiles are also expected according to Kakalios and Jackson (1988) for a time-dependent diffusion coefficient. Yet various types of deviations of diffusion profiles from the expected dependence have been reported.

a. Plateau-Like Depth Profiles

For *a*-Si:H/*a*-Si:D samples with a columnar microstructure, plateau-like deuterium diffusion profiles at low concentrations and stable H/D interfaces at high concentrations were observed by Street and Tsai (1988). Similar diffusion profiles were reported for *a*-Si with a columnar microstructure grown in a molecular beam epitaxy deposition system and for rf *a*-Si:H deposited at $T_s = 100°C$ (Petrova-Koch et al., 1987), for a-Si alloys, in particular for a-Si:C:H (Beyer, 1987; Shinar et al., 1994), as well as for certain microcrystalline Si samples (Beyer et al., 1997a). As shown in Fig. 33, one finds a similar diffusion behavior for the ECR a-Si:H samples (see Sec. II). The profile of implanted deuterium does not spread at high concentrations but shows a plateau of deuterium at low concentration that rises with annealing temperature (and time), while the total concentration of hydrogen and deuterium in the sample decreases. The nature of the fast deuterium diffusion process related to the plateau is not completely clear. Beyer (1987) attributed it to rapid surface diffusion. Street and Tsai (1988) gave two possible explanations, either a lower energy at the surface than in the bulk to excite hydrogen into mobile states or the diffusion of hydrogen molecules. For material of very high hydrogen concentration, Street (1991b) suggested the possibility of long-range hopping diffusion of hydrogen with a low activation energy. Recently, for deuterium-implanted μc-Si:H samples, a close agreement was found between the amount of molecular deuterium released in low-temperature effusion and the deuterium plateau level, suggesting that the fast diffusion process is indeed related to the motion of molecular hydrogen (deuterium) (Beyer et al., 1997a). Apparently, the

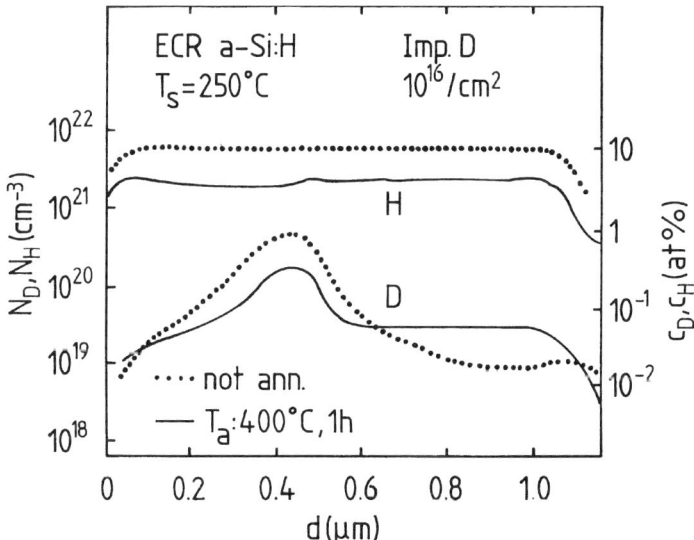

FIG. 33. SIMS depth profiles of hydrogen and deuterium prior to (dotted line) and after annealing at 400°C (solid line) in ECR plasma-grown a-Si:H with a columnar microstructure implanted with deuterium (dose 10^{16} cm^{-2}; energy: 30 keV).

deuterium plateau arises as molecular deuterium moves into the interconnected void network. The anneal-stable interfaces of H/D layer structures of void-rich material were explained by Shinar et al. (1989) in terms of a greatly reduced hydrogen diffusion coefficient. We consider it more likely, however, that anneal-stable interfaces indicate the presence of voids inhibiting long-range hydrogen diffusion, since effusion measurements of void-rich material generally show a decreased hydrogen stability compared with compact material (Beyer et al., 1991).

b. *Exponential Diffusion Profiles*

Branz et al. (1993a, 1993b, 1993c) reported exponential deuterium profiles for short-term tracer diffusion of deuterium in a-Si:H/a-Si:H:D/a-Si:H sandwich structures. Nearly exponential profiles also were observed by Beyer (1987) for in-diffusion of deuterium from a-Si:D source layers into hydrogenated amorphous silicon nitride films and tentatively attributed to diffusion enhanced by voids. In Fig. 34a, the results of SIMS depth profiling of an (rf1) a-Si:D/a-Si:H sample deposited at a substrate temperature of 150°C and annealed at 250°C for various time periods are plotted. This sample shows a behavior similar to that reported by Branz and coworkers. After annealing at $T_a = 250$°C for time periods of up to 20 hours, an exponential deuterium tail (wing) appears with an onset concentration (amplitude) rising with increasing annealing time. For extended annealing time (35 days), however, the deuterium profile changes to an error-function dependence. Similar results were obtained for higher annealing temperatures; the transition from exponential tail profiles to error-function profiles occurred then at shorter annealing times. Branz et al. (1993a) considered the appearance of exponential wings as compelling evidence for a trapping process. By solving the equations of trap-controlled tracer diffusion, Kemp and Branz (1993) described the exponential wing in terms of the detrapping time τ and the mean atomic displacement λ between deep trapping events. They suggest as a consistency check that the long-time diffusion coefficient D should equal the short-time diffusion coefficient $D = \lambda^2/\tau$. Following their analysis, we obtain for the exponential wings in Fig. 34a (short-time diffusion, annealing time $t < \tau$) $\tau \approx 20$ hours and $\lambda \approx 0.28\,\mu\text{m}$. Thus, with respect to τ, our results agree closely with those of Branz et al. (1993b, 1993c) ($\tau = 15 \pm 5$ hours and $\lambda = 100 \pm 30$ Å), but with respect to λ, our value is about a factor of 30 larger. Our short-time diffusion coefficient $D \approx 10^{-14}\,\text{cm}^2\,\text{sec}^{-1}$ is not consistent with the long-time diffusion coefficient ($D \approx 8 \times 10^{-17}\,\text{cm}^{-2}\,\text{sec}^{-1}$ for 35 days' annealing). Furthermore, we note that the short-time diffusion behavior in Fig. 34a differs completely from that of our standard samples.

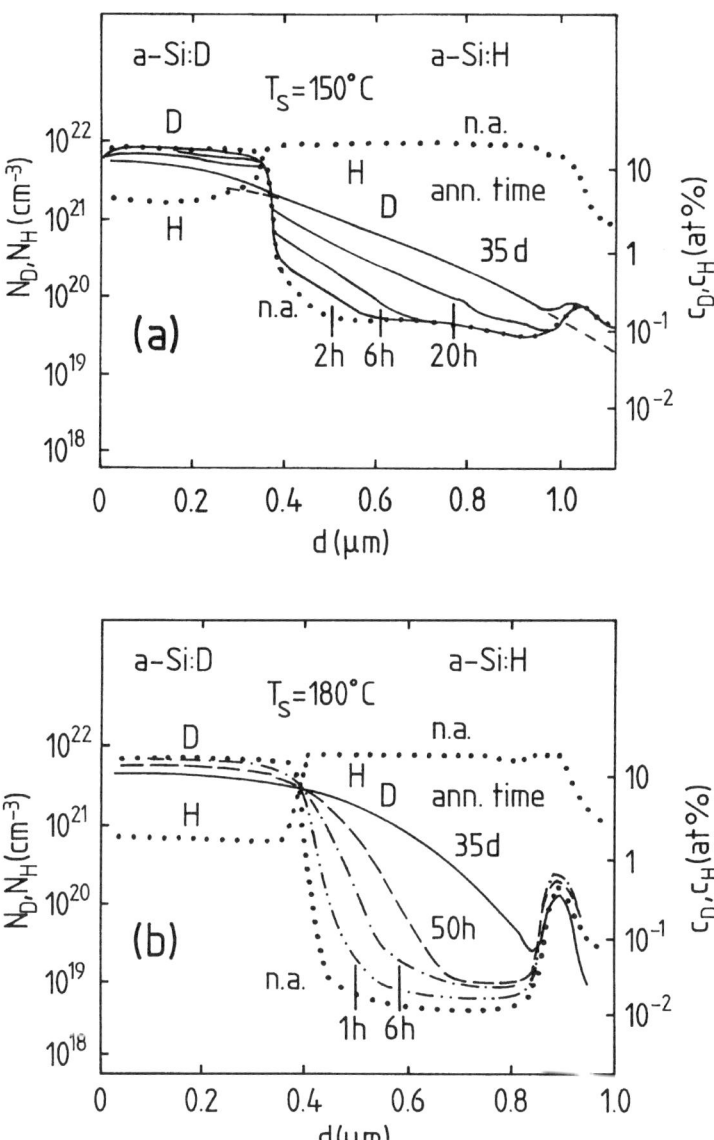

FIG. 34. SIMS depth profiles of deuterium and hydrogen in type A layer structures of (rf1) a-Si:H and a-Si:D films prior to (n.a.) and after annealing at 250°C for various annealing time periods for (a) $T_s = 150°C$ and (b) $T_s = 180°C$ samples.

Figure 34b shows typical SIMS depth profiles for such a sample, where all deuterium diffusion profiles, even for shortest annealing time, do not exhibit exponential wings. We note that although the two samples discussed in Fig. 34 were deposited at substrate temperatures only 30°C apart, both the hydrogen effusion and IR absorption microstructure parameters were quite different. For the $T_s = 150°C$ sample (Fig. 34a), $N_{LT}/N_{tot} = 0.11$ and $R^{IR} = 0.57$, whereas for $T_s = 180°C$ (Fig. 34b), $N_{LT}/N_{tot} = 0.06$ and $R^{IR} = 0.36$. Based on structural information from hydrogen effusion, we tentatively attribute exponential wings to a fast void-related diffusion, presumably by molecular deuterium emitted from the source layer and trapped in the diffusion layer. At long annealing times (or at increased annealing temperatures), the material is likely to reconstruct (see Secs. II and IV.8), resulting in a collapse of (interconnected) voids and in a change of hydrogen motion to diffusion of atomic hydrogen in a compact material. Thus we do not consider the results by Branz *et al.* (1993a, 1993b, 1993c) as evidence for H diffusion in *a*-Si:H to proceed generally by a trap-controlled mechanism.

c. Kinks in Error Function Profiles

Another type of deviation from error function diffusion profiles was first noted by Jackson *et al.* (1990, 1991a), who reported kinks for deuterium plasma in-diffusion profiles into material of low hydrogen concentration in agreement with a lower diffusion coefficient at low concentration than at high concentration. It was proposed that these kinks indicate the level of empty trapping sites within the amorphous material. For in-diffusion of deuterium below this level, the diffusing deuterium would see the empty traps and would diffuse differently from the situation in which the traps are filled by deuterium. In the early publications (Jackson *et al.*, 1990, 1991a), the level was associated with the concentration of dangling bonds, since, for the samples investigated, the ESR spin density agreed with the kink level within a factor of 3 to 4. In subsequent work it was recognized, however, that the kink level followed the hydrogen concentration rather than the dangling-bond concentration (Jackson *et al.*, 1991b, 1993; Jackson and Tsai, 1992). For explanation, hydrogen clusters were proposed as a possible trapping site for diffusing hydrogen. We note that Jackson's model would predict a different diffusion coefficient for *a*-Si:H with empty traps and with filled traps. As discussed in Section III.1 (see Beyer and Zastrow, 1996) and in Section IV.1, however, no significant influence of the hydrogen concentration (and thus of the trap density) on hydrogen diffusion is observed if the source concentration (using a deuterium plasma or an *a*-Si:D layer) is kept constant. An explanation of the kinks in the hydrogen diffusion profiles is possible also by the energy-band model of hydrogen diffusion. Here, kinks will arise if the chemical potential is not constant throughout the sample but drops at a certain depth from a higher value to a lower one for samples that are not in complete equilibrium. Origin can be in-diffusion of plasma-induced hydrogen at high concentration into material of low hydrogen concentration (and low chemical potential), shifting up the hydrogen chemical potential near the depth where the in-diffusing hydrogen concentration exceeds the original one. Thus the front of in-diffusing hydrogen (at low concentration) will diffuse with a lower diffusion coefficient than the majority of hydrogen atoms, with the result of a kink in the diffusion profile. Note that the energy-band model implies a kink position not determined by a concentration level of traps but rather by the depth where the in-diffusing hydrogen (deuterium) concentration exceeds the original concentration. This is in agreement with the plasma in-diffusion data by Jackson et al. (1991b). Also in agreement with the energy-band model is our observation of kinks in diffusion profiles for both layer diffusion and plasma in-diffusion samples if the source concentration exceeds the concentration in the probed layer (see Figs. 12b and 21). In the layer diffusion samples we associate the kink also with the depth where the chemical potential changes from its original low-level position in the probed layer to the increased position of the source layer. However, since small amounts of in-diffusing hydrogen can cause significant shifts in the chemical potential in these latter samples, SIMS measurements of very high accuracy would be needed to verify the correlation between kink position and concentration of in-diffusing hydrogen (deuterium).

Finally, we note that deviations from error function profiles must be expected for deuterium diffusion in a-Si:H layers if hydrogen concentration and solubility vary with depth (see Figs. 19b and 25).

6. PLASMA IN-DIFFUSION VERSUS LAYER DIFFUSION

A significantly different diffusion behavior of hydrogen for plasma in-diffusion and layer diffusion was first noted by Abeles et al. (1987a, 1987b), who studied deuterium plasma treatment of a-Si:H/a-Si:D multilayers. These authors estimated from their data a plasma in-diffusion activation energy of $E_D < 1$ eV and a plasma in-diffusion coefficient at $T = 160°C$ of 10^{-14} cm² sec⁻¹ and noted that this latter value exceeds the layer diffusion coefficient at the same temperature by many orders of magnitude. To explain this high diffusion coefficient for hydrogen plasma in-diffusion, Abeles et al. suggested interstitial diffusion of plasma-induced hydrogen (deuterium) interacting with bonded hydrogen by exchange. An enhanced diffusion coefficient for plasma in-diffusion also was reported by other authors (Nakamura et al., 1989; Santos and Jackson, 1992a, 1992b) and can be inferred from plasma in-diffusion data by Widmer et al. (1983). Both Nakamura et al. and Santos and Jackson suggested that hydrogen (deuterium) atoms injected from the plasma saturate deep traps (on the order of 10^{22} cm^{-3}) so that the hydrogen lifetime in mobile interstitial (bond-centered) sites is enhanced. The observed diffusion activation energy of about 0.5 eV was associated with the barrier height for interstitial hydrogen diffusion known from crystalline silicon data. However, compared with diffusion of interstitial hydrogen in crystalline Si, a diffusion prefactor lower by about 8 orders of magnitude is observed for plasma in-diffusion in a-Si:H (Santos and Jackson, 1992b). As a consequence, at typical plasma in-diffusion temperatures ($T = 250$–$400°C$), the diffusion coefficient for plasma in-diffusion is considerably lower than the data for diffusion of interstitial hydrogen in undoped c-Si. At $T = 250°C$, the extrapolated values by Van Wieringen and Warmoltz (1956) yield $D \approx 2 \times 10^{-7}$ cm^2 sec^{-1}, whereas Santos and Jackson (1992b) obtained less than $D = 10^{-15}$ cm^2 sec^{-1}. In view of this discrepancy in absolute values of the diffusion coefficient, we believe it more likely that the low diffusion energy for hydrogen plasma in-diffusion in a-Si:H has nothing to do with diffusion of interstitial hydrogen and is rather related to a strong temperature shift of hydrogen-related levels. Indeed, our plasma in-diffusion data as well as those of Santos and Jackson (1992b) fit well the Meyer-Neldel rule valid for layer diffusion (and diffusion with implanted deuterium source layers; see Fig. 15). In the equilibrium energy-band model of hydrogen diffusion, we explain the rather high plasma in-diffusion coefficient at low temperatures by a relatively high surface hydrogen chemical potential related to the high concentration of surface-adsorbed hydrogen (see Fig. 30b). While the diffusion process for plasma in-diffusion and layer diffusion seems to be basically the same, we note that plasma in-diffusion differs from layer diffusion significantly in another aspect; as discussed in Section V.1, the hydrogen concentration (and solubility) can be enhanced in the first case but not in the second.

7. Relation Between SIMS Diffusion Data and Hydrogen Effusion Data

As discussed in Section II, the low-temperature hydrogen effusion process observed for material with a significant microstructure is not rate limited by diffusion, and therefore, no direct correlation between effusion and diffusion data for this type of material can be expected. For samples with a pronounced LT effusion peak, rather anneal-stable H/D interfaces at high concentration and plateau-like or exponential diffusion profiles at low concentration are detected by SIMS profiling, as discussed in Section IV.5. In compact material, on the other hand, effusion is limited by hydrogen diffusion, and therefore, the H diffusion coefficient can be determined by both effusion and SIMS H/D depth profiling experiments. By solving the diffusion equation for out-diffusion from a film of thickness d, the temperature of maximum effusion T_M was found to be related to diffusion data by

$$(D/E_D) = (d/\pi T_M)^2(\delta/k) \tag{32}$$

where δ is the heating rate (Beyer and Wagner, 1982). In Fig. 35 we compare for the a-Si:D films of Fig. 1 the experimental data (exp.) for T_M with the calculated results (calc.) using Eq. (32) and the diffusion data of Fig. 14a and

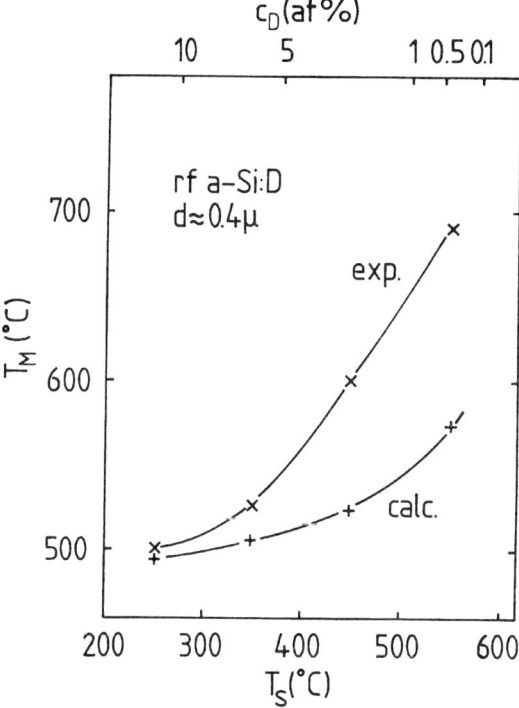

FIG. 35. Deuterium effusion peak temperature T_M versus substrate temperature T_s (and deuterium concentration c_D) for a-Si:D films of thickness $d \approx 0.4\,\mu$m; exp.: experimental data from Fig. 1; calc.: calculated from SIMS diffusion data (lines are guides to the eye).

assuming that these diffusion parameters determined in a temperature range between 250 and 500°C are still valid at T_M, i.e., up to 600–700°C. The results show for $T_s = 250$–350°C material a good agreement between experimental and calculated data, as reported previously (Beyer and Wagner, 1982; Chou et al., 1987; Beyer et al., 1988). At higher substrate temperatures, however, the experimental data for T_M exceed the calculated ones considerably. An explanation for this difference has been proposed by Beyer and Wagner (1994). It is based on the idea (Fig. 36) that diffusion in the energy-band model is determined by the chemical potential μ_H^{diff} situated at the upper edge of the Si-H density of states distribution $g_H(E)$ of a given material, whereas the (HT) effusion peak temperature T_M is determined by μ_H^{eff} valid when about half the original hydrogen concentration has been effused. An asymmetric Si-H density of states distribution for material of high hydrogen concentration with a maximum of $g_H(E)$ close to the upper edge of the density of states distribution can then explain $\mu_H^{diff} \approx \mu_H^{eff}$. For material of low hydrogen concentration, a more symmetric Si-H density of states distribution would cause a significant decrease of μ_H on hydrogen effusion. The experimental data of Fig. 35 suggest differences between μ_H^{eff} and μ_H^{diff} of up to 0.8 eV. Since this difference appears unrealistically high,

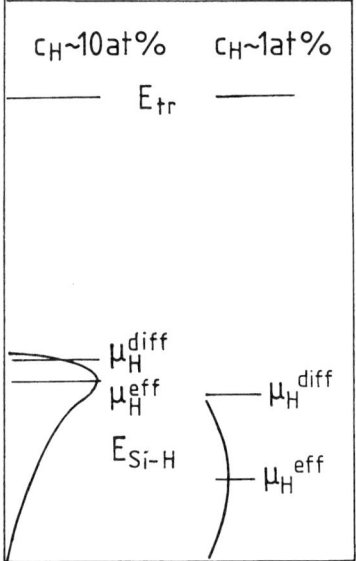

FIG. 36. Schematic energy-band diagram for hydrogen diffusion and effusion in a-Si:H of different hydrogen concentrations. (After Beyer and Wagner, 1994.)

other factors seem to play a role too. Conceivable is a contribution of molecular hydrogen trapped in voids to high-temperature effusion. Data for microcrystalline Si samples suggest that such hydrogen effuses at $T = 600$–$800°C$ (Beyer et al., 1997a). Data for deformed silicon crystals point to a broad effusion peak near 900°C of molecular hydrogen trapped in voids (Kisielowski-Kemmerich and Beyer, 1989). Thus an increasing concentration ratio of molecular hydrogen to silicon-bonded hydrogen could cause the increasing difference between T_M (exp.) and T_M (calc.) as the substrate temperature is increased.

8. INTERRELATION BETWEEN IR ABSORPTION SPECTRA AND EFFUSION TRANSIENTS

The concentration between hydrogen effusion transients and hydrogen-related infrared absorption was discussed extensively in the seventies and early eighties. The topic has been addressed again recently (Lusson et al., 1997). When the first hydrogen effusion and IR absorption measurements on variously deposited a-Si:H films were performed, the correlation between the appearance of an LT effusion peak (at 350–400°C) and the IR stretching absorption at $2100\,cm^{-1}$ on the one hand and between material with a single (HT) effusion peak and the IR stretching absorption at $2000\,cm^{-1}$ (see Figs. 1, 7, and 11) was soon recognized (Brodsky et al., 1977b; Fritzsche, 1977). Tentatively, the LT effusion was attributed by Fritzsche (1977) to the desorption of hydrogen from SiH_2 configurations considered to give rise to the $2100\,cm^{-1}$ stretching absorption, and the HT effusion was assigned to hydrogen desorption from SiH groups associated with the $2000\,cm^{-1}$ stretching absorption. Treating both HT and LT effusion peaks as surface desorption processes, several authors determined the kinetic parameters (Brodsky et al., 1977b; McMillan and Peterson, 1979a, 1979b). However, annealing experiments showing an increase of the $2100\,cm^{-1}$ absorption relative to $2000\,cm^{-1}$ when samples were heated to 370°C (Fritzsche et al., 1979) demonstrated that the direct association of the LT and HT effusion with dihydride and monohydride desorption, respectively, was an oversimplification. The thickness dependence of the HT effusion peak, recognized first by Biegelsen et al. (1979, 1980) and studied extensively by Beyer and coworkers (Beyer et al., 1981, 1982; Beyer and Wagner, 1981, 1982, 1983) demonstrated that in the HT effusion process hydrogen effusion is limited by diffusion in compact material. Analysis of H_2, HD, and D_2 effusion from layered structures of hydrogenated and deuterated material by Beyer and coworkers (Beyer et al., 1982; Beyer and Wagner, 1983) showed that the low-temperature effusion peak relates to out-diffusion of H_2 through an interconnected void structure, whereas the HT peak is due to diffusion of atomic hydrogen. For material showing both LT and HT peaks, a structural transformation from a void-rich material to a compact material was concluded (Beyer and Wagner, 1983). According to these latter results, no direct relation between the LT effusion and the $2100\,\text{cm}^{-1}$ absorption or between the HT effusion and the $2000\,\text{cm}^{-1}$ absorption can exist. The same conclusion was reached recently by Lusson et al. (1997). While hydrogen effusion measurements suggest quite generally (with the exception of our ECR samples) a densification of a-Si:H with rising annealing temperature, IR absorption data for hydrogen-effused a-Si:H samples are conflicting. Both an increase (Fritzsche et al., 1979; Ueda et al., 1985; Gleason et al., 1987) and a decrease (Biegelsen et al., 1979; Gleason et al., 1987; Lusson et al., 1997) in the $2100\,\text{cm}^{-1}/2000\,\text{cm}^{-1}$ absorption ratio was observed when hydrogen was effused near $300°C$ (i.e., in the LT effusion peak). In agreement with these findings are our results in Fig. 11 (see Sec. II.3.b), demonstrating that Si:H material that is compact according to hydrogen effusion ($N_{LT}/N_H < 0.2$) may have R^{IR} values between 0 and almost 0.7. Origin of this wide variation in R^{IR} values is presumably a varying concentration of isolated voids. While interconnected voids apparently show up in both IR absorption and effusion experiments, isolated voids barely seem to influence hydrogen effusion apart from a shift of the HT effusion peak to higher temperatures (see Sec. IV.7). Presumably, hydrogen trapped as H_2 in isolated voids returns to the bulk network via dissociative adsorption if H_2 pressure and temperature are high (Beyer et al., 1997a). The reported changes in IR absorption and hydrogen effusion on annealing of rf a-Si:H material can then be explained by a general disappearance of interconnected voids while the concentration of isolated voids is increased or decreased depending on the film structure.

V. Hydrogen Solubility Effects

With respect to the hydrogen incorporation in silicon, various solubility phenomena have been discussed in the literature. For (near-intrinsic) single-crystalline Si, a hydrogen solubility S decreasing strongly with falling temperature with an activation energy of $1.86\,\text{eV}$ has been reported by Van Wieringen and Warmoltz (1956). Their data refer to silicon in contact with molecular hydrogen at constant (1 atm) gas pressure and were taken at high temperatures. Extrapolation to low temperatures suggests very low values ($S < 2 \times 10^9\,\text{cm}^{-3}$) at temperatures $T < 500°C$. This (intrinsic) hydrogen solubility in silicon has been associated with the incorporation of hydrogen in interstitial sites (Van Wieringen and Warmoltz, 1956; Herring and Johnson, 1991). The H/D interdiffusion experiments in a-Si:H and hydrogen-implanted c-Si discussed in Section III.2, on the other hand, suggest the presence of a much higher solubility that essentially equals the hydrogen concentration and which seems to be rather independent of temperature if hydrogen effusion processes at elevated temperatures are omitted. This (extrinsic) solubility is associated with hydrogen incorporation in Si—H bonds. Both intrinsic and extrinsic solubility effects in compact Si:H will be discussed in Section V.1. The term *hydrogen solubility* also has been used in the literature in connection with the hydrogen-related void formation. This hydrogen solubility characterizes a hydrogen concentration threshold; up to this threshold, hydrogen is incorporated in compact material, whereas higher concentrations require a microstructure. This latter solubility will be discussed in Section V.2.

1. SOLUBILITY IN COMPACT MATERIAL

The energy-band model of hydrogen diffusion allows an estimate of the hydrogen concentration in the transport states. Since only hydrogen atoms in these states are expected to contribute to diffusion, their concentration n_{tr} can be calculated from the diffusion coefficient D and the hydrogen density N_H according to Eq. (9), provided the diffusion prefactor D_{H0} is known. Figure 37 shows n_{tr} versus $1/T$ for the type A a-Si:H samples of Fig. 13, calculated by assuming $D_{H0} = D_{00} \approx 10^{-3} \text{cm}^2 \text{sec}^{-1}$ (Eq. 6). Also included in Fig. 37 are (extrapolated) intrinsic solubility data for (near-intrinsic) single-crystalline Si (Van Wieringen and Warmoltz, 1956). Depending on the hydrogen density N_H, n_{tr} is found to attain a factor of 10–10,000 higher values than the intrinsic solubility in crystalline Si but shows a similar temperature dependence. A tempting conclusion from this result is that the temperature dependences of the intrinsic solubility in c-Si and of n_{tr} have both the same origin, namely, the equilibrium between mobile hydrogen in the transport states and hydrogen bound to silicon, governed by a hydrogen chemical potential. In case of c-Si, bound hydrogen is presumably located at the c-Si surface, arising from the reaction between hydrogen molecules and silicon (at high temperatures). Thus the intrinsic hydrogen solubility in crystalline Si would not be a property of the silicon network. This concept also would explain why in amorphous silicon n_{tr} can attain values considerably higher than reported for the intrinsic hydrogen solubility in crystalline Si.

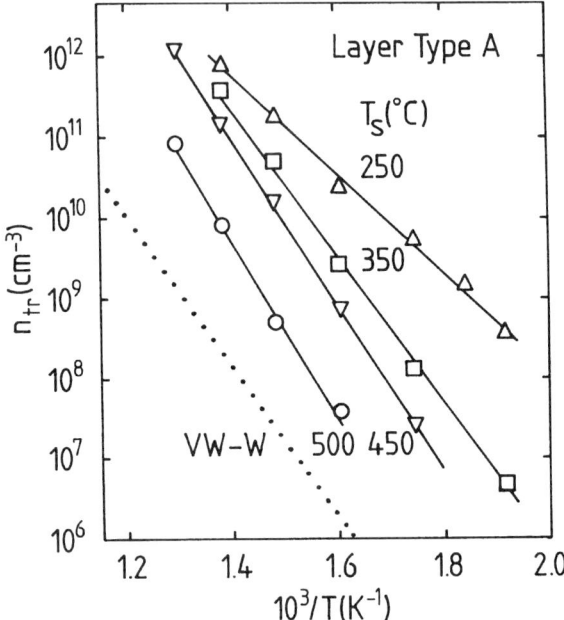

FIG. 37. Hydrogen concentration n_{tr} at the transport states versus $1/T$ for a-Si:H samples of Fig. 13a. Dotted line shows extrapolated literature data by Van Wieringen and Warmoltz (1956) of hydrogen solubility in (near-intrinsic) crystalline Si at 1 atm H_2 pressure.

The extrinsic hydrogen solubility in Si:H equals the sum of all hydrogen-bonding states that are filled for a given chemical potential, that is,

$$S = \int_{-\infty}^{\mu_H} g_H(E)\, dE \quad (33)$$

Accordingly, this extrinsic hydrogen solubility of a particular material will change in a hydrogen diffusion experiment only if the hydrogen density of states or the chemical potential (or both) change. Our results suggest that for layer H-D interdiffusion $g_H(E)$ remains essentially unchanged. We attribute the slight increase in deuterium penetration into a hydrogenated diffusion layer with rising source concentration (Fig. 17) not to an increase of g_H but rather to the increase in the hydrogen chemical potential. This assignment is supported by the simultaneous variation in the diffusion coefficient. Hydrogen diffusion within a material of fixed hydrogen solubility implies that the majority of Si—Si bonds (i.e., the strong Si—Si bonds) are

5 HYDROGEN PHENOMENA IN HYDROGENATED AMORPHOUS SILICON   233

not broken by in-diffusing hydrogen atoms. Diffusing hydrogen atoms confine, accordingly, to Si dangling bonds or weak Si—Si bonds that happen to be unhydrogenated during the diffusion jump. A rupture of strong Si—Si bonds at one place and the formation of a strong Si—Si bond at another, however, are not in disagreement with the experimental solubility data. If hydrogen and deuterium atoms are present, both compete for the same trapping sites. This explains the H-D exchange features of H-D interdiffusion profiles in Section III (Figs. 12b, 21, 24, and 25). The reason for the fixed solubility is presumably the fixed space for hydrogen to fit in. Since the Si—Si bond distance is 2.35 Å, two Si—H bonds with a bond length of about 1.5 Å cannot be inserted without significant dilatation of the Si-Si distance. The distance of two Si atoms with bonded hydrogen in between (three-center bond) is considered to exceed 3 Å (Van de Walle, 1991). With increasing substrate temperature and decreasing hydrogen concentration, hydrogen solubility in a-Si:H equals, according to Fig. 20, the hydrogen concentration down to $c_H \approx 1$ at.% and becomes rather constant at lower hydrogen concentration. This latter characteristic solubility level has been associated with the presence of growth-related weak bonds or dangling bonds defining a film density and thus the space for hydrogen to fit in (Beyer and Wagner, 1994; Beyer, 1997a). By hydrogen ion implantation, the characteristic solubility level is increased to about 5 at.%, presumably due to implantation defects (Beyer, 1996, 1997a). A material of a particularly low extrinsic hydrogen solubility is undoped single-crystalline silicon, where little spreading of implanted deuterium is observed down to concentration levels below 10^{18} cm^{-3} (Wilson, 1986).

According to the results in Figs. 21 to 23, the hydrogen solubility in a-Si:H is changed on out-diffusion of hydrogen and on hydrogen incorporation by hydrogen implantation or by plasma posthydrogenation. In the case of hydrogen out-diffusion, apparently the silicon network in rf a-Si:H densifies, and a material with a lower hydrogen concentration, solubility, and chemical potential results. The reverse occurs on hydrogen implantation and by plasma posthydrogenation. The (hydrogen) ion implantation results of Fig. 24 demonstrate that for crystalline Si, too, the (extrinsic) hydrogen solubility can be enhanced to values comparable with a-Si:H. The structure and diffusion of hydrogen within implantation-hydrogenated crystalline Si have been discussed recently (Beyer and Zastrow, 1998a, 1998b).

In view of the (almost) unchanged hydrogen solubility in the layer diffusion experiment, the increase in hydrogen solubility and hydrogen density of states by in-diffusion from a hydrogen plasma is surprising. Two factors seem to play a major role: (1) a high surface hydrogen chemical potential (see Fig. 30b) and (2) an easier expansion of the material next to the surface compared with bulk material when additional hydrogen is incorporated. More work on this interesting subject, however, is clearly desirable.

2. HYDROGEN-RELATED VOID FORMATION

The active role of hydrogen in void formation has been concluded from NMR, IR absorption, and hydrogen effusion measurements in the early eighties (Reimer, 1980; Cardona, 1983; Beyer and Wagner, 1983). Shanks et al. (1981) found that the 2000 cm^{-1} absorption band saturates with increasing hydrogen density at $N_H \approx 2 \times 10^{21}$ cm^{-3} (4 at.% hydrogen) and that the distributed phase in NMR saturates near 3×10^{21} cm^{-3} (6 at.% hydrogen). The latter value was associated by Cardona (1983) with a solubility limit of hydrogen in the bulk material, with hydrogen concentrations exceeding this threshold causing voids. Beyer and Wagner (1983) obtained from hydrogen effusion measurements a solubility limit of about 20 at.%. More recently, hydrogen-related void formation for $c_H > 3-4$ at.% was concluded from small-angle x-ray scattering (SAXS) measurements by Acco et al. (1996). The data in Figs. 6 and 10 suggest thresholds for standard rf-deposited samples of about 10 at.% from IR measurements and about 15–20 at.% from hydrogen effusion. The difference is presumably due to a different void sensitivity of the two experiments (Finger et al., 1992). Similar thresholds are obtained for hydrogen implantation into rf material (see Figs. 4a, b, and 10). For our ECR a-Si:H samples, the microstructure parameters depend on hydrogen concentration too but suggest a much lower threshold value for hydrogen-related void formation than observed for rf a-Si:H. However, this concentration dependence also could be due to a varying degree of coverage of the columnar microstructure by hydrogen. Much lower threshold hydrogen concentrations than in rf plasma a-Si:H have been reported for implantation-hydrogenated crystalline silicon (Beyer and Zastrow, 1998a, 1998b).

Since the statistical incorporation of hydrogen in SiH$_2$ bonding configurations remains rather low in the investigated concentration range (see Fig. 10), the hydrogen concentration seems not to be the only factor causing hydrogen clustering. The difference between hydrogen-related void formation in implantation-hydrogenated crystalline and amorphous silicon (Beyer and Zastrow, 1998a, 1998b) suggests that network strain may be involved; in an inflexible network, the hydrogenation of a particular Si—Si bond will likely cause the weakening or rupture of neighboring Si—Si bonds as well. Other mechanisms also may contribute to void formation. Acco et al. (1996) suggested that a defect-related trap concentration defines a solubility limit for hydrogen ($c_H \approx 3-4$ at.%) and that on annealing excess hydrogen atoms recombine to H_2 molecules that accumulate in nanoscale hydrogen complexes. Conceivable also is that a structural relaxation resulting in a drop in the hydrogen chemical potential and in the formation of a microstructure is triggered by a too high hydrogen chemical potential. Whatever reasons voids have, it seems that the method of hydrogen dilution can be effective in reducing the SiH_2 clustering and thus void formation, in particular for low substrate temperatures ($T_s \approx 50°C$). According to the results of Fig. 10, samples prepared from hydrogen-diluted silane showed the lowest degree of SiH_2 clustering of all investigated samples, with R^{IR} close to the statistical SiH_2 incorporation. The data of Fig. 3 demonstrate that material with a small N_{LT}/N_H ratio can be grown by the method of hydrogen dilution even at room temperature.

VI. Conclusions

Our results demonstrate the presence of a multitude of hydrogen phenomena in a-Si:H, like the influence of hydrogen on the frequency of hydrogen-related IR absorption spectra, the presence of a hydrogen-related or hydrogen-modified microstructure, and the effect of hydrogen concentration on the hydrogen diffusion coefficient and on hydrogen solubility. Our results confirm previous concepts concerning the structural information in IR absorption spectra and the role of voids for hydrogen stability. Concerning hydrogen diffusion in compact material, our data point against emission of hydrogen from distinct trapping levels as the rate-limiting step and favor a quasi-equilibrium process with both Si—H bond rupture and Si—Si bond reconstruction involved. Hydrogen diffusion phenomena can be well described in terms of an energy-band model with a hydrogen chemical potential located near the upper edge of a hydrogen density of states distribution and determining the concentration of mobile hydrogen atoms in hydrogen transport states. The time dependence of hydrogen diffusion is attributed to microstructural effects rather than to the motion of atomic hydrogen in a distribution of traps. Our data stress the importance of hydrogen solubility for hydrogen diffusion. Many aspects of hydrogen phenomena in a-Si:H, however, require further experimental and theoretical work. This refers in particular to the physics of the Meyer-Neldel rule of hydrogen diffusion, the nature of temperature shifts in hydrogen levels, details of the hydrogen density of states distribution, the interaction of hydrogen bonding and hydrogen motion with the electronic system, the influence of heterogeneity, and mechanisms of microstructure formation.

Acknowledgments

I profited greatly from the support by U. Zastrow in performing and evaluating SIMS measurements and by M. Gebauer who provided the ion implantations. Technical assistance by R. von de Berg, D. Lennartz, F. Pennartz, and R. Schmitz and helpful discussions with R. Carius, F. Finger, J. Herion, W. B. Jackson, A. H. Mahan, and H. Wagner are gratefully acknowledged. This work was supported by the Bundesministerium für Bildung, Wissenschaft, Forschung und Technologie (BMBF).

References

Abeles, B., Yang, L., Leta, D. P., and Majkrzak, C. (1987a). *M.R.S. Symp. Proc.*, **77**, 623.
Abeles, B., Yang, L., Leta, D., and Majkrzak, C. (1987b). *J. Non-Cryst. Solids*, **97–98**, 353.
Abo Ghazala, M., Beyer, W., and Wagner, H. (1991). *J. Appl. Phys.*, **70**, 4540.
Acco, S., Williamson, D. L., Stolk, P. A., Saris, F. W., Van den Boogard, M. J., Sinke, W. C., Van der Weg, W. F., Roords, S., and Zalm, P. C. (1996). *Phys. Rev. B*, **53**, 4415.
Amato, G., Della Mea, G., Fizzotti, F., Manfredotti, C., Marchisio, R., and Paccagnella, A. (1991). *Phys. Rev. B*, **43**, 6627.
Basrour, S., Bruyere, J. C., Bustarret, E., Gotet, C., Stiquert, J. P., and Sardin, G. (1989). In *Proceedings of the 9th European Photovoltaic Solar Energy Conference*. Dordrecht, Netherlands: Kluwer Academic, 1010.
Beyer, W., and Stuke, J. (1974). In *Amorphous and Liquid Semiconductors*, J. Stuke and W. Brenig, eds. London: Taylor and Francis, p. 245.
Beyer, W. (1985). In *Tetrahedrally Bonded Amorphous Semiconductors*, D. Adler and H. Fritzsche, eds. New York: Plenum Press, p. 129.
Beyer, W. (1987). *J. Non-Cryst. Solids*, **97–98**, 1027.
Beyer, W. (1991). *Physica B*, **170**, 105.
Beyer, W. (1996). *J. Non-Cryst. Solids*, **198–200**, 40.
Beyer, W. (1997a). *Phys. Stat. Solidi (a)*, **159**, 1997.
Beyer, W. (1997b). Unpublished work.
Beyer, W., and Wagner, H. (1981). *J. Phys. (France) Colloque*, **42**, C4–773.
Beyer, W., Wagner, H., and Mell, H. (1981). *Solid State Commun.*, **39**, 375.
Beyer, W., and Wagner, H. (1982). *J. Appl. Phys.*, **53**, 8745.
Beyer, W., and Wagner, H. (1983). *J. Non-Cryst. Solids*, **59–60**, 161.
Beyer, W., and Overhof, H. (1984). *Semiconductors and Semimetals*, **21C**, J. I. Pankove, ed. San Diego: Academic Press, p. 257.
Beyer, W., and Zastrow, U. (1993). *M.R.S. Symp. Proc.*, **297**, 285.
Beyer, W., and Wagner, H. (1994). *M.R.S. Symp. Proc.*, **336**, 323.
Beyer, W., and Zastrow, U. (1996). *M.R.S. Symp. Proc.*, **420**, 497.
Beyer, W., and Abo Ghazala, M. (1998). *M.R.S. Symp. Proc.*, **507**, 601.
Beyer, W., and Zastrow, U. (1998a). *J. Non-Cryst. Solids*, **227–230**, 880.
Beyer, W., and Zastrow, U. (1998b). *M.R.S. Symp. Proc.*, **507**, 679.
Beyer, W., Wagner, H., Chevallier, J., and Reichelt, K. (1982). *Thin Solid Films*, **90**, 145.
Beyer, W., Herion, J., Mell, H., and Wagner, H. (1988). *M.R.S. Symp. Proc.*, **118**, 291.
Beyer, W., Herion, J., and Wagner, H. (1989). *J. Non-Cryst. Solids*, **114**, 217.

Beyer, W., Herion, J., Wagner, H., and Zastrow, U. (1991). *Philos. Mag. B*, **63**, 269.
Beyer, W., Hapke, P., and Zastrow, U. (1997a). *M.R.S. Symp. Proc.*, **467**, 343.
Beyer, W., Hapke, P., and Zastrow, U. (1997b). Unpublished work.
Biegelsen, D. K., Street, R. A., Tsai, C. C., and Knights, J. C. (1979). *Phys. Rev. B*, **20**, 4839.
Biegelsen, D. K., Street, R. A., Tsai, C. C., and Knights, J. C. (1980). *J. Non-Cryst. Solids*, **35–36**, 285.
Boisvert, G., Lewis, L. J., and Yelon, A. (1995). *Phys. Rev. Lett.*, **75**, 469.
Boyce, J. B., and Ready, S. E. (1991). *Physica B*, **170**, 305.
Branz, H. M., Asher, S., and Nelson, B. P. (1993a). *Phys. Rev. B*, **47**, 7061.
Branz, H. M., Asher, S., Nelson, B. P., and Kemp, M. (1993b). *M.R.S. Symp. Proc.*, **297**, 279.
Branz, H. M., Asher, S., Nelson, B. P., and Kemp, M. (1993c). *J. Non-Cryst. Solids*, **164–166**, 269.
Branz, H. M., Yelon, A., and Movaghar, B. (1994). *M.R.S. Symp. Proc.*, **336**, 159.
Branz, H. M., Asher, S., Xu, Y., and Kemp, M. (1995). *M.R.S. Symp. Proc.*, **377**, 331.
Brodsky, M. H., Cardona, M., and Cuomo, J. J. (1977a). *Phys. Rev. B*, **16**, 3556.
Brodsky, M. H., Frisch, M. A., Ziegler, J. F., and Lanford, W. A. (1977b). *Appl. Phys. Lett.*, **30**, 561.
Cardona, M. (1983). *Physica Status Solidi (b)*, **118**, 463.
Carlson, D. E. (1986). *Appl. Phys. A*, **41**, 305.
Carlson, D. E., and Magee, C. W. (1978). *Appl. Phys. Lett.*, **33**, 81.
Chang, K. J., and Chadi, D. J. (1989). *Phys. Rev. B*, **40**, 11644.
Chou, S. F., Schwarz, R., Okada, Y., Slobodin, D., and Wagner, S. (1987). *M.R.S. Symp. Proc.*, **95**, 165.
Daey Ouwens, J., and Schropp, R. E. I. (1995). *M.R.S. Symp. Proc.*, **377**, 419.
Deák, P., Snyder, L. C., Heinrich, M., Ortiz, C. R., and Corbett, J. W. (1991). *Physica B*, **170**, 253.
Fang, C. J., Gruntz, K. J., Ley, L., Cardona, M., Demond, F. J., Müller, G., and Kalbitzer, S. (1980). *J. Non-Cryst. Solids*, **35–36**, 255.
Finger, F., Kroll, U., Viret, V., Shah, A., Beyer, W., Tang, X. M., Weber, J., Howling, A., and Hollenstein, C. (1992). *J. Appl. Phys.*, **71**, 5665.
Franz, A., Jackson, W. B., Jin, H.-C., Abelson, J. R., and Gland, J. L. (1998). *J. Non-Cryst. Solids*, **227–230**, 143.
Freeman, E. C., and Paul, W. (1978). *Phys. Rev. B*, **18**, 4288.
Fritzsche, H. (1977). In *Amorphous and Liquid Semiconductors*, W. E. Spear, ed. Edinburgh: Edinburgh University, p. 3.
Fritzsche, H., Tanielian, M., Tsai, C. C., and Gaczi, P. J. (1979). *J. Appl. Phys.*, **50**, 3366.
Gleason, K. K., Petrich, M. A., and Reimer, J. A. (1987). *Phys. Rev. B*, **36**, 3259.
Gracin, D., Desnica, U. V., and Ivanda, M. (1992). *J. Non-Cryst. Solids*, **149**, 257.
Herring, C., and Johnson, N. M. (1991). In *Hydrogen in Semiconductors*, J. I. Pankove and N. M. Johnson, eds. San Diego: Academic Press, p. 225.
Jackson, W. B. (1988). *Phys. Rev. B*, **38**, 3595.
Jackson, W. B. (1993). *J. Non-Cryst. Solids*, **164–166**, 263.
Jackson, W. B., and Tsai, C. C. (1992). *Phys. Rev. B*, **45**, 6564.
Jackson, W. B., Tsai, C. C., and Thompson, R. (1990). *Phys. Rev. Lett.*, **64**, 56.
Jackson, W. B., Zhang, S. B., Tsai, C. C., and Doland, C. (1991a). *AIP Conf. Proc.*, **234**, 37.
Jackson, W. B., Tsai, C. C., and Santos, P. V. (1991b). *J. Non-Cryst. Solids*, **137–138**, 21.
Jackson, W. B., Tsai, C. C., and Santos, P. V. (1992). *M.R.S. Symp. Proc.*, **258**, 319.
Jackson, W. B., Santos, P. V., and Tsai, C. C. (1993). *Phys. Rev. B*, **47**, 9993.
Jackson, W. B., Nickel, N. H., Johnson, N. M., Pardo, F., and Santos, P. V. (1994). *M.R.S. Symp. Proc.*, **336**, 311.

Johnson, N. M., Doland, C., Ponce, F., Walker, J., and Anderson, G. (1991). *Physica B*, **170**, 3.
Kakalios, J. (1991). In *Hydrogen in Semiconductors*, J. I. Pankove and N. M. Johnson, eds. San Diego: Academic Press, p. 381.
Kakalios, J., and Jackson, W. B. (1988). In *Amorphous Silicon and Related Materials*, H. Fritzsche, ed. Singapore: World Scientific, p. 207.
Kakalios, J., Street, R. A., and Jackson, W. B. (1987). *Phys. Rev. Lett.*, **59**, 1037.
Kemp, M., and Branz, H. M. (1993). *Phys. Rev. B*, **47**, 7067.
Kemp, M., and Branz, H. M. (1995). *Phys. Rev. B*, **52**, 13946.
Khait, Y. L., and Weil, R. (1995). *J. Appl. Phys.*, **78**, 6504.
Khait, Y. L., Weil, R., Beserman, R., Beyer, W., and Wagner, H. (1990). *Phys. Rev. B*, **42**, 9000.
Kirchheim, R., and Huang, X. Y. (1987). *Phys. Stat. Solidi*, **144**, 253.
Kirchheim, R., Sommer, F., and Schluckebier, G. (1982). *Acta Metall.*, **30**, 1059.
Kisielowski-Kemmerich, C., and Beyer, W. (1989). *J. Appl. Phys.*, **66**, 552.
Knights, J. C., Lucovsky, G., and Nemanich, R. J. (1978). *Philos. Mag. B*, **37**, 469.
Langford, A. A., Fleet, M. L., Nelson, B. P., Lanford, W. A., and Maley, N. (1992). *Phys. Rev. B*, **45**, 13367.
Le Comber, P. G., Loveland, R. J., Spear, W. E., and Vaughan, R. A. (1974). In *Amorphous and Liquid Semiconductors*, J. Stuke and W. Brenig, eds. London: Taylor and Francis, p. 245.
Lee, S. S., Kong, M. J., Bent, S. F., Chiang, C. M., and Gates, S. M. (1996). *J. Phys. Chem.*, **100**, 20015.
Lewis, A. J., Connell, G. A. N., Paul, W., Pawlik, J. R., and Temkin, R. J. (1974). *AIP Conf. Proc.*, **20**, 27.
Lucovsky, G. (1979). *Solid State Commun.*, **29**, 571.
Lusson, L., Lusson, A., Elkaim, P., Dixmier, J., and Ballutaud, D. (1997). *J. Appl. Phys.*, **81**, 3073.
Maeda, K., Kuroe, A., and Umezu, I. (1995). *Phys. Rev. B*, **51**, 10635.
Mahan, A. H., Raboisson, P., Williamson, D. L., and Tsu, R. (1987). *Solar Cells*, **21**, 117.
Mahan, A. H., Johnson, E. J., Crandall, R. S., and Branz, H. M. (1995). *M.R.S. Symp. Proc.*, **377**, 413.
Manfredotti, C., Fizzotti, F., Boero, M., Pastorino, P., Polesello, P., and Vittone, E. (1994). *Phys. Rev. B*, **50**, 18047.
Masson, D., Sacher, E., Yelon, A. (1987). *Phys. Rev. B*, **35**, 1260.
McMillan, J. A., and Peterson, E. M. (1979a). *J. Appl. Phys.*, **50**, 5238.
McMillan, J. A., and Peterson, E. M. (1979b). *Thin Solid Films*, **63**, 189.
Meyer, W., and Neldel, H. (1937). *Z. Tech. Phys.*, **18**, 588.
Mitra, S., Shinar, R., and Shinar, J. (1990). *Phys. Rev. B*, **42**, 6746.
Nakamura, M., Ohno, T., Miyata, K., Konishi, N., and Suzuki, T. (1989). *J. Appl. Phys.*, **65**, 3061.
Oguz, S., Anderson, D. A., Paul, W., and Stein, H. J. (1980). *Phys. Rev. B*, **22**, 880.
Pantelides, S. I. (1987). *Phys. Rev. Lett.*, **58**, 1344.
Petrova-Koch, V., Zeidl, H. P., Herion, J., and Beyer, W. (1987). *J. Non-Cryst. Solids*, **97–98**, 807.
Powell, M. J., and Deane, S. C. (1996). *Phys. Rev. B*, **53**, 10121.
Reimer, J. A., Vaughan, R. W., and Knights, J. C. (1980). *Phys. Rev. Lett.*, **44**, 193.
Reinelt, M., Kalbitzer, S., and Müller, G. (1983). *J. Non-Cryst. Solids*, **59–60**, 169.
Roth, J. A., Olson, G. L., Jacobson, D. C., and Poate, J. M. (1993). *M.R.S. Symp. Proc.*, **297**, 291.
Santos, P. V., and Jackson, W. B. (1992a). *M.R.S. Symp. Proc.*, **258**, 425.
Santos, P. V., and Jackson, W. B. (1992b). *Phys. Rev. B*, **46**, 4595.
Shanks, H., Fang, C. J., Ley, L., Cardona, M., Demond, F. J., and Kalbitzer, S. (1980). *Phys. Stat. Solidi*, **B106**, 43.

Shanks, H. R., Jeffrey, F. R., and Lowry, M. E. (1981). *J. Phys. Colloque*, **42**, C4-773.
Shinar, J., Shinar, R., Mitra, S., and Kim, J. Y. (1989). *Phys. Rev. Lett.*, **62**, 2001.
Shinar, J., Shinar, R., Wu, X. L., Mitra, S., and Girvan, R. F. (1991). *Phys. Rev. B*, **43**, 1631.
Shinar, R., Jia, H., Wu, X. L., and Shinar, J. (1992). *M.R.S. Symp. Proc.*, **258**, 419.
Shinar, R., Shinar, J., Jia, H., and Wu, X. L. (1993). *Phys. Rev. B*, **47**, 9361.
Shinar, R., Shinar, J., Subramania, G., Jia, H., Sankaranarayanan, S., Leonard, M., and Dalal, V. L. (1994). *M.R.S. Symp. Proc.*, **336**, 317.
Stavola, M., and Pearton, S. J. (1991). In *Hydrogen in Semiconductors*, J. I. Pankove and N. M. Johnson, eds. San Diego: Academic Press, p. 139.
Stein, H. J. (1975). *J. Electron. Mater.*, **4**, 159.
Stein, H. J. (1979). *Phys. Rev. Lett.*, **43**, 1030.
Street, R. A. (1991a). *Physica B*, **170**, 69.
Street, R. A. (1991b). *Solar Cells*, **30**, 207.
Street, R. A. (1991c). *AIP Conf. Proc.*, **234**, 21.
Street, R. A., and Tsai, C. C. (1988). *Philos. Mag. B*, 57, 663.
Street, R. A., Tsai, C. C., Kakalios, J., and Jackson, W. B. (1987). *Philos. Mag. B*, **56**, 305.
Street, R. A., Hack, M., and Jackson, R. A. (1988). *Phys. Rev. B*, **37**, 4309.
Tang, X. M., Weber, J., Baer, Y., and Finger, F. (1990a). *Phys. Rev. B*, **41**, 7945.
Tang, X. M., Weber, J., Baer, Y., and Finger, F. (1990b). *Phys. Rev. B*, **42**, 7277.
Tang, X. M., Weber, J., Baer, Y., and Finger, F. (1991). *Physica B*, **170**, 146.
Tatsumi, Y., and Ohsaki, H. (1989). *Properties of Amorphous Silicon*, EMIS Data Review Series 1. London: Inspec, p. 467.
Triska, A., Dennison, D., and Fritzsche, H. (1975). *Bull. Am. Phys. Soc.*, **20**, 392.
Tsu, D. V., and Lucovsky, G. (1987). *J. Non-Cryst. Solids*, **97–98**, 839.
Ueda, M., Chayahara, A., Nakashita, T., Imura, T., and Osaka, Y. (1985). *J. Non-Cryst. Solids*, **77–78**, 821.
Van de Walle, C. G. (1991). *Physica B*, **170**, 21.
Van de Walle, C. G. (1994). *Phys. Rev. B*, **49**, 4579.
Van de Walle, C. G., and Street, R. A. (1994). *Phys. Rev. B*, **49**, 14766.
Van de Walle, C. G., and Street, R. A. (1995). *M.R.S. Symp. Proc.*, **377**, 389.
Van Wieringen, A., and Warmoltz, N. (1956). *Physica*, **22**, 849.
Wagner, H., and Beyer, W. (1983). *Solid State Commun.*, **48**, 585.
Wert, C. A., and Zener, C. (1949). *Phys. Rev.*, **26**, 1169.
Widmer, A. E., Fehlmann, R., and Magee, C. W. (1983). *J. Non-Cryst. Solids*, **54**, 199.
Wilson, R. G. (1986). *Appl. Phys. Lett.*, **49**, 1375.
Yelon, A., and Movaghar, B. (1990). *Phys. Rev. Lett.*, **65**, 618.
Yelon, A., Movaghar, B., and Branz, H. M. (1992). *Phys. Rev. B*, **46**, 12244.
Zafar, S., and Schiff, E. A. (1989). *Phys. Rev. B*, **40**, 5235.
Zastrow, U. (1997). Unpublished work.
Zastrow, U., Beyer, W., and Herion, J. (1993). *Fresenius J. Anal. Chem.*, **346**, 92.
Zastrow, U., Szot, K., Beyer, W., and Speier, W. (1998). In *Secondary Ion Mass Spectrometry*, SIMS XI, G. Gillen, R. Lareau, J. Bennett, and F. Stevie, eds. New York: John Wiley & Sons, p. 393.
Zellama, K., Germain, P., Squelard, S., Monge, J., and Ligeon, E. (1980). *J. Non-Cryst. Solids*, **35–36**, 225.
Zellama, K., Germain, P., Squelard, S., Bourdon, B., Fontenille, J., and Danielou, R. (1981). *Phys. Rev. B*, **23**, 6648.
Zellama, K., Nedialkova, L., Bounouh, Y., Chahed, L., Benlahsen, M., Zeinert, A., Paret, V., and Theye, M. L. (1996). *J. Non-Cryst. Solids*, **198–200**, 81.

CHAPTER 6

Hydrogen Interactions with Polycrystalline and Amorphous Silicon—Theory

Chris G. Van de Walle

XEROX PALO ALTO RESEARCH CENTER
PALO ALTO, CALIFORNIA

I. INTRODUCTION 241
   1. Role of Hydrogen in Amorphous and Polycrystalline Silicon 241
   2. General Features of Hydrogen in Silicon 242
   3. Computational Approaches 245
II. HYDROGEN INTERACTIONS WITH AMORPHOUS SILICON 248
   1. Hydrogen Motion—Introduction 248
   2. Hydrogen Interactions with Dangling Bonds 253
   3. Hydrogen Interactions with Overcoordination Defects 256
   4. Hydrogen Interactions with Weak Si—Si Bonds 261
   5. Simulations of Amorphous Networks 262
   6. Hydrogen Diffusion and Metastability—Discussion 264
   7. Hydrogen Versus Deuterium for Passivation of Dangling Bonds 269
III. HYDROGEN IN POLYCRYSTALLINE SILICON 271
   1. Grain Boundaries 271
   2. Hydrogen Interactions with Grain Boundaries 272
   3. Hydrogen-Induced Generation of Donor-like Metastable Defects 273
   4. Hydrogen-Induced Generation of Acceptor-like Defects 277
IV. CONCLUSIONS AND FUTURE DIRECTIONS 277
   REFERENCES 278

I. Introduction

1. ROLE OF HYDROGEN IN AMORPHOUS AND POLYCRYSTALLINE SILICON

Hydrogen is known to interact with silicon in a wide variety of ways, including passivation of the surface, passivation of shallow as well as deep levels, generation of extended defects, etc. (Pankove and Johnson, 1991; Pearton et al., 1992). In amorphous and polycrystalline silicon, hydrogen passivates dangling bonds and eliminates deep levels from the bandgap, thereby dramatically improving the electronic quality of the material (Street, 1991a). However, hydrogen also has been established to play a role in creating metastable defects. Over the years, many models have been proposed to describe hydrogen diffusion and the creation and annihilation of defects. To put these models on a firmer basis, however, quantitative information is required on the energetics of the various processes and reactions. Such information is very hard to obtain directly from experiment, but it can be generated using computational techniques.

In this chapter I will focus on a review of theoretical approaches that include some computational component. I do not plan to give an exhaustive overview of the various computational approaches that have been used to generate amorphous networks; this could be the topic of a review chapter in its own right. I will refer to results from these approaches only when they address issues that are of direct relevance to the understanding of the interaction of hydrogen with the network.

In Section I.2 I will review some of the general features of hydrogen interactions with silicon. Many of these features actually apply to H interactions with semiconductors in general, but here the focus is on silicon. Most of the basic properties were first derived for hydrogen in a crystalline environment (see the reviews by Van de Walle, 1991a, 1991b; Estreicher, 1995), but the results also have bearing on amorphous materials.

The various computational techniques that have been used in the study of hydrogen interactions with polycrystalline or amorphous silicon (a-Si) will be discussed briefly in Section I.3.

2. GENERAL FEATURES OF HYDROGEN IN SILICON

a. *Isolated Interstitial Hydrogen*

Isolated interstitial hydrogen can assume different charge states in silicon. The impurity introduces a level in the bandgap, and the charge state depends on the occupation of that level. The charge state determines the most favorable location of hydrogen in the semiconductor lattice.

H^+ In the positive charge state (essentially a proton), hydrogen seeks out regions of high electronic charge density. In silicon, the maximum charge density is found at the bond-center (BC) site. The stability of the BC configuration arises from the formation of a three-center bond between the H atom and the two Si neighbors (Van de Walle, 1991b). H^+ is the preferred charge state in p-type material.

H⁻   In the negative charge state, hydrogen prefers regions of low electronic charge density, i.e., at the tetrahedral interstitial site. H⁻ is the preferred charge state in n-type material.

H⁰   The lattice location of neutral hydrogen (H⁰) is quite sensitive to the details of the charge density distribution. In crystalline Si, H⁰ is located at BC; however, the energy of H⁰ is always higher than that of H⁺ or H⁻; i.e., the neutral charge state is never the most stable state. This is the defining characteristic of a negative-U center.

The charge states of interstitial H in a-Si have not been established conclusively, but it is plausible that they would follow the same trends as for H in crystalline Si (c-Si).

Since H at the BC site is such an important configuration, it is useful to look in more detail at the basic atomic and electronic structure. Three-center bonds (Si—H—Si) were actually suggested for H in a-Si:H as early as 1978 by Fisch and Licciardello (1978). These bonds were studied in more detail by Zacher et al. (1986). As mentioned earlier, the stability arises because of Coulomb attraction between the proton and the high electron density at the bond center. At the same time, however, the stability is reduced due to the geometric constraints imposed by the surrounding network; in order to provide reasonable Si—H bonding distances, the Si atoms have to move outward from their regular lattice sites in c-Si, raising the elastic energy. Insertion of H into a bond that has a longer bond length than the equilibrium Si—Si bond length therefore increases the stability; this issue will be discussed in Sections II.4 and III.3.

b. Diffusion

Hydrogen, being a small and light impurity, is expected to move quite easily through the perfect lattice. The calculated migration barrier for the positive and neutral charge states in c-Si is 0.2 eV; for the negative charge state it is 0.25 eV (Van de Walle et al., 1989). Interstitial hydrogen in a-Si:H is expected to move with similar ease. However, in a-Si:H there is the possibility for hydrogen to become trapped at traps with various energy depths; a schematic of a possible diffusion process is shown in Fig. 1. The diffusion process therefore becomes quite complicated, with a measured activation energy of around 1.5 eV. A rigorous microscopic explanation for this phenomenon is not yet in place; various models will be discussed in Section III, along with attempts to associate specific energy values with the levels depicted in Fig. 1.

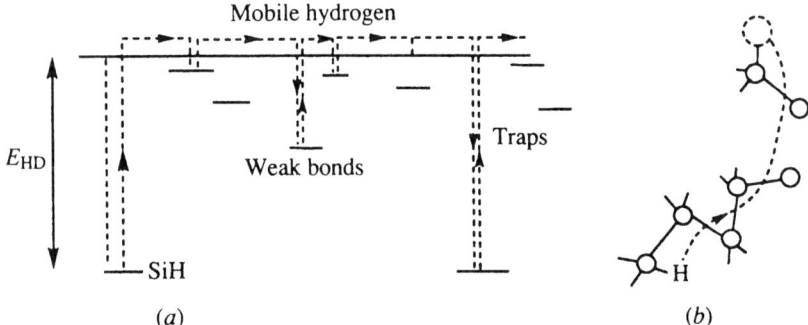

FIG. 1. Schematic diagram of the possible hydrogen diffusion mechanisms: (a) the potential wells corresponding to the trapping sites and the energy of the mobile hydrogen; (b) the motion of the hydrogen through the network. (From Street, 1991a.)

c. *Compensation and Passivation of Shallow Impurities*

Based on the characteristics of isolated interstitial hydrogen, one can immediately predict that hydrogen will interact with shallow dopants. Hydrogen always counteracts the electrical activity of the dopants. The presence of hydrogen in *p*-type material leads to formation of H^+, a donor, which compensates acceptors. H^+ can form a complex with the acceptor impurity, to which it is coulombically attracted. Similarly, H^- passivates donors.

d. *Interactions with Deep Levels*

Hydrogen, of course, also can interact with impurities or defects that form deep levels in the bandgap. A prime example is the interaction of hydrogen with dangling bonds in amorphous silicon, a process that significantly improves the electronic quality of the material. One expects hydrogen to interact with extended defects as well.

e. *Hydrogen Molecules and Molecular Complexes*

Theory predicts that H_2 molecules readily form in silicon; the binding energy is somewhat smaller than its value in free space but still large enough to make interstitial H_2 one of the more favorable configurations hydrogen can assume in the lattice. Another configuration involving two hydrogen atoms is the so-called H_2^* complex, in which a host-atom bond is broken and one hydrogen atom is inserted between the two atoms, in a BC position, while the other hydrogen occupies an antibonding (AB) site (Chang and Chadi, 1989). This configuration is somewhat higher in energy than the H_2 molecule, but it may play an important role in diffusion, in metastability, as well as in nucleation of larger hydrogen-induced defects.

f. Extended Hydrogen Complexes

It has been observed that hydrogenation can induce microdefects in a region within ~1000 Å of the surface (Johnson *et al.*, 1987, 1991; Ponce *et al.*, 1987). The defects, studied with transmission electron microscopy (TEM), have the appearance of platelets along {111} crystallographic planes and range in size from 50 up to 1000 Å. One or two H atoms per Si—Si bond are present. Several models have been proposed for the structure of these platelets (Ponce *et al.*, 1987; Van de Walle *et al.*, 1989; Zhang and Jackson, 1991); the model of Zhang and Jackson (1991) has been worked out in greatest detail. It consists of double-layer {111} H platelets resulting from clustering of diatomic H_2^* complexes; this clustering occurs because a certain amount of binding energy can be gained when bringing H_2^* complexes together. However, a definitive experimental microscopic identification of the platelet structure is still lacking. Zhang and Jackson (1991) suggested that similar structures exist in *a*-Si:H. Nickel *et al.* (1996) have observed hydrogen-induced platelets in polycrystalline Si.

g. Energies of Various Configurations of H in Si

First-principles calculations for a large number of configurations of H in Si were discussed by Van de Walle (1994). The results are summarized in Fig. 2. The calculated values were all obtained within a consistent theoretical framework (pseudopotential-density-functional theory) and provide direct information about the relative stability of different configurations.

3. COMPUTATIONAL APPROACHES

a. Pseudopotential-Density-Functional Calculations

This computational approach is now regarded as a standard for performing first-principles studies of defects or impurities in semiconductors. Density-functional theory (DFT) in the local density approximation (LDA) (Hohenberg and Kohn, 1964; Kohn and Sham, 1965) allows a description

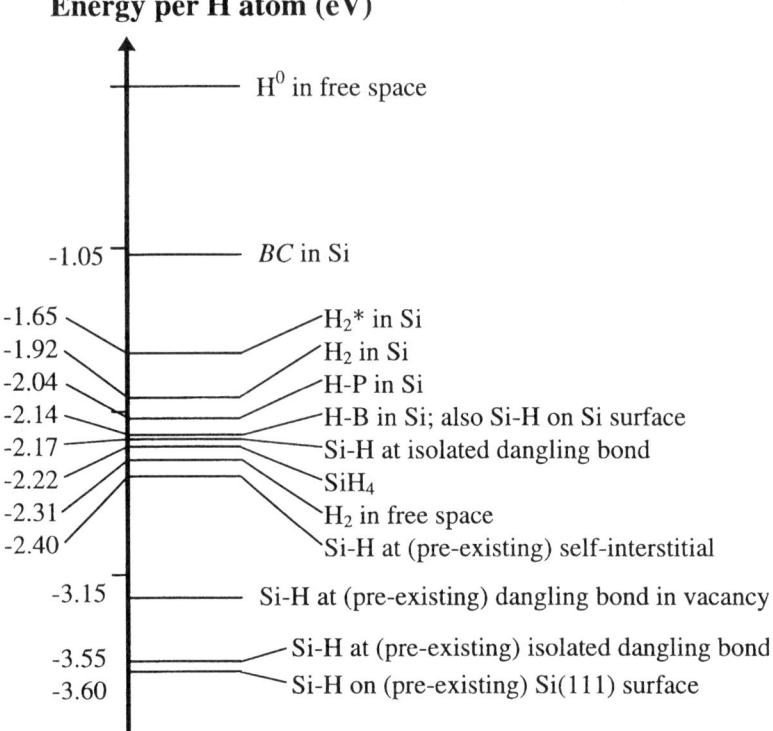

FIG. 2. First-principles energies for various configurations of H in Si. The zero of energy corresponds to a free H atom. The energy values were determined with first-principles pseudopotential-density-functional calculations and include zero-point energies.

of the ground state of the many-body system in terms of a one-electron equation with an effective potential. The total potential consists of an ionic potential due to the atomic cores, a Hartree potential, and a so-called exchange and correlation potential that describes the many-body aspects. This approach has proven very successful for a wide variety of solid-state problems. One shortcoming of the technique is its failure to produce reliable excited-state properties, widely referred to as the "bandgap problem." Many useful results of the calculations depend on ground-state properties and thus are not affected by this shortcoming. In cases where the bandgap enters the calculations either directly or indirectly, judicious inspection of the results may still allow extraction of reliable information.

Most properties of molecules and solids are determined by the valence electrons; the core electrons usually can be removed from the problem by representing the ionic core (i.e., nucleus plus inner shells of electrons) with a pseudopotential. State-of-the-art calculations employ nonlocal norm-conserving pseudopotentials (Hamann et al., 1979) that are generated solely based on atomic calculations and do not include any fitting to experiment.

The last ingredient commonly used in pseudopotential-density-functional calculations for defects or impurities is the supercell geometry. Ideally, one would like to describe a single isolated impurity in an infinite crystal. In the supercell approach, the impurity is surrounded by a finite number of semiconductor atoms, and this structure is periodically repeated. Maintaining periodicity allows continued use of algorithms such as Fast Fourier Transforms. One also can be assured that the band structure of the host crystal is well described (which may not be the case in a cluster approach). For sufficiently large supercells, the properties of a single, isolated impurity can be derived. Convergence tests have indicated that the energetics of impurities and defects are usually well described by using 32-atom supercells. The supercell approach also allows simulations of nonperiodic systems such as amorphous networks. Periodic boundary conditions still apply to the overall supercell, but within that cell an amorphous system consisting of tens or even hundreds of atoms can be simulated.

An alternative to the use of supercells for systems lacking periodicity is to employ a cluster geometry; a cluster of finite size is usually terminated with H atoms. This approach also can be used to model a defect or impurity in the solid. Whether using supercells (periodic boundary conditions) or clusters, convergence of the results with respect to cell size or cluster size always should be carefully checked.

Finally, I mention that the pseudopotential-density-functional method can be combined with a molecular dynamics approach, as pioneered by Car and Parrinello (1985). In molecular dynamics, atoms in the system move according to calculated forces; simulations can be performed as a function of temperature, mimicking, for instance, quenching of a system from the melt. The computational effort involved in first-principles molecular dynamics usually limits the size of the system, as well as the number of timesteps during which the evolution of the system can be studied.

b. *Approximate Pseudopotential-Density-Functional Calculations Using the Harris Functional*

To overcome the computational demands of the pseudopotential-density-functional method, Sankey and Niklewski (1989) and Sankey and Drabold (1991) developed an approximate method based on the Harris-functional approach (Harris, 1985) that eliminates the iterative approach typical for self-consistent calculations. Other approximations used by these authors include a limited basis set consisting of four local orbitals per site. The approach has been used by Fedders and Drabold (1993), Fedders (1995), and Tuttle and Adams (1996) to study amorphous systems containing up to several hundred atoms.

c. Empirical Potentials

As mentioned earlier, the computational complexity of first-principles calculations severely limits the size of systems that can be studied or the length of time over which molecular dynamics simulations can be performed. Various attempts have been made to describe the system of Si and H atoms with empirical potentials. For Si, such potentials have achieved a reasonable degree of reliability; modeling the Si-H interaction, however, has proved very difficult. Indeed, capturing the very different behavior of H in a covalent Si—H bond and in a three-center BC configuration is a challenge. One attempt was made by Biswas *et al.* (1991). Later, the same group went on to develop a tight-binding approach.

d. Tight-Binding Molecular Dynamics

Li and Biswas (1994a, 1995a, 1995b) have devised a tight-binding molecular dynamics approach that can be applied to systems containing several hundred atoms. First-principles calculations would not be feasible for systems of this size. The electronic states are approximated by a superposition of atomic orbitals. Parameters are obtained by fitting to first-principles calculations as well as to experimental values. Li and Biswas (1994a, 1994b) were able to successfully model a wide range of properties of crystalline Si, Si—H vibrational properties, and the total-energy surface for H in *c*-Si. They also found a good description of the electronic and structural properties of *a*-Si:H.

II. Hydrogen Interactions with Amorphous Silicon

1. HYDROGEN MOTION — INTRODUCTION

Hydrogen passivation of dangling-bond defects leads to a significant improvement in the electronic properties of amorphous Si (*a*-Si). The behavior of hydrogen in the amorphous material, however, is complex and still not fully understood. Hydrogen diffusion through the material is known to be affected by the doping level; in undoped samples, the activation energy E_A is 1.4–1.5 eV; in doped samples, the activation energy is smaller, 1.2–1.3 eV (Street et al., 1987). Hydrogen diffusion is also known to be affected by the presence of free carriers (Santos et al., 1991). In addition, hydrogen motion has been shown to be correlated with the *formation* of both metastable and equilibrium defects (Jackson and Kakalios, 1989).

A central process in these phenomena is the breaking of a silicon-hydrogen bond, in which hydrogen is promoted from a stable configuration (in which it is strongly bonded to a Si atom) to a higher-energy interstitial position (in which it is mobile and can easily move through the lattice). This basic framework was established by Street et al. (1987) and is illustrated in Fig. 3. The following reaction describes the breaking of a Si—H

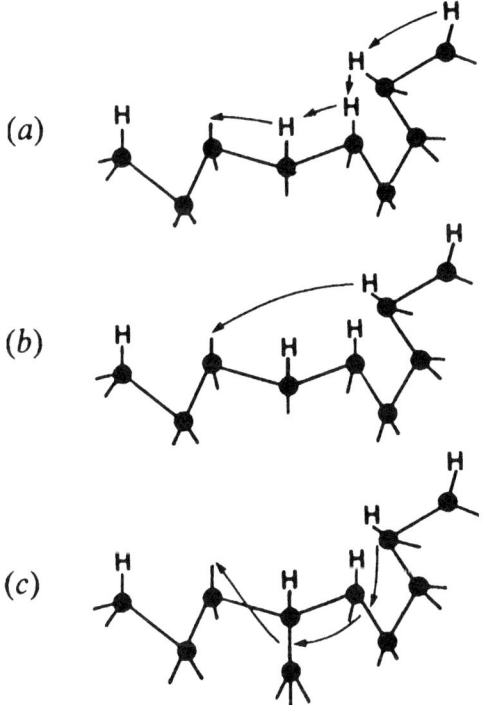

FIG. 3. Schematic diagrams showing three diffusion models for hydrogen in a-Si:H: (a) hydrogen moving to an immediately adjacent dangling bond; (b) hydrogen released into an interstitial site and diffusing until it is trapped by a dangling bond; (c) hydrogen released into an interstitial site and then breaking a weak Si—Si bond. (From Street et al., 1987.)

bond and insertion of H into a Si—Si bond, leaving a dangling bond (db) behind:

$$\text{Si—H} + \text{Si—Si} \leftrightarrow \text{Si—H—Si} + \text{db} \quad (1)$$

This reaction has been invoked in the context of defect creation; this picture, based on bond breaking and weak-bond to dangling-bond conversion, is commonly accepted now (Stutzmann et al., 1985; Smith and Wagner, 1987; Street and Winer, 1989; Street, 1991a). The process described by Eq. (1) has been suggested as the microscopic model for light-induced degradation (the Staebler-Wronski effect; Staebler and Wronski, 1977).

Equation (1) also has been used commonly to describe hydrogen diffusion, which proceeds by hydrogen being released from deep or shallow traps and moving in a transport state. A critical discussion of the use of Eq. (1) to describe diffusion can be found in Tuttle and Adams (1998).

Pantelides (1987a) suggested that diffusion and defect annihilation in a-Si:H could not be explained solely by invoking motion of hydrogen; he proposed that migration of the intrinsic defects themselves (in particular, floating bonds; Pantelides, 1986) played an important role. Pantelides (1987b) also proposed an explanation for the Staebler-Wronski effect based on floating bonds. Few additional investigations of this hypothesis have been performed; arguments against this proposal were discussed by Street (1991a).

Several groups contributed to the description of H diffusion through a-Si (Carlson and Magee, 1978; Street, 1991b, 1991c; Santos and Jackson, 1992; Kemp and Branz, 1993). The key energies that play a role in the diffusion process are illustrated in Fig. 4a. The quantity $E_{\text{Si-H}}$ represents the energy of a Si—H bond. μ_H is the chemical potential of H atoms. E_{BC} is the energy of an isolated interstitial H atom at its most stable site in the network; by analogy with crystalline Si (c-Si), this is taken to be at the bond-center (BC) site. E_S, finally, is the energy corresponding to the saddle point of migration of an interstitial H atom. Figure 4b illustrates the expected broadening of the energy levels in hydrogenated amorphous Si (a-Si:H) due to the atomic disorder (Street, 1991b, 1991c).

The hydrogen chemical potential μ_H determines the occupancy of various configurations; states with energies below μ_H are (mostly) occupied with hydrogen, and those above are (mostly) empty. The energy difference between mobile hydrogen (energy E_S) and μ_H determines the number of hydrogen atoms that can participate in diffusion. The diffusion activation energy E_A (1.4–1.5 eV in undoped material) therefore should be associated with $E_\text{S} - \mu_\text{H}$ (Street, 1991b).

Estimates about other energies in Fig. 4 also can be made. Hydrogen diffusion experiments in c-Si (Van Wieringen and Warmoltz, 1956) as well

6 HYDROGEN INTERACTIONS — THEORY

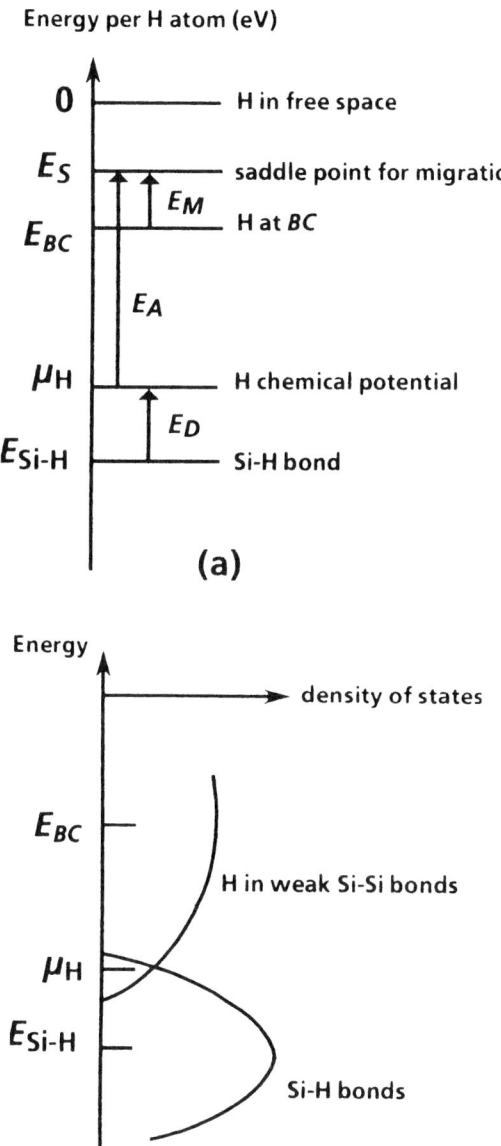

FIG. 4. (a) Energies relevant for H interactions with silicon. The various quantities are described in the text. (b) Schematic hydrogen density of states distribution in a-Si, corresponding to the level scheme in (a). The broadening of the levels is due to the disorder in the amorphous network. (From Van de Walle and Street, 1995.)

as H diffusion in a-Si when all deep traps are saturated (Santos and Jackson, 1992) indicate that the migration energy $E_M = E_S - E_{BC}$ is about 0.5 eV. On the other hand, first-principles calculations of the adiabatic energy surface (Van de Walle et al., 1989) indicate that E_M is only about 0.2 eV. I will therefore assume a range of values, 0.2–0.5 eV, for E_M. Analysis of solubility data in c-Si places E_{BC} at about 1 eV below the energy of a H atom in free space (Herring and Johnson, 1991), a value that agrees well with the theoretical value in Fig. 2 (Van de Walle, 1994). Finally, the energy difference between E_{Si-H} and μ_H can be associated with the dangling-bond formation energy, since the removal of H from Si—H results in a dangling bond; measurement of equilibrium-defect densities indicated that $E_D = \mu_H - E_{Si-H}$ is about 0.2–0.5 eV (Street and Winer, 1989; Street, 1991b).

Inspection of these values leads to a seemingly puzzling result: Based on the numbers quoted earlier, the energy difference between E_{BC} and E_{Si-H} should be around 1.5 eV or less; hence the energy E_{Si-H} would be less than 2.5 eV below the energy of a free H atom. This energy is much smaller (by more than 1 eV) than one would expect based on estimates of binding energies of Si—H bonds, e.g., in a silane (SiH_4) molecule. Indeed, it takes 3.92 eV to remove the first H atom from an SiH_4 molecule (i.e., SiH_3—H) (Walsh, 1981).

One explanation that was proposed assumes that H would be bonded predominantly in strong Si—H bonds on void surfaces and that transport would take place along these internal surfaces (Jackson and Tsai, 1992). The transport level would then be ~ 1.5 eV about the Si—H energy, meaning that it would lie significantly lower than in c-Si (as implied in Fig. 4). There are several problems with this picture. First, there is no evidence that the void structure in a-Si could allow for long-range diffusion in this fashion; this has been highlighted in the first-principles calculations for a-Si:H structures by Tuttle and Adams (1998). The theoretical results of Van de Walle and Street (1995) also do not support this mechanism.

In order to examine the validity of the framework described by Fig. 4, and to address the apparent inconsistent energetics, values for the various energies included in Fig. 4 can be calculated. In the following sections I will address the energetics of various configurations that hydrogen can assume in a-Si. In Section II.2 I focus on H interacting with dangling bonds, first in a crystalline environment and then in an amorphous network. In Section II.3 I discuss the interaction of H with overcoordination defects. Section II.4 describes calculations addressing the energetics of hydrogen insertion in weak Si—Si bonds. In Section II.5 I discuss simulations for amorphous networks that provide information about energy distributions for the various configurations. In Section II.6 I will return to the discussion of hydrogen diffusion and defect formation.

2. HYDROGEN INTERACTIONS WITH DANGLING BONDS

Until recently, little information has been available about the energetics of the Si—H bond in bulk (crystalline or amorphous) Si. It had been assumed that the bond strength would be similar to that in a silane (SiH_4) molecule, but then effects of the crystalline environment and possible distortions of the bonding configurations are ignored. As a first step, a study was performed for Si—H bonds in a crystalline environment (Van de Walle and Street, 1994). Even though that study did not explicitly take the amorphous nature of the network into account, it was argued that major features of the interaction would be mostly sensitive to the local environment, which is very similar in the crystalline and amorphous cases. Certain types of information, such as repulsion between neighboring H atoms and dissociation mechanisms of the Si—H bond, are also expected to be general in nature. These results are discussed in Section II.2.a. More recently, explicit calculations have been performed for H in an amorphous environment. These results are discussed in Section II.2.b.

a. Si—H Bonds Calculated in a Crystalline Environment

One way to define an energy for the Si—H bond is to address the question, How much energy is needed to remove the H atom from a Si—H bond, leaving a dangling bond behind and placing the hydrogen in an interstitial (BC) position? The resulting energy is a *binding energy*. Pseudopotential-density-functional theory (see Sec. I.3.a) was used to investigate the structure and energy of the Si—H bond in a number of distinct configurations (Van de Walle and Street, 1994). The first geometry was that of the hydrogenated vacancy; here the H atoms are close enough to significantly interact. The binding energy to remove one hydrogen from a fully hydrogenated vacancy was found to be 2.10 eV.

A second geometry placed a dangling bond in a larger void, created by the removal of four Si atoms. For the latter configuration it was found that the energy required to move the H atom from the Si—H bond to an interstitial position is 2.5 eV. The 0.4-eV difference between the two results was attributed to H-H repulsion effects. Additional calculations allowed a quantitative assessment of the hydrogen-hydrogen interaction energy U in the vacancy; it was found that $U \approx 0.11$ eV. We note that instead of choosing H at BC as the final state, one also can choose H in free space as the reference; this leads to the energy values shown in Fig. 2.

Another way of defining an energy for the Si—H bond is to assume that one starts from crystalline silicon and a hydrogen atom in free space and that the energy to create a dangling bond needs to be taken into account; this defines a *formation energy*. It is possible to extract a formation energy for an "ideal" Si—H bond (meaning it is located at a dangling bond that is isolated from other dangling bonds, with no H-H repulsion); the result is $-2.17\,\text{eV}$ (Van de Walle and Street, 1994). This value is very close to the calculated value for the SiH_4 molecule ($-2.22\,\text{eV}$), expressed with respect to bulk Si and a free H atom (see Fig. 2). This concept of a formation energy will be discussed in more detail in Section II.6.a.

It should be noted that all values of binding energies and formation energies mentioned here assume that both the dangling bond and the intestitial H atom occur in the neutral charge state; starting from these values, the corresponding values for other charge states can be derived if the energy levels of the dangling bond and of the interstitial H are known.

It is also interesting to investigate whether strain in the backbonds affects the strength of the Si—H bond. Strain effects might be important in the case of an amorphous network. To simulate strain, the three Si atoms to which the Si atom with the dangling bond is bonded were moved outward. The displacement was chosen to be in the plane of these three atoms and had a magnitude of $0.13\,\text{Å}$, representing a sizable strain (5% of the bond length). It was found (Van de Walle and Street, 1994) that the binding energy was only $0.05\,\text{eV}$ smaller than in the unstrained case, leading to the conclusion that strain effects have only a modest influence on the binding energy of the Si—H bond.

Finally, I discuss the phenomenon of hydrogen exchange at Si—H bonds. Various experiments have shown that this exchange can proceed with very low energy barriers; one way to investigate the exchange is to monitor substitution of hydrogen by deuterium (Street, 1987). Tuttle *et al.* (1999) have examined the hydrogen exchange mechanism at a Si—H bond. Initially, one hydrogen atom is bound at the Si—H bond, and the second H approaches it as an interstitial atom. In a first step, this second atom assumes a BC configuration in a Si—Si bond adjacent to the dangling bond; this configuration is only $0.15\,\text{eV}$ higher in energy than the configuration where hydrogen sits in a BC site far from the Si—H bond. In a second step, the two H atoms exchange places, a process that can proceed with a barrier of less than $0.2\,\text{eV}$. A detailed examination of this process reveals intermediate formation of an SiH_2-like complex. The overall low barriers for this process are in agreement with the experimental results of Branz *et al.* (1993) and Kemp and Branz (1995).

b. Si—H Bonds Calculated in an Amorphous Environment

Various simulations have addressed the atomic and electronic structure of Si—H bonds in an amorphous environment. Nelson *et al.* (1988) studied the effect of removing hydrogen atoms on the vibrational and electronic properties of a-Si:H using peudopotential-density-functional calculations and random-network models consisting of 54 Si atoms and 6 H. They found that dangling-bond states have an average energy 0.2 eV above the valence band, which is lower than indicated by experiment; however, the effect of LDA bandgap errors was not discussed. The Si—H bonding states were located 5.0 and 7.5 eV below the top of the valence band. For a simpler model, consisting of a hydrogenated vacancy, DiVincenzo *et al.* (1983) had found Si—H bonding character for states 4.5 eV below the valence band.

Additional information about binding of H to Si dangling bonds will be discussed in Section II.5.

c. Dissociation Mechanisms of Si—H Bonds

Several pathways can be considered for dissociation of Si—H bonds (Van de Walle and Street, 1994). The H atom can move along the direction of the Si—H bond away from the Si atom (Fig. 5); however, this is unlikely to be the most favorable path for two reasons: (1) the initial rise in energy in that direction is high, as indicated by the high vibrational frequency (around $2000\,cm^{-1}$) for the Si—H stretch mode, and (2) this path eventually leads

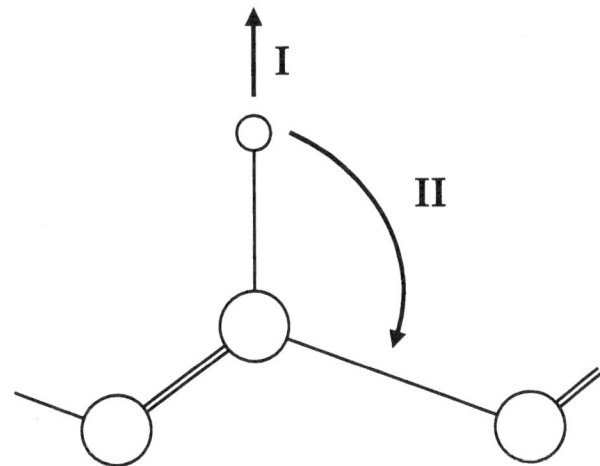

Fig. 5. Schematic representation (projected on a (110) plane) of a Si—H bond and the neighboring Si atoms. Si atoms are represented by large circles and the H atom by a small circle. Two paths for removing the H atom from the Si—H bond (leaving a dangling bond behind) are indicated; path *I* moves the H atom toward a tetrahedral interstitial site; path II moves the H atom toward a neighboring BC site. (From Van de Walle and Street, 1994.)

to a position of the H atom in the interstitial channel, which is not the lowest energy site for H in the neutral or positive charge state (in crystalline Si) (see Sec. I.2.a). Both these arguments actually favor a path (path II in Fig. 5) in which the H atom stays at an approximately constant distance from the Si atom to which it is bound: (1) the barrier in that direction is much lower, as indicated by the vibrational frequency (around $700\,\mathrm{cm}^{-1}$) for the Si—H wagging mode, and (2) this path leads to a H position near the BC site, which is the stable site for H^0 (and also for H^+) in crystalline Si.

The intermediate state, with the H atom in a BC site next to the dangling bond, is 1.5 eV higher in energy than the Si—H bond, to be compared with the 2.5 eV it costs to remove the H to a position far away from the dangling bond (Van de Walle and Street, 1994). Alternatively, the intermediate state may involve H in an AB site (Tuttle and Van de Walle, 1999). Even before the H atom reaches this intermediate position, however, two energy levels are introduced into the bandgap (near the valence band and near the conduction band), enabling the complex to capture carriers; after changing charge state, there is virtually no barrier to further dissociation. Experimentally, the involvement of carriers in the migration of H in a-Si:H was established by Santos and Johnson (1993); it was found that the motion of H is strongly suppressed in a depletion region under reverse bias, where no free carriers are available. The barrier for dissociation therefore can be reduced significantly in doped material or in the presence of light-induced carriers. An interesting application of the features of this dissociation path will be discussed in Section II.6.

3. HYDROGEN INTERACTIONS WITH OVERCOORDINATION DEFECTS

When discussing hydrogen passivation of deep levels in crystalline and amorphous silicon, it is often assumed that the deep levels are due to dangling bonds. Indeed, this type of interaction can be monitored explicitly, e.g., with electron spin resonance (ESR), which produces a characteristic signature of the dangling bond. However, one should not focus only on undercoordination defects (dangling bonds) but also contemplate the possible existence of overcoordination defects. Indeed, in crystalline Si it has been accepted for some time that vacancies are not the only type of intrinsic defect to play a role. The early first-principles calculations of Bar-Yam and Joannopoulos (1984) and Car *et al.* (1985) already indicated that self-interstitials have formation energies comparable with those of vacancies; this result has been confirmed by more recent state-of-the-art investigations by Blöchl *et al.* (1993).

In amorphous silicon, too, overcoordination defects (interstitials) may

6 HYDROGEN INTERACTIONS—THEORY 257

occur, in addition to undercoordinated atoms, as pointed out by Pantelides (1986). In crystalline silicon, the self-interstitials are known to play a role in self-diffusion (Blöchl et al., 1993), impurity diffusion (Cowern et al., 1994; Eaglesham et al., 1994), surface reconstructions (Dabrowski et al., 1994), planar interstitial defects (Kohyama and Takeda, 1992), and dislocation nucleation (Tan, 1981). However, because of their high mobility, isolated silicon self-interstitials have never been observed directly (Watkins, 1991). Despite the accepted importance of silicon self-interstitials and of hydrogen interactions with defects in silicon, the interaction between self-interstitials and hydrogen has been addressed only recently, in computational work by Jones et al. (see Bech Nielsen et al., 1995) and by Van de Walle and Neugebauer (1995). Experimental investigations were performed by Bech Nielsen et al. (1995).

Complexes consisting of one or two H atoms and a Si self-interstitial were investigated using first-principles calculations, addressing atomic structure, electronic structure, and vibrational frequencies (Van de Walle and Neugebauer, 1995). It was found that hydrogen interacts strongly with self-interstitials; while the calculated binding energy is smaller than for H interacting with a vacancy, it is large enough for the complexes to be stable at room temperature.

The lowest-energy structure of the isolated Si self-interstitial (Si_i), in the neutral charge state, consists of a pair of Si atoms, oriented in the $\langle 110 \rangle$ direction, sharing a substitutional lattice site (a split-interstitial configuration); the calculated geometry is illustrated in Fig. 6, showing a small cluster of atoms near the core of the defect. In the perfect crystal, this cluster would contain seven atoms, five of which lie in a $(\bar{1}10)$ plane (forming the characteristic zigzag chain) and two of which lie in a perpendicular (110) plane. One also can think of the cluster as consisting of a central atom surrounded by four nearest neighbors and two additional second-nearest neighbors. In the case of the self-interstitial, the lattice location at the center of the cluster is now shared by two atoms. Figure 6 shows the cluster viewed along the $[\bar{1}10]$ direction.

The structure of a complex between the self-interstitial and one H atom, as obtained by Van de Walle and Neugebauer (1995), is shown in Fig. 7. This configuration was found to have the lowest energy among many potential candidates obtained by adding the H atom (in various positions) to some of the basic configurations of the self-interstitial, including the $\langle 110 \rangle$ and $\langle 100 \rangle$ split interstitials and the tetrahedral interstitial site. The most stable configuration has all atoms at the core of the defect lying in the $(\bar{1}10)$ plane, with the Si atoms distorted from their positions in the $\langle 110 \rangle$ split interstitial (see Fig. 7). The calculated vibrational frequency for the stretch mode of the Si—H bond is 1870 cm^{-1}, slightly smaller than the values in SiH_4 or for Si—H bonds on an Si surface (Van de Walle, 1994).

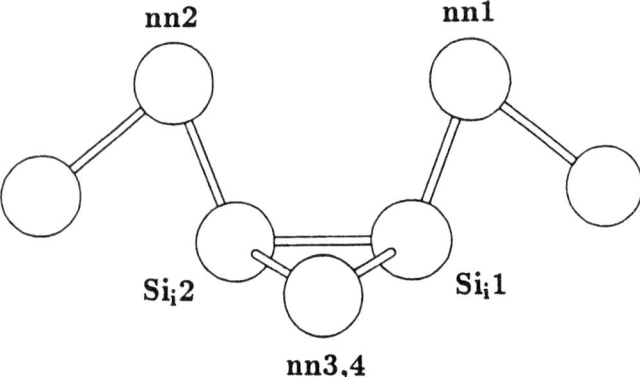

FIG. 6. Schematic representation of a cluster of Si atoms containing a Si self-interstitial in the split-interstitial configuration, oriented in the [110] direction. In the perfect crystal, the cluster consists of a central Si atom, surrounded by its four nearest neighbors, as well as two additional second-nearest neighbors. Now the central lattice site is shared by two Si atoms, labeled Si_i1 and Si_i2. The nearest neighbors nn1 and nn2, as well as the atoms to which they are bound, also lie in the plane of the figure. Nearest neighbors nn3 and nn4 lie in a plane perpendicular to the plane of the figure. (From Van de Walle and Neugebauer, 1995.)

The calculated binding energy, expressed with respect to an isolated H interstitial (at the BC site), is 1.34 eV. This is smaller than the binding energy for H to a dangling bond (2.50 eV) but larger than, for example, the binding energy of H to shallow impurities such as B or P (Van de Walle and Street, 1994). The electronic structure of the defect indicates that it is amphoteric

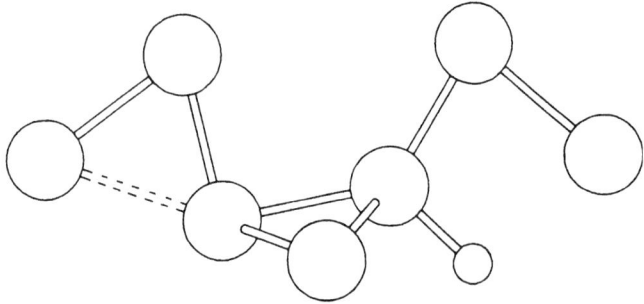

FIG. 7. Schematic representation of a cluster of Si atoms containing a complex between a Si self-interstitial and a single H atom. The Si atoms at the core of the defect are in positions that are distorted from the self-interstitial configuration illustrated in Fig. 6. (From Van de Walle and Neugebauer, 1995.)

in nature, with the acceptor level (transition level for the $0/-$ transition) located less than 0.1 eV above the donor level (the transition level for the $+/0$ transition) and both located around 0.4 eV above the valence band.

Complex formation between *two* H atoms and a Si self-interstitial also has been investigated (Van de Walle and Neugebauer, 1995). The structure of the complex is illustrated in Fig. 8. Now the two Si atoms at the core of the interstitial have twisted out of the $(\bar{1}10)$ plane. The bond between the two Si atoms attached to H is oriented 28° off $\langle 110 \rangle$ and only 17° off $\langle 100 \rangle$, so the configuration is actually closer to a $\langle 100 \rangle$ interstitial (Jones, 1997). The distortion allows these atoms to assume nominally fourfold coordination; i.e., they are bonded to four other atoms (three Si and one H), although the bond angles are significantly distorted (by as much as 35°) from the tetrahedral bond angle of 109°. The calculated vibrational frequency for the Si—H stretch mode is 1915 cm^{-1}. The calculated binding energy of this complex is 2.40 eV per H atom (with respect to the free H atom), essentially the same as for the complex with one H atom. The complex with two hydrogens has no levels in the bandgap, consistent with all the bonds being satisfied. These results for the atomic and electronic structure indicate that a single self-interstitial is unlikely to bind more than two H atoms.

Bech Nielsen *et al.* (1995) reported pseudopotential-density-functional calculations for complexes between the self-interstitial and two hydrogens. Their calculated structure is close to the one obtained by Van de Walle and Neugebauer (1995), with the bond between the two Si atoms attached to H 30° off $\langle 110 \rangle$ and only 12° off $\langle 100 \rangle$ (Jones, 1997). The calculated vibrational frequencies reported by Bech Nielsen *et al.* (1995) for the complex are 2106 and 2107 cm^{-1}. Bech Nielsen *et al.* (1995) also carried out vibrational spectroscopy for hydrogen interacting with native defects in *c*-Si and tentatively ascribed two modes, at 1987 and 1990 cm^{-1} to a $\langle 100 \rangle$ split interstitial saturated by two hydrogens. These values are somewhat closer to those calculated by Van de Walle and Neugebauer (1995), lending additional support to the assignment to a structure based on a $\langle 110 \rangle$ split interstitial.

The structures described here were derived for a crystalline environment; still, the general features also may apply to overcoordination defects in *a*-Si. First-principles calculations of Si—H bonds in amorphous networks by Tuttle and Adams (1998) included one example where removing a H atom from a Si—H bond leaves a fivefold coordinated Si atom behind. The binding energy in this case was calculated to be 1.5 eV, referenced to H at the BC in *c*-Si; this value is in good agreement with the results quoted earlier for H binding to a self-interstitial in *c*-Si. I also note that the calculated vibrational frequencies for the Si—H stretch modes, which are somewhat lower than for Si—H bonds at dangling bonds, could help explain why the

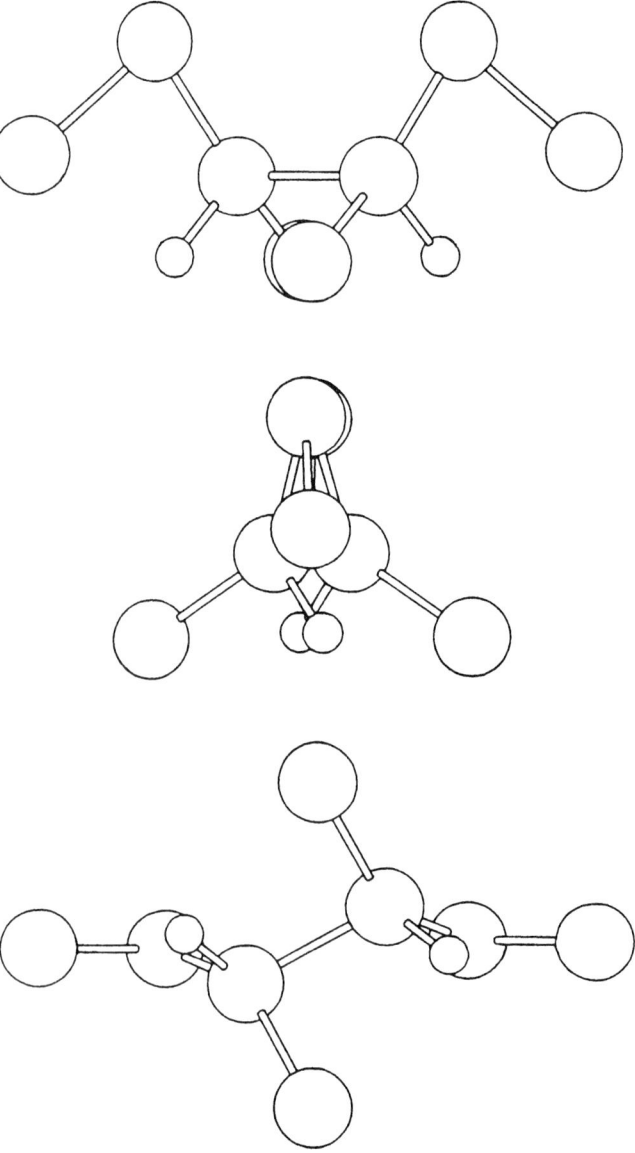

FIG. 8. Schematic representation of a cluster of Si atoms containing a complex between a Si self-interstitial and two H atoms. The Si atoms at the core of the defect are in positions that are distorted from the self-interstitial configuration illustrated in Fig. 6. In order to show the distortion out of the ($\bar{1}10$) plane, the cluster is viewed from three different directions: along the [$\bar{1}10$] direction (top), along the [$\bar{1}\bar{1}0$] direction (center), as well as along the [001] direction (bottom). (From Van de Walle and Neugebauer, 1995.)

absorption band corresponding to Si—H stretch modes extends to lower frequencies. The electrical activity of the complex described earlier is also relevant for understanding deep levels in a-Si. Furthermore, this complex may be a candidate for the H-induced acceptors observed by Nickel *et al.* (1995) in polycrystalline Si, as described in Section III.4.

4. HYDROGEN INTERACTIONS WITH WEAK Si—Si BONDS

Li and Biswas (1995a, 1995b) used a tight-binding molecular dynamic study (see Sec. I.3.d) to derive formation energies for insertion of hydrogen into weak Si—Si bonds. First, they generated a model of a-Si:H containing 272 atoms with periodic boundary conditions, containing both SiH and SiH_2 species. They calculated the formation energy of the reaction described by Eq. (1) at more than 70 different sites. In order to increase the likelihood of weak-bond occurrence, they systematically created strained a-Si:H configurations by dilating the supercells. Their results are shown in Fig. 9.

They arrived at the conclusion that the defect formation energy scales almost linearly with the bond-length deviation:

$$E = E_0 - \alpha \Delta R \qquad (2)$$

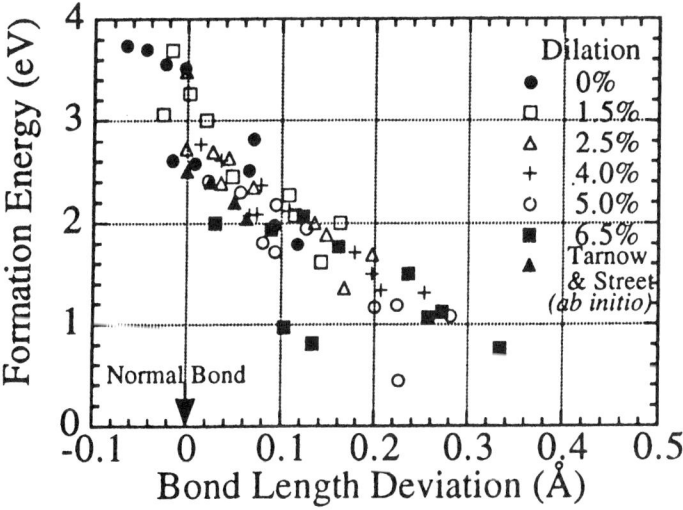

FIG. 9. Formation energy for the reaction described by Eq. (1) as a function of the Si—Si bond-length deviation. To enhance the likelihood of weak-bond occurrence, calculations were performed in cells with varying amounts of dilation. First-principles results from Tarnow and Street (1992) are also included. (From Li and Biswas, 1995a.)

where ΔR is the deviation of the Si—Si bond length from the average value. The calculations of Li and Biswas (1995a, 1995b) produced $E_0 = 2.5\,\text{eV}$ and $\alpha \approx 6.3\,\text{eV}/\text{Å}$. In Section III.3 I will discuss calculations (Van de Walle and Nickel, 1995) that were performed in a crystalline environment and produced a value for α between 4 and $5\,\text{eV}/\text{Å}$. The agreement is quite reasonable, with the larger value for a-Si:H indicative of an environment in which the atoms can relax more easily to accommodate the hydrogen-induced defect. It is possible, however, that the larger value of α obtained by Li and Biswas (1995a) is affected to some extent by their procedure, which involves dilation of the a-Si:H model (Tuttle and Adams, 1998).

5. SIMULATIONS OF AMORPHOUS NETWORKS

In this section I discuss computations for hydrogenated amorphous networks, which provide information about larger-scale atomic structure and/or energetics of the Si—H interactions.

Buda et al. (1991, 1992) performed first-principles molecular dynamics simulations, based on the pseudopotential-density-functional method and the Car-Parrinello approach (see Sec. I.3.a), for supercells containing 64 Si atoms and 8 H atoms. They obtained the amorphous structure by cooling from the melt; the cooling rate is inevitably much faster than in real experiments. Only monohydride groups occurred in the sample, supporting the idea that SiH_2 and other polyhydride groups have a low probability of occurring. A pronounced peak was found in the H-H pair correlation function around 2.5 Å, indicating a tendency for H to cluster.

Lee and Chang (1994) performed first-principles molecular dynamics simulations based on the density-functional-pseudopotential approach for supercells containing 64 atoms. Rather than using a fast quench from a liquid phase, they followed the procedure of Drabold et al. (1990), in which disordered networks are produced by a brief heating of c-Si, characterized by an incompletely melted sample. Their structural parameters were in reasonable agreement with experiment. Lee and Chang (1994) found that their amorphous model contained more dangling bonds than previously generated samples from liquid-quench simulations; this agrees with an assessment by Tuttle and Adams (1996) to be discussed below. Lee and Chang (1994) analyzed the electronic structure of the defects and found that the wave function of the midgap state has p-like character and is more localized on a threefold coordinated site than on a fivefold coordinated defect, in agreement with ESR (electron spin resonance) data.

Tuttle and Adams (1996) created a 242-atom model of a-Si:H using the approximate first-principles-pseudopotential approach based on the Harris functional (see Sec. I.3.b). They carried out a critical assessment of models derived from molecular dynamics; models produced from a liquid tend to be overcoordinated, whereas models produced from a crystal tend to lack disorder. They observed clustering of monohydrides in two distinct fashions: on the one hand, microclusters of two to four H, and on the other hand, H passivation of larger cavities.

Fedders and Drabold (1993) and Fedders (1995) used the approximate first-principles method described in Section I.3.b to create and study amorphous networks in supercells containing 62 Si atoms and varying numbers of H atoms. The cells contained 0, 1, or 2 well-localized defects. Fedders (1995) obtained properties of hydrogen configurations, in different charge states, for H at dangling bonds, in BC sites, and in other interstitial sites. The main result was a large spread of energies for a given type of defect; both H at BC and H at a dangling bond exhibited a spread of about 1 eV in binding energies, accompanied by considerable variations in the structural properties. This led Fedders to the conclusion that the distinction between BC hydrogen and a dangling bond plus a Si—H bond was not clear-cut.

More recently, first-principles results by Tuttle and Adams (1998) have contradicted the conclusions of Fedders (1995). For their study, Tuttle and Adams used self-consistent pseudopotential-density-functional theory on the a-Si:H model created by Guttman (1981) and Guttman and Fong (1982). They examined the energies of H inserted into Si—Si bonds and found the BC position to have an average energy 0.2 eV lower than BC in c-Si. For H bound in Si—H bonds, they found energies between 1.7 and 2.1 eV below H at BC in c-Si; the higher-lying energies are due to H-H interactions. They also identified configurations where removal of a H atom leads to reconstruction of a Si—Si bond; these Si—H bonds have energies in the range of 1.2–1.3 eV below H at BC in c-Si. The range over which the energies of the "shallow" and "deep" traps vary is much smaller than found by Fedders (1995); Tuttle and Adams attributed the differences with the work of Fedders (1995) to the more approximate nature of Fedder's calculational method.

Tuttle and Adams (1997) also examined the geometric and vibrational properties of clusters formed by hydrogen passivating a weak bond. They performed molecular-dynamics simulations for four different models of a-Si:H: one containing 54 atoms (Guttman, 1981; Guttman and Fong, 1982), one containing 68 atoms (Fedders and Drabold, 1993), one with 72 atoms (Buda *et al.*, 1991), and finally a 242-atom model developed by Tuttle and Adams themselves (1996). They proposed that these complexes are responsible for hydrogen clustering in disordered Si.

Finally, I note that the simulations of Fedders and Drabold (1996) revealed the existence of a quasi-continuous manifold of nearly degenerate and conformationally distinct metastable minima that are accessible to each other at moderate simulation temperatures. They suggested that these are the states associated with two-level systems. Low-energy excitations were observed, such as elastic modes with anomalously low frequencies ("floppy modes").

6. HYDROGEN DIFFUSION AND METASTABILITY — DISCUSSION

a. *Quantifying Energy Levels*

Now that we have accumulated a host of quantitative information on configurations and energetics of H in *a*-Si:H, we can return to the issues raised in Section II.1 and address the identification of the energies depicted in Fig. 4. One approach was given by Van de Walle and Street (1995), based on energy values included in Fig. 2. The zero of energy in that figure was chosen to correspond to the energy of a neutral H atom in free space. The level at -3.55 eV corresponds to the energy required to remove a hydrogen atom from an ideal Si—H bond, leaving a dangling bond behind. The theoretical definition of the level indicates that it is equivalent to the E_{Si-H} level in Fig. 4.

A key energy value (Van de Walle and Street, 1995) is the *formation energy* of a Si—H bond, referenced to bulk Si. This energy is listed in Fig. 2 at -2.17 eV and is derived by calculating the energy for the Si—H bond embedded in a Si environment and subtracting the energy of the H atom and of the Si atoms, assuming that all Si atoms are part of a bulk environment. Van de Walle and Street (1995) argued that this value is connected to the hydrogen chemical potential μ_H. The difference with the energy at -3.55 eV discussed earlier is that one includes the formation of a dangling bond, whereas the other does not. Note that the difference between the binding energy and the formation energy is $-2.17 - (-3.55) = 1.38$ eV; this value constitutes an estimate for the formation energy of a dangling bond in *c*-Si.

To understand the connection between the formation energy of Si—H bonds and μ_H, we need to describe the significance of μ_H in the case of *a*-Si:H. First of all, it is important to realize that μ_H should *not* be located at E_{Si-H}, which is the energy level corresponding to Si—H bonds. Instead, the chemical potential is located at an energy that corresponds to a minimum in the density of states. The density of states illustrated in Fig. 4b has contributions from two types of states: Around E_{Si-H} there is a distribution of Si—H bond energies; at higher energies we find a distribution of energies for interstitial hydrogen associated with distortions in weak Si—Si bonds. Hydrogen can move between these two types of configurations: Starting from a Si—H bond, hydrogen can move into a weak Si—Si bond, leaving a dangling bond behind, as described by the reaction in Eq. (1). The Si—H—Si bond can, in turn, convert to a Si—H bond plus a dangling bond; this does not affect the discussion here.

The chemical potential μ_H has to be located at a minimum in the density of states, between the weak-bond and Si—H bond levels. This follows from an energy-minimization argument: If μ_H were higher, a larger number of H atoms in weak Si—Si bonds would be formed, costing energy; if μ_H were lower, additional H atoms would be removed from Si—H bonds, creating dangling bonds, which also costs energy. The minimum-energy situation therefore occurs for μ_H located at the minimum in the density of states. This location is consistent with the observation that a-Si:H contains a relatively low number of dangling bonds; most bonds in the structure are satisfied, either because they are part of Si—Si bonds or by forming Si—H bonds. Furthermore, this location of μ_H corresponds to the condition that the formation energy of a Si dangling bond in a-Si:H is zero. Indeed, μ_H lies at the level where the formation energy of a Si—H—Si bond is equal to the formation energy of a Si—H bond; if we associate an energy with each of the terms in Eq. (1), this leads to the condition that the formation energy of the Si dangling bond in a-Si:H must be zero.

This discussion about μ_H immediately provides a connection with Fig. 2. The arguments in the preceding paragraphs show that the hydrogen chemical potential μ_H corresponds to the level where the formation energy of the dangling bond in a-Si:H is zero. We thus find that μ_H should be identified with the level corresponding to the formation energy of a Si—H bond, which was calculated to be at -2.17 eV. This allows derivation of a theoretical value for the activation energy E_A in Fig. 4: $E_A = E_S - \mu_H = (E_S - E_{BC}) + (E_{BC} - \mu_H) = [0.2 \cdots 0.5] + (1.12) = [1.32 \cdots 1.62]$ eV, where the range takes into account the uncertainty in $E_S - E_{BC}$. This value agrees well with experimental numbers for the diffusion activation energy E_A.

The formation energy of the Si—H bond is actually very close to the calculated formation energy of the Si—H bond on the Si(111) surface (-2.14 eV), as well as to the formation energy of SiH_4 (-2.22 eV), all expressed with respect to Si bulk (see Fig. 2). The similarity of these values indicates that there is no single microscopic structure that gives rise to this energy value; the key is that no energy cost needs to be paid for breaking any Si—Si bonds. Specific microscopic "implementations" also could include, for example, the extended H_2^*-based defects proposed by Zhang and Jackson (1991). The consistency of all these values strengthens our identification of this level with the hydrogen chemical potential. Placing μ_H at

−2.17 eV is also in good agreement with a rough estimate made by Street (1991c) based on an entirely different approach.

One remaining problem with the assignment of energy levels proposed by Van de Walle and Street (1995) is that the energy difference between μ_H and E_{Si-H} becomes equal to $-2.17 - (-3.55) = 1.38$ eV, i.e., the formation energy of a dangling bond in c-Si. This value is unphysically large for a-Si for two reasons: (1) In order for the assumption of thermal equilibrium to be valid, hydrogen needs to be able to redistribute among the various states at the temperatures of interest; an energy of 1.38 eV would be too large to allow exchange of hydrogen between Si—H bonds and the chemical potential at typical diffusion temperatures. (2) The dangling-bond creation process in a-Si involves excitation of a H atom out of an Si—H bond, which once again would cost 1.38 eV; however, experimental observations give a value below 0.5 eV (Street and Winer, 1989).

Both concerns can be addressed by establishing that the E_{Si-H} level is *higher* in a-Si than in c-Si. Indeed, explicit simulations for Si—H bonds in amorphous networks (see, e.g., Tuttle and Adams, 1998) show that the Si—H bond energies can be significantly higher than the value for H at an ideal, isolated dangling bond due to hydrogen clustering and H-H repulsion effects. In addition, mechanisms that facilitate the interchange between states at E_{Si-H} and μ_H can be invoked. For instance, exchange of diffusing H atoms with H bonded in Si—H atoms may occur with lower activation barriers than needed for excitation of a single H out of an Si—H bond; evidence for such a mechanism was found in experiments where deuterium was exchanged with hydrogen (Street, 1987), as discussed in Section II.2.a. Finally, the barrier for dissociation of a Si—H bond may be reduced significantly in the presence of free carriers (see Sec. II.2.c).

b. Explicit Calculation of Energy Distributions

Li and Biswas (1996) used the formation energies they obtained for weak Si—Si bonds (see Sec. II.4, Eq. 2) to derive an energy distribution for the formation energy of metastable defects under the assumption of a Gaussian distribution of the bond length. This allowed them to obtain a thermal equilibrium defect density, as well as a light-induced defect density, using a rate equation. Their calculated distribution of formation energies is shown in Fig. 10. This distribution depends on the value of α derived by Li and Biswas (1995a) (see Sec. II.4); if α is overestimated, as argued by Tuttle and Adams (1998), the distributions may be affected.

Li and Biswas (1996) also used their calculated energy distribution to examine the annealing dynamics of the defects. They found that (except at

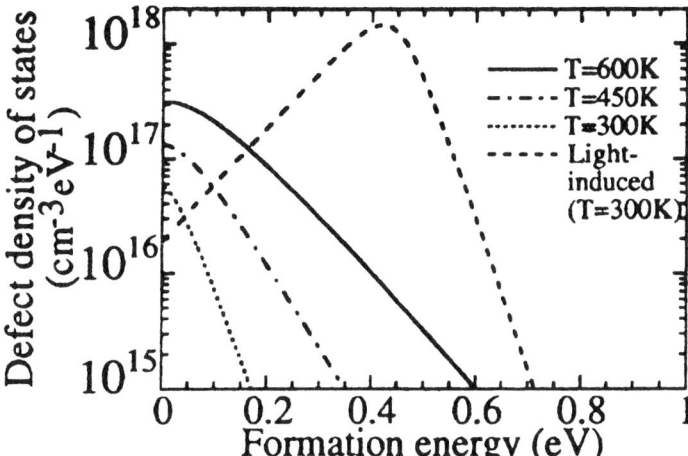

FIG. 10. Distributions of formation energies for thermal equilibrium and for light-induced defects. (Reprinted with permission from Li and Biswas, 1996. Copyright © 1996 American Institute of Physics.)

long times) the relaxation curves are well fit by a stretched exponential, in agreement with experiment (Kakalios et al., 1987). An alternative, simple explanation for stretched-exponential-type relaxations was proposed recently (Van de Walle, 1996a, 1996b). This model is based on a description of the release and retrapping of hydrogen at trap sites with a *single* energy level; it was found that this model provides a fit to experimental data that is at least as good as the traditional stretched exponential, without requiring a distribution of trap energies. This finding indicates that a stretched-exponential-like relaxation is not necessarily indicative of the presence of a distribution of trap energies.

c. *Models for Metastability*

Jones and Lister (1990) (see also Jones, 1991) performed first-principles pseudopotential calculations, in a cluster approach, to study H in various positions in c-Si, including configurations where the Si—Si bond was stretched. They found that as the Si—Si distance exceeds 3.8 Å, a transition takes place from the neutral symmetric BC configuration to a configuration with a Si—H bond plus a dangling bond. The positively charged BC center was found to be more stable.

Jones and Lister (1990) found that metastable states with an energy as low as 0.2 eV can form if the Si—Si bond is prestrained by 40%. They suggested that the SW effect occurs as follows: Upon illumination, electron-hole pairs are created at strained Si—Si bonds (~20% strain), which are consequently weakened further (>40%), enabling spontaneous hydrogen migration into them before electron-hole recombination occurs. Alternatively, electron-hole pairs are created at 40% strained bonds, and nonradiative recombination promotes H migration over the barrier into the bond center. Jones and Lister acknowledged that the density of strained bonds required for this mechanism to explain the experimental observations would exceed the equilibrium density; they suggested the strained bonds would be introduced athermally during the deposition process.

An alternative model was proposed by Biswas et al. (1991). These authors identified a bridge-bonded hydrogen interstitial as the initial or annealed state; this configuration consists of a hydrogen in a three-center bond between two Si atoms at a distance of about 3.2 Å. In the light-soaked state, the Si—Si separation increases, and the hydrogen forms a Si—H bond with one of the Si atoms, leaving a dangling bond on the other Si. One problem with this model is that the initial state is already electrically active.

Jackson (1990) suggested that two-hydrogen-atom complexes were involved in the metastability. These neutral complexes, the so-called H_2^* (Chang and Chadi, 1989; see Sec. I.2.e), consist of one H in a BC-like position and another H in an AB position. The complex can dissociate either thermally or through light-induced carriers into isolated atomic H configurations that diffuse rapidly to other weak Si—Si bond sites, creating midgap defects. Annealing occurs when these isolated H atoms form complexes either with the same or different H atoms. This proposal was backed up by first-principles calculations by Zhang et al. (1990), which demonstrated an explicit reaction pathway of carrier-induced changes through dissociation of paried-H complexes. The complex was shown to dissociate through the capture of charge carriers, with barriers that were found to be in agreement with the experimentally observed activation energies for single-carrier defect creation in a-Si:H.

Finally, I note that Li and Biswas (1996) recently used their tight-binding total-energy calculations to describe the changes in local structure following either electron or hole capture by a neutral dangling bond. They found significant changes in bond angles (by up to 10°) and displacements of nearest-neighbor Si and nearby H atoms (by more than 0.2 Å). Structural relaxations around a dangling-bond site, as a function of charge state, were investigated previously in a crystalline environment by Northrup (1989). The molecular dynamics simulations of Li and Biswas (1996) showed these relaxations to be fast, probably happening on a time scale less than 1 ms. These results do not support the suggestion that experimentally observed anomalously slow relaxations are associated with charge-state changes of dangling bonds (Cohen et al., 1992).

7. HYDROGEN VERSUS DEUTERIUM FOR PASSIVATION OF DANGLING BONDS

Recently, some very exciting results were published on the different characteristics of hydrogen and deuterium (D) for passivation of Si dangling bonds. The phenomena were first studied in the context of passivation of dangling bonds on Si surfaces; specifically, the Si(100)-(2 × 1) surface can be passivated perfectly by hydrogen (or deuterium). It was demonstrated that desorption of hydrogen from this surface can be achieved, with atomic resolution, by irradiation with electrons emitted from a scanning tunneling microscope (STM) tip (Lyding et al., 1994). When the experiment was repeated with deuterium (Avouris et al., 1996a), a very strong isotope effect was observed; the deuterium desorption yield was much lower (up to 2 orders of magnitude) than the hydrogen desorption yield.

Lyding et al. (1996) built on these observations to conduct an experiment in which the difference between H and D was studied in the context of passivation of defects at the Si/SiO_2 interface. This interface lies at the heart of Si metal-oxide-semiconductor (MOS) transistor technology. The interface exhibits a high degree of perfection, but any remaining defects (usually considered to be dangling bonds) can be passivated by hydrogen, which is accomplished in an intentional hydrogenation during processing. Unfortunately, some degradation occurs over time, mostly due to hot-electron effects. Lyding et al. (1996) observed significant improvements in the lifetime of (MOS) transistors when deuterium, rather than hydrogen, was incorporated at the Si/SiO_2 interface. Similarly to the surface desorption experiments, this indicates that the Si—D bond is more resistant to hot-electron excitation than the Si—H bond.

At first sight, it is very difficult to understand why the mass difference between H and D should cause such a large difference. From an electronic point of view, the isotopes are equivalent, and the static electronic structure of the Si—H and Si—D bonds is identical. The difference therefore must be attributed to dynamics—but it is hard to come up with an explanation that yields such a remarkable difference in desorption rates (up to 2 orders of magnitude). For instance, one can invoke the difference in zero point energies between the Si—H and the Si—D bond; however, this difference is on the order of 0.1 eV, whereas the dissociation barrier that needs to be overcome is on the order of several electronvolts; hence the small difference in zero-point energies has a negligible effect on tunneling rates, etc. Various other theoretical approaches have been suggested (Shen et al., 1995; Avouris et al., 1996b). Recently, a mechanism was proposed (Van de Walle and Jackson, 1996) that builds on these theoretical descriptions but proposes a specific pathway for the dissociation of Si—H and Si—D bonds, providing a natural explanation for the difference in dissociation rates.

It has been proposed that (at least in the low-voltage regime) the dissociation of Si—H bonds from Si(100) proceeds via a multiple-vibrational excitation by tunneling electrons (Shen et al., 1995)—a mechanism that would apply both to desorption from the surface and to the Si/SiO$_2$ interface. Electrons excite Si—H vibrational transitions with a rate proportional to the tunneling current. The extent to which vibrational energy can be stored in the bond depends on the lifetime, i.e., on the rate at which energy is lost by coupling to phonons. Because the lifetime of H on Si is long (Guyot-Sionnest et al., 1990, 1995), efficient vibrational excitation is expected. In the quantitative analysis by Shen et al. (1995) it was assumed that the vibrational energy is deposited in the *stretch* mode of the Si—H bond, which has a frequency around 2100 cm^{-1}. The same assumption is usually made implicitly in discussions of dissociation of Si—H bonds.

Van de Walle and Jackson (1996) pointed out that both the vibrational lifetime and carrier-enhanced dissociation mechanisms are most likely controlled by the Si—H *bending* modes, as discussed in Section II.2.c. The vibrational frequency of the bending mode for Si—H is around 650 cm^{-1}, and the estimated frequency for Si—D is around 460 cm^{-1}. This value is close to the frequency of bulk TO phonon states at the X point (463 cm^{-1}) (Madelung, 1991). Therefore, one can expect the coupling of the Si—D bending mode to the Si bulk phonons to result in an efficient channel for deexcitation. While it is quite possible to reach a highly excited vibrational state in the case of Si—H, this will be more difficult for Si—D. Deuterium therefore should be much more resistant to STM-induced desorption and hot-electron-induced dissociation due to the relaxation of energy through the bending mode.

Furthermore, as discussed in Section II.2.c, the bending mode provides a likely pathway for dissociation in the presence of carriers. Two conclusions mentioned in Section II.2.c have direct consequences for the vibrational excitation mechanism: (1) In the bulk, the bending-mode pathway for dissociation of a single Si—H involves an intermediate metastable state with the H atom located adjacent to the dangling bond (from which H has been removed). The activation barrier to reach this intermediate state is 1.5 eV, which is 1 eV lower than the energy difference between the Si—H bond and an isolated interstitial H. In this intermediate state, the system has electrically active levels in the bandgap; on capture of a carrier, the charge state of the hydrogen can change, and subsequent dissociation will proceed with no additional barrier. (2) Even before this intermediate metastable state is reached, levels are introduced into the bandgap. When the displacement of H from its equilibrium position reaches about 0.8 Å, gap levels emerge from the conduction and the valence bands. Again, capture of carriers in a gap level results in immediate dissociation. The bending-mode excitation pathway is therefore attractive because of the potential for carrier-enhanced dissociation and because the overlap of the Si—D bending-mode frequency with Si bulk phonons provides a natural explanation for its reduced dissociation rate.

It is tempting, of course, to suggest that the enhanced stability of Si—D compared with Si—H at Si surfaces and Si/SiO$_2$ interfaces also would apply to Si—D bonds in *a*-Si and could be used to address metastability and degradation problems. To my knowledge, no conclusive experiments on this issue have been performed. To observe an effect, it may be necessary to have most or all of the hydrogen in the sample replaced by deuterium; as long as any Si—H bonds are present, they may be the first to dissociate, leading to severe degradation even in the presence of deuterium. Further work in this area should be worthwhile.

III. Hydrogen in Polycrystalline Silicon

1. GRAIN BOUNDARIES

The grain boundaries are the most distinguishing feature of polycrystalline silicon. Inside the grains, the material is essentially bulk crystalline silicon, and hydrogen will display its usual range of behaviors there, including fast diffusion, formation of H$_2$ molecules, passivation of shallow and deep levels, formation of platelets (Nickel *et al.*, 1996), etc. Near the grain boundaries, however, the structure of the material is qualitatively different, and new phenomena can be expected to occur. The grain boundary is a two-dimensional extended defect (Bourret, 1985). When two grains are joined at a boundary, there is a strong driving force for reconstruction (Queisser, 1983); however, some point defects remain, which have been detected by electron spin resonance (ESR) and identified as Si dangling bonds (Johnson *et al.*, 1982). In addition, shallow localized states occur that form band tails; these states have been attributed to strained Si—Si bonds (Jackson *et al.*, 1983).

Detailed microscopic information on the atomic structure of grain boundaries is only gradually becoming available. A small number of computational studies have been performed; the following overview is not intended to be exhaustive but mainly to highlight the main features I will be using here. Thomson and Chadi (1984) performed tight-binding calculations of a high-angle tilt boundary in Si; they found that no dangling bonds were present, illustrating the large degree of reconstruction that takes place. Some deviations in bond length and bond angle were found, however. Similar conclusions were found by DiVincenzo et al. (1986) for a twin boundary in Si. More recently, Kohyama and Yamamoto (1994) performed semiempirical tight-binding calculations of twist boundaries in Si. They found that twist boundaries exhibit greater structural disorder and larger interfacial energies than tilt boundaries. The twist boundaries contain more coordination defects (dangling bonds or floating bonds) and also exhibit larger bond distortions (with bond-length deviations up to 10%). Recently, empirical-potential molecular-dynamics simulations produced the result that in thermodynamic equilibrium all high-energy grain boundaries are highly disordered, with a layer of about 5 Å around the boundary that is amorphous in nature (Keblinski et al., 1997).

Detailed information has been obtained from first-principles studies for grain boundaries in Ge (Tarnow et al., 1989, 1990). For the Σ5 and Σ5* twist boundaries, an *ab initio* molecular dynamics approach produced optimized geometries, for which the distribution of bond lengths was then examined. Atoms in the first layer near the boundary were found to exhibit nearest-neighbor bond lengths ranging from 2.2 Å to more than 2.8 Å; the equilibrium bond length in Ge is 2.45 Å. The distribution was asymmetric, with stretched bonds outnumbering compressed bonds.

In Section III.2 we will be particularly interested in the bond-length deviations that occur near grain boundaries. One may expect that bond distortions around dislocations bear some similarity to those around grain boundaries. I note that first-principles calculations also have been carried out for 90° partial dislocations in Si (Bigger et al., 1992); one may expect that the features of the atomic geometry around dislocations bear some similarity to those around grain boundaries. The results of Bigger et al. (1992) indicate that, once again, there is a tendency for bonds to be stretched, by amounts that can exceed 5% of the bond length (i.e., larger than 0.1 Å).

I conclude that (1) coordination defects (such as dangling bonds) can occur near grain boundaries, although they seem to occur mostly near higher energy (e.g., twist) boundaries, and (2) deviations in bond length and bond angle are common near grain boundaries, with stretched bonds occurring more frequently than compressed bonds and distortions of up to 10% of the bond length.

2. Hydrogen Interactions with Grain Boundaries

It is well known that hydrogen can passivate deep levels associated with the grain boundaries in poly-Si. Unlike the *a*-Si case, where H is incorporated during growth, dangling-bond defects in poly-Si are passivated by posthydrogenation at elevated temperatures. Exposure of poly-Si to monatomic hydrogen (from a remote plasma) at temperatures between 250 and 450°C leads to a decrease in the dangling-bond density as well as a reduction in the density of band-tail states (Johnson et al., 1982; Jackson et al., 1983). The details of the hydrogenation procedure can have a strong effect on the remaining spin density (Nickel et al., 1993a, 1993b); for more specific information, see Section IV of Chap. 4 in this volume. The passivation effects due to hydrogen are probably quite similar to those in amorphous silicon, as discussed in Section II. Not surprisingly, light-induced metastability has been observed (Nickel et al., 1993b), which has been associated with the interaction between hydrogen, dangling bonds, and strained Si—Si bonds.

One key experiment was performed on undoped samples that had been hydrogenated until the defect density saturated at a minimum value (Nickel et al., 1994). It was found that annealing around 160°C followed by a quench to low temperature produced an enhancement in the conductivity. The enhancement was metastable and decayed with time. The temperature dependence of the decay rate indicated that the activation energy for the decay process is $E_A \approx 0.74\,\text{eV}$. This activation energy allows for the metastable state to have a lifetime on the order of hours at room temperature. These observations are consistent with a model in which quenching leads to the trapping of H in an electrically active metastable state. The decay of the conductivity is then associated with the release of H from the metastable configuration. The H that is released from the traps returns to a lower-energy state (Fig. 11); this "reservoir" of H was found to have an energy 0.35 eV below that of the metastable state (Nickel et al., 1994). The experimental aspects are described in more detail in Section V.2 of Chap. 4 in this volume. I will examine the microscopic nature of this metastable state in the next section.

3. HYDROGEN-INDUCED GENERATION OF DONOR-LIKE METASTABLE DEFECTS

In the Introduction (Sec. I.2) I discussed the main features of the interactions between H and Si. The configurations that have energies in the relevant range are shown in Fig. 11 (configurations with energies below $-2.15\,\text{eV}$ are not included in the figure). An isolated, neutral H interstitial in Si resides at the BC site, with an energy 1.05 eV below the energy of H in free space. In this configuration, hydrogen acts as a donor, with a level about 0.2 eV below the conduction band (Bech Nielsen, 1991; Johnson et al., 1994; Van de Walle et al., 1989). The activation energy E_A mentioned earlier

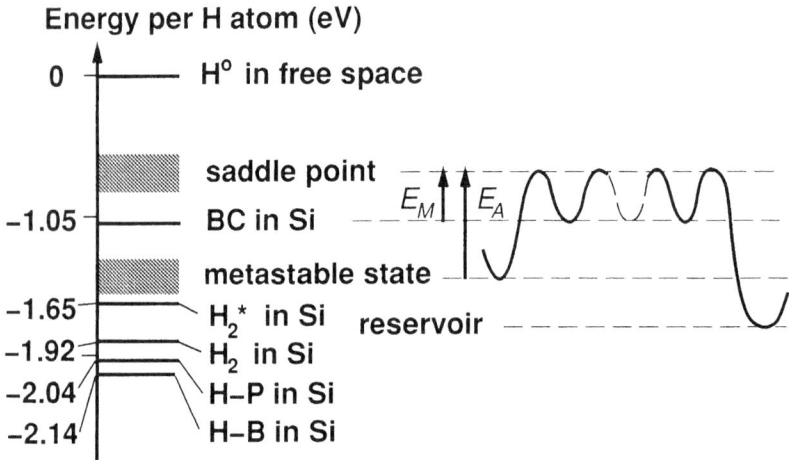

FIG. 11. Schematic diagram depicting first-principles energies of various configurations of hydrogen in silicon, as well as energy as a function of position. The saddle point for migration lies $E_M = 0.2 - 0.5$ eV above the level of H at BC. The activation energy $E_A \approx 0.74$ eV, obtained from the experiments by Nickel et al. (1994), is the energy difference between the metastable configuration and the saddle point. The energy of the metastable state therefore lies between -1.6 and -1.3 eV. (From Van de Walle and Nickel, 1995.)

corresponds to the energy difference between the metastable state and the saddle point of the migration path of an interstitial H atom. This saddle point occurs at an energy E_M above the level of the H interstitial at BC. The activation barrier for migration can be estimated to be between 0.2 and 0.5 eV, the lower number resulting from a first-principles calculation of an adiabatic energy surface (Van de Walle et al., 1989) and the higher number from high-temperature diffusion experiments (Van Wieringen and Warmoltz, 1956). We thus find that the energy of the metastable state is between -1.6 and -1.3 eV.

Figure 11 shows that stability of most Si—H configurations is such that activation energies much higher than 0.74 eV would be required to release hydrogen. In fact, the only configuration that allows for hydrogen to escape at room temperature or below is that of an isolated interstitial H atom at BC. In crystalline Si, the stability of BC hydrogen has been derived from DLTS (deep-level transient spectroscopy) measurements. It was found that the BC configuration is only stable at temperatures below 100 K (Bech Nielsen, 1991). However, variations on the basic BC configuration are possible that provide a higher stability and lower the energy of this configuration to fall in the range of -1.6 to -1.3 eV. As discussed in Section I.2.a, the formation energy of the BC configuration involves a balance between energy gain due to Coulomb interaction between the proton and the electron density in the bond and energy cost due to elastic energy required to move the Si atoms outward. Hypothetically, if the Si atoms initially were spaced further apart, the energy *gain* due to the formation of the three-center bond would still be the same, but the energy *cost* involved in moving the Si atoms would be lower, leading to a net increase in stability of the configuration. While this hypothesis is incompatible with the nature of pure c-Si, it is an attractive possibility in the case of poly-Si, where bond distortions near the grain boundaries are known to provide a variety of Si—Si bond distances.

This increase in stability of the BC configuration as a function of bond distortion was examined by Tarnow and Street (1992) and by Van de Walle and Nickel (1995); both studies employed first-principles pseudopotential-density-functional theory. Tarnow and Street (1992) investigated one aspect of the problem, namely, the effect of shear strain. Van de Walle and Nickel (1995) addressed bond-length variations as well as bond-angle variations.

Figure 12 summarizes the results for bond-stretching distortions (Van de Walle and Nickel, 1995). In this case, the Si atoms are displaced along the original bond direction. The circles in Fig. 12 indicate results for the change in formation energy expressed as a function of the *initial* distortion in the

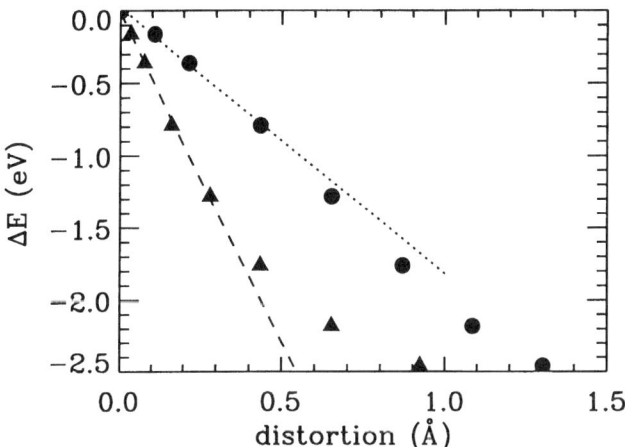

FIG. 12. Lowering in formation energy of the BC Si—H—Si configuration, for a bond-length distortion, as a function of *initial* distortion (circles) and as a function of the change in the *relaxed* bond length (triangles). For small distortions, the calculated points can be described by linear fits: 0.18 eV per 0.1 Å (dotted line) and 0.46 eV per 0.1 Å (dashed line).

Si—Si bond length. If relaxation is allowed, the Si atoms around the interface, of course, tend to move closer together, in order to try to restore the original Si—Si bond length. After relaxation, the change in bond length therefore will be smaller than given by the *initial* distortion; the triangles in Fig. 12 indicate results plotted as a function of the distortion in the *relaxed* bond length. For the configuration with H inserted in the bond, the Si atoms are, of course, also relaxed. The change in formation energy ΔE is always calculated using energy values for relaxed configurations, both in the absence and in the presence of hydrogen.

Figure 12 shows that for displacements up to 0.3 Å (which is the physical range of interest; see Sec. III.1), the behavior can be described adequately by a linear fit. Within the accuracy of the calculations, one finds a change in formation energy of 0.46 eV per 0.1 Å of increase in the (relaxed) Si—Si bond length. This value agrees quite well with a calculation by Li and Biswas (1995a, 1995b) for similar distortions in *a*-Si, as discussed in Section II.4. For bond-angle distortions, the results of Van de Walle and Nickel (1995) are in general agreement with the work of Tarnow and Street (1992). When the results were analyzed as a function of the increase in bond length that accompanies the bond-angle distortion, the behavior was found to be remarkably similar to that for pure bond-length distortions. The interesting conclusion was reached that for small displacements, no matter what the origin of the increase in bond length (be it pure bond stretching or bond bending), the formation energy of the Si—H—Si bond will be lowered by ~ 0.4–0.5 eV per 0.1 Å of distortion in the Si—Si bond.

These results can now be combined with the information about bond-length distortions near grain boundaries that was discussed in Section III.1. The microscopic atomic structure near grain boundaries includes a significant number of bonds with bond lengths exceeding the bulk bond length. Changes in the bond length of 0.1 Å seem common; the theoretical results then indicate that the formation energy of H atoms inserted in such a stretched bond would be lowered by 0.4–0.5 eV; according to Fig. 11, this is precisely the amount needed to bring the formation energy in the range where it could explain the experimental observations of Nickel *et al.* (1994).

As discussed earlier, the Si—H—Si configuration has been identified as a donor in *c*-Si, with an energy level approximately 0.2 eV below the conduction band. Not surprisingly, this configuration continues to behave as a donor state if the Si—Si bonds are strained prior to inserting H. However, the donor level moves away from the conduction band as the bond becomes distorted, at a rate of ~ 0.1 eV per 0.1 Å change in the bond length of the Si—Si bond (Van de Walle and Nickel, 1995). The experiments of Nickel *et al.* (1994) are not sensitive to the precise position of the donor level; indeed, the modification in the conductivity probably occurs through the introduction of charged defects at the grain boundary, which affect the depletion layers and hence the potential barriers at the boundaries (see Harbeke, 1985).

4. HYDROGEN-INDUCED GENERATION OF ACCEPTOR-LIKE DEFECTS

Prolonged exposure of poly-Si to monatomic hydrogen at elevated temperatures causes the generation of acceptor defects (Nickel *et al.*, 1995) (see also Sec. VI.1 of Chap. 4 in this volume). Type conversion was observed, and the acceptors were clearly established to be hydrogen induced. In contrast to the donor-type defects, for which a detailed microscopic model was discussed in the preceding subsection, the identification of the acceptor-type defects is more speculative at this stage. One explanation has been proposed, in terms of formation of hydrogen complexes with self-interstitial-type defects. These complexes, which were discussed in Section II.3, behave as acceptors, and it is plausible that their formation could be facilitated in the neighborhood of grain boundaries, which may have a tendency to form overcoordination defects (Kohyama and Yamamoto, 1994).

IV. Conclusions and Future Directions

In this chapter I have attempted to review the current theoretical understanding of hydrogen interactions with amorphous and polycrystalline silicon, focusing on studies that build on computational approaches. Over the past decade, a wealth of information has been generated about the structural and energetic aspects of the interaction of hydrogen with a-Si. Some of this information was obtained from calculations on hydrogen interacting with c-Si; many interesting results can be produced in this fashion. Of course, explicit simulations of the amorphous networks are essential to provide information about energy distributions, H-H interactions and clustering, etc. Unfortunately, calculations for amorphous structures require large systems and are computationally very costly. Such calculations therefore are currently difficult to accomplish with first-principles techniques. Progress in this area can proceed along two paths: (1) improvements in algorithms (e.g., order-N methods) as well as computational efficiency and (2) development of computationally less demanding tight-binding or empirical-potential methods. Both approaches would allow calculations for increasingly larger and more realistic systems. They also would allow molecular dynamics simulations on longer time scales. While much progress has been made, I do not think we can claim to have a full understanding of the phenomenon of metastability in a-Si. It is my feeling that progress in this area will not come exclusively from being able to simulate increasingly larger systems. Advances also should focus on devising new experiments, accompanied by new theoretical analyses, for instance, on the effect of substituting hydrogen with deuterium.

For polycrystalline silicon, many challenges remain as well. Since most of the interesting features revolve around the grain boundaries, more accurate experimental and theoretical information about the structure of grain boundaries needs to be generated. This, in turn, will enable a microscopic picture of the interaction of hydrogen with the grain boundaries. In particular, the nature of the acceptor-like defects introduced during prolonged hydrogenation (Nickel et al., 1995) requires further investigation.

ACKNOWLEDGMENTS

Thanks are due to W. B. Jackson, J. Neugebauer, N. H. Nickel, and R. A. Street for productive collaborations, to R. Biswas, R. Jones, and B. Tuttle for constructive comments, and to N. Greenleaves for a critical reading of the manuscript.

REFERENCES

Avouris, Ph., Walkup, R. E., Rossi, A. R., Shen, T.-C., Abeln, G. C., Tucker, J. R., and Lyding, J. W. (1996a). *Chem. Phys. Lett.*, **257**, 148.
Avouris, Ph., Walkup, R. E., Rossi, A. R., Akpati, H. C., Nordlander, P., Shen, T.-C., Abeln, G. C., and Lyding, J. W. (1996b). *Surface Science*, **363**, 368.
Bar-Yam, Y., and Joannopoulos, J. D. (1984). *Phys. Rev. Lett.*, **52**, 1129.
Bech Nielsen, B. (1991). *Phys. Rev. Lett.*, **66**, 2360.
Bech Nielsen, B., Hoffmann, L., Budde, M., Jones, R., Goss, J., and Öberg, S. (1995). *Mater. Sci. Forum*, **196–201**, 933.
Bigger, J. R. K., McInnes, D. A., Sutton, A. P., Payne, M. C., Stich, I., King-Smith, R. D., Burd, D. M., and Clarke, L. J. (1992). *Phys. Rev. Lett.*, **69**, 2224.
Biswas, R., Kwon, I., and Soukoulis, C. M. (1991). *Phys. Rev. B*, **44**, 3403.
Blöchl, P. E., Smargiassi, E., Car, R., Laks, D. B., Andreoni, W., and Pantelides, S. T. (1993). *Phys. Rev. Lett.*, **70**, 2435.
Bourret, A. (1985). In *Polycrystalline Semiconductors: Physical Properties and Applications*, **57**, G. Harbeke, ed. Berlin: Springer, p. 2.
Branz, H. M., Asher, S., Nelson, B. P., and Kemp, M. (1993). *J. Non-Cryst. Solids*, **164–166**, 269.
Buda, F., Chiarotti, G. L., Car, R., and Parrinello, M. (1991). *Phys. Rev. B*, **44**, 5908.

Buda, F., Chiarotti, G. L., Stich, I., Car, R., and Parrinello, M. (1992). *J. Non-Cryst. Solids*, **114**, 7.
Car, R., and Parrinello, M. (1985). *Phys. Rev. Lett.*, **55**, 2471.
Car, R., Kelly, P. J., Oshiyama, A., and Pantelides, S. T. (1985). *Phys. Rev. Lett.*, **54**, 360.
Carlson, D. E., and Magee, C. W. (1978). *Appl. Phys. Lett.*, **33**, 81.
Chang, K. J., and Chadi, D. J. (1989). *Phys. Rev. Lett.*, **62**, 937.
Cohen, J. D., Leen, T. M., and Rasmussen, R. J. (1992). *Phys. Rev. Lett.*, **69**, 3358.
Cowern, N. E. B., van de Walle, G. F. A., Zalm, P. C., and Vandenhoudt, D. W. E. (1994). *Appl. Phys. Lett.*, **65**, 2981.
Dabrowski, J., Müssig, H.-J., and Wolff, G. (1994). *Phys. Rev. Lett.*, **73**, 1660.
DiVincenzo, D. P., Bernholc, J., and Brodsky, M. H. (1983). *Phys. Rev. B*, **28**, 3246.
DiVincenzo, D. P., Alerhand, O. L., Schlüter, M., and Wilkins, J. W. (1986). *Phys. Rev. Lett.*, **56**, 1925.
Drabold, D. A., Fedders, P. A., Sankey, O. F., and Dow, J. D. (1990). *Phys. Rev. B*, **42**, 5135.
Eaglesham, D. J., Stolk, P. A., Gossmann, H.-J., and Poate, J. M. (1994). *Appl. Phys. Lett.*, **65**, 2305.
Estreicher, S. K. (1995). *Mater. Sci. Engr. Reps.*, **14**, 319.
Fedders, P. A. (1995). *Phys. Rev. B*, **52**, 1729.
Fedders, P. A., and Drabold, D. A. (1993). *Phys. Rev. B*, **47**, 13277.
Fedders, P. A., and Drabold, D. A. (1996). *Phys. Rev. B*, **53**, 3841.
Fisch, R., and Licciardello, D. C. (1978). *Phys. Rev. Lett.*, **41**, 889.
Guttman, L. (1981). *Phys. Rev. B*, **23**, 1866.
Guttman, L., and Fong, C. Y. (1982). *Phys. Rev. B*, **26**, 6756.
Guyot-Sionnest, P., Dumas, P., Chabal, Y. J., and Higashi, G. S. (1990). *Phys. Rev. Lett.*, **64**, 2156.
Guyot-Sionnest, P., Lin, P. H., and Miller, E. M. (1995). *J. Chem. Phys.*, **102**, 4269.
Hamann, D. R., Schlüter, M., and Chiang, C. (1979). *Phys. Rev. Lett.*, **43**, 1494.
Harbeke, G., ed. (1985). *Polycrystalline Semiconductors: Physical Properties and Applications*, Springer Series in Solid-State Sciences, **57**, part II, M. Cardona, ed. Berlin: Springer-Verlag.
Harris, J. (1985). *Phys. Rev. B*, **31**, 1770.
Herring, C., and Johnson, N. M. (1991). In *Semiconductors and Semimetals*, **34**, J. I. Pankove and N. M. Johnson, eds. San Diego: Academic Press, p. 279.
Hohenberg, P., and Kohn, W. (1964). *Phys. Rev.*, **136**, B864.
Jackson, W. B. (1990). *Phys. Rev. B*, **41**, 10257.
Jackson, W. B., and Kakalios, J. (1989). In *Advances in Disordered Semiconductors*, **1A**, H. Fritzsche, ed. Singapore: World Scientific, p. 247.
Jackson, W. B., and Tsai, C. C. (1992). *Phys. Rev. B*, **45**, 6564.
Jackson, W. B., Johnson, N. M., and Biegelsen, D. K. (1983). *Appl. Phys. Lett.*, **43**, 882.
Johnson, N. M., Biegelsen, D. K., and Moyer, M. D. (1982). *Phys. Rev. Lett.*, **40**, 882.
Johnson, N. M., Ponce, F. A., Street, R. A., and Nemanich, R. J. (1987). *Phys. Rev. B*, **35**, 4166.
Johnson, N. M., Herring, C., Doland, C., Walker, J., Anderson, G., and Ponce, F. (1991). In *Proceedings of the 16th International Conference on Defects in Semiconductors*, G. Davies, G. G. DeLeo, and M. Stavola, eds. Zürich: Trans Tech, p. 33.
Johnson, N. M., Herring, C., and Van de Walle, C. G. (1994). *Phys. Rev. Lett.*, **73**, 130.
Jones, R. (1991). *Physica B*, **170**, 181.
Jones, R. (1997). Private communication.
Jones, R., and Lister, G. M. S. (1990). *Philos. Mag. B*, **61**, 881.
Kakalios, J., Street, R. A., and Jackson, W. B. (1987). *Phys. Rev. Lett.*, **59**, 1037.
Keblinski, P., Philpot, S. R., Wolf, D., and Gleiter, H. (1997). *J. Am. Cer. Soc.*, **80**, 717.

Kemp, M., and Branz, H. M. (1993). *Phys. Rev. B*, **47**, 7067.
Kemp, M., and Branz, H. M. (1995). *Phys. Rev. B*, **52**, 13946.
Kohn, W., and Sham, L. J. (1965). *Phys. Rev.*, **140**, A1133.
Kohyama, M., and Takeda, S. (1992). *Phys. Rev. B*, **46**, 12305.
Kohyama, M., and Yamamoto, R. (1994). *Phys. Rev. B*, **50**, 8502.
Lee, I.-H., and Chang, K. J. (1994). *Phys. Rev. B*, **50**, 18083.
Li, Q., and Biswas, R. (1994a). *Phys. Rev. B*, **50**, 18090.
Li, Q., and Biswas, R. (1994b). *Mater. Res. Soc. Symp. Proc.*, **336**, 219.
Li, Q., and Biswas, R. (1995a). *Phys. Rev. B*, **52**, 10705.
Li, Q., and Biswas, R. (1995b). *Mater. Res. Soc. Symp. Proc.*, **377**, 407.
Li, Q., and Biswas, R. (1996). *Appl. Phys. Lett.*, **68**, 2261.
Lyding, J. W., Shen, T. C., Hubacek, J. S., Tucker, J. R., and Abeln, G. C. (1994). *Appl. Phys. Lett.*, **64**, 2010.
Lyding, J. W., Hess, K., and Kizilyalli, I. C. (1996). *Appl. Phys. Lett.*, **68**, 2526.
Madelung, O., ed. (1991). *Data in Science and Technology: Semiconductors*. Berlin: Springer-Verlag.
Nelson, J. S., Fong, C. Y., Guttman, L., and Batra, I. (1988). *Phys. Rev. B*, **37**, 2622.
Nickel, N. H., Johnson, N. M., and Jackson, W. B. (1993a). *Appl. Phys. Lett.*, **62**, 3285.
Nickel, N. H., Jackson, W. B., and Johnson, N. M. (1993b). *Phys. Rev. Lett.*, **71**, 2733.
Nickel, N. H., Johnson, N. M., and Van de Walle, C. G. (1994). *Phys. Rev. Lett.*, **72**, 3393.
Nickel, N. H., Johnson, N. M., and Walker, J. (1995). *Phys. Rev. Lett.*, **75**, 3720.
Nickel, N. H., Anderson, G. B., and Walker, J. (1996). *Solid State Commun.*, **99**, 427.
Northrup, J. E. (1989). *Phys. Rev. B*, **40**, 5875.
Pankove, J. I., and Johnson, N. M., eds. (1991). *Hydrogen in Semiconductors, Semiconductors and Semimetals*, **34**. San Diego: Academic Press.
Pantelides, S. T. (1986). *Phys. Rev. Lett.*, **57**, 2979.
Pantelides, S. T. (1987a). *Phys. Rev. Lett.*, **58**, 1344.
Pantelides, S. T. (1987b). *Phys. Rev. B*, **36**, 3479.
Pearton, S. J., Corbett, J. W., and Stavola, M. (1992). *Hydrogen in Crystalline Semiconductors*. Berlin: Springer-Verlag.
Ponce, F. A., Johnson, N. M., Tramontana, J. C., and Walker, J. (1987). In *Proceedings of the Microscopy of Semiconductivity Materials Conference*, A. G. Cullis, ed. *Inst. Phys. Conf. Ser.* **87**. London: Adam Hilger, Ltd., p. 49.
Queisser, H. J. (1983). *Mater. Res. Soc. Symp. Proc.*, **14**, 323.
Sankey, O. F., and Niklewski, D. J. (1989). *Phys. Rev. B*, **40**, 3979.
Sankey, O. F., and Drabold, D. A. (1991). *Bull. Am. Phys. Soc.*, **36**, 924.
Santos, P. V., and Jackson, W. B. (1992). *Phys. Rev. B*, **46**, 4595.
Santos, P., and Johnson, N. M. (1993). *Appl. Phys. Lett.*, **62**, 720.
Santos, P., Johnson, N. M., and Street, R. A. (1991). *Phys. Rev. Lett.*, **67**, 2686.
Shen, T.-C., Wang, C., Abeln, G. C., Tucker, J. R., Lyding, J. W., Avouris, Ph., and Walkup, R. E. (1995). *Science*, **268**, 1590.
Smith, Z. E., and Wagner, S. (1987). *Phys. Rev. Lett.*, **59**, 688.
Staebler, D. L., and Wronski, C. R. (1977). *Appl. Phys. Lett.*, **31**, 292.
Street, R. A. (1987). *Mater. Res. Soc. Symp. Proc.*, **95**, 13.
Street, R. (1991a). In *Hydrogenated Amorphous Silicon*. Cambridge: Cambridge University Press.
Street, R. (1991b). *Physica B*, **170**, 69.
Street, R. (1991c). *Phys. Rev. B*, **43**, 2454.
Street, R. A., and Winer, K. (1989). *Phys. Rev. B*, **40**, 6236.
Street, R., Tsai, C. C., Kakalios, J., and Jackson, W. B. (1987). *Philos. Mag. B*, **56**, 305.

Stutzmann, M., Jackson, W. B., and Tsai, C. C. (1985). *Phys. Rev. B*, **32**, 23.
Tan, T. Y. (1981). *Philos. Mag. A*, **44**, 101.
Tarnow, E., and Street, R. A. (1992). *Phys. Rev. B*, **45**, 3366.
Tarnow, E., Bristowe, P. D., Joannopoulos, J. D., and Payne, M. C. (1989). In *Atomic Scale Calculations in Materials Science,* J. Tersoff, D. Vanderbilt, and V. Vitek, eds. Materials Research Society Symposia Proceedings, Vol. 141. Pittsburgh: Materials Research Society, p. 333.
Tarnow, E., Dallot, P., Bristowe, P. D., Joannopoulos, J. D., Francis, G. P., and Payne, M. C. (1990). *Phys. Rev. B*, **42**, 3644.
Thomson, R. E., and Chadi, D. J. (1984). *Phys. Rev. B*, **29**, 889.
Tuttle, B., and Adams, J. (1996). *Phys. Rev. B*, **53**, 16265.
Tuttle, B., and Adams, J. (1997). *Phys. Rev. B*, **56**, 4565.
Tuttle, B., and Adams, J. (1998). *Phys. Rev. B*, **57**, 12859.
Tuttle, B., and Van de Walle, C. G. (1999). *Phys. Rev. B*, in press.
Tuttle, B., Van de Walle, C. G., and Adams, J. (1999). *Phys. Rev. B*, in press.
Van de Walle, C. G. (1991a). In *Hydrogen in Semiconductors*, **34**, J. I. Pankove and N. M. Johnson, eds. San Diego: Academic Press, p. 585.
Van de Walle, C. G. (1991b). *Physica B*, **170**, 21.
Van de Walle, C. G. (1994). *Phys. Rev. B*, **49**, 4579.
Van de Walle, C. G. (1996a). *Phys. Rev. B*, **53**, 11292.
Van de Walle, C. G. (1996b). In *Amorphous Silicon Technology*, M. Hack, R. Schropp, E. A. Schiff, A. Matsuda, and S. Wagner, eds. Pittsburgh: Materials Research Society, p. 533.
Van de Walle, C. G., and Street, R. A. (1994). *Phys. Rev. B*, **49**, 14766.
Van de Walle, C. G., and Street, R. A. (1995). *Phys. Rev. B*, **51**, 10615.
Van de Walle, C. G., and Nickel, N. H. (1995). *Phys. Rev. B*, **51**, 2636.
Van de Walle, C. G., and Neugebauer, J. (1995). *Phys. Rev. B*, **52**, R14320.
Van de Walle, C. G., and Jackson, W. B. (1996). *Appl. Phys. Lett.*, **69**, 2441.
Van de Walle, C. G., Denteneer, P. J. H., Bar-Yam, Y., and Pantelides, S. T. (1989). *Phys. Rev. B*, **39**, 10791.
Van Wieringen, A., and Warmoltz, N. (1956). *Physica*, **22**, 849.
Walsh, R. (1981). *Acc. Chem. Res.*, **14**, 246.
Watkins, G. D. (1991). In *Materials Science and Technology*, **4**, Cahn, Haasen, and Kramer, eds. Weinheim: VCH, p. 107.
Zacher, R., Allen, L. C., and Licciardello, D. C. (1986). *J. Non-Cryst. Solids*, **85**, 13.
Zhang, S. B., and Jackson, W. B. (1991). *Phys. Rev. B*, **43**, 12142.
Zhang, S. B., Jackson, W. B., and Chadi, D. J. (1990). *Phys. Rev. Lett.*, **65**, 2575.

CHAPTER 7

Hydrogen in Polycrystalline CVD Diamond

Karen M. McNamara Rutledge

DEPARTMENT OF CHEMICAL ENGINEERING
WORCESTER POLYTECHNIC INSTITUTE
WORCESTER, MASSACHUSETTS

I. INTRODUCTION 283
II. SOLID-STATE CHARACTERIZATION TECHNIQUES 286
   1. *Fourier Transform Spectroscopy* 286
   2. *Nuclear Magnetic Resonance* 288
   3. *Electron Paramagnetic Resonance* 290
   4. *Other Analysis Techniques* 293
III. RESULTS OF SOLID-STATE ANALYSIS 294
   1. *Covalent Bonding Environments* 294
   2. *Quantitative Hydrogen Concentrations* 297
   3. *Local Hydrogen Distribution* 298
   4. *Proximity to Paramagnetic Defects* 302
   5. *Macroscopic Hydrogen Distributions* 304
IV. EFFECTS OF HYDROGEN ON OBSERVED PROPERTIES 305
   1. *Infrared Transmission* 305
   2. *Thermal Conductivity* 306
V. SUMMARY 308
   REFERENCES 309

I. Introduction

The physical properties of natural diamond are in many cases a unique set of extremes that leads to their potential for use in a variety of applications. For example, the high thermal conductivity and resistivity of most type IIa natural diamonds make the material suitable for heat dissipation in electronic devices. Likewise, optical transparency combined with rigidity and hardness makes diamond an attractive material of fabrication for optical windows. However, the low availability of this precious natural material has resulted in little exploitation in these applications.

   Chemical vapor–deposited (CVD) diamond films with similar properties to those exhibited by natural diamond therefore have great economic potential. Exploitation of diamond's extreme properties depends primarily on our ability to reproduce the properties of the natural material. However, many CVD diamond samples do not exhibit some or all of the properties of natural diamond. The primary cause of this deviation is the polycrystalline nature of most synthetic materials and the presence of defects and impurities that are incorporated during the growth process.

Like many CVD processes for depositing thin films, the growth precursor for diamond deposition typically contain hydrogen. Hydrocarbon gases often are used as the carbon source. In addition to the small amount of hydrogen introduced with the carbon precursor, the remainder of the gas phase in CVD diamond growth is most often composed of hydrogen as well. In fact, most deposition environments contain hydrogen in excess of 95%. (The noted exception to this is the combustion synthesis of CVD diamond, which will not be discussed in detail here.) Thus hydrogen is incorporated as an impurity in many chemical vapor–deposited (CVD) semiconductor materials, and it is not surprising that in CVD diamond films it is the most abundant impurity.

In addition to a hydrogen-rich gas phase, the diamond growth process requires energy. The primary difference between most of the techniques used to grow diamond at low pressure is the manner by which energy is introduced into the system. Although a number of techniques have been used, the most commonly implemented methods are the arc jet, the microwave source, or a simple hot filament. The hot-filament system uses a resistively heated tungsten or tantalum filament to transfer thermal energy to the system, while the arc-jet and microwave sources, which typically have much higher power densities, cause the formation of plasmas that contain not only the radical species created in the thermal method but also ionized species. Studies, however, indicate a significant similarity in the quantity and bonding configurations of hydrogen in diamond produced by all three of these techniques.

It is proposed that hydrogen plays a number of critical roles in the deposition and processing of CVD diamond. In the low-pressure range (subatmosphere), where diamond is deposited, graphite is known to be the stable form of carbon, yet deposits containing tetrahedrally bonded, sp^3 carbon atoms are obtained routinely. Diamond can be deposited under such conditions as a result of the small free-energy difference between diamond and graphite. Hydrogen, however, is attributed with stabilizing the newly formed diamond by forming a carbon-hydrogen bond on the surface (Fig. 1), thus preventing the collapse to a graphitic or sp^2 bonded structure. This would not be possible were it not for the superequilibrium of hydrogen radicals present under growth conditions.

7 HYDROGEN IN POLYCRYSTALLINE CVD DIAMOND   285

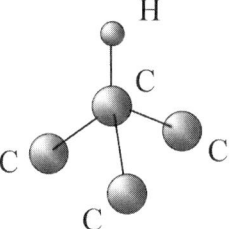

FIG. 1. Hydrogen-terminated sp^3 bonded carbon group.

A fully hydrogen-terminated sp^3 carbon surface does not allow for the continued growth and propagation of the diamond lattice. Hydrogen radicals in the gas phase are again thought to be crucial for the reaction to continue. The hydrogen atom now stabilizing the diamond surface must be removed by abstraction with a gas-phase radical (predominantly H), leaving an active site for carbon deposition. Of course, this open site may recombine with a carbon-containing radical, as shown in Fig. 2, or with another gas-phase hydrogen radical. The latter process is statistically more favorable, resulting in a relatively low efficiency of diamond growth. This critical role and the abundance of atomic hydrogen at the growth interface highlight the importance of hydrogen not only as a bulk defect but also as a significant surface and grain boundary defect in these polycrystalline materials. Finally, hydrogen is thought to contribute to the creation of carbon radicals in purely gas-phase reactions and to provide some of the energy required to maintain the temperature of the growth surface at approximately 1000°C through recombination on the surface.

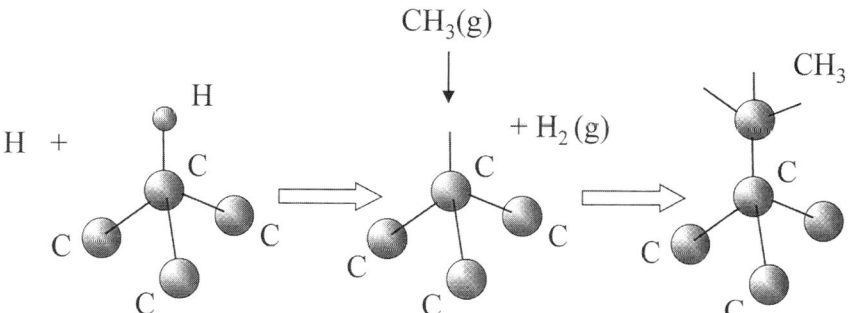

FIG. 2. Hydrogen recombination on a hydrogen-terminated diamond surface, followed by recombination with a hydrocarbon radical.

Hydrogen incorporation in CVD diamond has been investigated by a number of investigators using a variety of complementary techniques. These techniques include Fourier transform infrared spectroscopy (FTIR), nuclear magnetic resonance spectroscopy (NMR), electron paramagnetic resonance (EPR), nuclear reaction analysis (NRA), and Rutherford backscattering. In addition, there have been several studies of hydrogen defects associated with boron and nitrogen in single-crystal diamond. These single-crystal studies will not be discussed in detail, but rather, the reader is referred elsewhere (Sellschop *et al.*, 1974; McNamara Rutledge and Gleason, 1996).

Specifically discussed are the presence of hydrogen-related defects associated with lattice vacancies in polycrystalline CVD diamond as well as the probable distribution of these and other hydrogen environments in CVD diamond. These environments include both rigidly held static hydrogen and hydrogen-containing groups with significant mobility. The relative and absolute concentrations of each defect type are discussed. Although these results consider defects in diamond specifically, they can provide insight into phenomena observed in other semiconductor materials, such as hydrogen activation at grain boundaries in silicon.

II. Solid-State Characterization Techniques

1. FOURIER TRANSFORM SPECTROSCOPY

FTIR is an attractive technique for analyzing hydrogen in synthetic diamond because it is nondestructive and may be performed relatively quickly. In addition, FTIR can be used to identify specific hydrocarbon groups responsible for given IR absorptions. For example, it can readily distinguish moieties such as CH, CH_2, CH_3, etc. It is important to note, however, that FTIR will not identify noncovalently bonded hydrogen, such as H atom interstitials. H_2 inclusions, as have been observed in amorphous silicon, would not be observed in the absence of a local electronic field gradient as well.

In the simplest experiment, the incident beam of IR irradiation enters the sample at an angle of 90° with respect to the surface, and the transmitted energy is collected on the opposite side. Such measurements are made over a frequency range of 900–4000 cm^{-1}, in order to obtain an IR spectrum. The IR spectrum of a typical CVD diamond sample is shown in Fig. 3. Samples with a smooth, polished surface yield the most accurate results, since signal can be lost due to scattering on rough surfaces. If rough surfaces are used, scattering must be taken into account. Combined contributions of

7 HYDROGEN IN POLYCRYSTALLINE CVD DIAMOND 287

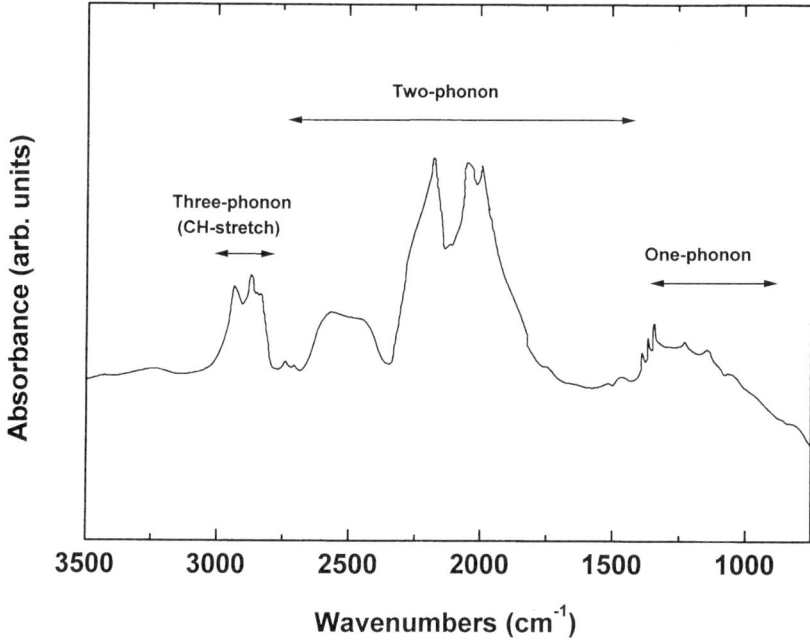

FIG. 3. FTIR absorption of a diamond film demonstrating absorptions in the one-, two-, and three-phonon regions.

internal and surface scattering are often estimated by an appropriate polynomial baseline fit. The measure transmittance T is often converted to absorbance A by Eq. (1), where T_0 represents the polynomial fit. If free-standing films cannot be obtained, samples on their foreign substrates may be examined using reflection techniques.

$$A = \log(T_0/T) \qquad (1)$$

Quantitative information also may be obtained for materials with similar absorption coefficients. Absorption coefficients observed in the literature for systems that may be similar to synthetic diamond, such as alkanes (Francis, 1950), paraffinic hydrocarbons (Angus et al., 1984), and a-Si:CH (Nakawaza et al., 1982), show only small variations, indicating that it may be possible to identify an effective absorption coefficient characteristic of polycrystalline diamond and obtain quantitative information from the FTIR spectrum of these materials. While some studies report absorption coefficients for the CH stretch comparable with those systems listed above (McNamara et al., 1992a), others have noted discrepancies (Hartnett and Miller, 1990). Thus complementary techniques such as NMR (McNamara et al., 1992a, 1992c) and Rutherford backscattering (Derry et al., 1983) are often used for a quantitative study.

2. Nuclear Magnetic Resonance

Quantitative measurements of impurities such as ^{1}H, averaged over the entire sample volume, can be obtained using solid-state NMR. Proton NMR is nondestructive and particularly valuable for measuring hydrogen concentrations, since this mobile, low-Z element is difficult to quantify by other spectroscopies. In cases where paramagnetic impurity concentrations are low and spin-lattice relaxation times are fast (which I demonstrate is the case for CVD diamond films), each nuclei in the magnetic field will give rise to the same integrated signal intensity, regardless of its local bonding environment. Thus comparison of the integrated area under the NMR line shape with that of a known standard provides a quantitative overall hydrogen count. Typical ^{1}H NMR spectra yield overall hydrogen concentrations ranging from <0.017 to ~1 at.% H for CVD films (McNamara Rutledge and Gleason, 1996).

As mentioned, NMR experiments are quantitative, regardless of the local bonding environment of the nuclei, only for samples in which the concentration of paramagnetic and ferromagnetic defects is small in comparison with the total concentration of nuclei of interest. In samples containing such impurities, a small fraction of nuclei in close proximity to such defects will remain undetected as a result of extreme line broadening due to interaction with the unpaired electron of the paramagnetic impurity (Henrichs et al., 1984). That this represents only a small fraction (<0.05%) of the total hydrogen in CVD diamond films will be demonstrated in the EPR section below. As additional supporting evidence, the ability to quantitatively correlate IR spectroscopy and NMR results confirms that paramagnetic defects do not shield a large number of protons from NMR detection in films examined to date (McNamara et al., 1994).

For quantitative measurements, it is also important to ensure that sufficient spin-lattice relaxation occurs between signal acquisitions (McNamara et al., 1992a, 1992c). This relaxation is characterized by a time constant T_1 known as the *spin-lattice relaxation time constant* and determined by saturation recovery techniques and subsequent application of the Bloch equation (Abragham, 1983):

$$\frac{M(t) - M_{eq}}{M_0 - M_{eq}} = \exp\left(-\frac{t}{T_1}\right) \qquad (2)$$

where $M(t)$ is the net magnetization parallel to the external magnetic field at time t, M_{eq} is the equilibrium magnetization, and M_0 is the initial magnetization at $t = 0$. The spin-lattice relaxation for hydrogen in CVD diamond is relatively fast, between 5 msec and 1 sec at room temperature (Levy and Gleason, 1992; Lock et al., 1993).

NMR spectroscopy also can provide information on the distribution and local environment of an element. The room-temperature and low-temperature ^{1}H NMR spectra for a typical CVD diamond film is shown in Fig. 4. Such spectra often contain two components, a narrow Lorentzian and a broad Gaussian, indicating at least two different bonding configurations for hydrogen (McNamara et al., 1992a). The ratio of integrated intensities of these components shows that the majority of the hydrogen contributes to the Gaussian component, whereas the Lorentzian component can be attributed to only a small fraction of the hydrogen present. The actual integrated intensity under each line shape is directly proportional to the number of nuclei in its environment. In addition, the average interproton spacing within the Gaussian component can be determined through application of the van Vleck equation (van Vleck, 1948), which reduces for a polycrystalline material with closely spaced hydrogen nuclei to

$$\text{FWHM}_G = 189.6(\Sigma r_{ij}^{-6})^{1/2} \text{ Å}^3 \text{kHz} \tag{3}$$

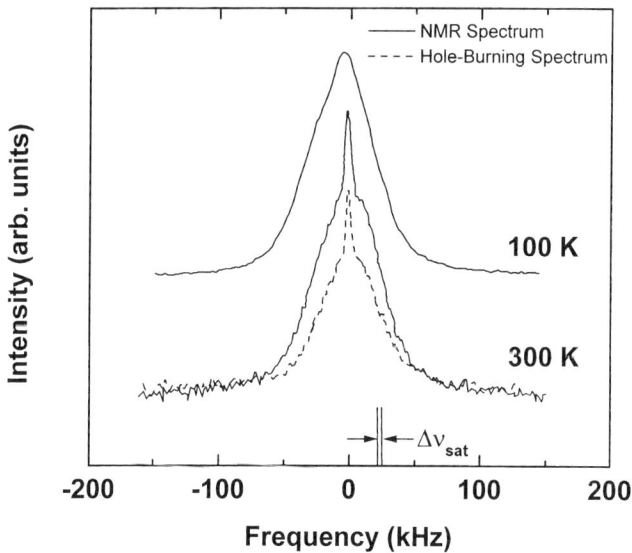

FIG. 4. Solid-state hydrogen NMR spectrum of a typical CVD diamond film demonstrating the characteristic Lorentzian and Gaussian line shapes.

where $FWHM_G$ is the full-width at half-maximum intensity of the Gaussian line shape, and r_{ij} is the interproton spacing.

In addition, the dimensional distribution of hydrogen on a greater length scale can be explored by ^{1}H multiple-quantum (MQ) NMR studies (Levy and Gleason, 1992). Up to 20 hydrogen atoms within a 10-Å radius were correlated in CVD diamond films despite the very low (<0.1 at.% H) overall bulk concentrations, indicating large-scale clustering of hydrogen. The initial MQ growth for the films can be compared with that for known bulk (three-dimensional) hydrogen distributions such as for CaH_2 and with known two-dimensional proton distributions such as intentionally hydrogenated diamond powder. Such comparisons lead to conclusions on the dimensionality of clustered hydrogen in diamond films.

3. ELECTRON PARAMAGNETIC RESONANCE

Electron paramagnetic resonance (EPR) spectroscopy detects defects that have an unpaired electron. In diamond, unpaired electrons often are localized at impurity atom sites, such as substitutional nitrogen, or at lattice defects, such as vacancies. Like NMR, EPR can give quantitative and distributional information about the unpaired electrons. In addition, detailed information can be obtained about the identities of the atoms and symmetry that surround the unpaired electron (Loubser and van Wyck, 1978). Nuclei having angular momentum that are in the proximity to the paramagnetic defect often can be identified through EPR as a result of hyperfine interactions. The strength and orientation dependence of hyperfine couplings depend on the location of the nuclear spin (or spins) with respect to the unpaired electron. Since the predominant isotope of hydrogen, ^{1}H, is a spin-1/2 nucleus, its interactions with nearby paramagnetic defects can be observed in this manner.

In contrast to natural and HPHT diamonds, which often contain significant nitrogen impurities, dangling-bond defects often are assumed to dominate the EPR spectra of CVD diamond films. Their concentration N_S ranges from 10^{17} to 10^{19} cm^{-3} (Watanabe and Sugata, 1988; Fabisiak et al., 1993; Fanciulli and Moustakas, 1993), significantly lower than observed hydrogen contents. Previous studies have identified an $S = 1/2$ electron paramagnetic resonance (EPR) at $g = 2.0028$ with a pair of weaker, partially resolved satellites in CVD diamond films produced from methane and hydrogen gases (Holder et al., 1994; Hoinkis et al., 1991; Portis, 1953; Trammell et al., 1958). One of these studies, using samples grown from deuterated reactants, has shown that the satellites arise from the hyperfine interaction with nearby hydrogen. A second study, combining observations at 9.8 and 35 GHz, additionally has proposed that they arise from the forbidden $\Delta m \pm 1$ nuclear spin flips of hydrogen during the EPR $\Delta M \pm 1$ transitions. Line widths of dangling-bond defects in polycrystalline diamond range from 1.9 to 12 G (Holder *et al.*, 1994; Hoinkis *et al.*, 1991; Portis, 1953; Trammell *et al.*, 1958; Zhang *et al.*, 1992; Erchak *et al.*, 1992). The smaller line widths are observed in films with lower dangling-bond densities, consistent with the reduction of dipole-dipole interactions between randomly distributed paramagnetic centers (Erchak *et al.*, 1992).

Zhou *et al.* (1996) reported the study of this EPR signal at the intermediate frequencies of 14 and 20 GHz. These results confirm in detail the $\Delta m \pm 1$ origin of the satellites. In addition, I find the result that the relative intensities of the central and satellite components, as well as their overall line shapes, can be accurately matched at all four frequencies by the assumption of a single unique defect with hyperfine interaction $A_{\parallel} = +28$ MHz and $A^{\perp} = -6$ MHz with a single hydrogen atom. From analysis of the hyperfine anisotropy, the separation between the electronic and nuclear spins is determined, leading to a microscopic model of the defect. An additional resonance is also observed in the spectra of samples containing relatively low hydrogen contents, and two possible interpretations for it are discussed.

Results at microwave frequencies $n = 14$ and 20 GHz are shown in Fig. 5, along with the results of Holder *et al.* (1994) at 9 and 35 GHz. Taken together, they clearly demonstrate a monotonically increasing separation and decreasing intensity of the satellites as the microwave frequency increases. That this is the characteristic signature for forbidden $\Delta m \pm 1$ nuclear spin flip transitions (Holder *et al.*, 1994; Trammell *et al.*, 1958; Halliburton *et al.*, 1979) can be seen as follows: The spin Hamiltonian for an electronic spin $S = 1/2$, with isotropic g, coupled to a nuclear spin $I = 1/2$ is

$$H = g\mu_B \mathbf{S} \cdot \mathbf{B} + \mathbf{S} \cdot \mathbf{A} \cdot \mathbf{I} - g_N \mu_N \mathbf{I} \cdot \mathbf{B} \qquad (4)$$

from which the nuclear Hamiltonian, to first order in $A/g\mu_B B$, becomes

$$H_N = -g_N \mu_N \mathbf{I} \cdot [\mathbf{B} - (\mathbf{A} \cdot \mathbf{B} M / g_N \mu_N B)] \qquad (5)$$

where g and g_N are the electronic and nuclear g values, respectively, μ_B and μ_N are the corresponding Bohr magnetrons, $\mathbf{S}$ and $\mathbf{I}$ are the corresponding spin operators, $\mathbf{A}$ is the nuclear hyperfine tensor, and $M \pm 1/2$ is the azimuthal quantum number for $\mathbf{S}$ quantized along $\mathbf{B}$, the applied magnetic field. The bracketed term represents an effective magnetic field seen by the nucleus, and when A is anisotropic, its two vector components, B and $-\mathbf{A} \cdot \mathbf{B}/2g_N, \mu_N B$, can point in different directions, as illustrated schemati-

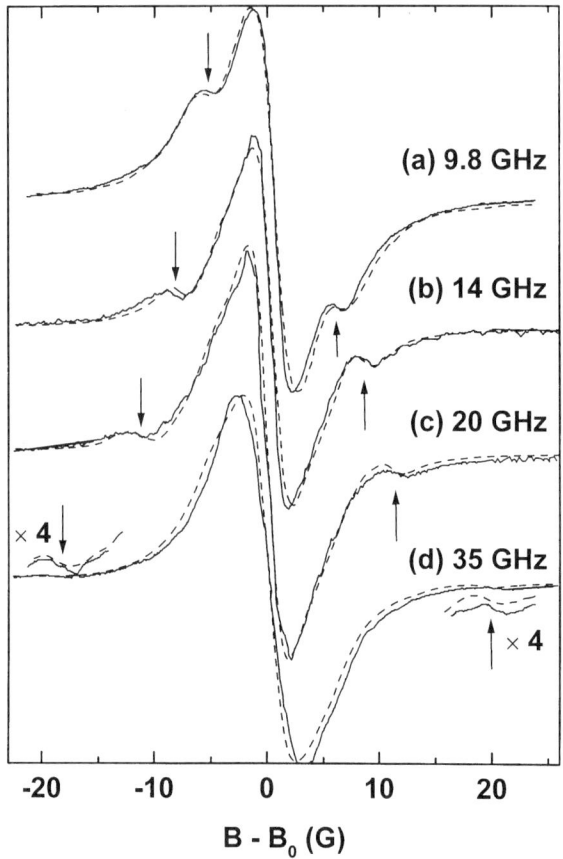

FIG. 5. Magnetic field dependence of EPR spectra of a defect in polycrystalline CVD diamond.

cally in Fig. 6b. Thus the quantization axis for the nucleus differs for the two quantization states, and as a consequence, the orthogonality of the $\Delta m \pm 1$ nuclear quantization states is destroyed. The intensity of the EPR transitions are proportional to

$$I(+1/2, m\langle - \rangle - 1/2, m') \propto |\langle +1/2|S + |-1/2\rangle|^2 |\langle m|m'\rangle|^2 H_1^2 \qquad (6)$$

where $|\langle m|m'\rangle|^2 = \cos^2(\Theta/2)$ for the $\Delta m = 0$ transitions and $\sin^2(\Theta/2)$ for the $\Delta m = \pm 1$ transitions and Θ is the angle between the nuclear quantization axis for $M = -1/2$ and that for $M = +1/2$, and H_1 is the amplitude of the microwave magnetic field. As illustrated in the figure, Θ can depart

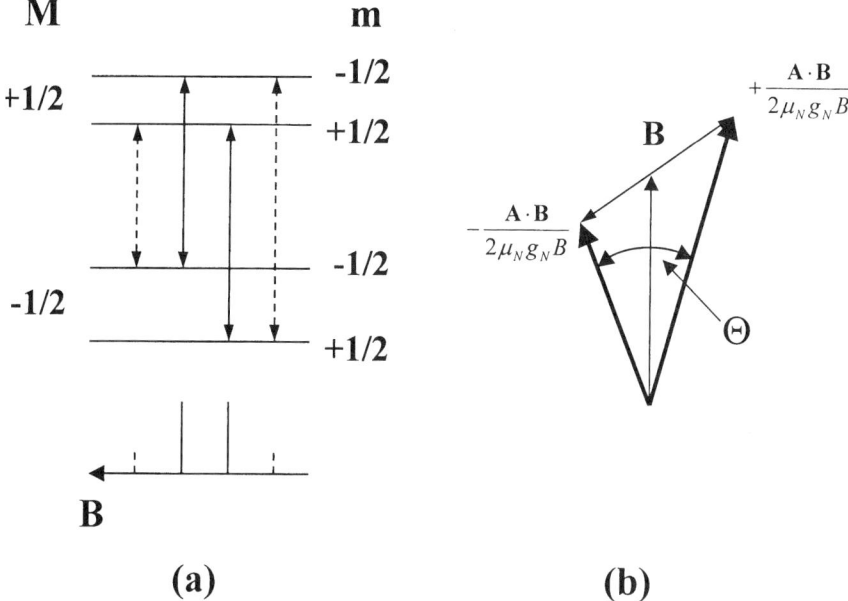

FIG. 6. (a) Energy-level diagram and EPR transitions for an $S = 1/2$ electron spin coupled to an $I = 1/2$ nucleus. (b) The effective magnetic field seen by the nucleus for the two M states.

substantially from zero when the anisotropy in $A \approx 2g_N\mu_N B$, leading to significant intensity in the forbidden $\Delta m = \pm 1$ transitions (Zhou et al., 1996). As B increases, Θ decreases as B^{-1}; the forbidden intensities decrease, therefore, as B^{-2}, where $B \approx B_0 = h\nu/g_B$, and the satellite positions, given by Eq. (1), approach field values of $\pm(g_N\mu_N/g\mu_B)B_0$ to either side of the central line.

4. OTHER ANALYSIS TECHNIQUES

Elastic recoil spectrometry (ERS) and nuclear reaction analysis (NRA) also have been used to obtain information on hydrogen in CVD diamond. Both can provide quantitative information on hydrogen contents in these materials. ERS typically detects only atoms that are lighter than the incident helium ions and has been used to study both hydrogen in diamond produced with normal isotopic compositions and deuterium in isotopically enriched samples. Hydrogen is also well suited to study by NRA because nuclear reactions involving ^{15}N and ^{19}F give interference-free hydrogen signatures (Sellschop et al., 1992, 1994; Windischmann et al., 1994).

III. Results of Solid-State Analysis

1. COVALENT BONDING ENVIRONMENTS

The IR spectrum of a natural type IIa diamond is shown in Fig. 7, along with the spectrum of a high-quality CVD diamond sample. The spectra are well matched and show ~70% transmission over nearly the entire IR portion of the spectrum. Both show the intrinsic two-phonon absorption between 1333 and 2666 cm^{-1} that is present in all diamond. It is important to note the transparency of both samples in the 8- to 12-μm region (850–1250 cm^{-1}) and in the CH-stretch region between 2750 and 3300 cm^{-1}, since these are areas of commercial interest. Most CVD diamonds, however, show absorptions in these regions as a result of defects. A more typical spectrum obtained from a CVD diamond film is shown in Fig. 3 (McNamara et al., 1994). The CH-stretch absorptions are particularly prevalent in CVD diamond and have been related directly to the hydrogen content in these films.

A large number of recent IR studies on CVD diamond have concentrated on this absorption in the CH-stretch region of the IR spectrum (McNamara Rutledge and Gleason, 1996). The stretching vibrations of carbon-hydrogen

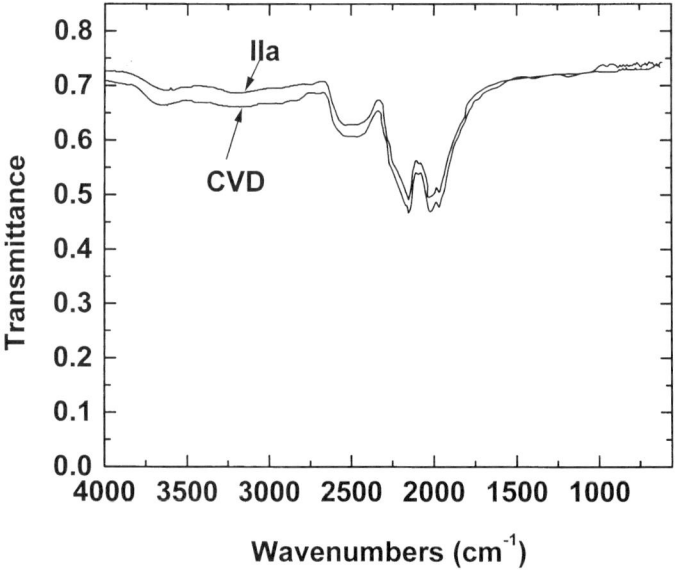

FIG. 7. The IR spectrum of a natural type IIa diamond window compared with that of a state-of-the-art polycrystalline CVD sample.

bonds occur in this region (2750–3300 cm^{-1}), providing information about hydrogen that is covalently bound to carbon. (Note again that free hydrogen atoms or H$_2$ molecules do not give rise to absorptions in the CH-stretch region.)

The frequency of the C—H bond vibrations shift slightly in response to different local environments. For example, the absorption due to hydrogen bonded to sp^2 bonded carbon appears above 2950 cm^{-1}, whereas that associated with sp^3 bonded carbon will appear below 3000 cm^{-1} (Bellamy, 1980; Socrates, 1980). Thus the CH-stretch region of the IR spectrum yields information about the carbon-bonding environment as well as that of hydrogen. Indeed, absorptions at 3025 cm^{-1}, related to sp^2 bonded carbon, have been observed in diamond films described as being "of poor quality" (McNamara et al., 1992a; Dischler et al., 1993; Zhang et al., 1992).

The number of hydrogen atoms bonded to a single carbon atom also will cause slight variations in the frequency of the CH-stretch vibrations (Table I) (McNamara Rutledge and Gleason, 1996). The dominant absorptions in the CH-stretch region typically appear near 2850 and 2920 cm^{-1} and are indicative of the symmetric and asymmetric stretching of sp^3-CH$_2$ groups. In many diamond films, the absorption intensity for the symmetric and asymmetric stretches of sp^3-CH$_2$ are clearly unequal as a result of steric effects, making quantitative deconvolution of these absorptions difficult (McNamara et al., 1994). The somewhat less intense absorptions present in the spectrum at 2880 and 2960 cm^{-1} are similarly related to the symmetric and asymmetric stretching vibrations of sp^3-CH$_3$ groups.

Two other absorptions also have been observed in CVD diamond, but their origin remains the subject of debate (McNamara et al., 1994; Dischler et al., 1993; Zhang et al., 1992). These absorptions appear at frequencies below those normally observed for CH-stretching vibrations in alkanes and amorphous carbon, specifically at 2820 and 2833 cm^{-1}. They have been

TABLE I

CHARACTERISTIC CH-STRETCHING VIBRATION FREQUENCIES

Frequency (cm^{-1})	Characteristic group
2850	sym. sp^3-CH$_2$
2880	sym. sp^3-CH$_3$
2920	asym. sp^3-CH$_2$
2960	asym. sp^3-CH$_3$
2980	sym. sp^2-CH$_2$
3025	sp^2-CH
3080	asym. sp^2-CH$_2$

observed in films produced by microwave, hot-filament, and dc arc-jet deposition. One possibility is that the peaks are related to the hydrogen-terminated $\langle 111 \rangle$ surface of diamond (Dischler et al., 1993). Others suggest that the absorptions at 2820 and 2833 cm^{-1} can be attributed to nitrogen- and oxygen-related defects such as N—CH$_3$ and O—CH$_3$, respectively (McNamara et al., 1994; Zhang et al., 1992). Observation of these groups is well documented in the organic chemistry literature and is considered positive identification of such a group.

Based on the peak positions in Table I, least-squares fits to the experimentally measured CH-stretch absorptions, such as that shown in Fig. 8, can determine which CH$_i$ species are present (McNamara et al., 1994; Dischler et al., 1993). However, it should be noted that there are a large number of fitted parameters, and fits do not provide quantitative information. It is possible to reduce the number of fitted parameters by fixing positions, peak widths, and/or intensity ratios of the individual components. However, the ratio of symmetric to asymmetric absorptions should not be

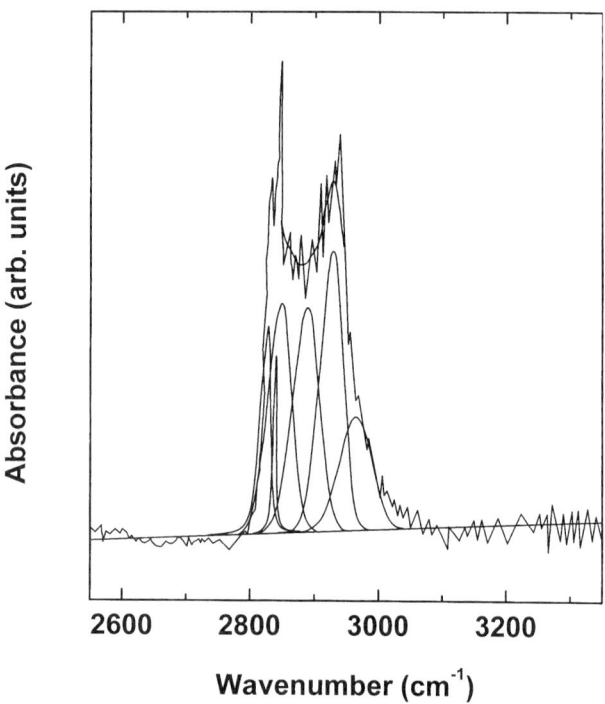

FIG. 8. Least-squares fit to the experimentally measured CH-stretch absorption of a CVD diamond film showing groups with both sp^3 and sp^2 carbon bonding.

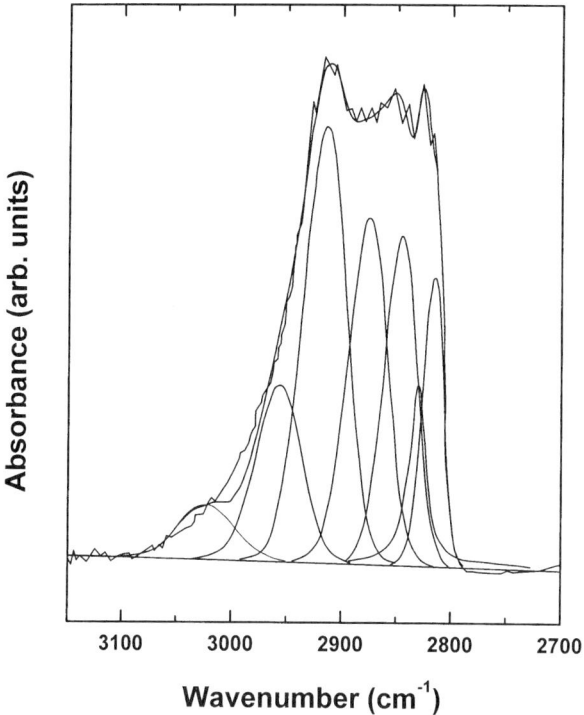

FIG. 9. Least squares fit to the experimentally measured CH-stretch absorption of a CVD diamond film showing groups with both sp^3 and sp^2 carbon bonding.

assumed, since the diamond environment is ill-defined, and it is possible for this ratio to vary considerably (McNamara et al., 1994). Figure 9 shows the spectra of a second film, where not only the groups with sp^3 bonded carbon but also some groups with sp^2 bonded carbon are observed.

2. QUANTITATIVE HYDROGEN CONCENTRATIONS

The total hydrogen content determined from the total integrated area under the composite line shape for a number of diamond films, whose ^{1}H NMR spectra contain both Lorentzian and Gaussian components, as shown in Fig. 4, is listed in Table II. Comparing the integrated area under each individual line shape shows that the majority (>95%) of the hydrogen in CVD diamond films contributes to the Gaussian component of the line shape, and only a small fraction of the hydrogen in the samples is related to

TABLE II

ESTIMATED DEFECT CONCENTRATION MEASURED BY NMR AND EPR

Sample	Total H content (at.%)	H1 (10^{16} cm^{-3})	H2 (10^{16} cm^{-3})
1	0.021	0	3.2
2	0.025	0	4.8
3	—	0	≤0.4
4	0.027	0	≤0.5
5	0.031	≤3	0
6	0.043	50	0
7	0.045	16	0
8	0.085	≤6	0
9	0.150	67	0
10	0.290	119	0
11	0.320	92	0
12	0.330	78	0

the Lorentzian component. The low-temperature NMR experiment at 100 K (Fig. 4) (McNamara et al., 1994) shows that the Lorentzian component is broadened, indicative of reduced molecular motion (McNamara et al., 1994). It is unlikely that this peak is a result of trapped H_2 as observed in amorphous silicon, since the motion of H_2 would not be significantly reduced at 100 K. The Gaussian component remains unchanged at 100 K, indicating rigidly held hydrogen (McNamara et al., 1992a, 1992c). As the hydrogen content increases, it can sometimes become difficult to resolve the narrow feature from the dominant broad line. A "hole burning" NMR experiment where magnetization is transferred from the Gaussian environment to the Lorentzian environment (Fig. 4) shows that these two hydrogen environments are located within 5 Å of each other (McNamara Rutledge and Gleason, 1996).

3. LOCAL HYDROGEN DISTRIBUTION

Although the average hydrogen concentration in the CVD films is low, the proton homonuclear dipolar line broadening in the Gaussian component is large (~ 60 kHz). This discrepancy indicates that locally high hydrogen concentrations exist, requiring significant segregation of hydrogen in polycrystalline diamond (McNamara et al., 1992a, 1992c). Randomly dispersed CH or CH_2 groups would provide too little homonuclear broadening to account for the observed line width. However, a fit to the van Vleck equation (Eq. 2) shows that the broadening is consistent with the areal densities for hydrogen-passivated diamond surfaces, typically 1–3 × 10^{15} H/cm^3.

Also, consistent with this interpretation is the correlation of the observed NMR hydrogen contents with the absorption of the films in the CH-stretch region (2750–3050 cm^{-1}) of the IR spectra for films with <0.2 at.% total hydrogen (Fig. 10). This agreement suggests that the majority of hydrogen in these CVD diamonds is covalently bonded to the lattice. An effective absorption coefficient for the overall CH-stretch region can be calculated if the intrinsic two-phonon absorption is calibrated with respect to natural diamond. The absorption coefficient can be defined as

$$A = \frac{1}{c} \alpha(v)\, dv \qquad (7)$$

where c is the concentration of C—H bonds in mol liter^{-1}, $\alpha(v)\, dv$ is the integrated intensity of the CH-stretch region in cm^{-2}, and A_{eff} is in liters mol^{-1} cm^{-2}. The average calculated absorption coefficient for the CH-stretch region in diamond is 4.1 ± 1.4 × 10^{-3} liters mol^{-1} cm^{-2}, which is in good agreement with literature values for alkanes (Francis, 1950) and

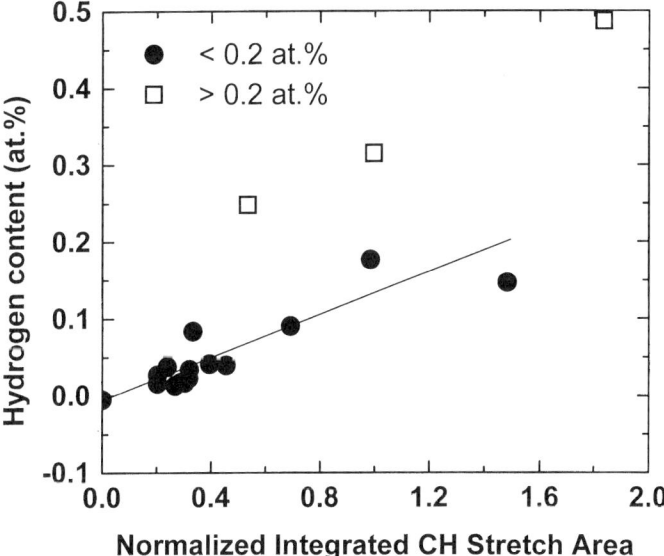

FIG. 10. Correlation of the normalized absorption in the CH-stretch with total hydrogen contents measured by NMR.

paraffinic hydrocarbons (Angus et al., 1984), 3.6×10^{-3} liters mol^{-1} cm^{-2} and 3.75×10^{-3} liters mol^{-1} cm^{-2}, respectively.

Multiple quantum NMR experiments also have confirmed the two-dimensional nature of the hydrogen distribution in CVD diamond (Levy and Gleason, 1992). These experiments show correlation growth rates for CVD diamonds that are similar to those observed for a two-dimensional hydrogen surface on natural diamond powder and do not agree with correlation growth rates observed in three-dimensional hydrogen environments. Thus it is likely that the majority of the hydrogen in CVD diamond films is involved in surface passivation. Both internal surface area (e.g., grain boundaries) and the top growth surface are potential sites for such passivation. Internal voids also may contribute to the surface area (McNamara et al., 1992c) and influence the thermal conductivity of diamond films (McNamara et al., 1995).

It is reasonable to ask, then, if the measured hydrogen contents are consistent with grain boundary coverage. Figure 11 shows the atomic hydrogen concentration required for full coverage of 3×10^{15} H/cm^2 (McNamara et al., 1992a). For films with fewer hydrogen than the detection limit of the NMR experiment, 0.001 at.%, the corresponding grain size is in excess of 100 μm. For films with a grain size of approximately 1 μm, the surface is

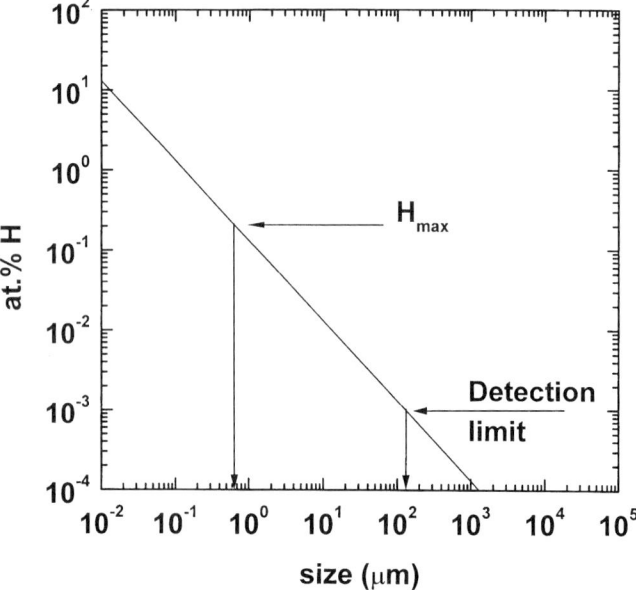

FIG. 11. Average crystallite size versus hydrogen content required for full surface coverage.

completely saturated by ~0.2 at.% hydrogen. The experimentally observed grain sizes in CVD diamond films are consistent with this range and coverage, although the relationship between grain size and hydrogen content is only qualitative. For films with hydrogen contents in excess of 0.2 at.%, the correlation no longer holds. It is interesting that the quantitative correlation shown in Fig. 10 also breaks down for films with greater than 0.2 at.% hydrogen. It has been suggested that the excess hydrogen in these films may be incorporated differently than that for films with lower hydrogen contents. One reason for the difficulty in quantitatively relating grain size and hydrogen content is the large disparity in grain sizes within individual films. This makes it difficult to estimate the grain boundary

FIG. 12. SEM of a diamond film showing a large disparity in grain size.

surface area accurately, as is demonstrated in Fig. 12 (McNamara et al., 1994). The variation in grain size with film thickness also will contribute to this problem.

4. PROXIMITY TO PARAMAGNETIC DEFECTS

Quantitative values for the concentration of two dangling-bond hydrogen defects, H1 and H2, respectively, are also listed in Table II (Zhou et al., 1996). Note these values are 1 to 2 orders of magnitude smaller than the total hydrogen content observed by NMR. To simulate the EPR spectra, axial symmetry is assumed for **A** with a single hydrogen atom, which, for any angle Φ between the symmetry axis and **B**, gives four lines, two for the $\Delta m = 0$ transitions and two for the $\Delta m = \pm 1$, whose intensities are given by Eq. (6) (Abragham, 1983). This is illustrated in Fig. 6a. These are convoluted with a random distribution function for the defect axis orientation, $N(\Phi)\, d\Phi = \frac{1}{2}\sin\Phi\, d\Phi$, and then convoluted again with the derivative of a Lorentzian line shape versus **B**. There are thus only three adjustable parameters, $A^{\|}$, $A^{\wedge}$, or the isotropic part $a = \{A^{\|} + 2A^{\perp}\}/3$, the anisotropic part $b = \{A^{\|} - 2A^{\perp}\}/3$, and the Lorentzian line width, which must give a good fit to the experimental derivative spectra at all four microwave frequencies.

In Fig. 5, the fit for all four frequencies using the values of $a = 5.5$ MHz and $b = 11.0$ MHz is shown. [The Lorentzian used for the best match shown in the figure for the sample used in my 14- and 20-GHz results had a peak-to-peak derivative width of PPDW = 2.8 G. For the sample used by Holder et al. (1994) at 9.8 and 35 GHz, the PPDW was taken to be 3.6 G. Studies reported in previous work display different degrees of resolution of the satellites consistent with sample-dependent breadth difference (Holder et al., 1994; Hoinkis et al., 1991; Portis, 1953; Trammell et al., 1958), and such an adjustment, which is one of resolution only, is therefore reasonable.] By adjusting a and b, it was concluded that the excellent fit at all four frequencies shown in the figure requires $b = +11.0$ MHz and $a = +5.5$ MHz, with an uncertainty of ± 2.5 MHz for each. The relative intensities of the satellites are determined primarily by b (i.e., $\sim b^2$), and the shape of the central line, which arises from the normally allowed hyperfine structure, depends strongly on the ratio of a to b. Because of this, the range of values for a and b beyond which a satisfactory fit cannot be obtained is so narrowly defined. The absolute signs of a and b cannot be determined but have been taken as those appropriate for a dipole-dipole origin of the anisotropy.

The narrow allowable range of the hyperfine parameters strongly suggests that the resonance results from a single well-defined defect. The isotropic part of the hyperfine interaction, $a = 5.5\,\text{MHz}$, is small, corresponding to $<0.5\%$ of the atomic value (1420 MHz), so the hydrogen atom must be considered only a neighbor of the paramagnetic site. Therefore, I treat the anisotropic part, $b = +11.5\,\text{MHz}$, as arising from the dipole-dipole interaction between separated electronic and nuclear dipoles, $b = g\mu_B\mu_N/r^3$, which gives for the separation $r = 1.9\,\text{Å}$ (and determines the sign of b to be positive). This is an interesting result; the distance between the nearest C—C diamond lattice separation and the next-nearest separation are 1.54 and 2.51 Å, respectively.

These are similar to what one might expect for distances between a carbon atom next to a lattice vacancy and a hydrogen atom bonded to one of the other carbon neighbors of the vacancy (1.74 Å with a typical C—H bond distance of 1.09 Å and the carbon atoms in their normal unrelaxed lattice sites. On breaking the weak next-nearest neighbor reconstructed bond, one expects relaxation backward, and this distance should increase). Many vacancy-like sites are expected in such polycrystalline films — in the bulk at dislocation cores but particularly at or near the many grain boundaries, which, of necessity, contain vacancy-like reconstructed bonds adjacent to the extra atom planes that support the angle between the lattice planes between two crystalline grains. Such stretched bonds are vulnerable to single hydrogen atoms that can enter and break the bond, attaching to one of the carbon atoms and activating a vacancy-like dangling bond on the other as the two relax backward. A schematic of the defect is shown in Fig. 13. If we had a hydrogen in an isolated bulk vacancy, we would expect to see easy hopping of the dangling bond among the three remaining carbon neighbors as well as motion of the hydrogen itself, as is commonly observed for such defects (Watkins and Corbett, 1964, 1965; Elkin and Watkins, 1968). For all the EPR active defects observed here, however, no such motion is observed in experiments from 1.5 K to room temperature. Therefore, I expect to find these highly distorted vacancy-like defects at grain boundaries, surfaces, and other areas of misfit resulting from the polycrystalline growth. The results of molecular modeling are in agreement with this interpretation, predicting $r = 1.96\,\text{Å}$.

Finally, let me mention that in a few samples that contained significantly lower hydrogen contents, as determined by NMR and IR spectroscopy, the H1 signal described above is missing, and a weaker but distinguishably different hydrogen-related center, H2, which has not been reported previously, is observed. It is narrower and with relative intensity factors of $\sim 2\text{--}3$ weaker than the H1 defect. For it, I estimate $b \approx 6.7\,\text{MHz}$, which corresponds to a hydrogen separation of $\sim 2.3\,\text{Å}$. It is shown elsewhere that this is a distinct defect from the H1 defect and that its EPR Gaussian line shape supports the notion of nearby regions of extended hydrogenated structural

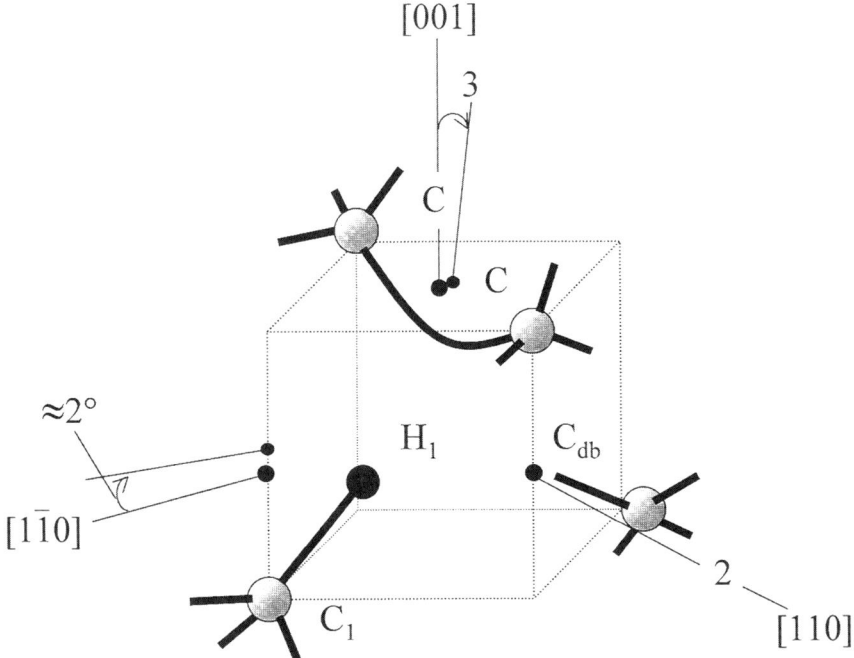

FIG. 13. Model showing approximate atom positions for a single hydrogen atom in a vacancy and the principal axis for the hydrogen hyperfine interaction.

defects, such as the hydrogen-terminated grain boundaries and surfaces indicated by previous NMR results (McNamara Rutledge and Gleason, 1996).

5. MACROSCOPIC HYDROGEN DISTRIBUTIONS

Hydrogen NMR also has been used to study the macroscopic variations in the hydrogen content of CVD diamond films. McNamara and Gleason (1992) examined samples produced by both hot-filament and dc arc-jet deposition for radial variations in the hydrogen content of the films. Hydrogen contents at the center of the two samples were comparable, being 0.031 and 0.023 at.%, respectively. This may indicate a common mechanism by which hydrogen is incorporated during diamond film growth in the two systems. Figure 14 (McNamara and Gleason, 1992) shows the radial variation of both films as a function of normalized radius. In both cases, the hydrogen content in the films increases as a function of radial position,

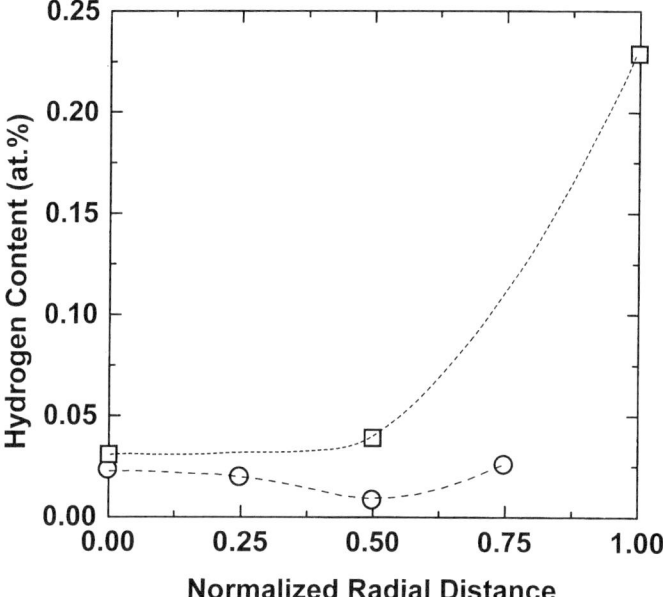

FIG. 14. Hydrogen content of dc arc-jet and hot-filament CVD diamond films as a function of normalized radial position.

although the hot-filament case is more pronounced. The presence of these variations in hydrogen content may result from fluctuations in the gas-phase concentrations above the growth surface and nonuniformities in surface temperature, orientation of the crystallite growth surface, and surface area of these crystallites. The observation of variations in different reactor environments serves to highlight the importance of surface chemistry and the control of local growth conditions.

IV. Effects of Hydrogen on Observed Properties

1. INFRARED TRANSMISSION

As noted earlier, hydrogen contents, measured by NMR spectroscopy, correlate quantitatively with the CH-stretch absorption in the IR spectrum (Fig. 10). This also can be seen in Fig. 15 (McNamara et al., 1992a), which shows the detailed CH-stretch absorption ($2750-3050\,\text{cm}^{-1}$) for films containing 0.136, 0.078, and 0.017 at.% hydrogen, respectively. It is also ob-

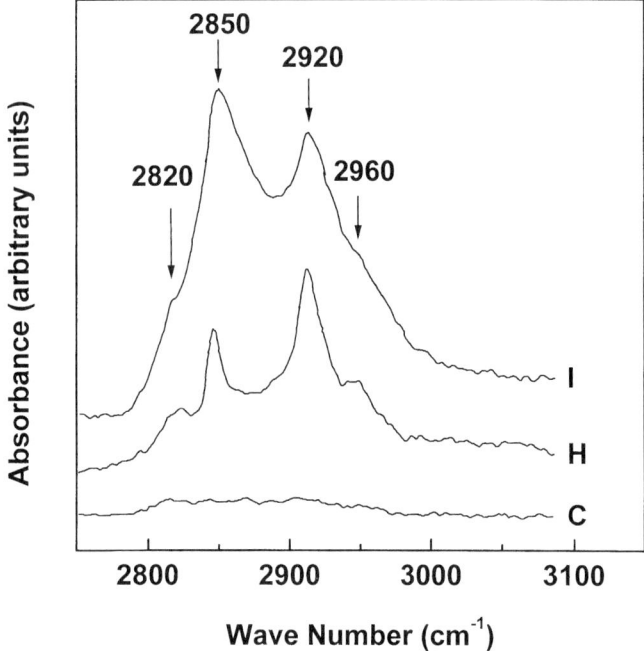

FIG. 15. The CH-stretch absorption for three CVD diamond films with different total hydrogen concentrations.

served that the IR absorption in the one-phonon region (below $1332\,\text{cm}^{-1}$) of the IR spectrum correlates qualitatively with hydrogen content. These absorptions are of great interest because many commercial and military applications of CVD diamond require transparency in this region. These one-phonon absorptions are normally disallowed as a result of the symmetry of the diamond lattice. However, if that symmetry is disturbed by the presence of impurities or defects, absorptions may be observed. The qualitative correlation of these absorptions with hydrogen content may indicate a link between hydrogen and the disruption of symmetry of the diamond lattice. This is important to our understanding and control of processes designed to produce materials for optical applications.

2. THERMAL CONDUCTIVITY

The room-temperature thermal conductivity of diamond κ has been found to generally decrease with hydrogen content (Fig. 16) (McNamara et al.,

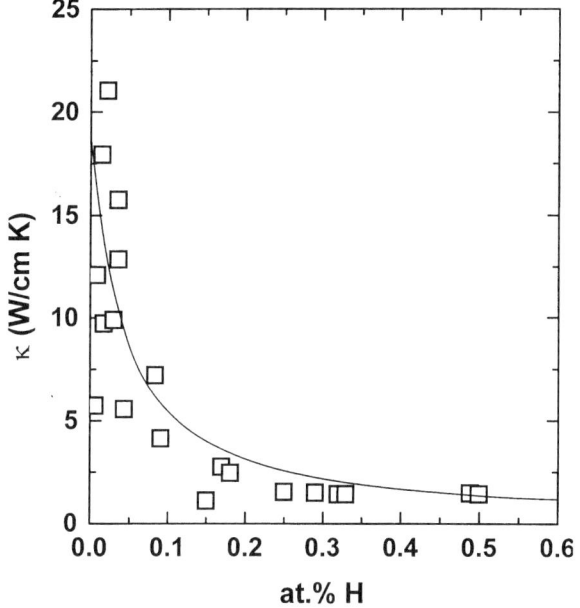

FIG. 16. Thermal conductivity versus hydrogen concentration.

1995), consistent with the hypothesis that hydrogenated defects act as phonon-scattering centers. The expected functional form of this relationship is shown in Eq. (8):

$$\kappa(\%H) = \frac{C_V^2 V/3S_i}{S_{D,H}/S_i} + \frac{S_{D,NH}}{S_i} + 1 \qquad (8)$$

where C_V is the constant volume heat capacity of diamond, V is the phonon velocity, S_i is the intrinsic phonon-phonon scattering rate, and $S_{D,H}$ and $S_{D,NH}$ are the scattering rates due to hydrogenated and nonhydrogenated defects, respectively.

Regardless of the type of defect occupied by hydrogen (point, lines, or surface), $S_{D,H}$ is expected to be proportional to the atomic percentage of hydrogen in the film. The upper limit of κ at any given hydrogen content will occur when $S_{D,NH} = 0$. This prediction is shown in the solid line in Fig. 16, taking $S_{DH}/S_i = 30$ (at.% H). As expected, the observed values of κ fall on or to the left of the predicted line within the experimental uncertainty. Films having a thermal conductivity lower than this upper limit may have significant phonon scattering from nonhydrogenated defects.

It is also interesting that at hydrogen concentrations in excess of 0.1 at.% in Fig. 16 (McNamara et al., 1995), the thermal conductivity is relatively insensitive to further increases in hydrogen content. This is approximately the same concentration range where the correlation between the total hydrogen content measured by NMR and the IR absorption in the CH-stretch region breaks down. This may indicate a change in the bonding structure in films with higher hydrogen concentrations that could complicate the hyperbolic relationship between thermal conductivity and hydrogen content.

V. Summary

Hydrogenated defects in polycrystalline diamond films have been quantitatively observed using solid-state characterization techniques. Results show that the overall hydrogen content in these materials is typically <0.2 at.%. The majority of this hydrogen ($>99\%$) is spatially removed from paramagnetic sites, covalently bound, and rigidly held. A small fraction ($<1\%$) of the hydrogen observed is associated with covalently bound motional groups, such as rotating surface methyl groups, while <0.0005 at.% is associated with dangling-bond defects. Evidence indicates that all these defects are likely to be located within the same region in the sample, specifically near highly distorted defects found at grain boundaries, surfaces, and other areas of misfit resulting from polycrystalline growth. Because of their potential electrical activity, even very low concentrations of such defects are expected to effect the electrical, thermal, and optical performance of these materials.

ACKNOWLEDGMENTS

I would like to acknowledge the contributions of many collaborators with whom I have worked over the years, particularly George Watkins at Lehigh University and Richard P. Messmer of General Electric Company, for their efforts in the study of electron paramagnetic defects and Karen K. Gleason of MIT for her work in the nuclear magnetic resonance studies. I would also like to thank the many who have encouraged, advised, and enriched my work and all those who have contributed to the understanding of hydrogen in CVD diamond.

REFERENCES

Abragham, A. (1983). *The Principles of Nuclear Magnetism.* New York: Oxford University Press.
Angus, J. C., Shultz, J. E., Shiller, P. J., MacDonald, J. R., Mitrich, M. J., and Domitz, S. (1984). *Thin Solid Films*, 311.
Bellamy, L. J. (1980). *The Infrared Spectra of Complex Molecules*, Vol. 2. New York: Chapman & Hall.
Derry, E., Madiba, C. C. P., and Sellschop, J. P. F. (1983). *Nuclear Inst. Methods Phys. Res.*, **218**, 559.
Dischler, B., Wild, C., Mulle-Serbet, W., and Koidl, P. (1993). *Physica B*, **185**, 217.
Elkin, E. L., and Watkins, G. D. (1968). *Phys. Rev.*, **174**, 881.
Erchak, D. P., Ulyashin, A. G., Glefand, R. B., Penian, N. M., Zaitsev, A. M., Varichenko, V. S., Efimov, V. G., and Stelmakh, V. F. (1992). *Nucl. Instrum. Methods Phys. Res.* **B69**, 271.
Fabisiak, K., Maar-Stumm, M., and Blank, E. (1993). *Diamond Rel. Mater.*, **2**, 722.
Fanciulli, M., and Moustakas, T. D. (1993). *Phys. Rev. B*, **48**, 14982.
Francis, S. A. (1950). *J. Phys. Chem.*, **18**, 861.
Halliburton, L. E., Perlson, B. D., Weeks, R. A., Weil, J. A., and Wintersgill, M. C. (1979). *Solid State Commun.*, **30**, 575.
Hartnett, T. M., and Miller, R. P. (1990). *SPIE Proc.*, **1307**, 60.
Henrichs, P. M., Cofield, M. L., Young, R. H., and Hewitt, J. M. (1984). *J. Magn. Reson.*, **58**, 85.
Hoinkis, M., Weber, E. R., Landestrass, M. I., Plano, M. A., Han, S., and Karris, D. R. (1991). *Appl. Phys. Lett.*, **59**, 1870.
Holder, S. L., Rowan, I. G., and Krebs, J. J. (1994). *Appl. Phys. Lett.*, **64**, 1091.
Levy, D. H., and Gleason, K. K. (1992). *J. Phys. Chem.*, **31**, 8125.
Lock, H., Wind, R. A., Maciel, G. E., and Johnson, C. E. (1993). *J. Chem. Phys.*, **99**, 3363.
Loubser, J. H. N., and van Wyk, J. A. (1978). *Rep. Prog. Phys.*, **41**, 1201.
McNamara, K. M., and Gleason, K. K. (1992). *J. Appl. Phys.*, **71**, 2884.
McNamara, K. M., and Gleason, K. K. (1994). *Chem. Mater.*, **6**, 39.
McNamara, K. M., Gleason, K. K., and Robinson, C. J. (1992a). *J. Vac. Sci. Technol. A*, **10**, 3143.
McNamara, K. M., Gleason, K. K., and Butler, J., and Vestyck, D. J. (1992b). *Diamond Rel. Mater.*, **1**, 1145.
McNamara, K. M., Levy, D. H., Gleason, K. K., and Robinson, C. J. (1992c). *Appl. Phys. Lett.*, **60**, 580.
McNamara, K. M., Williams, B. E., Gleason, K. K., and Scruggs, B. E. (1994). *J. Appl. Phys.*, **76**, 2466.
McNamara, K. M., Scruggs, B. E., and Gleason, K. K. (1995). *J. Appl. Phys.*, **77**, 1459.
McNamara Rutledge, K. M., and Gleason, K. K. (1996), *Chem. Vap. Deposition*, **2**, 37.
Nakazawa, K., Udea, S., Kumeda, M., Morimoto, A., Shinuzu, T. (1982). *Jpn. J. Appl. Phys*, L176.
Portis, A. M. (1953). *Phys. Rev.*, **91**, 1071.
Pruski, M., Lang, D. P., Hwang, S.-J., Jia, H., and Shinar, J. (1994). *Phys. Rev. B*, **49**, 10635.
Sellschop, J. P. F., Bibby, D. M., Erasmus, C. S., and Mingay, D. W. (1974). *Indust. Diamond Dist.*, **43**.
Sellschop, J. P. F., Connell, S. H., Madiba, C. C. P., Haddad, E. S., Stemmet, M. C., and Ram, K. B. (1992). *Nucl. Inst. Methods Phys. Res.*, **B68**, 133.
Socrates, G. (1980). *Infrared Characteristic Group Frequencies.* New York: Wiley.
Trammell, G. T., Zeldes, H., and Livington, R. (1958). *Phys. Rev.*, **110**, 630.

van Vleck, J. H. (1948). *Phys. Rev.*, **74**, 1168.
Watanabe, I., and Sugata, K. (1988). *Jpn. J. Appl. Phys.*, **27**, 1808.
Watkins, G. D., and Corbett, J. W. (1964). *Phys. Rev.*, **134**, A1359.
Watkins, G. D., and Corbett, J. W. (1965). *Phys. Rev.*, **138**, A543.
Windicshmann, H., Epps, G. F., Cong, Y., and Collins, R. W. (1991). *J. Appl. Phys.*, **69**, 2231.
Zhang, W., Zhang, F., Wu, Q., and Chen, G. (1992). *Mater. Lett.*, **15**, 292.
Zhou, X., Watkins, G. D., McNamara Rutledge, K. M., Messmer, R. P., and Chawala, S. (1996). *Phys. Rev. B*, **54**, 7881.

CHAPTER 8

Dynamics of Muonium Diffusion, Site Changes, and Charge-State Transitions

Roger L. Lichti

DEPARTMENT OF PHYSICS
TEXAS TECH UNIVERSITY
LUBBOCK, TEXAS

I. INTRODUCTION 311
II. EXPERIMENTAL TECHNIQUES 316
   1. *Transverse-Field Methods* 318
   2. *Longitudinal-Field Methods* 319
III. IDENTIFICATION AND CHARACTERIZATION OF MUONIUM STATES 328
   1. *Neutral Paramagnetic Centers* 329
   2. *Charged Diamagnetic Centers* 334
IV. DYNAMICS OF MUONIUM TRANSITIONS 341
   1. *Silicon: The Basic Model* 342
   2. *Germanium* 354
   3. *Gallium Arsenide* 359
   4. *Other III-V Materials* 364
V. RELEVANCE TO HYDROGEN IMPURITIES 366
   REFERENCES 369

I. Introduction

The behavior of hydrogen in semiconductors has been studied extensively since the early 1980s following the discovery that H can passivate shallow donors and acceptors in Si. A large amount of work over the intervening years has demonstrated much more generally that hydrogen modifies the properties of dopants, other impurities, and more extended defects in a wide range of semiconducting materials by forming complexes with those defects which move the related energy levels into or out of the bandgap. A great deal of experimental information has been accumulated on these hydrogen-related complexes, but far less direct data exist on isolated atomic hydrogen impurities. This book and earlier books specifically on hydrogen in semiconductors provide an excellent overview of hydrogen-related defects and their properties (Pankove and Johnson, 1991; Pearton et al., 1992; Pearton, 1994). Isolated hydrogen can exist in three charge states (H^0, H^+, and H^-) in a semiconductor depending on the level of the Fermi energy with respect to the hydrogen-related levels. The interstitial location of isolated H and its diffusive properties are to a great extent directly related to its charge state. Furthermore, the reactivity of H with other impurities is significantly impacted by charge states and diffusive motion. Since directly studying isolated hydrogen has proven to be extremely difficult, it is fortunate that an analogous impurity exists in muonium (Mu = $\mu^+ e^-$), for which obtaining the relevant properties is much more straightforward. Muonium is effectively a very light, short-lived pseudoisotope of hydrogen with a mass of 0.1125 amu ($\sim\frac{1}{9}$ that of H) and a mean lifetime of 2.197 μsec. Following a brief introduction to the static properties of Mu centers and the methods used to study Mu in semiconductors, I devote the bulk of this chapter to describing results from recent studies of motional properties of the Mu centers and the dynamics of transitions between the various muonium states. Detailed discussion will be confined to research over the last 5 to 7 years in which I was directly involved and to muonium in group IV and III-V materials, which has been the major focus of that effort.

Muonium is formed when positive muons (μ^+) are implanted into a target material, in this case a semiconductor. The states formed by the implanted muons have been thoroughly investigated, and their static properties are well documented. Most of this work was accomplished during the 1980s and has been reviewed exhaustively (Patterson, 1988; Kiefl and Estle, 1991). For the most part, muonium is expected to form the same charge states as hydrogen and to occupy the same interstitial locations. For a discussion of theoretical considerations, the reader may consult the extensive literature on modeling of hydrogen in semiconductors, recently reviewed (Estreicher, 1995), or Chapter 2 in this book. Because of the large mass difference, quantum effects such as zero-point energy and tunneling will be much more important for muonium than for hydrogen. However, with due regard for these differences, the results from investigations of muonium centers can be transferred to isolated hydrogen. The static properties of the centers should be very similar, and dynamic properties will be at least qualitatively comparable.

One of the largest differences in experiments on hydrogen and muonium is related to the short muon lifetime. As a result, Mu nearly always occurs as an isolated center, and one seldom directly observes interactions with other impurities; thus Mu and H experiments probe opposite ends of the reaction sequence leading from isolated H or Mu to impurity complexes. Even at the highest intensities of available muon beams, only a few muons exist within a sample at any instant. A second major difference that needs to be kept in mind when comparing the literature on Mu and H is that muonium experiments essentially always explore highly nonequilibrium situations, especially at low temperatures, initiated by the high-energy muon implantation and the thermalization processes. The final muon stopping sites are well away from any damage from the implantation, and the Mu centers observed in any experiment are determined primarily by the basic properties of the semiconductor, dopant type and concentration, and temperature. The transitions that are seen at low temperatures are those which move the system toward thermal equilibrium, whereas at sufficiently high temperatures, well above room temperature in most cases, they represent dynamics within the equilibrium mix of states.

Four types of muonium centers are well established in the tetrahedrally coordinated semiconductors (Patterson, 1988; Kielf and Estle, 1991). These include the three charge states and two types of interstitial locations, one labeled as bond-centered (BC), in which the muon resides at or near the center of a stretched bond, and the second labeled as tetrahedral (T), where the muon is situated within the largest open region of the diamond or zincblende lattice, the center of which has tetrahedral symmetry. Each of these sites can support two separate charge states in general. The known stable and metastable centers are Mu_{BC}^0, Mu_{BC}^+, Mu_T^0, and Mu_T^- using the now common notation that denotes both the charge state and general location. The early literature used notations of *Mu* and *Mu** for the *normal* and *anomalous* paramagnetic Mu^0 centers now identified with the T and BC sites, respectively.

The T-site neutral center Mu_T^0 is highly mobile at all temperatures, particularly in Si, Ge, and GaAs, where the evidence is most convincing (Kadono *et al.*, 1990, 1994; Schneider *et al.*, 1992), and quite likely in the other semiconductors as well. The lowest-energy site within the tetrahedral cage region is off center at the so-called antibonding (AB) location in many of the theoretical models; however, due to rapid motion, the resulting signal observed in any experiment reflects the averaged symmetry of the central T site. The electron wave function for this neutral center is much like a slightly constricted atomic *s* state, and the resulting hyperfine interaction is isotropic at roughly half the free muonium value. The lattice constant or physical size of the cage appears to be the primary material parameter that correlates with variations in the Mu_T^0 hyperfine constant among the various materials. The T site is a metastable configuration for Mu^0 in diamond and silicon but may be the neutral ground state in Ge and GaAs, as I will discuss. Because this region of the lattice has the lowest electronic charge density, the T site is the stable location for a Mu^- center. With a second electron, the cage represents much more of a size constriction, thereby greatly reducing the mobility of the negative ion compared with that of the neutral center. In

III-V compounds there are two T sites that are no longer equivalent, one with its four nearest neighbors all of group III and the other surrounded by group V atoms. Because of partially ionic bonding in these compounds the T_{III} site will be strongly favored for Mu^- and becomes the lower-energy region for Mu_T^0 by a much smaller margin. The T_{Ga} cage has been established experimentally as the Mu^- location in GaAs (Chow et al., 1995), providing the only spectroscopically determined local structure for an isolated charged Mu or H center in any semiconductor.

The identification of the BC site as the ground-state configuration for Mu^0 in Si (Symons, 1984; Cox and Symons, 1986; Estle et al., 1986; Estreicher, 1987) and the experimental determination of the physical and electronic structure of this center (Kiefl et al., 1988) represent one of the more significant contributions of muonium research to the field of impurities in semiconductors. Comparison of the Mu_{BC}^0 hyperfine parameters, properly scaled, with those of the AA9 EPR center for isolated H^0 (Gorelkinski and Nevinnyi, 1987, 1992; Bech Neilsen et al., 1994; see also Chap. 3) shows the complete equivalence of the Mu and H electronic structures in the only case where similar data exist for hydrogen and muonium impurities. Comparison of parameters for Mu_{BC}^0 and H_{BC}^0 in silicon remains the one available experimental reference point for translating muonium results over to the analogous isolated hydrogen centers. Mu_{BC}^0 is observed in Si, Ge, diamond, GaAs, and GaP, whereas the isotropic Mu_T^0 center is seen in many semiconductor materials (Patterson, 1988). The Mu_{BC}^0 hyperfine interaction is small and anisotropic, axially symmetric about a $\langle 111 \rangle$ bond direction. Detailed measurements of the muon hyperfine parameters for Mu_{BC}^0 and the nuclear hyperfine parameters for interactions with near neighbors in Si, GaAs, and GaP show that the muon is near the center of a bond with the two nearest neighbors moving significantly outward, stretching the host bond length by roughly a third (Kiefl et al., 1987, 1988; Schneider et al., 1993). The wave function for the unpaired electron of this center is best described as predominantly made up of a host antibonding orbital, with almost zero charge density at the muon site and 40% to 70% located on the two nearest host atoms and the rest spread out over more distant atoms. This structure is asymmetric in the compounds, as should be expected, with the muon somewhat off center, inequivalent outward lattice relaxation, and the Mu_{BC}^0 electron density on the nearest host atoms unequally distributed. In all these materials, Mu_{BC}^0 appears to be stationary, remaining at a single BC site over the roughly 10-μsec time scale of the experiments. The BC site represents the expected configuration for the positively charged center, although no direct spectral data exist from which details of the Mu^+ local structure can be determined. For the more ionic III-V compounds, arguments similar to those noted earlier for Mu^- imply that the T_V cage region may become competitive with the BC site as a possible location for Mu^+; thus an additional metastable Mu^+ center needs to be considered in the compound semiconductors.

The more recent work has focused on understanding the dynamics of transitions between muonium states. The collaboration of which I am a member has identified and characterized a total of eight separate transition processes for Mu in silicon and has developed a configuration coordinate model based on these data (Lichti, 1995; Kreitzman et al., 1995). Because of the similarities in observed Mu centers in the tetrahedrally coordinated semiconductors, the results for Si can serve as a starting point for understanding Mu behavior in all these materials. Comparisons with the few available measurements of similar experimental data on isolated hydrogen in silicon show very similar energy parameters (Gorelkinski and Nevinnyi, 1992; Bech Neilsen et al., 1994; Holm et al., 1991), indicating that the dynamic features of muonium and isolated hydrogen are quite comparable and that conclusions from the muonium investigations should be transferable to hydrogen at least at the semiquantitative level (Lichti, 1995; Kreitzman et al., 1995; Lichti et al., 1994a, 1995).

Ongoing investigations into muonium dynamics in other semiconductors are aimed at testing the general application of the basic model of Mu dynamics that resulted from the silicon data and determining how the details need to be modified for individual materials. In Si, the various dynamic features were conveniently well separated, greatly simplifying identification of specific processes and allowing the dynamic parameters for individual transitions to be determined fairly accurately. Even with this fortuitous isolation of most processes, at least five or six different samples were examined before a consistent picture of the transition identities emerged, and several additional samples were required to check specific identifications. There remain a few questions regarding Mu behavior in Si and the modeling of data related to the various transition processes; however, the model has held up very well to this continued scrutiny, and we are quite confident that the general features are correct. A single set of dynamic parameters for muonium along with the basic description of dopants and electronic properties for silicon satisfactorily reproduces experimental results from a wide range of dopant concentrations and different sample orientations. Similar detailed study of Ge, GaAs, InP, and other materials is at a much earlier stage, but important tentative conclusions already can be drawn. A full description of all this work and the various experimental checks is beyond the scope of this chapter. I will concentrate on our present understanding of the dynamic behavior of muonium in several specific semiconductors and use selected experimental results to illustrate how we have arrived at this still-emerging picture.

II. Experimental Techniques

All the experimental techniques used in muonium research are variants of nuclear magnetic resonance (NMR) methods. They make use of the fact that the positive muon has a spin of $\frac{1}{2}$ with a magnetic moment that is roughly three times larger than that of its proton counterpart: $\gamma_\mu = 135.54$ MHz/T compared with $\gamma_p = 42.58$ MHz/T. These techniques are known collectively as μSR for *muon-spin research* when used in the generic sense. The *R* often has been assigned other designations such as *rotation, relaxation,* or *resonance* depending on the specific technique or particular application. Much of the terminology and many of the standard theoretical treatments associated with continuous-wave (cw) NMR apply directly to μSR, for example, the typical spin Hamiltonian treatments of the various magnetic interactions. The major differences between cw-NMR and μSR are related to the formation and decay processes for the muon. For our purposes, the important facts are (1) that the muon creation process and beam-collection techniques yield nearly 100% spin-polarized muon beams and (2) that in the muon decay process a positron is emitted preferentially along the direction of the muon's spin. These are a consequence of parity violation in pion decay that creates a muon and in the decay process of the muon itself. The first of these results in muon beams that have spin polarization characteristics that are ideal for a magnetic resonance probe particle. The sources we have commonly used are so-called surface beams of muons collected from decay of pions stopped near the surface of a primary target. The beamlines are tuned to a momentum of about 28 MeV/c, for which the muon spin direction is opposite to the muon momentum or beam propagation direction. The spin polarization is maintained during implantation into a sample and subsequent thermalization; thus the initial muonium state is formed with a known direction of the muon's spin. This yields a maximum signal strength independent of temperature or magnetic field and means that μSR techniques can operate at any field (including zero) with no spectrometer tuning requirements.

The second feature, asymmetric positron emission in the μ^+ decay, provides a convenient signal-detection scheme. Directional rate measurements for positrons emitted from muons stopped in the sample provide the raw data in a μSR measurement. More specifically, the asymmetry in the decay positron rates along two directions 180° from each other provides a measure of the muon spin polarization along that axis. Figure 1 displays a typical experimental setup for the longitudinal geometry in schematic form. One set of positron counters is shown providing emission-rate measurements along the initial spin (beam) direction; a second set along a perpendicular axis is often used when operating in the transverse mode. A

8 DYNAMICS OF MUONIUM DIFFUSION 317

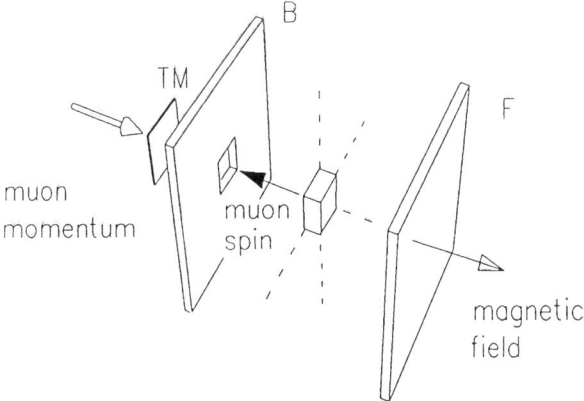

FIG. 1. Typical setup for μSR measurements in a longitudinal-field geometry. The back (B) and front (F) decay positron counters monitor the muon decay asymmetry, which is proportional to the polarization. For the transverse-field mode, the field is applied perpendicular to the beam, and a second set of counters may be used.

magnetic field is applied either parallel to the initial spin direction (longitudinal) or perpendicular (transverse) to both the initial spin and second measurement directions. In actual practice, the muon spin may be rotated 90° prior to implantation to give a transverse field condition rather than changing the applied field direction; the net effect is essentially identical. Data may be taken in time-differential (TD) mode, in which one muon at a time, or a tighly bunched pulse of muons, is allowed into the sample and the time between implantation as obtained from the TM counter and decay as detected in one of the positron counters (F or B) is used to generate a time histogram of the rates in each positron counter. A single data point typically is obtained from time histograms containing 5 to 20 million muon decay events. The back/front asymmetry as a function of time is constructed from the raw count rates as a first step in data analysis using

$$A_{BF} = \frac{N_B - \alpha N_F}{N_B + \alpha N_F} \quad (1)$$

for each time bin in the histogram. Here N_B and N_F are the counts in the back and front counters, respectively, with any background removed, and α is a parameter that corrects for differences in solid angle acceptances and efficiencies of the two counters. As defined here, the asymmetry is positive in the initial spin direction. The time-dependent asymmetry, which has the muon lifetime effectively removed, is then further analyzed to yield information on the muonium states and their behavior using analysis equations briefly discussed below for common experimental conditions. If the time history is not especially important, data can be collected in a time-integral (I) mode, in which the maximum beam intensity can be used and the raw data simply consist of positron rates along appropriate directions normalized to the beam rate. This is often used in a longitudinal field geometry if one wishes to measure the average asymmetry or polarization as a function of field, and details of the depolarization function are unimportant. I comment briefly on the techniques most important to this work and refer the interested reader to a chapter in a recent volume of this series where μSR methods are discussed in more detail (Chow et al., 1997b).

1. TRANSVERSE-FIELD METHODS

One of the earliest and most commonly used μSR techniques is known as TF-μSR for transverse-field muon-spin rotation. A magnetic field is applied normal to the initial muon polarization, and the spin then precesses about that field. The asymmetry obtained from Eq. (1) is analyzed to obtain the precession frequency and the relaxation function and associated rate constant. For a single precession frequency, one has

$$A^{TF}(t) = A_0 G^{TF}(t) \cos(\omega t + \phi) \tag{2}$$

where A_0 is the asymmetry at time zero, ω is the angular precession frequency, ϕ is a phase angle that depends on the counter directions, and $G^{TF}(t)$ is the transverse relaxation function. The relaxation function is typically Gaussian when the muonium center is stationary and switches over to exponential if the center is diffusing rapidly. A Gaussian rate constant gives information on the spread of internal dipolar magnetic fields due to neighboring nuclear moments, whereas an exponential rate constant is related to the hop rate, the lifetime of the observed muonium state, or the relevant parameter from some other dynamic process. For a static center, the transverse relaxation function is a measure of dephasing of the spin precession because various sites have different local dipolar fields, that is, spin-spin or T_2^{-1} relaxation in NMR terminology. When neighboring nuclei have quadrupole moments, the TF Gaussian rate constant as a function of applied field can be used to characterize the quadrupolar interactions of these nuclei in the electrical field gradient induced by the muonium impurity (Hartmann, 1977; Gamani et al., 1977; Chow et al., 1995). These quadrupolar decoupling curves obtained for different applied magnetic field directions also will provide symmetry information on the muonium site. For dynamic conditions with rapid fluctuations, an exponential transverse relaxation may reflect actual loss of polarization rather than dephasing; for these conditions, the dynamics are better determined by longitudinal field methods.

An alternative to the time analysis of TF-μSR data described by Eq. (2) is to perform a Fourier transform on the time-dependent asymmetry and examine the results as a function of frequency. In this case, the Gaussian or exponential time dependence of the transverse relaxation function translates into either a Gaussian or Lorentzian line-shape function, respectively. For a diamagnetic Mu$^{\pm}$ center the frequency is just $\omega_\mu = 2\pi\gamma_\mu B$, while for the paramagnetic Mu^0 centers there are four lines whose frequencies are controlled by the muonium hyperfine interaction and by the applied field direction if the center is anisotropic, such as Mu$^0_{BC}$. In recent experiments on Mu dynamics, TF-μSR has been used mostly to determine diamagnetic fractions, for a rough characterization of the internal fields and motional properties related to Mu$^{\pm}$ centers, and to obtain information on the coupling to neighboring quadrupolar nuclei.

2. LONGITUDINAL-FIELD METHODS

Specific μSR experiments utilizing longitudinal fields (LFs) can be separated into several distinct categories: spectroscopic methods (Lichti et al., 1992) that supplement TF-μSR, measurements of longitudinal relaxation rates, and characterization of hyperfine or dipolar fields by use of longitudinal field decoupling curves. Zero-field (ZF) measurements can be thought of as a special category that uses the longitudinal measurement geometry.

Because of the removal of phase-coherence constraints, methods using longitudinal fields are much more sensitive to final states of any transition than is TF-μSR. Delayed formation of a muonium state must occur within a small fraction of the precession period of the precursor for it to be observed in a transverse field; thus there is often a significant temperature region in which neither the precursor nor the final state is visible in TF-μSR. It is therefore extremely difficult to experimentally verify a simple direct transition between two states using only transverse-field measurements. Without the phase problems, it is often possible to examine both states over the transition temperature range using longitudinal-field spectroscopic techniques, thereby allowing much easier assignment of transitions. A combination of these LF spectroscopic methods and measurements of longitudinal relaxation characteristics provides for cross-checks on state and transition identities as well as yielding multiple measurements of dynamic parameters.

Two spectroscopic techniques will be discussed first, a standard magnetic resonance method using high-frequency ac fields to drive the resonance and loss of polarization due to state mixing to avoid the crossing of energy levels. In short form these are known as rf-μSR for *radio frequency* driving frequencies in the first case and *muon level-crossing resonance* (μLCR) in the second.

a. Muon Spin Resonance

Many of our results on muonium dynamics in Si, which forms the basic model we are now modifying for application to other semiconductors, came from rf-μSR measurements of the temperature-dependent amplitudes of resonances for muonium states (Kreitzman *et al.*, 1995). Basically, rf-μSR uses the traditional magnetic resonance technique of applying a dc magnetic field and matching the frequency for a perpendicular oscillating rf magnetic field to one of the muonium precession frequencies; for example, $\omega_{rf} = \omega_\mu = 2\pi \gamma_\mu B$ is the resonant condition for a diamagnetic Mu state. Signal detection is by the positron counting methods of μSR rather than measuring the absorption of rf power as in standard resonance methods. The dc field is applied in the longitudinal geometry, and the back/front asymmetry is measured. With the applied field parallel to the initial muon spin, the polarization remains at $P_z(t) = 1$ except when ω_{rf} is sufficiently near resonance. At resonance, the polarization precesses at a frequency governed by the magnitude of the rf field, consequently reducing the time-averaged longitudinal polarization and thus the measured asymmetry. We have used rf-μSR primarily in a time-integral mode, where the difference in asymmetry with the rf power on($+$) and off($-$) isolates the effects of the rf field and reduces systematic variations due to beam tuning, etc. The rf-related asymmetry change A_{rf} as defined by

$$A_{rf} = \frac{N_F^+ - N_F^-}{N_F^+ + N_F^-} - \frac{N_B^+ - B_B^-}{N_B^+ + N_B^-} \quad (3)$$

is the primary quantity analyzed in an rf-μSR experiment, where $N_{F/B}^{+/-}$ represents the integral positron counts in the appropriate counter with rf on/off. Data are taken in either a fixed-frequency, swept-field mode or vice versa depending on the characteristics of the resonance to be observed. A careful treatment of the spin dynamics under these rf conditions shows that the rf-μSR line shapes are Lorentzian (Kreitzman *et al.*, 1995; Chow *et al.*, 1997b; Kreitzman, 1990). Typically, the magnetic field (or ω_{rf}) is stepped through the region of a resonance to trace out that line shape, which is then fit to yield the on-resonance amplitude, line width, and resonant field or frequency. Additional analysis and normalization checks are required to yield the time-averaged fractional occupation of the observed muonium state.

While rf-μSR yields information similar to that from TF-μSR with regard to muonium state identification and electronic structure, it has definite advantages if there is a transition between states. Like other longitudinal techniques, rf-μSR does not require any phase coherence and thus is far more sensitive than TF methods to final states. rf-μSR can detect the resonance signal independent of when the state was formed so long as the polarization has been maintained. My colleagues and I have developed expressions based on a strong collision model for extracting transition rates from fits to the temperature and field dependences of the rf-μSR amplitudes of the various muonium states (Kreitzman et al., 1995; Hitti et al., 1997) and have successfully extracted the dynamics for eight separate transitions among the four states in silicon. These expressions are quite complicated and will not be reproduced here. The appendices in Kreitzman et al. (1995) provide details of an initial three-state version of this model using Mu_T^0, Mu_{BC}^0, and a combination of Mu^+ and Mu^- that are indistinguishable spectroscopically. We have been able to make assignments of the charged states based on dynamic features in the data.

b. Muon Level-Crossing Resonance

The basic principles of level-crossing resonance (LCR) have been reviewed previously in connection to its μSR application (Kiefl and Kreitzman, 1992), specifically its use in characterizing Mu_{BC}^0 (Kiefl and Estle, 1991). I will not go into the related theory here, but the basic idea is quite simple. Whenever two energy levels become nearly degenerate as an external parameter is varied, quantum state mixing will occur to avoid level crossing, provided some mechanism is available to mix these states. As used in muonium studies, these levels can be within the combined μ^+ and e^- spin system of Mu^0, the Mu^0 levels as modified by neighboring nuclear spins, or a system made up of just the μ^+ and neighboring nuclei for a diamagnetic Mu state. In semiconductors, μLCR played a crucial role in determining the electronic and physical structure of Mu_{BC}^0, with the initial data coming from GaAs (Kiefl et al., 1987) and Si (Kiefl et al., 1988) and others following shortly thereafter. LCR spectra for a paramagnetic center can provide locations of nearby nuclei and map the electronic wave function in a manner similar to double-resonance techniques such as ENDOR.

More recent use of the LCR technique has been to investigate the structures of diamagnetic muonium states using a version we have called

QLCR, for *quadrupolar LCR*. The levels involved are the muon Zeeman levels and the quadrupolar levels of neighboring nuclei. For this method to work, the nuclei must have spin greater than 1/2 and reasonably large dipole and quadrupole moments. Figure 2 shows a simulation of the levels for a diamagnetic muonium state and an $I = 3/2$ nucleus, along with the QLCR spectrum that results. The locations of these resonances are controlled primarily by the nuclear quadrupole splitting in an electrical field gradient induced by the presence of the nearby muon. Since the dipole-dipole interaction provides the state mixing in this situation, the amplitudes of QLCR spectra are related to the strength of the dipolar coupling between the muon and neighboring nucleus. QLCR spectral details identify the nearby nucleus and combined with quadrupolar decoupling curves help determine their locations.

In a slightly different picture of QLCR, the relevant levels would cross when the muon Zeeman splitting matches an energy difference in the nuclear quadrupolar levels of the neighbor involved. Energy-conserving spin-flip transitions can then occur in what is known as *cross-relaxation*. Observation of QLCR spectra is accomplished by scanning an applied longitudinal field to tune the muon Zeeman levels to the crossing condition. The QLCR

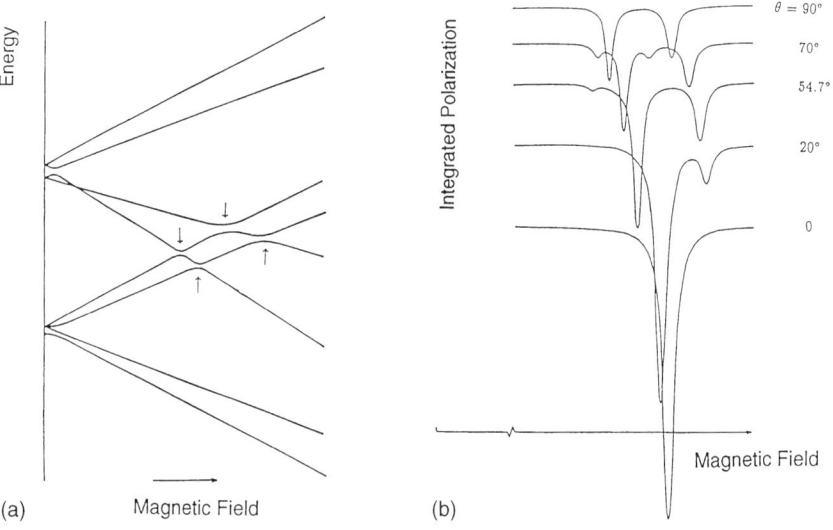

FIG. 2. Quadrupolar level-crossing resonance for a system involving the muon and a single spin 3/2 nucleus: (a) avoided crossings for energy levels of the combined spin system; (b) the resulting QLCR spectrum for different orientations of the applied field with respect to the principal axis for the electrical field gradient at the nucleus.

spectra are often collected as integral longitudinal asymmetry as a function of applied field. However, if time-differential data are collected, using a delayed time window increases the signal size because it takes some time for the state mixing to occur (Leon, 1992), and the related cross-relaxation can be observed as a peak in the longitudinal relaxation rate (Cox *et al.*, 1997a). The collaboration is currently using QLCR to examine the physical structure of a number of diamagnetic Mu states, especially in III-V compounds, where most host elements have quadrupolar isotopes. Much of this work is not yet complete, but this method was crucial in our determination of the Mu^- structure in GaAs (Chow *et al.*, 1995).

c. Longitudinal Muon-Spin Relaxation

In magnetic resonance, the longitudinal relaxation function, known traditionally as *spin-lattice relaxation* or T_1^{-1}, is commonly used to characterize dynamic fluctuations. For longitudinal-field muon-spin relaxation (LF-μSR), the magnetic field is applied parallel to the initial spin direction, and there is no spin precession and thus no dephasing. Under these conditions, any relaxation is due to dynamic processes that typically produce exponential relaxation functions. Strictly speaking, this is true only when the applied field is sufficiently large to dominate any local dipolar or hyperfine fields. Because of the selective sensitivity to dynamic processes, measurement of the longitudinal muon-spin relaxation rates, or more properly, the *depolarization* rates, has played an important part in the present work. An early application of LF relaxation to study muonium dynamics in semiconductors was in characterizing the motion of Mu_T^0 in GaAs (Kadono *et al.*, 1990; Schneider *et al.*, 1992). In this application, the fluctuations are in the effective dipolar fields from Ga and As nuclei, while the muon hops or tunnels from one T site to another as it diffuses through the lattice.

The LF relaxation can yield information on muonium transition rates. The number of different processes are far too many to give detailed expressions characteristic of each type of transition here. In many cases of a one-way transition between two muonium states, the longitudinal relaxation shows two components. The relative asymmetries and rate constants associated with the two components, particularly their field and temperature dependences, often allow a determination of the precise transition, the initial fractions of the two states, and the transition rate. Many of the relevant expressions for specific transitions have been worked out in full detail (Smilga and Belousov, 1994); however, in practice, simplifying assumptions often give satisfactory results. As an example, a transition from an initial paramagnetic state having nonoscillating visible polarization p_o and depolarization rate λ to a diamagnetic final state with a conversion rate κ gives rise to two-component longitudinal relaxation (Cox et al., 1997b). The function used to fit the relaxation data, normalized to yield the measured polarization, is then

$$P(t) = p_1 e^{-\lambda_1 t} + p_2 e^{-\lambda_2 t} \tag{4}$$

Under fairly general conditions, the simple model for this transition yields an observable longitudinal polarization of

$$P(t) = p_i(t)e^{-\kappa t} + \kappa \int_0^t p_i(\tau)e^{-\kappa \tau} d\tau \tag{5}$$

which, along with $p_i(t) = p_o e^{-\lambda t}$ describing the relaxation in the initial state, yields

$$p_1 = p_o \frac{\lambda}{\lambda + \kappa} \quad \lambda_1 = \lambda + \kappa \tag{6}$$

$$p_2 = p_o \frac{\kappa}{\lambda + \kappa} \quad \lambda_2 = 0 \tag{7}$$

as parameters for the measured fast and slow components, respectively. The slow rate constant in this simple model is zero because a diamagnetic final state normally does not show any longitudinal relaxation. Both signal amplitudes p_1 and p_2 should show the field dependences characteristic of p_o from the initial paramagnetic Mu^0 state, as discussed in Section II.2.d. The expected temperature dependence of the conversion rate κ provides an extra experimental check and helps to identify the conversion process. Other unidirectional transitions will have their own unique signatures. Usually both the temperature and field dependences are necessary to identify the states and dynamic processes involved, even when it is a simple case, as in this example.

In many of the semiconductors, rapid cyclic transitions between one of the Mu^0 centers and one of the charged $Mu^\pm$ centers occur at high temperatures (Chow et al., 1993; Lichti et al., 1994b). Because of the hyperfine oscillations while in the Mu^0 state, up to half the existing polarization is lost in each charge cycle depending on the Mu^0 lifetime compared with the hyperfine period. An approximate expression for the expected single-exponential longitudinal relaxation rate for these muonium charge cycles gives results nearly identical to the full-blown theory (Chow, 1995). For an isotropic neutral center like Mu_T^0, this expression, used extensively to fit the temperature and field dependence of the rate constants from longitudinal depolarization due to cyclic charge-state transitions, becomes

$$T_1^{-1} \simeq \frac{1}{2}\left(\frac{\lambda_0 \lambda_d}{\lambda_0 + \lambda_d}\right)\left(\frac{\omega_0^2}{\lambda_0^2 + \omega_{24}^2}\right) \tag{8}$$

where $\lambda_{0,d}$ represents the transition rate from $Mu^0 \to Mu^\pm$ and $Mu^\pm \to Mu^0$, respectively, $\omega_o = 2\pi A_\mu$, $\omega_{24} = \omega_o\sqrt{1 + x^2}$, and $x = B/B_o$ with $B_o = A_\mu/(\gamma_e + \gamma_\mu)$ defines a field scaling parameter. If the Mu_{BC}^0 center is the active relaxing species, then this expression is much more complicated; however, for very rapid cycle rates, the function looks essentially equivalent with a small average A_μ characteristic of the BC configuration. For slow cycle rates involving Mu_{BC}^0, the longitudinal relaxation rate will have a relatively sharp peak at $B = (A_\parallel \cos^2\theta + A_\perp \sin^2\theta)/2\gamma_\mu$, where θ is the angle between the BC symmetry axis and the applied field (Chow et al., 1994a). At this field, the *effective field* for half the polarization becomes fully transverse due to the anisotropy, thereby reducing the visible longitudinal polarization at a maximum rate. Observation of this longitudinal relaxation peak is a signature that Mu_{BC}^0 is definitely involved in the relaxation processes.

d. Hyperfine Decoupling Curves

Because of the lack of a precession frequency for the LF geometry, the decoupling of local fields as characterized by the field dependence of the longitudinal asymmetry has been important in determining which muonium state is active in a particular dynamic feature and in helping to assign the specific dynamic process. While the field-dependent longitudinal asymmetry can be used to characterize dipolar fields, the main use of this particular application of longitudinal μSR measurements is to identify which Mu^0 center is present under conditions where the identifying spectra cannot be observed. The field dependence of the longitudinal asymmetry follows a specific form that is scaled to the Mu^0 hyperfine parameter A_μ and is proportional to the parallel, or z, component of the muon polarization. Using x and ω_{24} as defined in Eq. (8), the time- and field-dependent longitudinal polarization function has the form

$$P_z(x, t) = \frac{1 + 2x^2}{2(1 + x^2)} + \frac{1}{2(1 + x^2)}\cos(\omega_{24}t) \tag{9}$$

Under most circumstances, ω_{24} is too high a frequency to be observed directly except with extremely high timing resolution. Thus ordinarily the

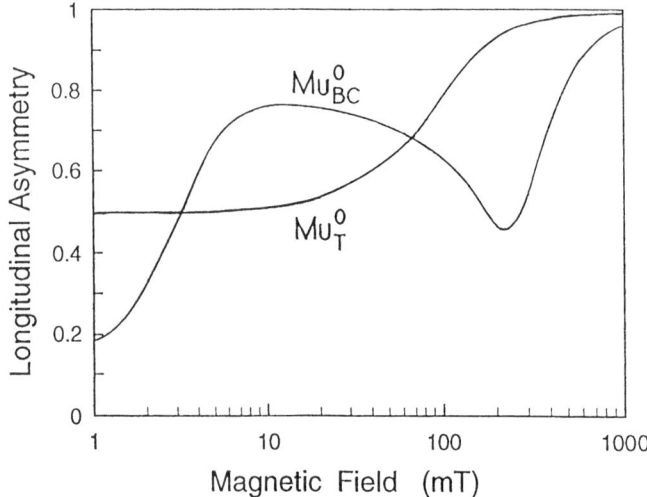

FIG. 3. The hyperfine decoupling, or repolarization, curves for the Mu^0 centers shown as the nonoscillating polarization observed in a longitudinal field. The Mu^0_{BC} curve is for $\mathbf{B} \| \langle 100 \rangle$, and A_μ values are for silicon.

first term, that represents the effective decoupling or the *repolarization* curve, is all that is observed. For an isotropic paramagnetic center like Mu^0_T, assuming no rapid dynamic depolarization that might further reduce the observed signal, this curve goes from a relative asymmetry of 1/2 at low fields to 1 at high fields with a characteristic decoupling field of B_o. Since Mu^0 has vastly different hyperfine parameters at the T and BC sites, such curves can quickly distinguish between the two neutral muonium states. Figure 3 shows the hyperfine decoupling curves for Mu^0_T and for Mu^0_{BC} with the applied field in the $\langle 100 \rangle$ direction, where the four BC symmetry axes are all equivalent. Details of the curve for Mu^0_{BC} are considerably more complicated because of the anisotropy in the hyperfine interaction (see, for example, Patterson, 1988, for details) and the precise shape for various crystallographic orientations can yield both $A_\|$ and $A_\perp$ for that center. Our use of hyperfine decoupling curves has been primarily for Mu^0 state identification rather than for detailed characterization.

e. Zero-Field Muon-Spin Relaxation

The muon-spin relaxation function in a zero magnetic field can be used to obtain a measure of the local dipolar fields at a static diamagnetic muonium center and provides a very sensitive measurement of slow hopping motion. Zero-field muon-spin relaxation (ZF-μSR) data are taken in the longitudinal geometry. Random local dipolar fields cause slow spin precession about different directions; thus the measured polarization initially decreases following the Gaussian envelope of the local field distribution. However, on average, a third of the local field components lie parallel to the initial spin direction; thus at long times a polarization of 1/3 is observed for a static state. Overall, the polarization drops to near zero and then recovers to 1/3 at long times. The standard model for the ZF relaxation is the Kubo-Toyabe function, which assumes that a large number of randomly oriented nuclear moments contribute to these local fields. Expressed as the time-dependent longitudinal polarization, the static Gaussian Kubo-Toyabe expression used to fit ZF-μSR data (Hayano et al., 1979) for a static center is

$$p_z(t) = \frac{1}{3} + \frac{2}{3}(1 - \Delta^2 t^2) \exp\left(-\frac{\Delta^2 t^2}{2}\right) \qquad (10)$$

where $\Delta = \gamma_\mu \langle B_z^2 \rangle^{1/2}$ is the Gaussian width parameter related to the second moment of the local field distribution. The minimum in this static zero-field relaxation function, shown as the dashed curve in Fig. 4, is at $t = \sqrt{3}/\Delta$.

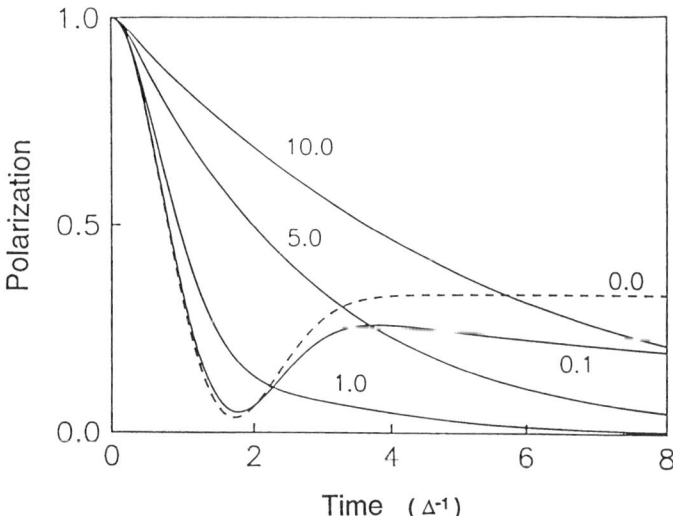

FIG. 4. The zero-field Kubo-Toyabe relaxation functions normalized to the static Gaussian line-width parameter Δ. Labels on the curves are the hop rates in units of Δ.

Thus far µSR has provided the only technique whereby this unique zero-field relaxation function can be demonstrated.

For the dynamic situation where a muonium center hops among equivalent sites, each with its own local dipolar field, this expression needs to be modified (Hayano et al., 1979). The result is not analytical but can be used to accurately extract the rate for (assumed instantaneous) jumps between equivalent muon sites when the static width parameter Δ has been measured previously and is fixed in the fits. The dynamic Kubo-Toyabe function is obtained from the static result of Eq. (10) by

$$P_z(t) = p_z(t)e^{-vt} + v \int_0^t p_z(\tau)e^{-v\tau}P(t - \tau)\,d\tau \tag{11}$$

where the additional variable v is the mean hop rate. Figure 4 shows the dynamic zero-field relaxation function of Eq. (11) appropriately normalized with time in units of $1/\Delta$ for several different hop rates that are given in units of Δ. This function varies from the static result as v increases, first by depressing the long-time 1/3 tail and then by washing out the dip at $\sqrt{3}/\Delta$, after which it takes an Abragamian form and eventually becomes exponential at very large hop rates. We have made extensive use of zero-field data to characterize diamagnetic muonium states, especially in III-V compounds, where all the host atoms have magnetic moments and thus contribute to local dipolar fields. Determination of the Kubo-Toyabe Δ is therefore a means to distinguish different diamagnetic Mu states even if a specific identification of the charge and microscopic structure is not possible.

III. Identification and Characterization of Muonium States

The identification of muonium centers traditionally has been accomplished using the muon-spin precession signals from TF-µSR measurements. These techniques and the characteristic frequencies for the four Mu states observed in tetrahedral semiconductors are well established (Patterson, 1988). In this section I am more concerned with how one can identify the centers present under conditions where the traditional methods fail to distinguish between specific states. One such case is nearly always present. Diamagnetic Mu centers, where the precession signal arises only from the muon moment, all contribute to a single precession signal. Second, for a number of materials or under certain conditions in others, the TF-µSR signals from the paramagnetic Mu^0 states are not visible due to late formation, very rapid transverse relaxation, or dynamic depolarization. During the course of studying muonium dynamics, a number of methods have been used to determine whether an invisible paramagnetic center may be present and to obtain a rough characterization of its properties.

Whether a state is paramagnetic or diamagnetic, in order to fully determine its properties, one would like to know both the electronic and physical structures. An early step in doing so is to obtain characteristic local fields, either the dipolar field for a diamagnetic center or the hyperfine field for a neutral state. The finer structural details come from determining interactions with specific neighbors that are typically obtained from level-crossing spectra. Beyond a center's static properties, information on its motion while maintaining its identity is necessary for a full characterization of that Mu center and in order to understand its interactions with other impurities or dopants.

1. NEUTRAL PARAMAGNETIC CENTERS

The first experimental step in characterizing the muonium states in a new sample is to take TF-μSR data at a nominal field of roughly 10 mT, where one typically sees just the diamagnetic fraction oscillating at the standard muon precession frequency. Common practice is to use silver data as a normalization for the maximum signal from any experimental setup because Ag has a 100% diamagnetic Mu fraction and minimal relaxation. For most semiconductors at low temperature, the diamagnetic fraction formed promptly during implantation is considerably less than 100% (Patterson, 1988), with the remainder presumably forming neutral centers. In samples containing a large fraction of nuclei with magnetic moments and also at μSR facilities with a pulsed-beam structure, it is quite challenging to observe the TF precession frequencies from these Mu^0 states. Spectroscopic techniques other than TF-μSR are rather time-consuming, even when they will work; thus nonspectroscopic methods of identifying Mu^0 centers are often required.

In the absence of spectral information, the most common method of identifying Mu^0 centers has been hyperfine decoupling curves, that is, measurements of the asymmetry as a function of applied field in the longitudinal geometry, as described by the first term in Eq. (9). Figure 3 in the preceding section displayed such curves for each of the two neutral states using typical parameters. The exact location of the features for Mu_{BC}^0 depends on the relative orientation of the bond axis and the applied longitudinal field, reflecting the center's axial symmetry. The differences between the two sites are readily apparent in the decoupling data, and a quick visual inspection of these curves allows a decision as to whether the neutral state is Mu_{BC}^0 or Mu_T^0, or if both states are present.

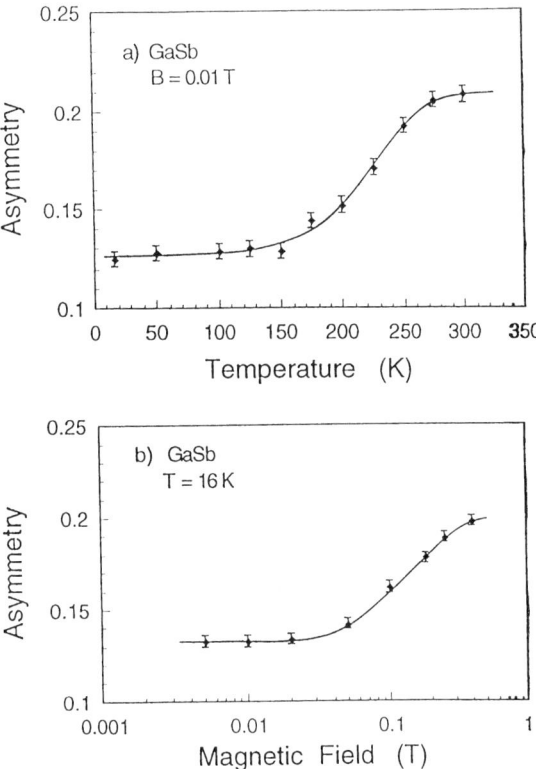

FIG. 5. Data related to the Mu^0 center in n-type GaSb: (a) the diamagnetic TF-μSR asymmetry where a missing fraction indicates a Mu^0 state; (b) the repolarization curve that indicates this center is Mu$_T^0$ with a hyperfine constant of $A_\mu \simeq 3100$ MHz.

Figure 5 shows the diamagnetic amplitude as a function of temperature and the low-temperature decoupling curve for an n-type GaSb sample. Roughly 40% of the muons are not observed in the diamagnetic signal below about 150 K. These muons are shown to produce a Mu$_T^0$ state by the hyperfine decoupling curve that yields a characteristic field of about 110 mT. This translates into a hyperfine constant of $A_\mu \simeq 3100$ MHz for the T site neutral in GaSb, somewhat larger than the value of 2884 MHz for GaAs or 2914 MHz for GaP (Patterson, 1988). The temperature dependence of the diamagnetic TF-μSR amplitude suggests that a transition from Mu$_T^0$ to a diamagnetic state occurs very rapidly by 200 K, where the full fraction is seen in that signal.

As a second example of the use of longitudinal decoupling curves to identify possible neutral states, I show data from a polycrystalline sample of cubic-BN in Fig. 6. The high-field part of this curve, as extended by the dashed line, represents the polarization visible at time zero in an applied longitudinal field, whereas the data points and solid line are for the *residual* or nonrelaxing asymmetry. The initial visible signal strongly implies that a Mu_T^0 state is formed in cubic-BN. The low-field data are less clear. Comparing the solid line in Fig. 6 with the Mu_{BC}^0 curve in Fig. 3 might suggest that a BC neutral state is also present. However, the region between the solid and dashed lines represents an exponentially relaxing longitudinal signal, and the diamagnetic TF-μSR signal from this sample shows the same amplitude and an exponential relaxation with a very similar rate constant. The tentative conclusion is that the relaxing signal in low-field longitudinal data is from the diamagnetic fraction undergoing some dynamic process. The low-field LF data would then represent decoupling of dipolar fields responsible for the fluctuations leading to both the LF and TF relaxation. A very rapidly moving diamagnetic center is the likely explanation for this particular combination of data.

In *n*-type materials, an e^- spin-flip scattering process, sometimes called *spin exchange*, can occur between electrons in the conduction band and Mu^0. This process randomizes the spin orientation of the Mu^0 electron, which is in turn transferred to the muon by the hyperfine coupling, thus

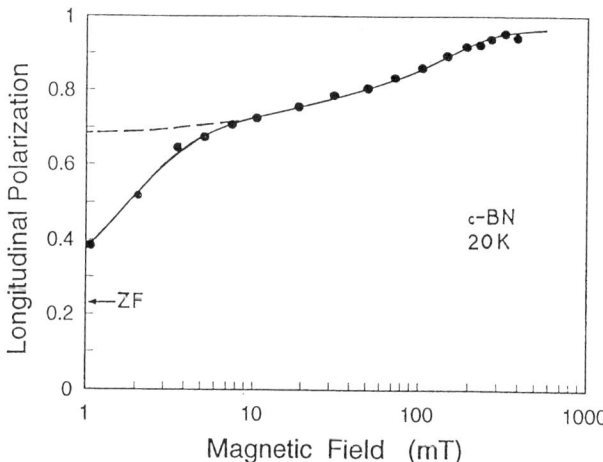

FIG. 6. The nonrelaxing longitudinal polarization for polycrystalline cubic-BN showing the presence of at least two different states, one of which is Mu_T^0. The asymmetry in zero field is shown by the arrow and represents a background signal from the sample holder.

causing rapid depolarization. Consequently, the Mu^0 spin-precession signals broaden and become undetectable. In a longitudinal field, this process leads to relaxation that appears very similar to that discussed for muonium charge cycles. For both spin-exchange and charge-state cycles, the shape of the field dependence for the longitudinal relaxation rate provides some information as to the identity of the Mu^0 center involved. A firm identification of Mu_{BC}^0 can be made in these cases if there is a peak in the relaxation rate versus field (Chow et al., 1994a). Figure 7 shows such data at 120 K for an n-type GaAs(Si) sample doped at 3×10^{15} cm^{-3}. This peak is present from roughly 30 K, where the Si donors ionize, to about 200 K, showing conclusively that a BC neutral muonium state is present up to 200 K in this GaAs sample. Similar data have been used in more heavily doped GaAs and in n-type Si to demonstrate that Mu_{BC}^0 is in fact present under conditions where the TF-μSR signals are not visible. Typical conditions for this to occur is in n-type materials where spin exchange can take place at temperatures too low for ionization or a BC → T site change (Chow et al., 1994a).

I turn briefly to characterization of the diffusive motion of Mu_T^0, where the data on GaAs is by far the most complete and convincing (Kadono et al., 1990; Schneider et al., 1992). In a material like GaAs, where all the nuclei have fairly large magnetic moments, when a muonium state moves rapidly through the lattice, the local fields are modulated by the different dipolar states for the near neighbors of each visited site. For a Mu^0 center, this fluctuating field acts as a random term in the nuclear contributions to the

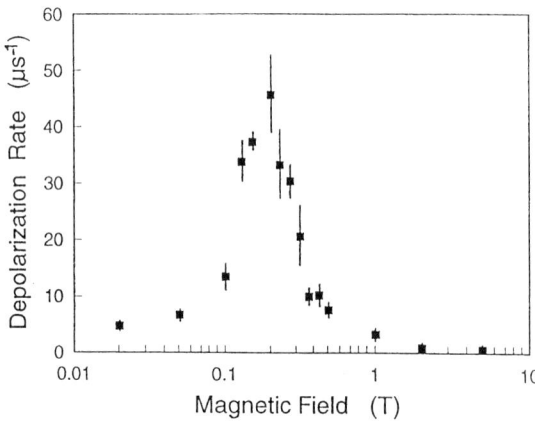

FIG. 7. Typical longitudinal depolarization signature of spin or charge exchange involving Mu_{BC}^0. Data are from a weakly doped n-type GaAs(Si) sample for which this signal is present up to 200 K.

hyperfine fields. The net effect is depolarization of the muon spin, which can be measured as longitudinal relaxation. A simple model (Celio, 1987) assumes a random term for the nuclear hyperfine contributions of the form $H_n = \delta_n \mathbf{S} \cdot \mathbf{T}(t)$, where δ_n is the effective strength of the nuclear hyperfine interactions, $\mathbf{S}$ is the muonium electron spin, and $\mathbf{T}(t)$ is a unit vector having a randomly fluctuating direction with a correlation time τ_c. The inverse of τ_c can be viewed as the Mu^0 hop rate v. A fluctuating interaction of this type introduces spin flips of the muonium electron that are transferred to the muon via hyperfine coupling, thereby producing muon-spin depolarization. While the longitudinal relaxation function is complicated for this model (Yen, 1988), an approximate expression given in Eq. (12) for the dominant exponential rate in the depolarization function can be used to extract the correlation time (Kadono, 1990). The raw LF relaxation data at several different fields are fit to exponential functions, and the resulting rate constants at each temperature are then fit to

$$T_1^{-1} \approx \left(1 - \frac{x}{\sqrt{1 + x^2}}\right) \frac{\delta_n^2 \tau_c}{1 + \omega_{12}^2 \tau_c^2} \qquad (12)$$

to yield a value for τ_c at that temperature. The x in this expression is the field-scaling variable defined earlier, and ω_{12} is the angular frequency for a transition between the relevant Mu^0 energy levels.

Figure 8 shows hop rates from the preceding model as a function of temperature for Mu_T^0 in high-resistivity GaAs (Schneider et al., 1992). These data show a minimum hop rate near 90 K that represents a change from a classic thermally activated or phonon-assisted hopping process at higher temperatures to a quantum tunneling process at lower temperatures. The increase in hop rate as the temperature decreases below 90 K implies incoherent tunneling where the basic motion remains hoplike, that is, to a near-neighbor T site. An additional change in the character of the motion takes place near 10 K, below which the tunneling becomes coherent and the mean free path is longer than the T-site spacing (Kagan and Prokof'ev, 1990). While the study of these quantum tunneling processes is extremely interesting in its own right, of primary interest in the present context is the fact that the Mu_T^0 center is very rapidly moving at all temperatures. As seen in Fig. 8, the minimum hop rate in GaAs is greater than 10^8 sec^{-1}, on the order of 10^2 jumps during an average muon lifetime. Evidence in the other semiconductors strongly suggests that Mu_T^0 is highly mobile in all these materials, with estimates for the minimum rate in intrinsic Si and Ge above 10^{10} sec^{-1} (Patterson, 1988).

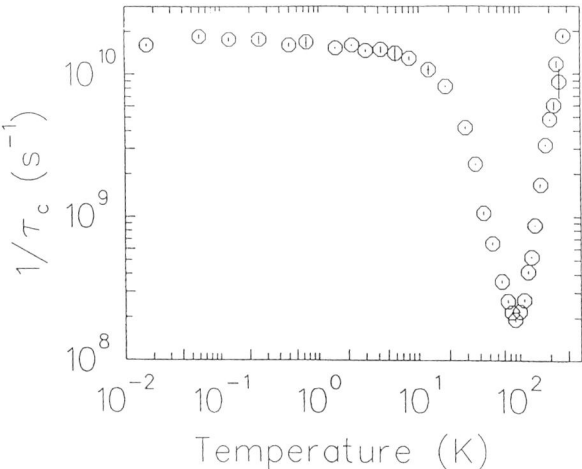

FIG. 8. The temperature-dependent hop rate for the fast-diffusing Mu_T^0 center in semi-insulating GaAs. These data imply quantum tunneling motion below 90 K. (From Schneider et al., 1992.)

2. CHARGED DIAMAGNETIC CENTERS

Identification of the charge state for diamagnetic Mu centers has been a serious problem since the beginning of μSR. In most materials, any diamagnetic signal is assumed to come from a bare μ^+ and usually has carried that label. However, a negatively charged center Mu^- or any bound muonium state with no unpaired e^- such as a Mu-dopant pair is also diamagnetic. A method to experimentally determine exactly what state is responsible for the observed diamagnetic signals would clearly fill a major hole in the available information. We have made some significant progress on this problem recently using relaxation functions to provide additional characterization of diamagnetic centers and QLCR spectra to yield some details of their local structures.

GaAs provides a case study in use of these techniques in an attempt to identify the diamagnetic centers. My colleagues and I have managed to fully characterize the Mu^- center in GaAs, determining its structure and motional properties (Chow et al., 1995, 1996). Work on the states related to Mu^+ is incomplete; two separate centers can be demonstrated, but neither is yet conclusively identified (Chow et al., 1997a). Efforts to categorize the diamagnetic centers in GaAs began with the data of Fig. 9, which shows the TF-μSR Gaussian line width or relaxation rate as a function of temperature for very heavily doped n-type GaAs(Si) and p-type GaAs(Zn) doped to

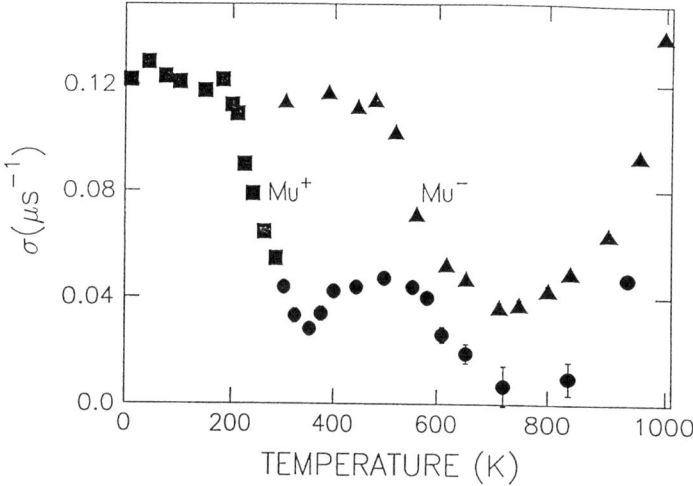

FIG. 9. The TF-μSR Gaussian relaxation rate σ for the diamagnetic signal in heavily doped GaAs; the triangles are for Mu$^-$ in n-type GaAs(Si), and the squares and circles are for Mu$^+$ in p-type GaAs(Zn). (From Chow et al., 1997.)

metallic levels of 5×10^{18} and $3 \times 10^{19}\,\text{cm}^{-3}$, respectively (Adams et al., 1995; Chow et al., 1997a). These data demonstrate clear differences in the diamagnetic muonium states in these two samples. Isolated Mu centers should be Mu$^-$ in the n-type sample and Mu$^+$ in the p-type case, but a Mu-Si or Mu-Zn pair also could occur. The data for heavily doped GaAs(Te), which is n-type but with the donor substituting on the opposite sublattice, are essentially the same as for GaAs(Si) up to 700 K. Since the increase in relaxation rate at the highest temperatures is from muonium charge cycles and is expected to shift with concentration, these data suggested that the same state is seen in both n-type materials implying an isolated Mu$^-$.

Two additional types of experiments, QLCR and quadrupolar decoupling curves taken at room temperature, allowed the structure of Mu$^-$ to be determined (Chow et al., 1995). Figure 10 shows the results for GaAs(Si). Essentially identical data were again obtained for the GaAs(Te) sample, which verifies that the center is indeed an isolated Mu$^-$. The combination of these two types of data, with decoupling curves taken for applied fields along each of the three high-symmetry crystallographic directions and QLCR spectra for $\mathbf{B}\| \langle 100 \rangle$, allowed the quadrupolar and dipolar coupling constants to be determined for both the nearest and second shells of neighbors. The QLCR spectrum (Fig. 10b) has four higher-field lines that

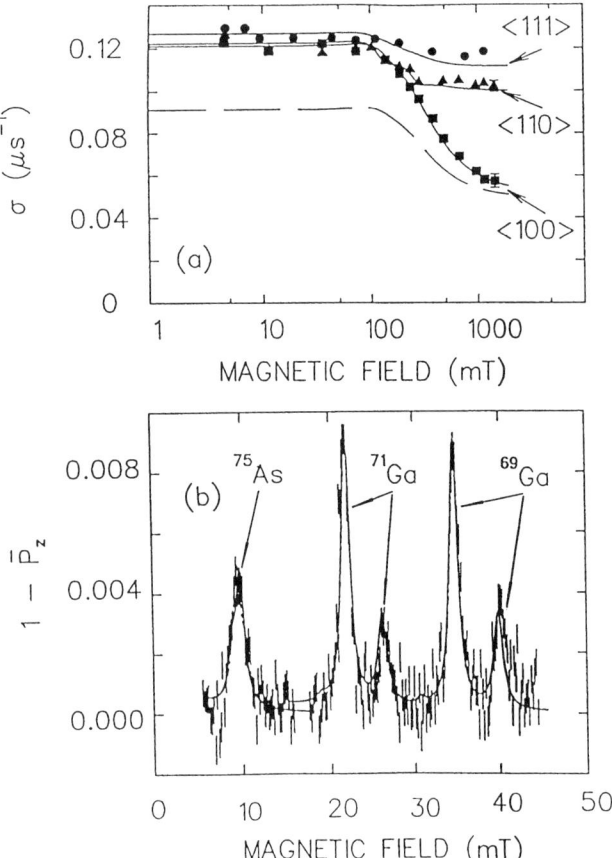

FIG. 10. Data used to determine the structure of Mu⁻ in heavily doped n-type GaAs(Si): (a) quadrupolar decoupling curves shown as the field-dependent TF-μSR relaxation rates for fields along the labeled crystal axes; (b) the QLCR spectrum for $\mathbf{B}\|\langle 100\rangle$ implying equivalent Ga nearest neighbors along $\langle 111\rangle$ directions from the muon. (From Chow et al., 1995, © 1995 American Physical Society.)

are identified as from Ga atoms along the equivalent $\langle 111\rangle$ directions and a lower-field line with unresolved structure due to more distant As neighbors. Duplicate two-line features at fields in the ratio of the Ga quadrupole moments and intensities with the Ga dipole moment ratio make the Ga identification unmistakable, whereas the lack of such structure for the lower-field feature implies As. The decoupling curves yield a better determination of the dipolar constants, and modeling of these curves with assumed structures independently suggested a T site having Ga nearest neighbors (Chow et al., 1994b). The combined data on both n-type samples imply a T_{Ga} location for Mu$^-$ in GaAs with a 9–10% inward relaxation of the Ga atoms and a slight outward relaxation of the second-neighbor As atoms (Chow et al., 1995). A theoretical calculation (Adams et al., 1995) using a small cluster gave about a 3% inward relaxation for the Ga atoms when the second neighbors were kept fixed and implied that the minimum-energy location is at a Ga antibonding location, but with a very small barrier for AB_{Ga} to AB_{Ga} motion through the central T_{Ga} site. The muonium zero-point energy is several times as large as the energy differences in this calculational model; thus motion would yield the averaged higher symmetry as observed, but perhaps with a shorter Mu—Ga distance than if the location were actually at the T site, depending on the details of the averaging.

The motional properties of this Mu_T^- center are qualitatively implied by the narrowing feature just above 500 K in the TF-μSR line-width data of Fig. 9. The motion is better characterized by the relaxation function in zero field as modeled by the dynamic Kubo-Toyabe function of Eq. (11) with the Gaussian width parameter Δ fixed to the static value. For Mu_T^-, the lower-temperature zero-field relaxation data fit to Eq. (10) for a stationary center yielded $\Delta = 0.165(5)$ MHz. Figure 11 displays the hop rate v extracted from fits of the dynamic function to the zero-field data between 475 and 625 K. These hop rates are well described by an Arrhenius law $v(T) = v_0 e^{-E_a/k_B T}$, with $v_0 = 5.6(5) \times 10^{12}$ sec^{-1} and $E_a = 0.73(1)$ eV. The dramatic

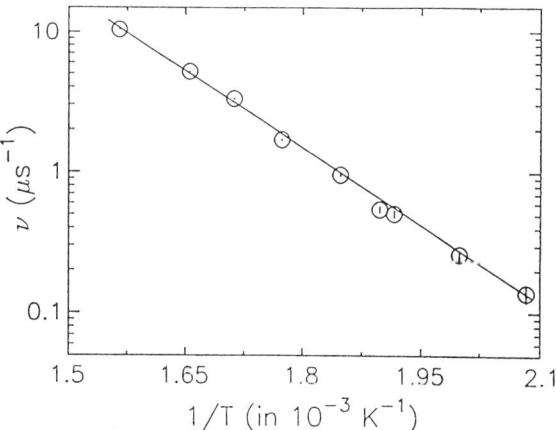

FIG. 11. An Arrhenius plot of the hop rates for Mu$^-$ in GaAs(Si) yielding a barrier height of 0.73 eV for motion between nearest T_{Ga} sites. (From Chow et al., 1996, © 1996 American Physical Society.)

difference in the motional properties of Mu_T^- compared with those of Mu_T^0 is illustrated by the approximately 10 orders of magnitude difference in hop rates at room temperature. While the larger radius expected for a Mu^- ion compared with an atomic-like Mu^0 would lead to reduced motion, the more likely origin for this huge difference lies with the partial ionicity of GaAs bonds. The unequivalent T sites have nearly the same energy for Mu_T^0, but the slightly positive charge for Ga lowers the T_{Ga} site energy for Mu^- and negative charge associated with As raises the energy for T_{As} (Adams et al., 1995). Thus Mu^- motion proceeds from T_{Ga} to T_{Ga}, with the intervening T_{As} location representing a significant barrier rather than a nearly equivalent minimum as for Mu_T^0.

While a detailed picture exists of both the structure and motional properties of Mu^- in heavily doped GaAs, understanding of Mu^+ states in p-type samples is not nearly as advanced. The TF-μSR line-width data for Mu^+ in Fig. 9 show the onset of motion near 200 K as a drop in the width; however, the width increases again near 300 K and shows a broad peak between 400 and 650 K (Chow et al., 1997a). Such peaks are typical of muonium centers trapped at or near an impurity. Here it might indicate the formation of a Mu-Zn pair, but this remains to be verified. Zero-field relaxation data show features consistent with the picture arrived at from the transverse data. An attempt to fit the ZF data with a single dynamic Kubo-Toyabe signal failed to yield good fits but did indicate that the motion between 200 and 300 K occurs in two steps rather than as a single activated process. Using results from several attempts to fit these data with slightly different models, I suggest the following scenario: At low temperatures, the state is most likely Mu_{BC}^+, which is expected theoretically to be the most stable Mu^+ center. As the temperature is raised, the weaker bond in this bridging structure breaks, and the Mu^+ hops locally among the four equivalent BC directions around one common partner, although it is not clear whether this should be Ga or As since it depends on whether ionic or covalent features dominate for the stronger bond. At a slightly higher temperature, the stronger bond also will break, allowing global migration of the Mu^+. The double dynamic feature at roughly 200 and 280 K that shows up in the zero-field data is assumed to represent this asymmetric release from the BC configuration. When the Mu^+ is able to migrate throughout the lattice, it can then find a more stable location, such as paired with a Zn acceptor, thereby forming the higher-temperature trapped state. Finally, at sufficiently high temperatures this state dissociates, and full diffusive motion sets in. While this picture is feasible, no single piece of it has been confirmed by a structural determination. We have demonstrated that the two quasi-stationary states, below 200 K and near 520 K, are definitely different by looking at the quadrupolar decoupling curves. QLCR data were recorded at 520 K, although the partial spectrum is not sufficiently detailed at present to draw any conclusion regarding the local structure.

A number of other III-V materials show trapping peaks similar to that seen in p-type GaAs. Both the n- and p-type samples of InP and GaSb have such features, which in each material are different for the two doping types. We have taken extensive zero-field data on both these materials and have characterized the local fields and motional properties of several of these centers. Quadrupolar decoupling curves have shown clear differences in the low-temperature centers and the trapped states in each case, but as for GaAs, firm structural results do not yet exist from which to confirm any tentative model of the sequence of muonium states and transitions that may occur. However, in n-type InP we have managed to fit the zero-field data on the diamagnetic centers to a three-signal dynamic Kubo-Toyabe relaxation to obtain the Δ parameter characterizing the static local fields and an activation energy associated with the motion for each state (Lichti et al., 1997a). Because the low-temperature states (two in each sample) are identical in n- and p-type InP, and since no Mu$^-$ state has ever been observed in any p-type material, we have suggested that these states are two separate Mu$^+$ centers in the BC and T$_V$ configurations and that the third state formed at higher temperatures after Mu^0 disappears is most likely Mu$_T^-$. As in GaAs, these assignments must remain tentative until one or more of these identities and local structures have been confirmed by QLCR spectra. The fitted asymmetries and hop rates for the three states are shown in Fig. 12, and the tentative assignments and associated parameters are listed in Table I. Similar fits to ZF data for the other samples of InP and GaSb have produced less consistent results thus far, and work continues to identify the various diamagnetic muonium states in III-V materials and to determine their respective local structures.

TABLE I

TENTATIVE ASSIGNMENTS AND PARAMETERS FOR MUONIUM STATES IN InP FROM DYNAMIC KUBO-TOYABE FITS TO ZERO-FIELD μSR DATA

Muonium center	Low-T f_μ (%)		KΓ width (μs^{-1})		Hop barrier (eV)	
	n-type	p-type	n-type	p-type	n-type	p-type
Mu$_{BC}^+$	68	62	Δ_1 0.3041(12)	0.3065(11)	0.167(8)	0.148(15)
Mu$_{TP}^+$	16	12	Δ_2 0.096(8)	0.108(12)	0.270(11)	
Mu$_T^0$	16	26	Missing			
Mu$_{T_{In}}^-$			Δ_n 0.282(7)		0.75(11)	
Mu-A (?)			Δ_p	0.147(5)		

From Lichti et al., 1997.

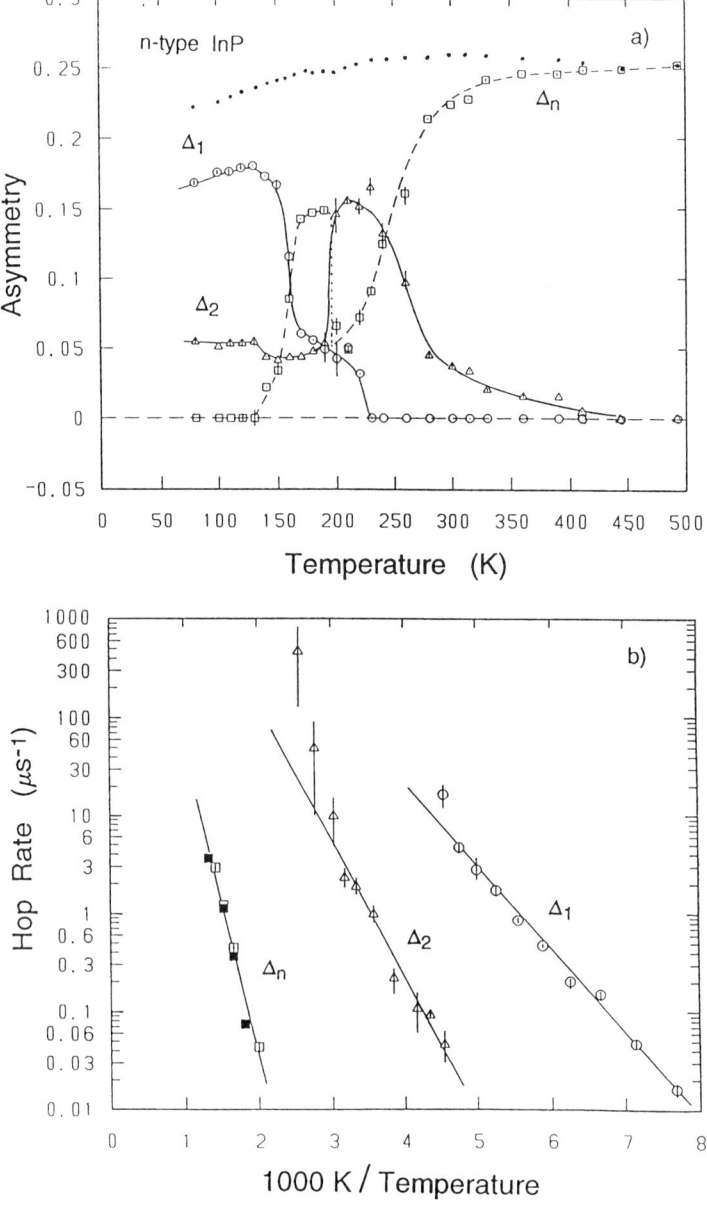

FIG. 12. Results of a three-signal dynamic Kubo-Toyabe analysis of zero-field data on n-type InP: (a) the amplitudes; (b) the hop rates for observed states. The resulting parameters and tentative assignments are listed in Table I. (From Lichti et al., 1997.)

IV. Dynamics of Muonium Transitions

The main emphasis of recent efforts has been to investigate the transitions among muonium states in semiconductors, initially in silicon and more recently in germanium and GaAs. Very recently this work has begun to be extended to other III-V materials. A configuration coordinate diagram was developed from the transitions required to fit rf-μSR data for Si (Lichti, 1995; Kreitzman *et al.*, 1995), and results for the other materials have been used to modify specific features of this diagram and adjust the energy parameters for each material. Figure 13 shows this diagram roughly to scale for the silicon parameters. In building up a model for muonium transitions, only the four known Mu states were considered along with the various site changes and charge-state transitions with $\Delta q = \pm e$ that could occur among those states. The primary data in silicon were temperature-dependent rf-μSR amplitudes of the Mu$^\pm$ diamagnetic states, with the rf results for the neutral centers and longitudinal relaxation measurements as qualitative checks on the developing model of Mu transition dynamics. Thus far the main data for the other materials have been relaxation measurements in all three field conditions, transverse, longitudinal, and zero field, with the rf resonance data playing a secondary role. One of the remaining tasks to demonstrate that the basic picture is in fact correct is to fully integrate the relaxation and rf results within a single model for at least one material. In the rest of this section I discuss the data for each material separately with reference back

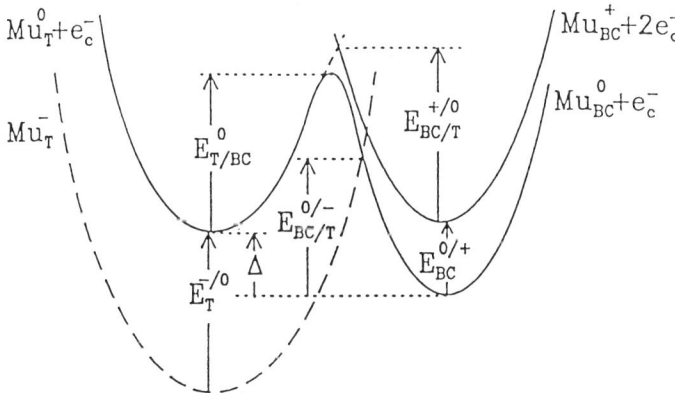

FIG. 13. The configuration coordinate diagram representing an operational model of muonium transitions in silicon. All energies in the diagram except Δ have been obtained from rf-μSR data. This diagram serves as a general reference for muonium transition dynamics in other semiconductors. (From Kreitzman *et al.*, 1995, © 1995 American Physical Society.)

to the diagram of Fig. 13. Each transition that has thus far been clearly identified will be discussed, and where the transition-rate parameters have been determined, the current best fit values are given in the accompanying tables.

1. SILICON: THE BASIC MODEL

For silicon, the rf-μSR data are cleanest in the p-type materials, where only three of the four muonium states participate. These are the two neutrals Mu_T^0 and Mu_{BC}^0 along with the positively charged center Mu_{BC}^+. The rf data on a p-type Si sample doped with boron at a level of 10^{13} cm^{-3} is shown in Fig. 14, which displays all the main transition features observed in p-type silicon. Before discussing the transitions, we need to briefly examine the starting conditions immediately after implantation.

During the muon implantation process, a large number of electron-hole pairs are produced along with some atomic displacements. The main result of this stopping spur on the muonium states that are formed initially is to provide a source of electrons. In fitting the rf-μSR data for silicon, an assumption that works quite well is that each μ^+ captures an electron to form Mu^0 and that these neutrals form Mu_T^0 and Mu_{BC}^0 centers in a fixed ratio independent of temperature, doping type, and concentration. The low-temperature diamagnetic Mu fraction in the most intrinsic Si sample is

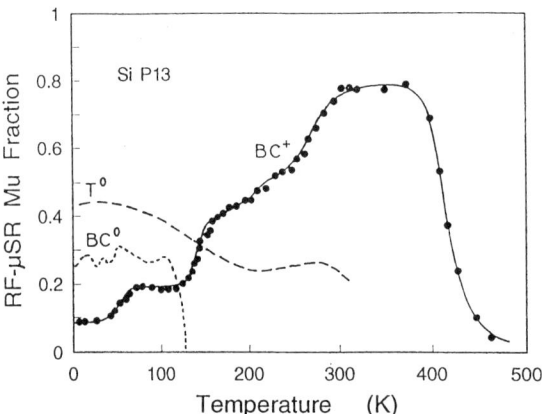

FIG. 14. Time-integrated rf-μSR amplitudes for the three muonium centers in p-type Si(B) doped at 10^{13} cm^{-3}. The diamagnetic amplitudes (data points) were fit to extract rate parameters for transitions as discussed in the text. (From Kreitzman et al., 1995, © 1995 American Physical Society.)

less than 3%. At the very lowest temperatures, where dopants should not be ionized, the rf data show an increasing diamagnetic fraction with increased concentration of either dopant type (Kreitzman et al., 1995), a trend that is also observed in TF-μSR data (Patterson, 1988). In backtracking results on transition dynamics for various processes, we found that this increase in diamagnetic states always appears to come from the highly mobile Mu_T^0 fraction. Therefore the low-temperature increase in quasi-prompt diamagnetic states in doped silicon has been assigned to a charge-exchange scattering interaction between Mu_T^0 and neutral dopants, either donors or acceptors. Because this takes place extremely rapidly, little additional information can be obtained. I propose the following interactions as the most likely processes to form diamagnetic states out of the Mu_T^0 initial fraction:

$$Mu_T^0 + D^0 \rightarrow Mu_T^- + D^+(\rightarrow[Mu, D]^0) \quad (13)$$

and

$$Mu_T^0 + A^0 \rightarrow Mu_{BC}^+ + A^-(\rightarrow[Mu, A]^0) \quad (14)$$

where it is unclear experimentally whether or not the final step occurs. Since the initial charge exchange results in oppositely charged impurities that are spatially close together, the likelihood of Coulomb capture to form a bound pair should be relatively high. Because the time frame for these interactions is extremely short based on the TF-μSR data, I suspect that the process is completed during the later stages of the initial thermalization.

Recent experiments show that applying large electrical fields to drive the electrons in the spur ionization cloud along or opposite the μ^+ implantation direction can modify the diamagnetic/neutral ratio (Storchak et al., 1997), demonstrating that the availability of free electrons during the later stages of thermalization is important to the initial mix of states, particularly to the charge-state mix at the BC site. These electrical field data imply that the Mu_T^0 state is formed early in the thermalization, while most of the Mu_{BC}^0 formation involves delayed capture of an e^- by a fully thermalized μ^+ already at the BC location.

As the temperature is increased, the time-integrated diamagnetic fraction, as shown in Fig. 14 for weakly doped p-type Si, displays several steps, each of which indicates that an additional muonium transition process has become active on a microsecond time scale. The lowest-temperature transition is seen at about 50 K, where the $Mu^\pm$ amplitude increases. This is the temperature region in which donors or acceptors ionize. A check of samples with different acceptor species indicated a temperature shift in accordance with the acceptor ionization energy, verifying that this feature is associated with the increase in majority carrier concentrations due to dopant ionization. The specific muonium transition assigned to this feature in p-type samples is

$$\text{Mu}_{\text{BC}}^0 + h^+ \to \text{Mu}_{\text{BC}}^+ \tag{15}$$

The transition rate was modeled as $v_c^h(T) = n_h(T)v_h(T)\sigma_c^h$, where we have used standard expressions for the temperature dependence of the carrier density $n_h(T)$ and mean thermal velocity $v_h(T)$ within an appropriate statistical model and assumed a temperature-independent hole capture cross section σ_c^h.

Some evidence exists for hole capture by Mu_T^0 in the decrease of its rf-μSR signal intensity beginning at 50 K but showing a much weaker temperature slope (long dashed curve in Fig. 14) compared with hole capture at the BC site. Much stronger evidence that the BC process dominates in p-type Si is the fact that for higher concentrations the diamagnetic amplitude change occurs in two steps that are correctly modeled by the BC anisotropy (Kreitzman et al., 1995), and the net amplitude change correlated well with the initial BC fraction. Hole capture by the T-site neutral does occur under other conditions. The T-site process has been proposed to explain a low-temperature T $\to$ BC site change induced by optical excitation of n-type samples (Kadono et al., 1997), although in that experiment the final state was apparently Mu_{BC}^0, implying that hole capture to form Mu_{BC}^+ at low temperatures will quickly be followed by capture of an electron if the density n_e is sufficiently large. The rf evidence clearly indicates that the primary low-temperature hole-capture process in p-type silicon occurs predominantly at the BC site.

In the most intrinsic Si sample that was investigated (p-type at $\sim 10^{11}$ cm^{-3}), the 50-K step is not observed, and the lowest-temperature feature is near 140 K, that is, the second feature observed with increasing temperature in Fig. 14. This increase in the diamagnetic signal is identified as thermal ionization of the BC neutral center:

$$\text{Mu}_{\text{BC}}^0 \to \text{Mu}_{\text{BC}}^+ + e^- \tag{16}$$

The ionization was treated as a simple activated process with a rate expression of the form $v_i(T) = v_o^i e^{-E_i/kT}$. The diamagnetic amplitude increase related to this feature also occurs in two steps properly modeled by the anisotropy of the Mu_{BC}^0, unambiguously identifying the initial state. The strong correlation of the drop in Mu_{BC}^0 amplitude shown as the dotted curve in Fig. 14 and the initial increase in diamagnetic amplitude associated with

Mu_{BC}^+ confirm that this is a simple direct transition. The best value for the ionization energy as obtained from the rf data is from the near-intrinsic sample where the ionization step is cleanest, yielding 0.21(1) eV compared with 0.17 eV from temperature-dependent TF-μSR relaxation rates for Mu_{BC}^0 (Patterson, 1988) and an average just below 0.20 eV obtained from temperature-dependent longitudinal field quenching curves on several samples (Scheuermann et al., 1995). The rf-μSR value comes from features related to the final state, whereas the other two are associated with the initial state. For the data shown in Fig. 14, the net amplitude change associated with the two lowest-temperature transitions adds up to the initial BC fraction consistent with the preceding process assignments. At concentrations beyond the middle 10^{14} cm^{-3} range, h^+ capture dominates with only a very small ionization contribution, and below roughly 10^{12} cm^{-3}, ionization completely dominates as the Mu_{BC}^0 to Mu_{BC}^+ transition process in p-type silicon.

As the temperature is raised above 230 K, the diamagnetic fraction continues to increase beyond what can be associated with the initial BC fraction. This feature and the rapid decrease in the Mu_T^0 rf-μSR signal amplitude near room temperature is assigned to a T $\rightarrow$ BC site change for Mu^0 as the rate-limiting step in a two-step process leading to the Mu_{BC}^+ final state, with either of the two processes just discussed providing the second step depending on n_h and temperature. The site-change dynamics are modeled as a simple thermally activated process giving $v_a(T) = v_o e^{-E_a/kT}$ as the transition rate. The overlapping temperature region in which both the Mu_T^0 and Mu_{BC}^+ rf signals are observed is related to the different rf frequencies used in the two resonance measurements and thus different sensitivities to the transition time scale. Vastly different longitudinal decoupling characteristics for Mu_T^0 and Mu_{BC}^0 and dependences on field direction for Mu_{BC}^0 allow transitions from the two paramagnetic initial states to be cleanly separated by varying the magnitude of the field and the sample orientation. Thus the assignment of diamagnetic features to T and BC initial Mu^0 states has been verified experimentally using several different fields and orientations for several samples. These checks on transition assignments and the associated dynamic parameters were expecially useful in situations where features related to the two Mu^0 sites overlapped significantly.

At room temperature, all the existing information on the muonium states in p-type silicon agree that essentially 100% of the muons are in a diamagnetic state and located at the BC site. The presumed center is an isolated Mu_{BC}^+, although if some fraction of these were a Mu-acceptor pair, the theoretical structure would place the muon in a near-BC location next to the dopant atom. We have searched for a signature of such a bound state in Si(B) samples and have not found convincing evidence of pair formation within a few muon lifetimes at any temperature. A small increase in the TF-μSR diamagnetic line width roughly associated with acceptor ionization could be construed as evidence of interaction with the boron magnetic moments; however, a clear indication of a pair, such as the boron-related QLCR spectra, that should be present for a static Mu-B pair, has not yet been found despite several attempts.

Returning to the muonium dynamics observed in p-type silicon, as the temperature is raised above 300 K, the rf-μSR amplitude decreases to zero starting just below 400 K in the most intrinsic sample and shifting to higher temperatures with increasing p-type doping levels. Longitudinal relaxation measurements show the onset of rapid depolarization, and the TF-μSR relaxation rates increase at these same temperatures. All these observed features are related to cyclic muonium charge-state transitions. The shift to higher onset temperatures with increasing acceptor concentration N_A implies that the cycles are controlled by the conduction electron concentration n_e. Detailed fits of the drop in diamagnetic rf-μSR signal intensity assuming e^- capture by Mu_{BC}^0 require the capture process to be thermally activated beyond effects in n_e (Kreitzman et al., 1995). Fits to the high-temperature longitudinal relaxation data for the near-intrinsic Si sample, the results of which are shown in Fig. 15, strongly imply that the neutral center active in the charge-cycle depolarization is Mu_T^0. We are thus left with an initial Mu_{BC}^+ state and an active relaxing state of Mu_T^0. The simplest additional transition that satisfies such a situation is for the charge cycle to be initiated by a combined e^- capture and site-change transition into the metastable T-site neutral muonium state:

$$Mu_{BC}^+ + e^- \to Mu_T^0 \qquad (17)$$

The dynamics of this transition were modeled using an activated capture process as the simplest type of rate expression consistent with the data; $v_{ac}^e(T) = n_e(T)\,v_e(T)\sigma_o e^{-E_c/kT}$, where $\sigma(T) = \sigma_o e^{-E_c/kT}$ is considered to be an activated total-process cross section. The resulting energy of 0.40(2) eV for this capture process is presumably related to the site change. A check of the relative rates for various transitions out of the three states seen in p-type samples, using the dynamic parameters from fits to the rf data, verifies that a three-state cycle should occur involving Mu_{BC}^0 ionization to Mu_{BC}^+, then the activated e^- capture to Mu_T^0, and finally the T to BC site change as Mu^0 back to the first state. I shall return to a discussion of the muonium charge cycles after looking at the low-temperature Mu transition dynamics in n-type samples.

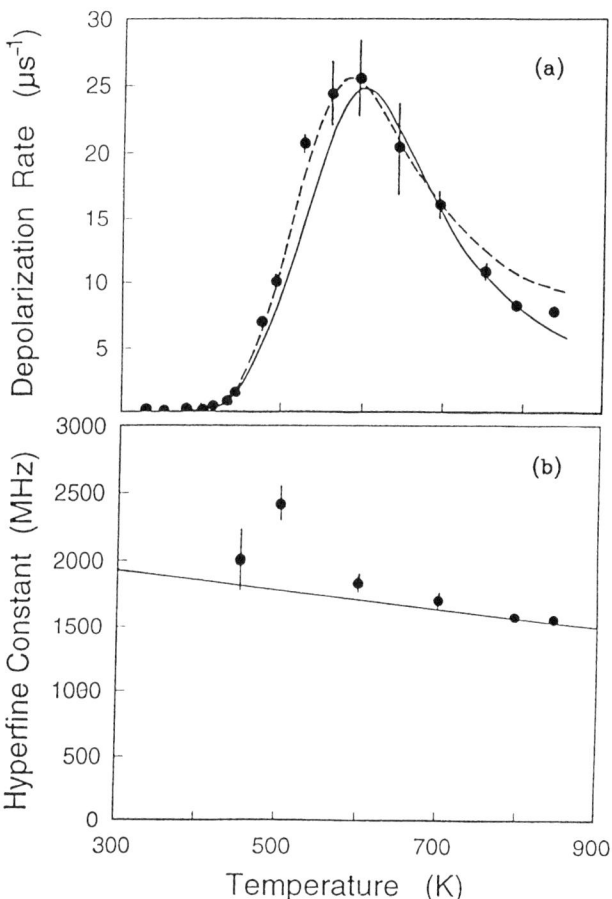

FIG. 15. (a) Low-field longitudinal depolarization rates due to rapid muonium charge cycles in near-intrinsic p-type Si. (b) The Mu^0 hyperfine constant extracted from a two-state fit to the field dependences (data not shown). The line in (b) is extrapolated from low-temperature data on Mu_T^0. Solid curve in (a) used parameters from the field dependence, and the dashed line is a fit to these data for $T < 650$ K. (From Chow et al., 1993, © 1993 American Physical Society.)

In n-type silicon, the fourth muonium state Mu_T^- is important, and additional depolarization mechanisms for the Mu^0 centers become active (Kreitzman et al., 1995; Hitti et al., 1997). Several interactions involving the conduction electrons must be included to obtain consistent fits to the rf-μSR data on the n-type samples. Spin-exchange scattering between conduction

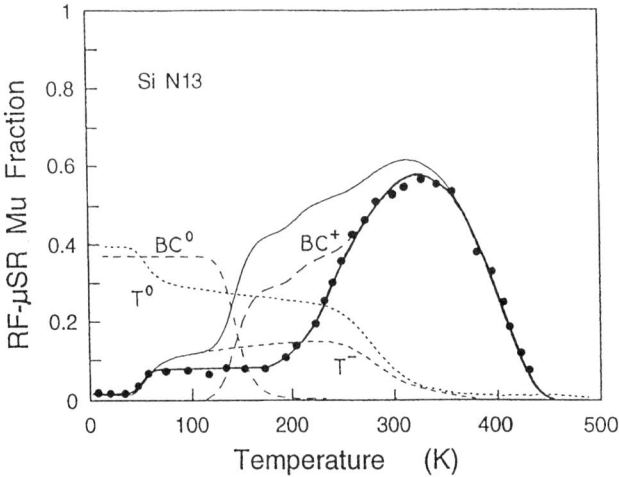

FIG. 16. Results of a four-state model for the rf-μSR amplitudes in weakly doped n-type Si(P) with net $N_D = 2 \times 10^{13}$ cm^{-3}. The thin solid line is the sum of Mu$_T^-$ and Mu$_{BC}^+$ fractions, while the thicker line is a fit to the diamagnetic amplitude (points) including depolarization of Mu^0 precursor states. (From Lichti et al., 1995.)

electrons and Mu$_{BC}^0$ and a charge-state cycle with the muon remaining at the BC site cause significant depolarization for the initial BC fraction at low and intermediate temperatures even in the most weakly doped n-type samples.

Figure 16 shows the rf-μSR data for an n-type Si sample doped with phosphorous for a net donor concentration of 2×10^{13} cm^{-3}. This figure also shows fractional occupations of the Mu^0 states and contributions to the diamagnetic signal for each charged state as a function of temperature, as modeled with transition dynamics alone, that is, without the additional depolarization mechanisms (Lichti et al., 1995). The heavier solid line through the data represents a fit with the extra Mu$_{BC}^0$ depolarization included. Note the region between about 140 and 240 K, where the sum of Mu$_{BC}^+$ and Mu$_T^-$ curves shows the largest deviation from the data. This difference is the diamagnetic signal loss due to the muon depolarization in the precursor states, primarily from Mu$_{BC}^0$. Figure 17 displays the low-temperature longitudinal relaxation data at 200 mT for the same sample. The depolarization rate increases near 50 K when the donors ionize, thereby providing electrons for the spin-exchange scattering processes. There is a second increase in the depolarization rate above 150 K from Mu$_{BC}^0$ ionization and subsequent recapture of an electron by Mu$_{BC}^+$ resulting in a rapid

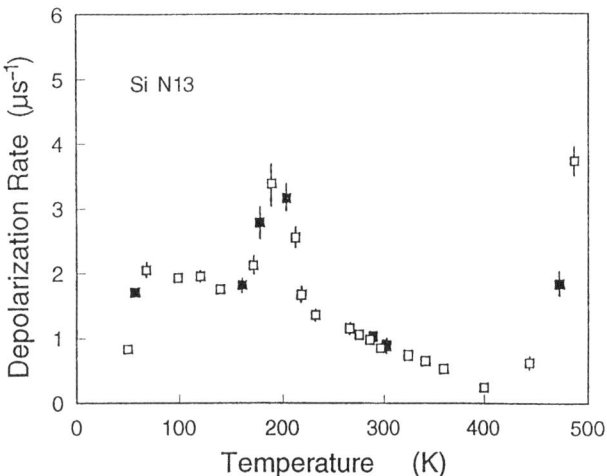

FIG. 17. Longitudinal depolarization rates at 200 mT for the 10^{13} cm^{-3} n-type Si(P) sample showing spin-exchange scattering effects, a peak around 200 K from a $0/+$ charge cycle at the BC site, and the onset of the high-temperature charge cycle. (From Lichti et al., 1995.)

charge cycle at the BC site. A previously unlisted e^- capture transition, namely,

$$\mathrm{Mu}_{BC}^+ + e^- \to \mathrm{Mu}_{BC}^0 \qquad (18)$$

is implied by this relaxation peak and is necessary to fit the rf data (Hitti et al., 1997). The depolarization rate from this BC charge cycle stays low because the hyperfine constant is small for Mu_{BC}^0, and the rapid increase in ionization rate with temperature soon overtakes the hyperfine frequency, thereby reducing the net depolarization. Near 200 K, the Mu_{BC}^0 center also can capture an e^- and change sites in competition with the other BC transitions, as will be demonstrated in more heavily doped samples to be discussed shortly. In the current sample, several things occur in this temperature range, making it difficult to obtain specific information on a given process.

In the 10^{13} cm^{-3} n-type silicon sample of Figs. 16 and 17, the identity of the diamagnetic signal is almost entirely determined by what transition dominates the exit route from the metastable Mu_T^0 initial fraction. When the Mu^0 site change to the more stable BC configuration dominates, the eventual diamagnetic state is Mu_{BC}^+ as in p-type silicon. However, if electron capture occurs faster than the site change, then the diamagnetic state is Mu_T^-. The latter situation prevails up to 230 K for this sample based on a comparison of fitted transition rates (Kreitzman et al., 1995). Most of the rapid rise in diamagnetic fraction between 200 and 280 K is due to this change in identity of the final diamagnetic state originating from Mu_T^0. The e^- capture by Mu_T^0 gives the small step in diamagnetic amplitude in Fig. 16 near 50 K and is related to ionization of the P donors, which rapidly increases the electron concentration at that point. The muonium transition is

$$Mu_T^0 + e^- \to Mu_T^- \qquad (19)$$

as stated earlier. The rate was fitted to a standard carrier capture expression analogous to that presented for hole capture. A considerably more accurate determination of the cross section can be obtained with higher donor concentrations, where the size of this step increases, as seen in Fig. 18 for a 10^{15} cm^{-3} n-type sample (Hitti et al., 1997). This figure also demonstrates the increase in the Mu_T^- fraction at very low temperatures assigned to the charge-transfer interaction between the mobile Mu_T^0 center and neutral donors, listed earlier as Eq. (13).

Figure 18 shows two other features related to Mu_T^-. Recent checks of the rf data on n-type silicon samples in the 10^{15} and 10^{16} cm^{-3} concentration range verify that nearly all the diamagnetic signals at these doping levels are from Mu_T^-. The feature at 200 K is assigned to electron capture by Mu_{BC}^0 and an accompanying site change:

$$Mu_{BC}^0 + e^- \to Mu_T^- \qquad (20)$$

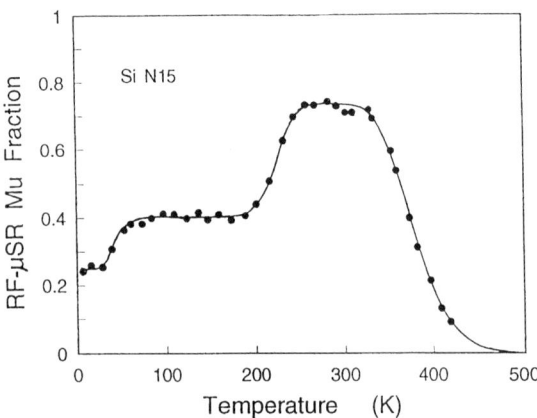

FIG. 18. The diamagnetic rf-μSR amplitude in n-type Si(P) samples doped at 10^{15} cm^{-3} is totally from Mu_T^-. The drop around 400 K is assigned to thermal loss of an e^- from Mu^- locating the T($-$/0) electronic level at 0.56 eV below the conduction band. (From Kreitzman et al., 1995, © 1995 American Physical Society.)

The rate expression used for this transition is the activated capture form given earlier. This process is active in the initial part of the 200-K amplitude rise in Fig. 16 and shows a small shift to lower temperatures as the donor concentration increases, providing one of the clues to its identity. Several checks, including the disappearance of the Mu_{BC}^0 longitudinal relaxation peak at the same temperature, verify this transition assignment.

Perhaps the most important information with respect to the analogous H^- center that can be obtained from the Mu^- dynamics is extracted from the drop in amplitude starting near 350 K in the 10^{15} cm^{-3} and higher concentration samples. At lower n-type doping levels and in p-type Si, the disappearance of the rf-μSR diamagnetic signal is associated with the Mu_{BC}^+ to Mu_T^0 transition that signals the start of the high-temperature charge cycle. In all samples with n_e less than the middle 10^{14} cm^{-3} range, this drop is assigned to the same transition, and its onset shifts to lower temperature with increasing n_e, as expected for an electron-capture process. At the upper end of this electron concentration range, the signal loss has shifted to below rom temperature. However, for higher-concentration samples, this feature occurs at a significantly higher temperature than expected from this trend and does not vary with concentration. Such behavior can be expected for thermal promotion of an electron from Mu_T^- to the conduction band in an ionization-like process, that is,

$$Mu_T^- \rightarrow Mu_T^0 + e^- \qquad (21)$$

and presumably starts a T-site charge cycle between Mu^0 and Mu^-. The calculated Mu^- to Mu^+ crossover point is certainly above this diamagnetic intensity drop for the 10^{16} cm^{-3} sample, lending strong support to this assignment. Several additional checks were performed to verify this assignment, but the abrupt change in the onset temperature versus concentration, more specifically the constant nature of this feature at the two highest concentrations, provides the best experimental clue. The energy parameter obtained from fitting the decrease in diamagnetic intensity for the higher donor concentrations is 0.56(3) eV, which places the $T(-/0)$ muonium energy level much deeper in the gap than the $BC(0/+)$ level of about 0.2 eV. The difference in Mu^0 configuration energy between the T and BC sites (the parameter Δ in Fig. 13) is required along with these single-site electronic energy levels in order to confirm that Mu is indeed a negative-U impurity, as suggested by this result. The value of Δ has not been obtained experimentally because the reverse Mu^0 site change from BC to T is not observed. This implies that Δ is probably somewhat larger than 0.3 eV, which is the activation energy for e^- capture to form Mu_T^- in n-type silicon for the highest-temperature competing exit route from Mu_{BC}^0.

TABLE II
Transitions Identified for Mu in Silicon Along with the Dynamic Parameters from Best Fits to rf-μSR Data with the Complete Model

Mu transition	Rate parameters	
BC charge-state transitions		
$Mu_{BC}^0 \rightarrow Mu_{BC}^+ + e^-$	$v_{BC}^{0/+} = 3.1 \times 10^{13} \text{sec}^{-1}$	$E_{BC}^{0/+} = 0.21 \pm 0.01 \text{ eV}$
$Mu_{BC}^0 + h^+ \rightarrow Mu_{BC}^+$	$\sigma_{BC}^{0/+} \simeq 20 \text{ Å}^2$	
$Mu_{BC}^+ + e^- \rightarrow Mu_{BC}^0$	$\sigma_{BC}^{+/0} \simeq 3300 \text{ Å}^2$	
T charge-state transitions		
$Mu_T^0 + e^- \rightarrow Mu_T^-$	$\sigma_T^{0/-} \simeq 12 \text{ Å}^2$	
$Mu_T^- \rightarrow Mu_T^0 + e^-$	$v_T^{-/0} = 4.2 \times 10^{14} \text{ sec}^{-1}$	$E_T^{-/0} = 0.56 \pm 0.03 \text{ eV}$
Site transitions		
$Mu_T^0 \rightarrow Mu_{BC}^0$	$v_{T/BC}^0 = 9.5 \times 10^{12} \text{ sec}^{-1}$	$E_{T/BC}^0 = 0.39 \pm 0.04 \text{ eV}$
Combined charge/site transitions		
$Mu_{BC}^+ + e^- \rightarrow Mu_T^0$	$\sigma_o^{+/0} = 1.2 \times 10^{-10} \text{ cm}^2$	$E_{BC/T}^{+/0} = 0.40 \pm 0.02 \text{ eV}$
$Mu_{BC}^0 + e^- \rightarrow Mu_T^-$	$\sigma_o^{0/-} = 5.0 \times 10^{-10} \text{ cm}^2$	$E_{BC/T}^{0/-} = 0.32 \pm 0.01 \text{ eV}$

Table II provides a summary of the muonium transitions and related dynamic parameters that were obtained from fitting the rf-μSR data on at least 12 different Si samples. This is a single set of parameters that satisfactorily reproduces the data for the full set of samples ranging from middle 10^{15} cm^{-3} p-type to middle 10^{16} cm^{-3} n-type. Slightly better fits can be obtained for each individual sample; however, the range of the resulting differences in energy parameters was at most 5–10%. Instead, energies were kept fixed at values from fits to data where the relevant transition feature had the largest amplitude or the least overlap with effects from other transitions. Prefactors and cross sections were then adjusted for the best overall result on the full sample set using a single set of transition-rate parameters. Initial fractions for the various states also were adjusted sample by sample for the best result within this compromise. The main effect is a transfer from T-site species to BC initial states for the more heavily doped samples, somewhat stronger for n-type than for p-type. This is most likely a consequence of more frequent interactions between the mobile Mu_T^0 center and dopants, including a reduction of the average hop rate for this state due to the impurity strain fields. Since these state modifications occur within the first few nanoseconds, minimal details can be extracted from the data, and any assignment of specific transition sequences would have to be based on feasibility arguments and speculation.

One of the biggest open questions in the analysis of muonium dynamics in silicon is related to the high-temperature longitudinal depolarization measurements and our modeling of the cyclic charge-state transitions. We have confirmed that at least in the onset region the same three-state cycle is active in producing the polarization loss for conduction electron concentrations up to nearly 10^{15} cm^{-3}; thus all the p-type and the lower-concentration n-type samples appear to show identical muonium states and processes at high temperatures. For more heavily doped n-type silicon, the depolarization rates become too fast to obtain good data, and the only recourse is to rely on a comparison of competing transition rates and calculated effects; however, extrapolation and questions concerning the completeness of the model pose a serious problem.

The difficulties related to an incomplete calculational model and extrapolation of the transition rates from low to high temperatures are present even for the most intrinsic Si sample. When the rate parameters obtained from fits for individual transitions in the rf data below 400 K are used to calculate the depolarization at higher temperatures, the result yields much faster rates than are actually measured. The reason for this discrepancy is not immediately obvious. One effect is certainly present; the calculation only includes the transition cycle that dominates the depolarization. Other cycles should be present but do not significantly contribute to the measured depolarization. However, they will effectively reduce the net time spent in the relevant transition cycle. This is especially important because the electron capture and ionization cycle at the BC site, that is, $\mathrm{Mu}_{BC}^{0} \rightleftharpoons \mathrm{Mu}_{BC}^{+}$, occurs extremely rapidly compared with the three-state cycle through Mu_{T}^{0} that dominates the depolarization. The direct contribution of the $0/+$ charge cycle at the BC site to the depolarization is almost zero because the cycle rate is extremely rapid compared with the BC hyperfine frequency. We have yet to extract the measurable longitudinal depolarization rates from a full-model calculation that includes all the muonium transitions and all the depolarization mechanisms for each center. The temperature and field dependence of the measured longitudinal depolarization data barely support full fits to the most simple two-state charge cycle, and even for the active three-state cycle alone, several parameters must be held constant. Therefore, no independent confirmation of any complicated model can be expected from these data in terms of fitted parameter agreement.

A second likely possibility for differences between the high- and low-temperature regimes is that many of the dynamic parameters can be expected to have some temperature dependence, whereas the model has thus far assumed constant values. Although this assumption works well up to 400 K, at some point it shoud break down. In addition to likely quantitative changes in dynamic parameters, one also should expect to see qualitative changes in muonium behavior as the motion of the Mu centers becomes too rapid for the heavier lattice to respond. As a result of such a breakdown of adiabatic muonium motion, the nature of the high-temperature Mu states may become quite different from the states observed and studied at low temperatures. Again, the sensitivity of the main high-temperature data to additional parameters is marginal at best; thus an experimental answer to any of these questions will be extremely difficult to obtain.

2. GERMANIUM

The data on muonium transitions in germanium are much less extensive than for silicon. Our collaboration's efforts have been confined to high-purity undoped samples thus far, with work just commencing on doped material. Several other groups have concentrated more heavily on muonium behavior in Ge, and a few of these results will be discussed briefly where they impact conclusions drawn from the collaboration's experiments. Some differences are expected theoretically in Mu (or H) behavior for Si and Ge (Estreicher, 1995). Most important, the configurational energy for the two Mu^0 centers is predicted to be nearly identical in Ge, while the BC site is clearly more stable in models for silicon. The older transverse and longitudinal μSR relaxation data (Patterson, 1988) on germanium produced a number of questions that have not been resolved. The TF-μSR signals from both Mu^0 centers broaden and disappear between 80 and 120 K, yielding very small energy parameters for transitions out of these states. These were assumed to be Mu^0 ionization; however, the increase in diamagnetic TF-μSR amplitude seen above 160 K yields much larger energies (Patterson, 1988). The difference in temperature for features associated with a simple ionization is typical of TF data, but the energy discrepancy raises questions of exactly what the transition sequence may be.

The initial rf-μSR examination of an undoped Ge sample showed the loss of both the Mu_T^0 and Mu_{BC}^0 amplitudes just above 100 K, quite consistent with the TF relaxation data (Lichti et al., 1992). However, only a small fraction of the initial Mu^0 states contributes to a rise in diamagnetic rf signal strength in the same temperature region. Furthermore, the rf-μSR intensities indicate a time-integrated occupation of diamagnetic states $>50\%$ below 100 K compared with $<20\%$ in the TF data. Since the latter measures only the promptly formed states, while the rf data are sensitive to both prompt and slowly formed states, these results imply a slow reaction leading to a diamagnetic Mu state even at the lowest temperatures (5–10 K). The rf-μSR diamagnetic fractions in an ultrapure Ge sample are shown in Fig. 19. Two additional higher-temperature steps of decreasing intensity are present in these data, one near 200 K back to what appears to be a smooth extrapolation of the intensity below 100 K and a second decrease above 320 K. The rf data on doped Ge from other groups (Iwanowski et al., 1994) show very

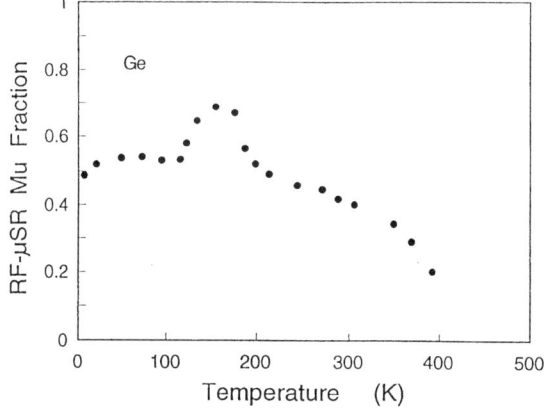

FIG. 19. The diamagnetic fraction in rf-μSR measurements on an ultrapure Ge sample. Comparison with TF-μSR suggests the broad underlying feature represents a slowly formed state, stable to 350 K.

large variations depending on doping type. More samples will need to be studied to arrive at any comprehensive model.

Longitudinal depolarization rates were examined up to 900 K in ultrapure germanium to obtain additional information on dynamic features. There is very slow longitudinal relaxation at low temperatures associated with up to 40% of the muon polarization, and significant depolarization involving a much larger fraction is seen above 200 K, as shown in Fig. 20. A combination of the temperature dependence at low fields (Fig. 20) and field dependences at several temperatures between 250 and 600 K implies a cyclic charge-exchange depolarization process roughly similar to that observed in silicon. Attempts to fit these data with a single transition cycle were not successful, and we conclude that two separate cycles occur for Ge (Lichti et al., 1994b, 1997b). The field-dependent rates show a slightly different behavior at 250 and 290 K compared with the curves above 350 K, consistent with a change in the transition cycle, as suggested in Fig. 20. The higher-temperature broad feature initially was fit to a two-state model. The parameters from this fit indicate an effective cycle between Mu_T^0 and Mu^+, with all other cycle options yielding unrealistic parameters. The high-temperature fit was then subtracted from the depolarization data, and the remaining smaller peak near 280 K was fit to a separate two-state cycle. A much smaller hyperfine constant was required for this peak, and a $Mu_{BC}^0 \rightleftharpoons Mu^+$ cycle is implied by the results. The charge cycles in germanium thus occur separately through the two neutral centers, but both share

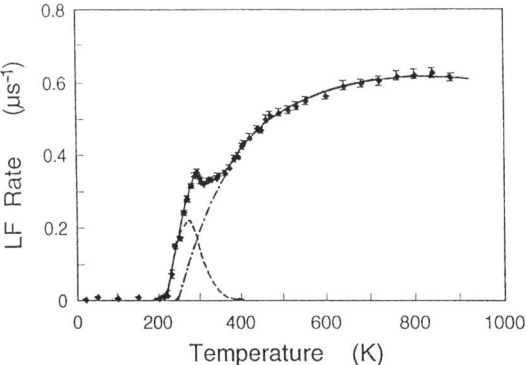

FIG. 20. The low-field longitudinal depolarization rates in Ge from muonium charge cycles. Two separate 0/+ cycles are present; the peak near 280 K involves Mu_{BC}^0, while the broad high-temperature peak is associated with Mu_T^0. (From Lichti et al., 1997.)

a common Mu^+ state that is expected to be at the BC location. Both cycles are best fit assuming e^- ionization and recapture, with a site change needed for capture into Mu_T^0. The Mu_{BC}^+ to Mu_T^0 transition should require activated capture if the process is analogous to that observed for silicon. However, using an activated cross section in the fits resulted in a negative activation energy. Separate capture cross sections were obtained for each of the constant-temperature field-dependence curves and we concluded that $\sigma(T) = \sigma_o T^{-2}$ works very well in describing the cross section for e^- capture to form Mu_T^0. The origin of this decrease in the effective capture rate is not completely understood but may reflect a decrease in the effective time spent in this cycle due to the competing cycle where the muon remains at the BC site. The 0/+ charge cycle at the BC location contributes very little to the depolarization above 350 K, but as discussed for silicon, it will continue to be present at higher temperatures.

The ionization energies for both neutrals, 0.17 eV for Mu_T^0 and 0.215 eV for Mu_{BC}^0, obtained from the charge-cycle data are much larger than the values from the low-temperature disappearance of the TF signals from the Mu^0 states. The energy from the rise in the TF-μSR diamagnetic amplitude in n-type Ge samples assuming a Mu_T^0 precursor was 0.17 eV (Andrianov et al., 1978), consistent with our charge-cycle results. The TF-μSR diamagnetic signal in the present sample shows a clear signature of this transition in the amplitude, relaxation rate, and the phase shift. A second step in the amplitude starts at 230 K, near the charge-cycle onset, and grows to 100% by 380 K. These TF amplitude changes are consistent with the larger ionization energies. The transverse relaxation rate joins the longitudinal depolarization rate near 360 K. Both rates involve all the muons and are due to the same dynamic process above that temperature. The TF relaxation rates show the effects of the charge cycles somewhat differently than the longitudinal data, in particular confirming that near 300 K more than one relaxation process is active.

The assignment of the larger energies to ionization leaves a major question as to the low-temperature transitions out of the neutral states. Reanalysis of composites of all the early published TF-μSR relaxation data on Mu_{BC}^0, and for Mu_T^0 in undoped samples, was undertaken in an attempt to understand the low reported energies. (These data are reviewed and a full set of references provided in Patterson, 1988.) We obtain energy parameters of 0.019(2) eV for Mu_T^0 and 0.030(5) eV for Mu_{BC}^0 with very low prefactors of 81 and 48 MHz, respectively. If the 100-K transitions are not due to ionization, then bidirectional Mu^0 site changes are good candidates. Since the minima for the two Mu^0 sites are nearly equal, any site change should be accompanied by a return transition with a similar barrier. Taking these low-energy parameters as site-change barriers implies that the BC site is slightly more stable than the T site, by 11 ± 6 meV, with a nearly flat potential surface between the two sites. Such a small barrier also would explain why the electron capture from Mu_{BC}^+ into the T-site neutral state does not appear to be activated. The low prefactors for these site changes imply that the phonon modes involved are different from the optical modes active in the Mu^0 site changes in Si and diamond, probably an acoustic mode. The low-temperature longitudinal measurements provide some support for the site-change assignments. The data fit well to a model of slow bidirectional transitions into and out of the Mu_T^0 state. The simple model used in this fit does not take into account the nature of the second state, assuming no contribution to the relaxation. The resulting energies are a bit higher than from the TF data on Mu^0 states, about 25 and 55 meV for transitions out of and into the T site, respectively, obtained from the relaxing amplitude between 16 and 150 K. A decoupling curve at 75 K identifies the relaxation as coming from Mu_T^0. The full set of transition assignments appears to be consistent with existing data on Mu in Ge but will require some additional checks for verification. Table III shows the parameters associated with muonium transitions in Ge as obtained from the charge-cycle depolarization and reanalysis of older data.

As a final point in discussing muonium in germanium, the longitudinal data on ultrapure Ge also show a 15% nonrelaxing fraction present up to about 350 K. The amplitude of this signal is consistent with the low-temperature diamagnetic amplitude in TF-μSR. Furthermore, the temperature at which this state joins the relaxing fraction matches the final drop in rf-μSR diamagnetic amplitude. A significant amount of early work on both

TABLE III

TRANSITIONS IDENTIFIED FOR MUONIUM IN GERMANIUM ALONG WITH THE DYNAMIC
PARAMETERS FROM LONGITUDINAL DEPOLARIZATION DATA

Mu transition	Rate parameters		
BC charge-state transitions			
$Mu_{BC}^0 \rightarrow Mu_{BC}^+ + e^-$	$v_{BC}^{0/+} = 2.1 \times 10^{13} \, sec^{-1}$		$E_{BC}^{0/+} = 0.215 \pm 0.010 \, eV$
$Mu_{BC}^+ + e^- \rightarrow Mu_{BC}^0$	$\sigma_{BC}^{+/0} = 2500 \pm 200 \, Å^2$		
Site transitions			
$Mu_T^0 \rightarrow Mu_{BC}^0$	$v_{T/BC}^0 = 8.1 \times 10^7 \, sec^{-1}$		$E_{T/BC}^0 = 0.019 \pm 0.002 \, eV$
$Mu_{BC}^0 \rightarrow Mu_T^0$	$v_{BC/T}^0 = 4.8 \times 10^7 \, sec^{-1}$		$E_{BC/T}^0 = 0.030 \pm 0.004 \, eV$
Combined charge/site transitions			
$Mu_{BC}^+ + e^- \rightarrow Mu_T^0$	$\sigma_o^{+/0} = 1.4 \times 10^5 \, T^{-2} \, Å^2$		$E_{BC/T}^{+/0} = 0.0$
$Mu_T^0 \rightarrow Mu_{BC}^+ + e^-$	$v_{T/BC}^{0/+} = 1.2 \times 10^{13} \, sec^{-1}$		$E_{BC/T}^{0/+} = 0.174 \pm 0.002 \, eV$

the longitudinal and transverse relaxation rates in doped Ge (see references in Patterson, 1988) strongly suggests interaction of the mobile Mu_T^0 center with dopants and other impurities, including isoelectronic Si. These interactions are strongest at the lowest temperatures when the dopants are neutral. The present data suggest that such interactions, similar to those of Eqs. (13) and (14), may be the origin of the slowly formed diamagnetic Mu state seen in rf resonance measurements. Figure 21 shows the nonrelaxing amplitude from longitudinal relaxation measurements through the temperature region where that state joins the relaxing fraction. A fit to this drop yields an energy

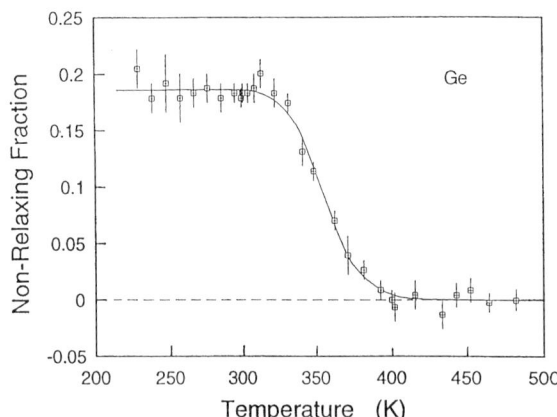

FIG. 21. Temperature dependence of the nonrelaxing signal in the longitudinal data on Ge in the region where this state joins the charge cycles. The resulting transition energy is 0.75 eV.

of 0.748(2) eV, which is significantly larger than the Ge bandgap of 0.66 eV near 350 K. This large energy strongly argues that this slowly formed diamagnetic Mu state is from a Mu-impurity complex, where the impurity is electrically inactive. Impurities such as Si, C, and O are possibilities and can be present at significant levels in electronically ultrapure Ge.

3. GALLIUM ARSENIDE

The experimental determination of muonium transition assignments and measurement of the transition-rate dynamics have not progressed as far in III-V compounds as it has in Si and Ge. Since the understanding of the static muonium states is more advanced for GaAs, our study of Mu transitions in III-V materials started with GaAs. The observed transitions vary significantly with doping type and concentration, more so than for the elemental semiconductors. For example, low-temperature transitions from BC to the T site are inferred from longitudinal data on n-type GaAs; however, the temperature at which this transition is seen varies over a range of nearly 150 K (Kadono et al., 1994; Lichti, 1995). The identity of the diamagnetic state in GaAs appears to be Mu^+ up to the 10^{16} cm^{-3} region for n_e (Estle et al., 1997) but is definitely Mu^- at metallic n-type concentrations (Chow et al., 1995). Longitudinal muon spin depolarization measurements in a semi-insulating sample and in weakly and heavily doped n-type GaAs show that there is a significant change in the high-temperature muonium charge cycle near a donor concentration of 10^{17} cm^{-3} (Estle et al., 1997). The charge cycle(s) for lower electron densities involve $0/+$ muonium transitions, with the primary cycle beginning near 550 K (Lichti et al., 1994b), while for higher donor concentrations the cycle is $-/0$ and starts above 700 K (Chow et al., 1996). I will discuss the details of the few transitions in GaAs that have been identified and dynamic parameters determined, beginning with the charge cycle in metallic n-type material.

For the n-type samples used to characterize the Mu^- structure ($N_D > 10^{18}$ cm^{-3}), the Mu_T^- center starts to move out of the T_{Ga} region above 500 K, as discussed earlier. This state undergoes a transition to Mu_T^0 at about 700 K, and a $-/0$ charge cycle begins at that temperature (Chow et al., 1996). The field dependence for longitudinal relaxation rates, measured at two temperatures for one of these samples, identified Mu_T^0 as the active relaxing center via the fitted hyperfine frequency, while Mu_T^- is the initial state at the cycle onset, thereby identifying the charged state. Figure 22 shows the transition rates extracted from fits of Eq. (8) to depolarization rates measured at two widely separated applied fields per temperature point (Chow et al., 1996). The nearly constant rate for the transition from the

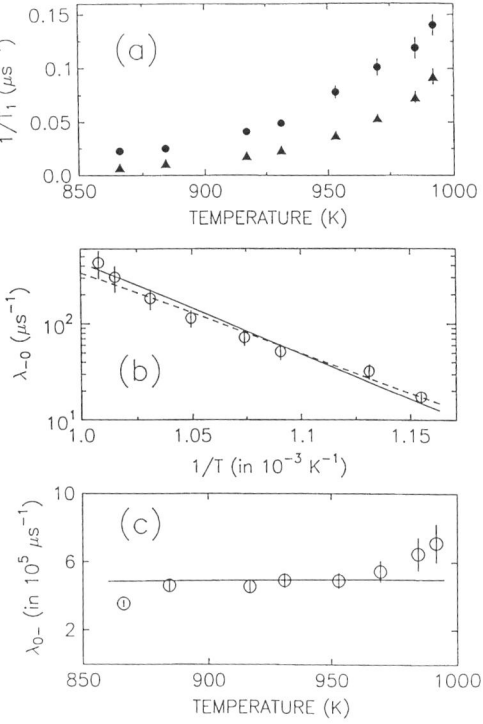

FIG. 22. Longitudinal depolarization rates in heavily doped GaAs(Si) shown in (a) for fields of 2.9 T (triangles) and 48 mT (circles) yield transition rates (b) and (c) for the $-/0$ charge cycle at the T site. The temperature dependences for these rates imply alternating carrier capture, i.e., electron-hole recombination at Mu_T. (From Chow et al., 1996, © 1996 American Physical Society.)

neutral to negative charge state implies an electron-capture process because n_e remains nearly constant over this temperature range at such high doping levels. The negative-to-neutral transition displays an activation energy of 1.66 eV, somewhat above the GaAs bandgap and consistent with thermal promotion of an electron from the valence band to the Fermi level within the conduction band, and therefore implies that hole capture is the relevant process. Fits of these temperature-dependent transition rates to standard carrier capture rate expressions yielded cross sections of 16(1) and 26(2) $\times 10^2$ Å^2, respectively, for e^- capture by Mu_T^0 and h^+ capture by Mu^- (Chow et al., 1996). These values are fully consistent with the expected geometric and Coulomb capture processes. The charge-cycle data for metallic n-type samples thus establishes Mu_T as a recombination center in

GaAs and strongly suggest that the muonium and hydrogen T($-/0$) energy level does not lie close to either band edge.

The charge-cycle data for less heavily doped n-type GaAs and semi-insulating samples show a dominant $0/+$ cycle above roughly 550 K (Lichti et al., 1994b; Estle et al., 1997). The field dependence of the depolarization rates are shown for several constant temperatures in Fig. 23. These data are typical of charge cycles involving Mu_T^0 neutral states. The most consistent parameters are obtained for transition-rate expressions representing ionization of Mu_T^0 and e^- capture by the positively charged center. Fits of the temperature and field dependences for the longitudinal depolarization rates yield an average ionization energy of 0.451(8) eV independent of what process is assumed for the return transition. However, the capture part of this cycle yields a small cross section, closer to that expected for carrier capture by a neutral center. However, attempts to fit the data to a $-/0$ charge cycle required hyperfine constants that were much too large and gave unreasonable capture rates as well. Recalling that a Mu_T^+ state located in the T_V cage region was postulated to explain some of the results in p-type III-V materials (Lichti et al., 1997), I suspect that the Mu^+ state participating in this high-temperature charge cycle may be this T-site Mu^+ species rather than the BC ground state. The neutral center for this $0/+$ cycle remains Mu_T^0, again identified by fits to the field-dependent depolarization rates at constant temperature (Estle et al., 1997).

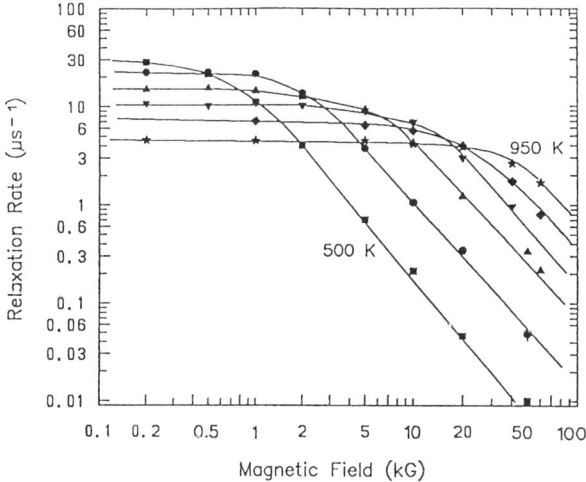

FIG. 23. Longitudinal depolarization rates as a function of field for constant temperatures of 500, 600, 700, 775, 850, and 950 K in weakly doped GaAs(Si) in the low 10^{15} cm^{-3} range. These data are typical of charge cycles involving Mu_T^0. (From Estle et al., 1997.)

The transition rates within a two-state $0/+$ cycle as extracted from field dependences at constant temperatures are displayed in Fig. 24 as an Arrhenius plot. Above 600 K, the slope for the ionization step shows the 0.45-eV ionization energy, while the return transition has a slope of 0.68 eV, half the bandgap roughly in the middle of the relevant temperature range (Estle *et al.*, 1997). This latter result appears consistent with carrier capture in the intrinsic limit. A check of the extrinsic-to-intrinsic crossover temperature for the listed donor concentration of this sample gives about 670–700 K, significantly higher than suggested by the data; however, GaAs is notorious for sample history dependences in electrical properties, especially at low dopant levels.

A different process seems to be present in the weakly doped n-type sample below ~ 600 K and involves up to about 60% of the muons (Estle *et al.*, 1997). Assuming that the corresponding transition rates in Fig. 24 represent a $0/+$ cycle in the extrinsic limit, the data show that the transition out of Mu^0 occurs with an energy parameter roughly equal to the bandgap, suggesting an h^+ capture process, while the return to Mu^0 occurs at a constant rate, implying e^- capture. However, other interpretations for the origin of the fast relaxing fraction below 600 K are possible, including spin-exchange scattering between conduction electrons and the initial Mu_T^0 fraction rather than a transition cycle.

The temperature dependence of low-field longitudinal data on the same sample (Lichti, 1995) are displayed in Fig. 25. This figure shows growth of

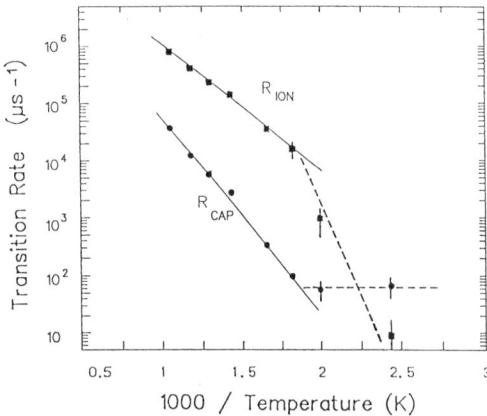

FIG. 24. Transition rates extracted from the data of Fig. 23. The temperature dependence above 600 K implies ionization and e^- capture in a $0/+$ cycle operating in the intrinsic limit, and below 600 K they suggest alternating h^+ and e^- capture for a $0/+$ cycle in the extrinsic limit. Both cycles involve Mu_T^0 and suggest a Mu_T^+ ionized state. (From Estle *et al.*, 1997.)

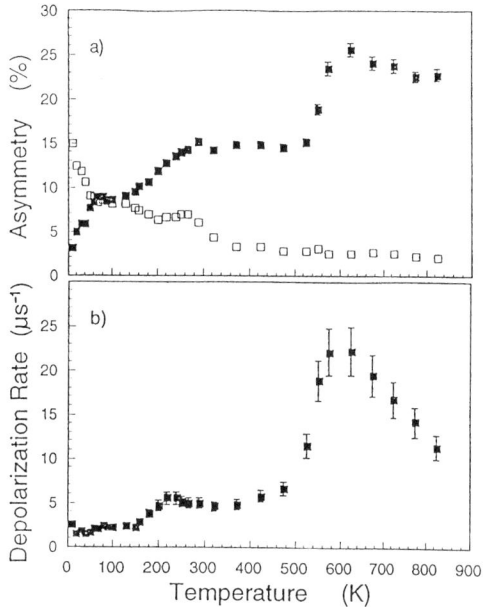

FIG. 25. The full temperature dependence for low-field longitudinal depolarization data on 10^{15} cm^{-3} n-type GaAs(Si) showing the onset of charge cycles at 500 K. The asymmetries in (a) are the relaxing (filled) and nonrelaxing (open) fractions. The feature at 200 K is assigned to a BC to T site change for Mu^0. (From Estle et al., 1997.)

a relaxing signal amplitude between 550 and 600 K that corresponds to the crossover between the two regimes observed in the field dependences discussed earlier. The low field data were taken at the ISIS facility, which has a pulsed beam structure, and there was a missing fraction below this crossover region for which the depolarization rates are too rapid to observe that part of the signal. The low-field limits for the field-dependent data from TRIUMF discussed earlier fall slightly above the ISIS values for $T > 600$ K and extend to higher rates at lower temperatures, where less than the full fraction participates in the very fast depolarization. The data for semi-insulating GaAs have very similar behavior above 550 K, but even more complicated dynamics are observed at lower temperatures. No diamagnetic TF-μSR precession signal is seen in the weakly n-type sample up to at least 350 K, while the diamagnetic signal in semi-insulating GaAs suggests competing ionization and site-change transitions for the initial Mu$^0_{BC}$ fraction, although details of the dynamics have not yet been obtained.

Overall, the muonium transition dynamics in GaAs are not well understood as in silicon; however, the fairly simple behavior in heavily doped

TABLE IV

TRANSITIONS TENTATIVELY ASSIGNED FOR MUONIUM IN GaAs ALONG WITH DYNAMIC PARAMETERS OBTAINED THUS FAR FROM LONGITUDINAL DEPOLARIZATION DATA

Mu transition	Rate parameters	
T charge-state transitions		
$Mu_T^0 + e^- \to Mu_T^-$	$\sigma_T^{0/-} = 16.0 \pm 0.6 \text{ Å}^2$	
$Mu_T^- + h^+ \to Mu_T^0$	$\sigma_T^{-/0} = 2600 \pm 200 \text{ Å}^2$	
$Mu_T^0 \to Mu_T^+ + e^-$	$v_T^{0/+} = 1.9 \times 10^{14} \text{ sec}^{-1}$	$E_T^{0/+} = 0.451 \pm 0.008 \text{ eV}$
$Mu_T^+ + e^- \to Mu_T^0$	$\sigma_T^{+/0} = 1.2 \times 10^{10} \text{ T}^{-3} \text{Å}^2$	
Site transitions		
$Mu_{BC}^0 \to Mu_T^0$	Sample dependent	

n-type samples has been fully characterized above about 200 K. Table IV shows the muonium transitions in GaAs that have been at least tentatively assigned and their dynamics characterized. A number of questions remain concerning the low-temperature states in GaAs in general, but more particularly, the assignments for the low-temperature processes and the wide range of dynamics apparently associated with specific transitions. These questions are being actively investigated at present, but sample dependence appears to be the rule rather than the exception for these low-temperature features, making it very difficult to draw any conclusions. Because sample history is apparently important, interactions with intrinsic defects or unknown impurities may very well dominate the dynamics of the relevant muonium transitions.

4. OTHER III-V MATERIALS

As indicated in Section III, our collaboration has begun to examine the muonium states and transitions in several additional III-V materials with the aim of finding systematic variations in Mu behavior with materials parameters such as bond ionicity, lattice spacing, bandgap, etc. While a few systematic changes do occur, such as a variation in charge-cycle onset temperatures with bandgap and the variations in Mu_T^0 hyperfine constant and the relative stability of Mu_T^0 and Mu_{BC}^0 with lattice constant, these are much more generally associated with muonium in crystalline solids. Although the static Mu states are very similar and some characteristics are generally predictable, for the most part it appears that each semiconducting material produces its own special features related to transition dynamics. Thus far we have investigated both n-type and p-type samples of InP and

GaSb in addition to GaAs and either semi-insulating or undoped samples of InP, InAs, and GaP. Each of these materials with the exception of GaP shows features in transverse and zero-field μSR data below room temperature that appear to be related to either motion or a site change for the Mu^+ charged state. Most of these samples show higher-temperature trapping peaks, indicating a more or less stationary diamagnetic Mu state that is different from the low-temperature center. In many of these cases these investigations are at an early stage, and precise transitions have not yet been determined, nor have the transition dynamics been extracted.

The data displayed in Section III for the diamagnetic amplitude versus temperature for TF-μSR measurements on GaSb (Fig. 5a) and in zero field for InP (Fig. 12a) show missing fractions at low temperatures. These nondiamagnetic states are most likely Mu^0 centers, although in both these cases no paramagnetic precession signals have been seen. For GaSb, decoupling curves did identify the Mu_T^0 center, while for InP, no recovery of the missing polarization is seen up to 0.5 T in longitudinal fields. The reasons for the lack of direct observation of the Mu^0 states are probably rapid depolarization of a mobile Mu_T^0 center interacting with the large nuclear moments of host nuclei. However, even if the Mu^0 center has not been identified, the growth of the diamagnetic signal with increasing temperature provides an estimate of the ionization temperature for whatever Mu^0 center is present. Some caution regarding assignment of the extracted energy needs to be exercised, since only the final state is monitored directly. The energy found from the growth of the TF-μSR precession amplitude is about 0.25 eV for GaSb, which tentatively can be assigned to Mu_T^0 ionization. Judging from the zero-field data on these samples, the resulting Mu^+ state is likely to be in the T_V interstitial region, since the signal-recovery temperatures are above the onset of Mu^+ motion and possible migration out of the BC site. Again, although the energy from growth of the diamagnetic signals should be valid, the transition assignment must be considered as somewhat speculative and based primarily on feasibility arguments rather than hard evidence.

At temperatures above 300 K, GaSb shows longitudinal depolarization characteristic of spin- or charge-exchange interactions. Because of the temperatures involved, the increased depolarization is assumed to be related to cyclic muonium charge-state transitions; however, the temperature and field dependences in these data are much more complicated than seen in GaAs or the elemental materials. Thus far we have not attempted to extract transition rates because it is clear that a simple model will be insufficient to deal with the complexity. No hint of any significant longitudinal depolarization is observed for InP samples up to fairly high temperatures. Thus, although Mu charge cycles are expected in InP based on observations in other semiconductors, there are no data to indicate their presence in either n- or p-type samples.

An important focus of the collaboration's investigations of III-V materials in the near term will be to identify the trapped Mu states seen in the transverse relaxation data. These states are the best candidates for the muonium analogue of hydrogen passivation complexes found to date. The second major thrust that has just begun is to investigate muonium in the group III nitrides, particularly GaN, as wide-gap semiconductors with developing technological importance. Very preliminary data on a typical n-type GaN thick film show a 100% diamagnetic signal that is likely to be Mu^-. We have observed a strong QLCR spectrum demonstrating that this state resides close to a Ga atom, but initial analysis suggests a site that does not appear to be energetically favorable from a theoretical viewpoint. Zero-field data show this state to be stationary up to above 500 K, where it begins slow hopping motion with an activation energy quite close to 1.0 eV. Considerable work will be required before details of Mu states and dynamics in the nitrides can be discussed with any confidence, but these initial data appear to be extremely promising in both respects.

V. Relevance to Hydrogen Impurities

The muonium results give an excellent qualitative picture of the behavior of isolated hydrogen defect centers in semiconductors, especially given the fact that only a few experimental measurements exist that directly examine the analogous hydrogen states. Comparison of the structure and various dynamic parameters for H^0_{BC} and Mu^0_{BC} in silicon currently provide the only overlap in measured energy parameters. These values are in reasonable agreement, with the differences between the various measurements on Mu or H of about the same magnitude as the disagreement between average values for the two. My colleagues and I have made detailed comparisons of the available H and Mu experimental results (Kreitzman et al., 1995; Lichti et al., 1995) and conclude that the muonium results should be considered as a good quantitative estimate for hydrogen, at least with respect to transitions that involve H^0_{BC}.

Barriers for motions between sites should be somewhat higher for H than for Mu based on the larger zero-point energy for muonium, while strictly electronic energies are expected to be very similar. As an example, we obtain 0.40 eV for the e^- capture transition from Mu^+_{BC} to Mu^0_T at the onset of the muonium charge cycles above room temperature, whereas a similar transition can be inferred from the annealing behavior of the DLTS (Bech Neilsen

8 DYNAMICS OF MUONIUM DIFFUSION 367

et al., 1994) and EPR (Gorelkinski and Nevinnyi, 1992) signals from H_{BC}^0 for which barriers of 0.44 and 0.48 eV were obtained. These values yield decent agreement when expected differences are taken into account, especially given the nearly 8 orders of magnitude difference in effective time scales between measurement methods. The DLTS measurements give an energy of 0.16 eV associated with the H_{BC}^0 to H_{BC}^+ transition (Holm *et al.*, 1991). When corrected for electrical field effects, this yields an ionization energy a bit above 0.18 eV, somewhat below the present best value of about 0.20 eV for ionization of Mu_{BC}^0 but within the range of results from the various μSR techniques that have been applied to this problem.

A third energy also may be compared. The muonium data in *n*-type samples identify the responsible transition as e^- capture by Mu_{BC}^0 resulting in a Mu_T^- final state. A barrier of 0.32 eV is obtained in fits using the full model. The corresponding energy related to capture of a second electron in the hydrogen counterpart gives 0.29 and 0.30 eV for ^1H and ^2H, respectively, from the DLTS annealing curves (Holm *et al.*, 1991). There remains some uncertainty whether the hydrogen result should be associated directly with the e^- capture process or a second step related to migration of the resulting center. The muonium data related to this transition show effects of Mu_{BC}^0 depolarization that complicate the fits, making the energy value somewhat less certain. Thus, even though these energies are oppositely ordered from what is expected based on simple arguments, the agreement is certainly satisfactory given the complications in analysis and interpretation of both the H and Mu data. Much more important than such detailed comparison of dynamic parameters, the basic behavior for the two BC impurities appears to be fully consistent with each other where the experimental information overlaps. This fundamental agreement provides a strong argument that the large volume of additional results from the muonium investigations should translate over to isolated hydrogen behavior in a similar semiquantitative manner.

The implications for hydrogen diffusive motion that arise out of the high-temperature muonium data are considerable and were not fully appreciated previously. The cyclic charge-state transitions that dominate the Mu dynamics at elevated temperatures in almost every semiconductor with a small or intermediate bandgap should occur for hydrogen as well. These transitions are rapid on a microsecond time scale and need to be included in any model of dynamic features related to isolated hydrogen impurities. The muonium results imply cyclic transitions in and out of the highly mobile Mu_T^0 state, which is metastable in silicon. For the Mu case, this center is many orders of magnitude more mobile than any of the other states and consequently can dominate diffusive motion even when it is not the expected state in thermal equilibrium. The muonium results emphasize in a very dramatic fashion that thermal equilibrium is not a static situation, especially at high temperatures.

There are at least two important results that are directly relevant to models of hydrogen diffusion. The first is the likely identification of the fast-diffusing state in silicon, namely, a metastable H^0 species that can easily hop among the tetrahedral interstices directly analogous to Mu_T^0 motion. The low-temperature quantum tunneling behavior that is important for Mu_T^0 should be less relevant for the hydrogen case; however, above 100 K, the motions should be very similar with some adjustment of the barrier height. Much of the hydrogen diffusion data obtained over the years require interrupted motion of a rapidly diffusing center, commonly interpreted as trapping and detrapping at another impurity. The muonium experiments certainly identify a good candidate for the fast-diffusing species in the T-site neutral center.

The muonium charge-cycle results are clearly relevant to hydrogen motion and suggest that diffusion could occur as repeatedly interrupted motion even without formation and dissociation of an H-impurity complex, even though there is no doubt that such complexes play a dominant role. In particular, measurements in Si, Ge, and GaAs where the details of the transitions in these charge cycles have been obtained suggest that diffusion of charged species may very well be dominated by charge-state transitions into the mobile H^0 center rather than motion of the H^+ or H^- centers themselves. This result should apply to hydrogen motion in bulk material but may be less directly relevant to measurements in depletion regions under the influence of electric fields, since the presence of charge carriers is crucial to cyclic charge-state transitions. The muonium data and the rapid transitions which they represent must be kept in mind when interpreting measurements on diffusion of hydrogen. Most important, any model that relies on only a single hydrogen species at elevated temperatures should be treated with some skepticism. Transfer of the muonium results to hydrogen, even in a very rough qualitative way, implies that transitions among the various isolated hydrogen states should occur extremely rapidly on the time scale of most of the measurement techniques applied to the study of hydrogen impurities.

In conclusion, the study of muonium has in the past provided a great deal of information regarding the nature of the isolated hydrogen defect centers in semiconductors. Our recent investigations of the site stability and dynamics of the motion of various muonium centers, and of transitions among the muonium states, has produced a detailed operational model of Mu dynamics in silicon. The configuration coordinate diagram associated with this model generally should be applicable to other tetrahedrally coordinated semiconductors. Ongoing investigations are aimed at determining specific modifications to this diagram and the dynamic parameters for muonium transitions in other materials. Significant progress has been made for Ge and GaAs, although many of the preliminary transition process assignments need to be verified by examining a variety of doped samples. The results for muonium transition dynamics apply qualitatively to the analogous isolated hydrogen impurities; thus these experiments yield information on hydrogen states that are extremely difficult to directly investigate. Muonium displays a very rich variety of dynamic features. In addition to its obvious application as an analogue to isolated hydrogen impurities, this system will continue to yield experimental data for comparison with the results of developing theoretical techniques for modeling the dynamic behavior of impurities in semiconductors.

ACKNOWLEDGMENTS

The contributions of the other members of the muonium in semiconductors collaboration to the work presented in this chapter is gratefully acknowledged. The participation of RLL in this effort is supported by the U.S. National Science Foundation (DMR-9623823) and the Robert A. Welch Foundation (D-1321).

REFERENCES

Adams, T. R., Robertson, M. A., and Lichti, R. L. (1995). *Philos. Mag. B*, **72**, 183.
Andrianov, D. G., Myasishcheva, G. G., Obukhov, Yu. V., Roganov, V. S., Savel'ev, G. I., Firsov, V. G., and Fistul', V. I. (1978). *Fiz. Tekh. Poluprovodn.*, **12**, 161.
Bech Neilsen, B., Bonde Neilsen, K., and Byberg, J. R. (1994). *Mater. Sci. Forum*, **143–147**, 909.
Celio, M. (1987). *Helv. Phys. Acta*, **60**, 600.
Chow, K. H. (1995). Ph.D. dissertation, University of British Columbia.
Chow, K. H., Kiefl, R. F., Schneider, J. W., Hitti, B., Estle, T. L., Lichti, R. L., and Schwab, C. (1993). *Phys. Rev. B*, **47**, 16004.
Chow, K. H., Lichti, R. L., Kiefl, R. F., Dunsiger, S., Estle, T. L., Hitti, B., Kadono, R., MacFarlane, W. A., Schneider, J. W., Schumann, D., and Shelley, M. (1994a). *Phys. Rev. B*, **50**, 8918.
Chow, K. H., Kiefl, R. F., Schneider, J. W., Estle, T. L., Hitti, B., Johnson, T. M. S., Lichti, R. L., and MacFarlane, W. A. (1994b). *Hyperfine Interact.*, **86**, 645.
Chow, K. H., Kiefl, R. F., MacFarlane, W. A., Schneider, J. W., Cooke, D. W., Leon, M., Paciotti, M. A., Estle, T. L., Hitti, B., Lichti, R. L., Cox, S. F. J., Schwab, C., Davis, E. A., Morrobel-Sosa, A., and Zavich, L. (1995). *Phys. Rev. B*, **51**, 14762.
Chow, K. H., Hitti, B., Kiefl, R. K., Dunsiger, S. R., Lichti, R. L., and Estle, T. L. (1996). *Phys. Rev. Lett.*, **76**, 3790.

Chow, H. K., Cox, S. F. J., Davis, E. A., Dunsiger, S. R., Estle, T. L., Hitti, B., Kiefl, R. F., and Lichti, R. L. (1997a). *Hyperfine Interact.*, **105**, 309.
Chow, K. H., Hitti, B., and Kiefl, R. F. (1997b). In *Identification of Defects in Semiconductors*, M. Stavola, ed., Semiconductors and Semimetals series. New York: Academic Press, p. 137.
Cox, S. F. J., and Symons, M. C. R. (1986). *Chem. Phys. Lett.*, **126**, 516.
Cox, S. F. J., Fuchslin, R., Meier, P. F., Estle, T. L., Cooke, D. W., Morrobel-Sosa, A., Lichti, R. L., Hitti, B., Cottrell, S. P., Chow, K. H., and Schwab, C. (1997a). *Hyperfine Interact.*, **106**, 57.
Cox, S. F. J., Cottrell, S. P., Hopkins, G. A., Kay, M., and Pratt, F. L. (1997b). *Hyperfine Interact.*, **106**, 85.
Estle, T. L., Estreicher, S. K., and Marynick, D. S. (1986). *Hyperfine Interact.*, **32**, 637.
Estle, T. L., Chow, K. H., Cox, S. F. J., Davis, E. A., Hitti, B., Kiefl, R. F., Lichti, R. L., and Schwab, C. (1997). *Mater. Sci. Forum*, **258–263**, 849.
Estreicher, S. K. (1987). *Phys. Rev. B*, **36**, 9122.
Estreicher, S. K. (1995). *Mater. Sci. Engr.*, **R14**, 319.
Gamani, M., Gygax, F. N., Rüegg, W., Schenck, A., and Schilling, H. (1977). *Phys. Rev. Lett.*, **39**, 836.
Gorelkinski, Yu. V., and Nevinnyi, N. N. (1987). *Pis'ma Zh. Tekh. Fiz.*, **13**, 105.
Gorelkinski, Yu. V., and Nevinnyi, N. N. (1992). *Physica B*, **170**, 155.
Hartmann, O. (1977). *Phys. Rev. Lett.*, **39**, 832.
Hayano, R. S., Uremura, Y. J., Imazato, J., Nishida, N., Yamazaki, T., and Kubo, R. (1979). *Phys. Rev. B*, **20**, 850.
Hitti, B., Kreitzman, S. R., Estle, T. L., Lichti, R. L., and Lightowlers, E. C. (1997). *Hyperfine Interact.*, **105**, 321.
Holm, B., Bonde Neilsen, K., and Bech Neilsen, B. (1991). *Phys. Rev. Lett.*, **66**, 2360.
Iwanowski, M., Maier, K., Major, J., Pfiz, T., Scheuermann, R., Schimmele, L., Seeger, A., and Hempele, M. (1994). *Hyperfine Interact.*, **86**, 681.
Kadono, R. (1990). *Hyperfine Interact.*, **64**, 615.
Kadono, R., Kiefl, R. F., Brewer, J. H., Luke, G. M., Pfiz, T., Riseman, T. M., and Sternlieb, B. J. (1990). *Hyperfine Interact.*, **64**, 635.
Kadono, R., Matsushita, A., Nagamine, K., Nishiyama, K., Chow, K. H., Kiefl, R. F., MacFarlane, A., Schumann, D., Fujii, S., and Tanigawa, S. (1994). *Phys. Rev. B*, **50**, 1999.
Kadono, R., Macrae, R. M., and Nagamine, K. (1997). *Hyperfine Interact.*, **105**, 327.
Kagan, Yu., and Prokof'ev, N. V. (1990). *Phys. Lett. A*, **150**, 320.
Kiefl, R. F., and Estle, T. L. (1991). In *Hydrogen in Semiconductors*, J. I. Pankove and N. M. Johnson, eds. New York: Academic Press, p. 547.
Kiefl, R. F., and Kreitzman, S. R. (1992). In *Prospectives in Muon Science*, T. Yamazaki, K. Nakai, and K. Nagamine, eds. New York: Elsevier, p. 265.
Kiefl, R. F., Celio, M., Estle, T. L., Luke, G. M., Kreitzman, S. R., Brewer, J. H., Noakes, D. R., Ansaldo, E. J., and Nishiyama, K. (1987). *Phys. Rev. Lett.*, **58**, 1780.
Kiefl, R. F., Celio, M., Estle, T. L., Kreitzman, S. R., Luke, G. M., Riseman, T. R., and Ansaldo, E. J. (1988). *Phys. Rev. Lett.*, **60**, 224.
Kreitzman, S. R. (1990). *Hyperfine Interact.*, **65**, 1055.
Kreitzman, S. R., Hitti, B., Lichti, R. L., Estle, T. L., and Chow, K. H. (1995). *Phys. Rev. B*, **51**, 13117.
Leon, M. (1992). *Phys. Rev. B*, **46**, 6603.
Lichti, R. L. (1995). *Philos. Trans. R. Soc.*, **A350**, 323.
Lichti, R. L., Lamp, C. D., Kreitzman, S. R., Kiefl, R. F., Schneider, J. W., Niedermayer, Ch., Chow, K. H., Pfiz, T., Estle, T. L., Dodds, S. A., Hitti, B., and DuVarney, R. C. (1992). *Mater. Sci. Forum*, **83–87**, 1115.

Lichti, R. L., Chow, K. H., Estle, T. L., Hitti, B., Kreitzman, S. R., and Schneider, J. W. (1994a). *Mater. Sci. Forum*, **143–147**, 915.
Lichti, R. L., Chow, K. H., Cooke, D. W., Cox, S. F. J., Davis, E. A., DuVarney, R. C., Estle, T. L., Hitti, B., Kreitzman, S. R., Macrae, R., Schwab, C., and Singh, A. (1994b). *Hyperfine Interact.*, **86**, 711.
Lichti, R. L., Schwab, C., and Estle, T. L. (1995). *Mater. Sci. Forum*, **196–201**, 831.
Lichti, R. L., Cox, S. F. J., Schwab, C., Estle, T. L., Hitti, B., and Chow, K. H. (1997a). *Hyperfine Interact.*, **105**, 333.
Lichti, R. L., Chow, K. H., Cox, S. F. J., Estle, T. L., Hitti, B., and Schwab, C. (1997b). *Mater. Sci. Forum*, **258–263**, 179.
Pankove, J. I., and Johnson, N. M., eds. (1991). *Hydrogen in Semiconductors*, **34**, R. K. Willardson and A. C. Beer, eds., *Semiconductors and Semimetals* series. New York: Academic Press.
Patterson, B. D. (1988). *Rev. Mod. Phys.*, **60**, 69.
Pearton, S. J., Corbett, J. W., and Stavola, M. (1992). *Hydrogen in Crystalline Semiconductors*. Berlin: Springer-Verlag.
Pearton, S. J., ed. (1994). *Hydrogen in Compound Semiconductors*. *Mater. Sci. Forum*, **174–178**.
Scheuermann, R., Schimmele, L., Seeger, A., Stammler, Th., Grund, Th., Hampele, M., Herlach, D., Iwanowski, M., Major, J., Notter, M., Pfiz, Th. (1995). *Philos. Mag. B*, **72**, 161.
Schneider, J. W., Kiefl, R. F., Ansaldo, E. J., Brewer, J. H., Chow, K. H., Cox, S. F. J., Dodds, S. A., DuVarney, R. C., Estle, T. L., Haller, E. E., Kadono, R., Kreitzman, S. R., Lichti, R. L., Niedermayer, Ch., Pfiz, T., Riseman, T. M., and Schwab, C. (1992). *Mater. Sci. Forum*, **83–87**, 569.
Schneider, J. W., Chow, K. H., Kiefl, R. F., Kreitzman, S. R., MacFarlane, W. A., DuVarney, R. C., Estle, T. L., Lichti, R. L., and Schwab, C. (1993). *Phys. Rev. B*, **47**, 10193.
Smilga, V. P., and Belousov, Yu. M. (1994). *The Muon Method in Science*. New York: Nova Science.
Storchak, V., Cox, S. F. J., Cottrell, S. P., Brewer, J. H., Morris, G. D., Arseneau, D. J., and Hitti, B. (1997). *Phys. Rev. Lett.*, **78**, 2835.
Symons, M. C. R. (1984). *Hyperfine Interact.*, **19**, 771.
Yen, H. K. (1988). M.Sc. thesis, University of British Columbia.

CHAPTER 9

Hydrogen in III-V and II-VI Semiconductors

*Matthew D. McCluskey**

XEROX PALO ALTO RESEARCH CENTER
PALO ALTO, CALIFORNIA

Eugene E. Haller

LAWRENCE BERKELEY NATIONAL LABORATORY AND UNIVERSITY OF CALIFORNIA AT BERKELEY
BERKELEY, CALIFORNIA

I. INTRODUCTION		373
II. HYDROGEN IN III-V SEMICONDUCTORS		376
1. Hydrogen in GaAs		376
2. Hydrogen in AlAs		390
3. Hydrogen in InP		396
4. Hydrogen in GaP		402
5. Hydrogen in AlSb		412
6. Hydrogen in GaN		422
7. Hydrogen in Other III-V Semiconductors		426
III. HYDROGEN IN II-VI SEMICONDUCTORS		426
1. Hydrogen in ZnSe		426
2. Hydrogen in CdTe		433
IV. SUMMARY AND DISCUSSION		434
REFERENCES		436

I. Introduction

Since the discovery of hydrogen passivation of acceptors (Johnson et al., 1985) and donors (Chevallier et al., 1985) in GaAs, a great deal of research has been performed on hydrogen in compound semiconductors. This chapter describes experimental and theoretical discoveries in the field of hydrogen in III-V and II-VI semiconductors that have been reported since

*Current address: Department of Physics, Washington State University, Pullman, Washington.

1991. The pre-1991 era of hydrogen in III-V semiconductors has been summarized by Pankove and Johnson (1991), and we focus on post-1991 developments.

In this chapter we concentrate on the microscopic structure of hydrogen-related complexes that can be obtained with experimental techniques such as infrared (IR) and Raman spectroscopy. Hydrogen-related local vibrational modes (LVMs) can be unambiguously identified through the substitution of deuterium, which reduces the LVM frequency by a factor of close to $\sqrt{2}$. The LVM frequencies and the isotopic frequency ratio $r = v_H/v_D$ provide information about the bonding and position of the hydrogen. Due to limited space, we will not describe the results of proton and deuteron implantation experiments. Interested readers are referred to Zavada and Wilson (1994).

To date, a large number of hydrogen-related complexes have been discovered in III-V semiconductors. C-H complexes in GaAs have been studied intensively, both theoretically and experimentally, over the past 5 years (Davidson et al., 1993). The four isotope combinations of these complexes (^{12}C-H, ^{13}C-H, ^{12}C-D, ^{13}C-D) each have 4 modes—one stretch (A_1), one longitudinal (A_1), and two transverse (E^- and E^+)—resulting in 16 LVMs, all of which have been been observed experimentally. In addition, the C_{As}-C_{As} split interstitial pair can be passivated, resulting in the formation of $(C_{As})_2H$ and $(C_{As})_2H_2$ complexes (Cheng et al., 1994; Davidson et al., 1994). C-H complexes also have been discovered in AlAs (Pritchard et al., 1994a), GaP (Clerjaud et al., 1991), and InP (Davidson et al., 1998).

Group VI donor-hydrogen complexes in GaAs give rise to sharp IR absorption peaks due to hydrogen and deuterium stretch and wag modes (Svensson and Weber, 1993; Rahbi et al., 1994; Vetterhöffer et al., 1994). In these complexes, hydrogen is believed to reside in an antibonding orientation. Pajot and Song (1992) have observed numerous IR absorption peaks between 2940 and 3500 cm^{-1} in oxygen-doped GaAs that they attribute to N-H and O-H stretch modes. Although hydrogen has long been suspected to form interstitial H_2 molecules in semiconductors, such molecules have been observed only recently in GaAs by Raman spectroscopy (Vetterhöffer et al., 1996b).

To explain the temperature dependence of hydrogen LVM frequencies and line widths, we discuss two competing models, both of which include an anharmonic interaction between LVMs and acoustical phonons. In the first model, the complex is assumed to interact weakly with a *single* acoustical phonon. In the second model, the complex is assumed to interact with *all* the phonons in the Brillouin zone. In addition to variable-temperature spectroscopy and the application of uniaxial stress (Veloarisoa et al., 1993) and hydrostatic pressure (McCluskey et al., 1997) have been used to perturb hydrogen-containing complexes, leading to shifts in the frequencies of the hydrogen LVMs.

For group II acceptor-hydrogen complexes in GaAs (Chevallier *et al.*, 1991), InP (Darwich *et al.*, 1993), and GaP (McCluskey *et al.*, 1995b), results from LVM spectroscopy show conclusively that hydrogen binds to the host anion in a bond-centered (BC) orientation, along a [111] direction, adjacent to the acceptor. By measuring overtones of hydrogen LVMs in InP, Darwich *et al.* (1993) have determined the anharmonicity of the hydrogen potentials. As the atomic number of the group II acceptor increases from Be to Cd, the stretch-mode frequency and isotropic frequency ratio $r = v_H/v_D$ increase. It is interesting to note that Mg is weakly passivated by H in GaAs (Rahbi *et al.*, 1993) but very effectively passivated in GaN (Amano *et al.*, 1989; Nakamura *et al.*, 1992). In addition, theoretical calculations (Neugebauer and Van de Walle, 1996) indicate that in Mg-doped GaN, H binds to a nitrogen in an *antibonding* orientation, in contrast to acceptor-hydrogen complexes in other III-V semiconductors. The results of LVM spectroscopy experiments (Götz *et al.*, 1996) have verified the theoretical prediction that hydrogen passivates the magnesium acceptor by forming a N—H bond.

In AlSb, the Se and Te DX centers are passivated by annealing in a hydrogen ambient at temperatures as low as 700°C (McCluskey *et al.*, 1996a). The second, third, and fourth harmonics of Se-H and Se-D wag modes have been observed and show splittings that are consistent with C_{3v} symmetry. Although the Se-D complex shows one stretch mode, the Se-H stretch mode, in contrast, splits into three peaks. This anomalous splitting is explained by a resonant interaction between the stretch mode and a phonon-wag combination mode (McCluskey *et al.*, 1998). As temperature or pressure is varied, the LVMs and extended lattice phonons exhibit anticrossing behavior.

In ZnSe, epilayers grown by MOCVD and doped with nitrogen form N-H complexes that give rise to stretch and wag mode peaks in IR and Raman spectra (Wolk *et al.*, 1993; Kamata *et al.*, 1993). N-D complexes were discovered in MBE-grown ZnSe (Yu *et al.*, 1996), with monatomic deuterium introduced via a thermal cracker in the chamber. Similarly, epilayers grown by MOCVD and doped with arsenic form As-H complexes (McCluskey *et al.*, 1996b). A comparison is made between GaAs:Zn,H and ZnSe:As,H complexes—in both cases, hydrogen resides in a BC position between a zinc and arsenic atom. It is clear that obtaining strongly *p*-type ZnSe grown by MOCVD is hindered at least in part by the formation of neutral acceptor-hydrogen complexes.

Unless stated otherwise, spectroscopic data were obtained at or near liquid helium temperatures (4–15 K).

II. Hydrogen in III-V Semiconductors

1. HYDROGEN IN GaAs

a. *C-H Complexes*

Of all hydrogen-related complexes in compound semiconductors, C-H complexes in GaAs have received the most theoretical and experimental attention. Carbon is a preferred *p*-type dopant in GaAs because it has a high solubility limit and a low diffusion coefficient. Hydrogen can passivate carbon acceptors during crystal growth, annealing in a hydrogen ambient (Kozuch et al., 1993), or during exposure to hydrogen plasma (Rahbi et al., 1993). To dissociate C-H pairs and thereby activate carbon acceptors, the samples are annealed at temperatures around 550°C.

Theoretical and experimental studies strongly indicate that a hydrogen binds directly to the carbon in a BC position, along a [111] axis (Fig. 1). This configuration is analogous to that of acceptor-hydrogen complexes in Si (Pankove, 1991). The stretch vibrational modes of the ^{12}C-H and ^{12}C-D complexes have frequencies of 2635 and 1969 cm^{-1}, respectively, with an isotopic frequency ratio $r = 1.3386$ (Chevallier et al., 1991). Clerjaud et al. (1990) provided evidence that the hydrogen binds directly to the carbon acceptor by observing an absorption line corresponding to ^{13}C-H at 2628 cm^{-1}. The ratio of the intensities of the ^{13}C-H to ^{12}C-H stretch modes matches the ratio of the natural abundances of ^{13}C and ^{12}C (1:99). In addition, the shift in the vibrational frequency can be explained by a diatomic C-H model. *Ab initio* calculations (Jones and Öberg, 1991; Bonapasta, 1993) show that the BC configuration minimizes the total energy. A calibration for the stretch mode gives an integrated absorption of 1 cm^{-2} for a concentration of $(9 \pm 2) \times 10^{15}$ C-H pairs/cm^2 (Davidson et al., 1996).

In addition to the stretch (A_1) mode, the C-H complex also has a low-frequency longitudinal mode (A_1) in which the hydrogen and carbon

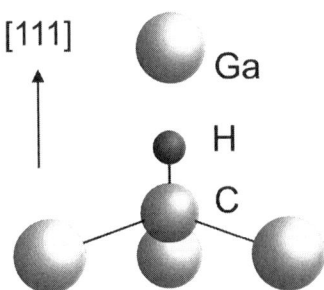

FIG. 1. Bond-centered model of the C-H complex in GaAs.

oscillate in phase. For the ^{12}C-H complex, this mode gives rise to an IR absorption peak at 453 cm^{-1}, commonly labeled X (Woodhouse et al., 1991). ^{13}C-H, ^{12}C-D, and ^{13}C-D complexes have X peaks at 438, 440, and 426 cm^{-1}, respectively (Davidson et al., 1993). Since the carbon and hydrogen move as one unit, the frequencies are approximately proportional to the square root of the C-H molecular mass. Based on Raman scattering, the symmetry of the X mode was shown definitively to be A_1 (Wagner et al., 1992).

The transverse modes (E) are classified as either E^- (antisymmetric) or E^+ (symmetric), depending on whether the hydrogen and carbon oscillate out of phase or in phase, respectively. In addition, a mode is classified as "hydrogen-like" or "carbon-like" if the hydrogen or carbon amplitude, respectively, is larger. Since carbon is a relatively light atom, there exists a significant interaction between the E^- and E^+ modes. Using a spring-and-ball model, Davidson et al. (1993) showed that as the mass of hydrogen is continuously varied 1 to 2 amu, an anticrossing occurs between the high (E^-) and low (E^+) frequency modes (Fig. 2). Whereas the E^- mode is hydrogen-like for the C-H complex, it is carbon-like for the C-D complex, and vice versa for the E^+ mode. Because the modes are qualitatively different for C-H and C-D complexes, the isotopic frequency ratios are not close to $\sqrt{2}$, a fact that caused a great deal of confusion for several years.

All LVMs related to carbon-hydrogen complexes except the ^{12}C-H and ^{13}C-H E^- modes have been observed by IR spectroscopy (Davidson et al., 1993). The absence of the C-H E^- (hydrogen-like) modes from IR spectra has been attributed to their unusually weak induced dipole moments, as shown by ab initio calculations (Jones and Öberg, 1991). The deuterium E^- and E^+ modes, on the other hand, both contain carbon-like components that have larger dipole moments. The ^{12}C-H and ^{13}C-H E^- modes were finally observed by Raman spectroscopy (Wagner et al., 1995a). Polarization-dependent Raman scattering measurements were used to verify the C_{3v} symmetry of the C-H complex as well as the symmetries of all the modes (Wagner et al., 1995b). The 4 modes—one stretch (A_1), one longitudinal (A_1), and two transverse (E^- and E^+)—combined with the 4 isotope combinations (^{12}C-H, ^{13}C-H, ^{12}C-D, ^{13}C-D) give rise to 16 LVMs, all of which have been observed experimentally (Table I). Ab initio calculations (Jones et al., 1994) agree with the experimental low-frequency modes but overestimate the frequencies of the stretch modes. The discrepancy between theory and experiment is attributed to the anharmonicity of the potential in which the hydrogen is bound. For some reason, the "improved" calculations gave worse predictions for the stretch modes than the "old" calculations, which predicted a ^{12}C-H stretch-mode frequency of 2605 cm^{-1} (Jones and Öberg, 1991).

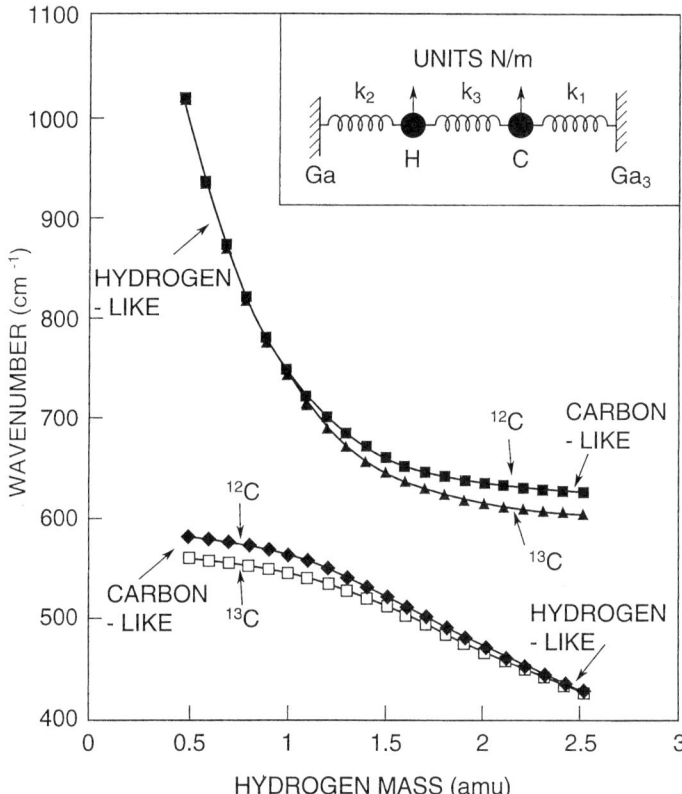

FIG. 2. Vibrational transverse E modes of C-H complexes in GaAs. As the hydrogen mass is continuously varied, there is an anticrossing between the modes. (From Davidson et al., 1993.)

In heavily carbon-doped GaAs epilayers grown by metalorganic chemical vapor deposition (MOCVD), weak IR absorption peaks were observed at 2643 and 2651 cm^{-1} (Stavola et al., 1992; Kozuch et al., 1993) along with the known ^{12}C-H stretch mode peak at 2636 cm^{-1}. In heavily carbon-doped GaAs epilayers grown by metalorganic molecular beam epitaxy (MOMBE), these three peaks were observed along with an additional peak at 2688 cm^{-1} (Kozuch et al., 1990, 1993; Stavola et al., 1992; Veloarisoa et al., 1992). Polarization-dependent IR absorption experiments showed that the peak at 2688 cm^{-1} is strongly polarized along a particular [110] direction (Fig. 3), whereas the other peaks have a random polarization (Cheng et al., 1994). The peak at 2688 cm^{-1} is attributed to the stretch mode of a $(C_{As})_2H$

TABLE I

EXPERIMENTAL AND THEORETICAL VALUES FOR LVMs OF
CARBON-HYDROGEN COMPLEXES IN GaAs

Mode	Observed frequency[a] (cm^{-1})	Theory[e] (cm^{-1})
Stretch (A_1)		
^{12}C-H	2635.2	2950
^{13}C-H	2628.5	2942
^{12}C-D	1968.6	2154
^{13}C-D	1958.3	2144
Longitudinal (A_1)		
^{12}C-H	452.7	456
^{13}C-H	437.8	440
^{12}C-D	440.2	442
^{13}C-D	426.9	428
Transverse (E^-)		
^{12}C-H	739[b]	888
^{13}C-H	730[b]	883
^{12}C-D	637.2	707
^{13}C-D	616.6	693
Transverse (E^+)		
^{12}C-H	562.6	553
^{13}C-H	547.6	536
^{12}C-D	466.2	495
^{13}C-D	463.8	487
Stretch		
(^{12}C$_{As}$)$_2$H	2688[c]	—
(^{12}C$_{As}$)$_2$H$_2$	2725[d]	—
	2775[d]	—

[a] Davidson et al., 1993.
[b] Wagner et al., 1995.
[c] Cheng et al., 1994.
[d] Davidson et al., 1994.
[e] Jones et al., 1994.

complex in which the hydrogen resides in a BC position between a carbon and a gallium, with a second carbon neighboring the gallium (Fig. 4). The complex is believed to orient in one particular [110] plane during growth, for a growth direction along the [100] axis.

Davidson et al. (1994) performed polarization-dependent IR absorption measurements on MOMBE-grown GaAs:C samples in which the 2 × 4 (001) surface reconstruction had been determined by reflection high-energy electron diffraction (RHEED). By correlating the polarization of the peak at 2688 cm^{-1} with the surface reconstruction pattern, they concluded that the

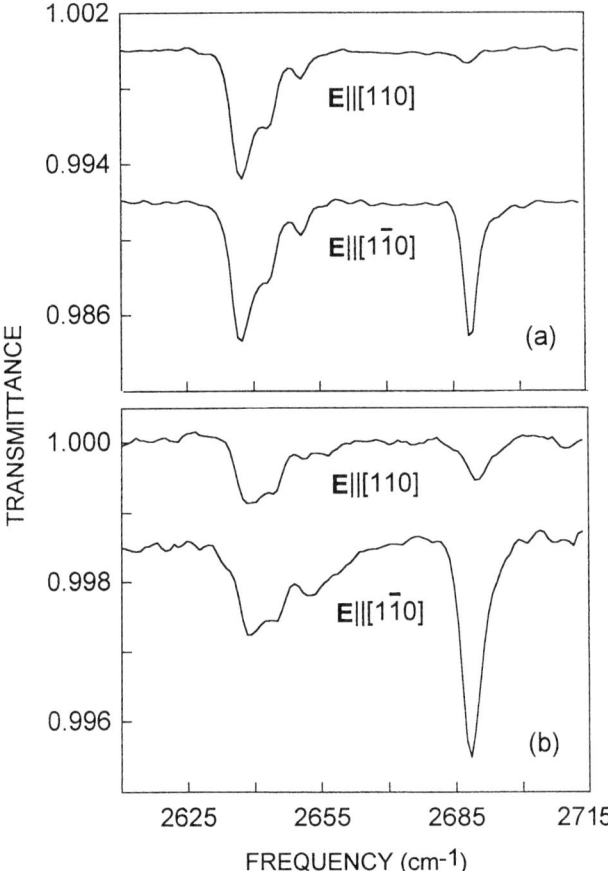

FIG. 3. Infrared absorption spectra of MOMBE-grown GaAs:C with (a) $N_A = 2 \times 10^{20}$ cm^{-3} and (b) $N_A = 5 \times 10^{20}$ cm^{-3}. The light is polarized along the [110] and [1̄10] directions. (From Cheng et al., 1994.)

$(C_{As})_2H$ complexes preferentially align along the [110] direction, perpendicular to the [1̄10] surface arsenic dimer bond. In samples grown by MOCVD, however, the surface reconstruction is $c(4 \times 4)$, and $(C_{As})_2H$ complexes align along the [1̄10] direction (Davidson et al., 1995). In both cases, the alignment is attributed to incomplete decomposition of trimethyl gallium (TMG) during growth.

In samples that are annealed in H_2 gas at temperatures above 525°C, the $(C_{As})_2H$ complexes dissociate and form new, randomly oriented $(C_{As})_2H$ complexes. This was shown experimentally by a loss in the dichroism of the

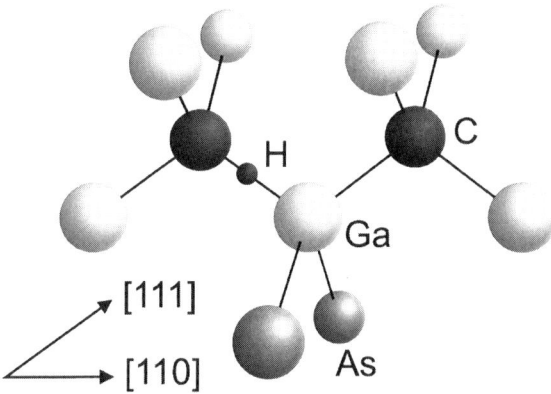

FIG. 4. Model of the $(C_{As})_2H$ complex in GaAs.

peak, which disappeared (random orientation) after annealing at temperatures above $\sim 600°C$ for a duration of 30 min (Cheng et al., 1994). At an annealing temperature of 700°C, the complex dissociated and the peak disappeared. When samples were annealed in He gas at $\sim 600°C$, the intensity of the peak at 2688 cm^{-1} decreased, indicating dissociation of the complexes without re-formation of new, randomly oriented complexes (Cheng et al., 1994).

In samples exposed to a hydrogen plasma at a temperature of 320°C, the peak at 2688 cm^{-1} disappeared and two new peaks appeared at 2725 and 2775 cm^{-1} (Davidson et al., 1994). The peaks at 2725 and 2775 cm^{-1} do not depend on polarization of the incoming light and are attributed to $(C_{As})_2H_2$ complexes that are formed via the capture of an additional hydrogen. Subsequent annealing in vacuum at the growth temperature (450°C) led to an increase in the peak at 2688 cm^{-1} at the expense of the peaks at 2725 and 2775 cm^{-1}, indicating partial dissociation of the $(C_{As})_2H_2$ complexes (Davidson et al., 1994). It would be interesting to test this model by exposing GaAs:C samples to a deuterium plasma, which could result in new peaks arising from $(C_{As})_2HD$ complexes.

Theoretical first-principles calculations (Lee and Chang, 1996) showed that both the $(C_{As})_2H$ and $(C_{As})_2H_2$ complexes are stable configurations, with the hydrogen atoms located at BC sites, in agreement with experimental observations. The C_{As}-C_{As} complex, however, is unstable and on annealing can dissociate into two isolated C_{As} acceptors. Thus annealing in an H_2 ambient causes the dissociation and re-formation of $(C_{As})_2H$ complexes, leading to a randomization of the alignment.

b. *Group VI Donor-Hydrogen Complexes*

Following the discovery of group IV donor-hydrogen complexes in GaAs (Chevallier *et al.*, 1991), a great deal of research has been focused on vibrational spectroscopy of group VI donor-hydrogen complexes. Svensson and Weber (1993) observed sharp IR absorption peaks in GaAs:Se exposed to a hydrogen or deuterium plasma at 1507.46 and 1084.82 cm^{-1}, respectively. These peaks are attributed to hydrogen stretch modes. In addition, wag modes were observed at 777.95 and 554.31 cm^{-1}, respectively. The hydrogen stretch and wag modes are shown in Fig. 5. Unlike group IV donor-hydrogen complexes, the integrated absorption of the Se-H wag mode is approximately 7 times that of the stretch mode.

Stretch and wag modes arising from Te-H and S-H complexes (Fig. 5) were observed by Rahbi *et al.* (1994) and Vetterhöffer *et al.* (1994). As shown in Table II, the frequency of the modes varies by only a few wavenumbers from S to Te. The insensitivity of the LVM frequencies to the donor species strongly suggests that the hydrogen resides in an antibonding orientation, along a [111] direction opposite the donor (Fig. 6). Polarized IR spectroscopy on GaAs:S,H samples under uniaxial stress verified the C_{3v} symmetry

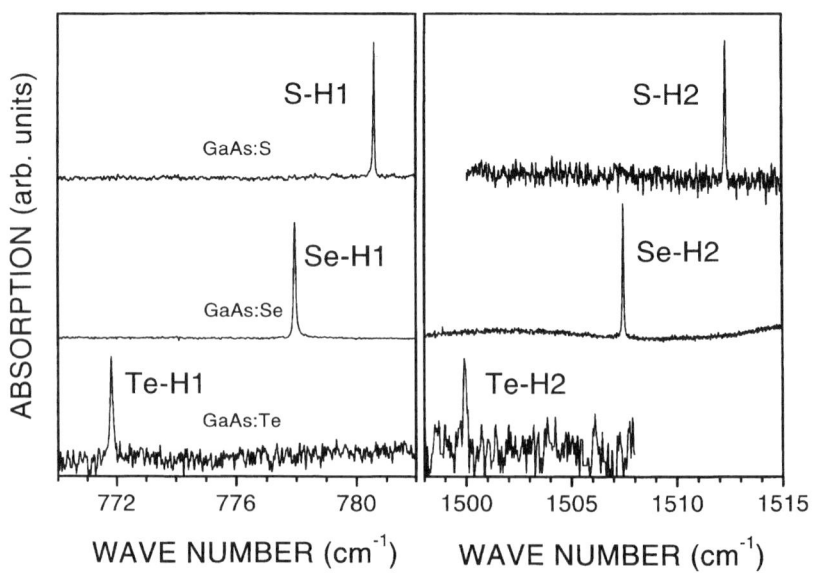

FIG. 5. Infrared absorption spectra of hydrogenated GaAs:S, GaAs:Se, and GaAs:Te. (From Vetterhöffer *et al.*, 1994.)

TABLE II
LVM FREQUENCIES OF GROUP VI DONOR-HYDROGEN COMPLEXES IN GaAs (FREQUENCIES IN cm^{-1})

Material	D-wag	D-stretch	H-wag	H-stretch
GaAs:S	556.1[c]	1088.4[c]	780.57[b,c]	1512.26[b,c]
GaAs:Se	554.31[a]	1084.82[a]	777.95[a]	1507.46[a]
GaAs:Te	550.0[c]	ND	771.81[b,c]	1499.85[b,c]

[a]Svensson and Weber, 1993.
[b]Rahbi et al., 1994.
[c]Vetterhöffer and Weber, 1996a.

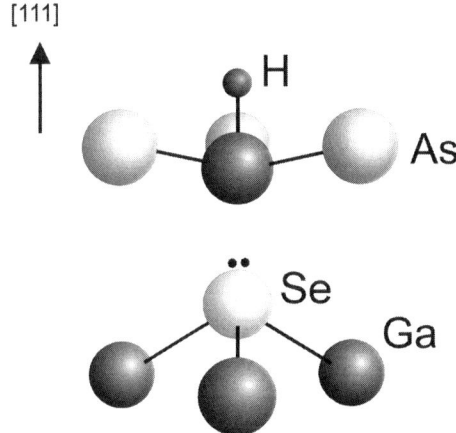

FIG. 6. Antibonding model of group VI donor-hydrogen complexes in GaAs.

of this complex (Vetterhöffer and Weber, 1996a). Ab initio calculations (Chang, 1991) showed that the antibonding orientation minimizes the total energy. The antibonding model is reminiscent of donor-hydrogen complexes in Si (Johnson et al., 1986).

c. *Group II Acceptor-Hydrogen Complexes*

Rahbi et al. (1993) performed a systematic study of hydrogen diffusion and the formation of acceptor-hydrogen complexes in GaAs. They observed IR absorption peaks corresponding to stretch modes of C-H, Zn-H, Cd-H, and Ge-H complexes, in agreement with previous observations (Chevallier et al., 1991). Hydrogenated and deuterated GaAs:Mg showed sharp IR absorption peaks at 2144 cm^{-1} and 1720 cm^{-1}, respectively, for an isotopic frequency ratio of $r = v_H/v_D = 1.3859$. These peaks exhibited a less pronounced temperature-dependent shift and broadening than Zn-H stretch modes (Bouanani-Rahbi et al., 1997), leading the authors to conclude that hydrogen resides in an antibonding location in Mg-H complexes. The antibonding orientation would reduce the interaction between the Mg and H. This model is similar to that which is proposed for the Mg-H complex in GaN (see Sec. II.6.a). In contrast, Bouanani-Rahbi et al. (1997) propose that hydrogen resides in a BC position in InP:Mg,H complexes (see Sec. II.3.a).

d. *O-H and N-H Complexes*

Pajot and Song (1992) have observed numerous IR absorption peaks between 2940 and 3500 cm^{-1} in oxygen-doped semi-insulating GaAs. The various peaks are attributed to N-H and O-H stretch modes. A strong peak at 3300.0 cm^{-1} and a weak peak at 2989.54 cm^{-1} are attributed to ^{16}O-H and ^{18}O-H stretch modes, respectively (Fig. 7). The ratio of the peak intensities agrees with the natural abundance of stable oxygen isotopes, and

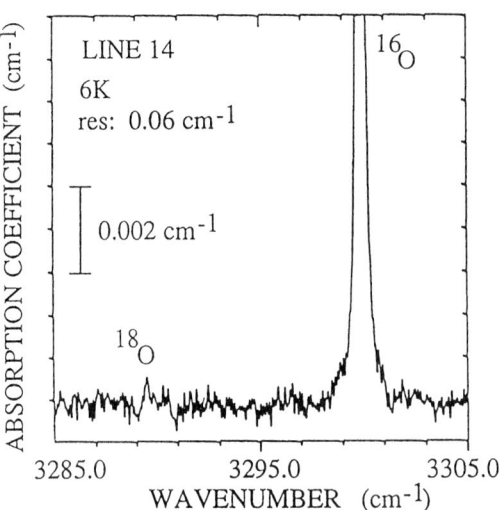

FIG. 7. Infrared absorption peaks corresponding to O-H complexes in GaAs. (From Pajot and Song, 1992.)

the frequency shift agrees with a diatomic molecule approximation. In addition, a peak at 2947.4 cm^{-1} was attributed to a ^{14}N-H complex with trigonal symmetry (Song, 1992). It has been proposed that the complex contains two hydrogen atoms, resulting in an additional hydrogen stretch mode at 1984.3 cm^{-1} and a transverse mode at 1010.1 cm^{-1}. In addition, a ^{15}N-H stretch-mode peak was found at 2947.4 cm^{-1}. The nitrogen-dihydrogen model in GaAs is similar to that in GaP (see Sec. II.4.c).

Clerjaud et al. (1990, 1997) made the argument that since C-H and N-H complexes are practically never observed in the same sample, neutral hydrogen atoms form N-H complexes, whereas protons form C-H complexes. By monitoring the transition from C-H to N-2H complexes as a function of Fermi level, they concluded that the hydrogen donor level H^0/H^+ is located 0.3 eV above the valence-band maximum.

e. *Isolated H_2 Molecules*

First-principles calculations by Pavesi and Giannozzi (1992) and Breuer et al. (1996) showed that while hydrogen enters GaAs as monatomic H, there exists a strong tendency to form H_2 molecules at tetrahedral sites. The H_2 molecules are confined in a deep potential well and therefore are immobile. Although the existence of H_2 molecules in crystalline semiconductors has long been suspected, they have only recently been observed directly by Raman spectroscopy (Vetterhöffer et al., 1996b) at a temperature of 77 K. Both p- and n-type GaAs samples exposed to a hydrogen plasma showed Raman-active peaks at 3925.9 and 3934.1 cm^{-1}, corresponding to H-H stretch modes with rotational quantum numbers $J = 1$ and $J = 0$, respectively (Fig. 8). The frequencies are ~ 200 cm^{-1} lower than that of H_2 gas (4161 cm^{-1}). Samples exposed to a deuterium plasma showed a D-D stretch-mode frequency at 2842.6 cm^{-1}, with a rotovibrational splitting that was presumably less than the instrumental resolution (2 cm^{-1}). Samples exposed to a hydrogen/deuterium plasma showed the H-H and D-D peaks as well as a H-D stretch-mode peak at 3446.5 cm^{-1}. The sharpness of the peaks as well as the observation of rotational splitting led the authors to conclude that the hydrogen molecules are only weakly perturbed by the GaAs host.

f. *Temperature Dependence of LVMs*

As the temperature of a semiconductor is increased, the lattice expands, and phonon modes are thermally populated. Although both these phenomena affect the vibrational modes of hydrogen-related complexes, lattice

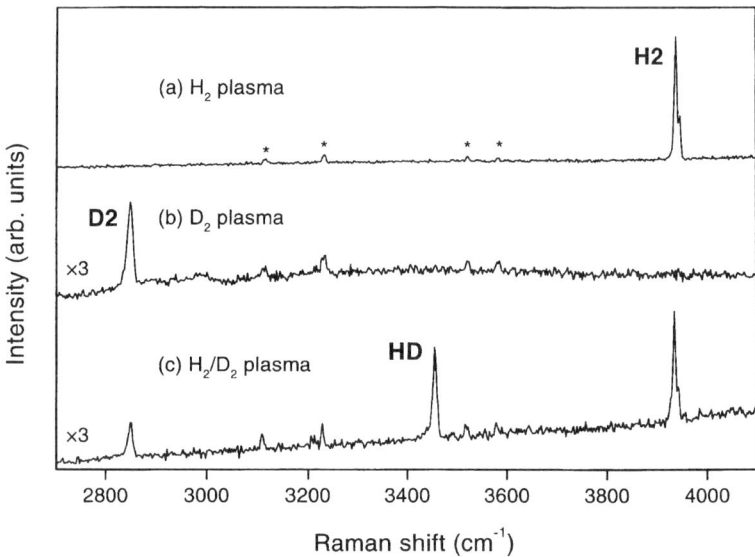

FIG. 8. Raman spectra ($T = 77$ K) of GaAs exposed to (a) hydrogen, (b) deuterium, and (c) hydrogen/deuterium plasma. Residual laser plasma lines are indicated by asterisks (*). (From Vetterhöffer et al., 1996b.)

expansion does not typically play a dominant role (McCluskey et al., 1995b; Vetterhöffer and Weber, 1996a). The temperature dependence of C-H (Clerjaud et al., 1992a), Si-H (Tuncel and Sigg, 1993), N-H (Darwich et al., 1994), Se-H, and Te-H (Vetterhöffer and Weber, 1996a) LVMs in GaAs is well described by a model in which the complex interacts with a single phonon. In the weak coupling limit, this interaction leads to a temperature-dependent frequency shift given by

$$\delta v = \frac{\delta v_0}{e^{hv_0/kT} - 1} \tag{1}$$

and line-width broadening given by

$$\delta \Gamma = \frac{2(\delta v_0)^2}{\eta} \frac{e^{hv_0/kT}}{(e^{hv_0/kT} - 1)^2} \tag{2}$$

where δv_0, v_0, and η are adjustable parameters. The temperature-dependent shifts δv and line widths $\delta \Gamma$ of several hydrogen-related LVMs in GaAs and AlAs are plotted in Figs. 9 and 10.

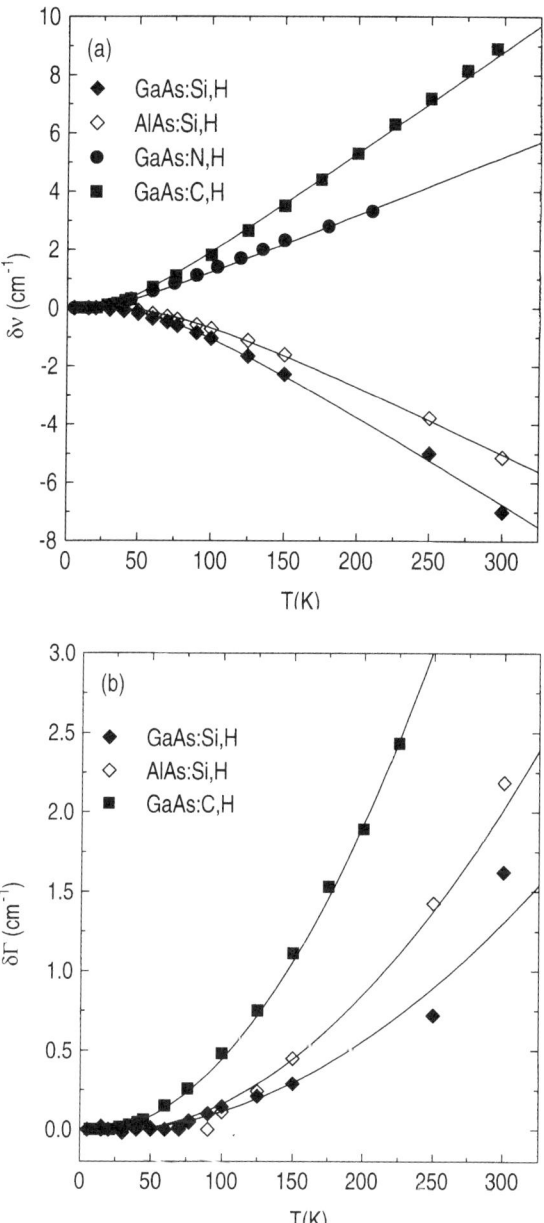

FIG. 9. Temperature-dependent shifts in the (a) frequency and (b) line width of hydrogen stretch modes in GaAs and AlAs. The solid lines are fits to Eqs. (1) and (2).

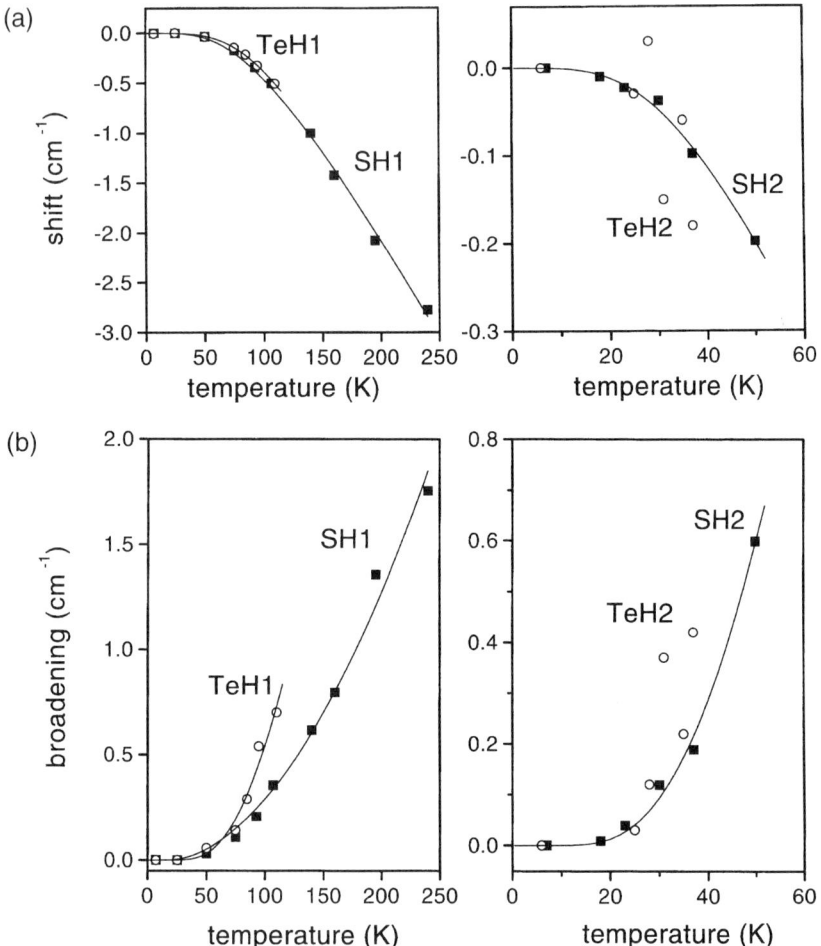

FIG. 10. Temperature-dependent shifts of the (a) frequency and (b) line width of hydrogen stretch and wag modes in GaAs:S and GaAs:Te. The solid lines are fits to Eqs. (1) and (2). (From Vetterhöffer and Weber, 1996a.)

It should be re-emphasized that the thermal expansion of the lattice does not contribute significantly to the temperature-dependent shifts of the hydrogen LVMs (McCluskey et al., 1995b; Vetterhöffer and Weber, 1996a). In the temperature range from 0 to 150 K, lattice expansion results in a shift of only $0.1\,\text{cm}^{-1}$.

The single-phonon model cannot explain the temperature-dependent shift of the C-H stretch mode frequency in GaP (Clerjaud et al., 1992b) or the

O-H stretch mode in GaAs (Darwich et al., 1994). In Section II.4.b, a different model is used to fit the temperature dependence of acceptor-hydrogen LVMs in GaP, in which the complexes are assumed to interact with all the phonons in the Brillouin zone.

g. Pressure Dependence of LVMs

Until recently, variable-pressure spectroscopy of hydrogen-related complexes has always used uniaxial stress (Stavola and Pearton, 1991). Using uniaxial stress, the symmetry and coupling of the ground-state energy to stress have been determined for Be-H and C-H complexes in GaAs (Veloarisoa et al., 1993).

IR spectroscopy of hydrogen LVMs under *hydrostatic* pressure was performed for the C-H, S-H, and Se-H complexes, as well as substitutional C_{As}, for pressures as high as 7 GPa (McCluskey et al., 1997). LVMs arising from $^{12}C_{As}$ and $^{13}C_{As}$ substitutional impurities vary linearly with pressure. The pressure-dependent shifts of the $^{12}C_{As}$-H and $^{13}C_{As}$-H stretch-mode frequencies have positive curvatures (Fig. 11), whereas the pressure depen-

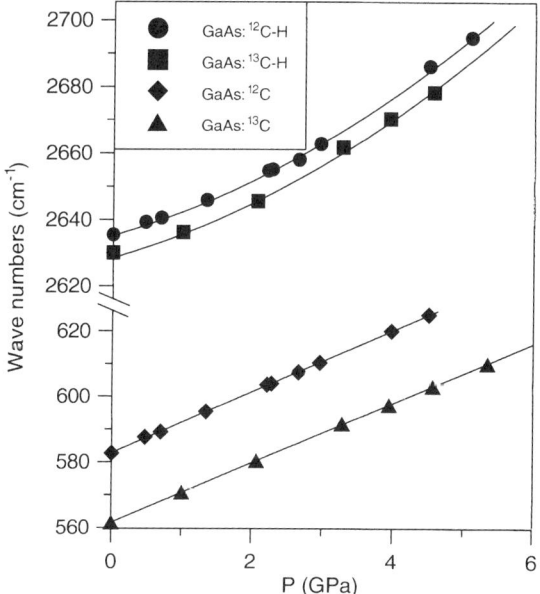

FIG. 11. Plot of C_{As} and C_{As}-H LVMs in GaAs as a function of hydrostatic pressure. (From McCluskey et al., 1997a.)

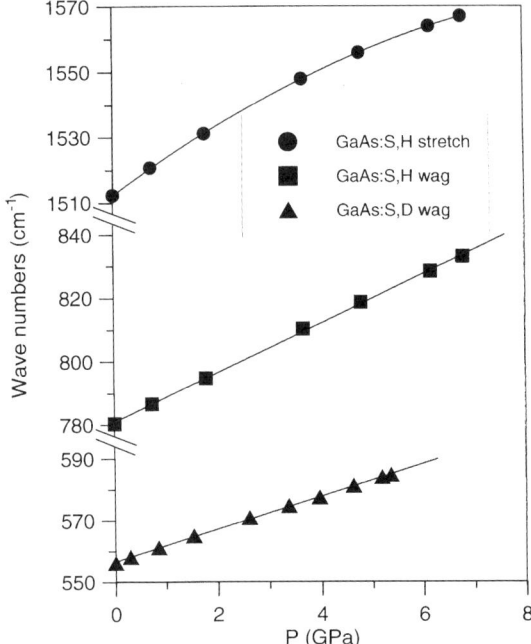

FIG. 12. Plot of S-H and S-D LVMs in GaAs as a function of hydrostatic pressure. (From McCluskey et al., 1997a.)

dent shift of the S-H stretch-mode frequency has a negative curvature (Fig. 12). This may be related to the fact that in the BC C-H complex, the hydrogen is compressed between the carbon acceptor and one gallium host atom, whereas in the S-H complex, the hydrogen occupies an interstitial position and is not crowded by neighboring atoms. The S-H and S-D wag-mode frequencies vary linearly with pressure (Fig. 12). Although *ab initio* calculations have been performed for the structures of the C-H (Jones et al., 1994) and S-H (Chang, 1991) complexes, the pressure dependence of LVMs awaits a detailed theoretical treatment.

2. HYDROGEN IN AlAs

a. *C-H Complexes*

Carbon is a preferred *p*-type dopant in AlAs, occupying As lattice sites (C_{As}). AlAs:^{12}C and AlAs:^{13}C epilayers grown by chemical beam epitaxy

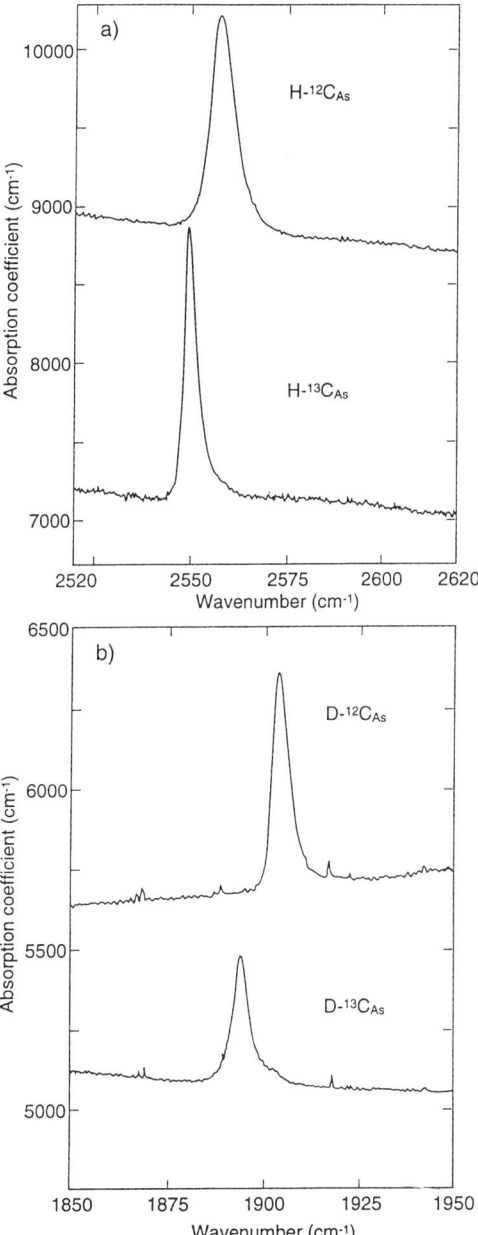

FIG. 13. Infrared absorption spectra of stretch modes of (a) $^{12}C_{As}$-H and $^{13}C_{As}$-H pairs and (b) $^{12}C_{As}$-D and $^{13}C_{As}$-D pairs in AlAs. (From Pritchard et al., 1994a.)

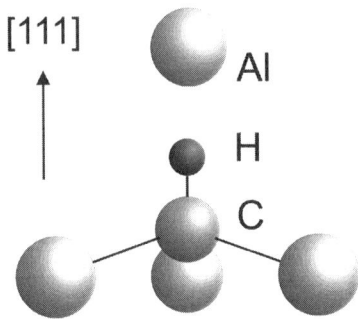

FIG. 14. Bond-centered model of C-H complexes in AlAs.

(CBE) showed absorption peaks at 2558 and 2550 cm^{-1}, respectively, which are attributed to C-H stretch modes (Pritchard et al., 1994a) (Fig. 13). Hydrogen binds directly to the carbon in a BC position, along a [111] axis (Fig. 14). After exposure to a hydrogen or deuterium plasma, additional peaks were observed. Analogous to the case in GaAs, the C-H complexes give rise to four vibrational modes: stretch (A_1), longitudinal (A_1), and transverse (E^- and E^+). For the four isotope combinations (^{12}C-H, ^{13}C-H, ^{12}C-D, ^{13}C-D), all but the E^+ modes have been observed by IR spectroscopy (Pritchard et al., 1994a). Polarized Raman scattering experiments (Wagner et al., 1995b) verified the symmetries of the modes. As with GaAs, ab initio calculations (Jones et al., 1994) agreed with the experimental low-frequency modes but overestimated the frequencies of the stretch modes. The experimental and theoretical frequencies are listed in Table III.

b. *C-H Complexes in AlGaAs*

IR absorption and Raman scattering measurements have been performed on $Al_xGa_{1-x}As$:C that contained C-H complexes that formed during growth or after exposure to a hydrogen plasma (Pritchard et al., 1994b). The ^{12}C-H and ^{12}C-D stretch (A_1) modes provide an excellent probe of the local host environment, with five peaks resulting from sites with zero, one, two, three, or four Al nearest neighbors. The ^{12}C-H stretch modes range in frequency from 2558 cm^{-1} (four Al neighbors) to 2636 cm^{-1} (zero Al neighbors). The longitudinal (X) ^{12}C-H mode, which has A_1 symmetry in GaAs and AlAs, also splits into five peaks in $Al_xGa_{1-x}As$. From the intensities of the peaks, it was determined that hydrogen preferentially occupies a BC position between gallium and carbon atoms, as opposed to a site between aluminum and carbon atoms. This observation was verified

TABLE III
EXPERIMENTAL AND THEORETICAL VALUES FOR LVMs OF CARBON-HYDROGEN COMPLEXES IN AlAs

Mode	Observed frequency[a,b] (cm^{-1})	Theory[c] (cm^{-1})
Stretch (A_1)		
^{12}C-H	2558.1	2885
^{13}C-H	2549.7	2877
^{12}C-D	1902.6	2111
^{13}C-D	1894.4	2100
Longitudinal (A_1)		
^{12}C-H	487.0	466
^{13}C-H	477.2	453
^{12}C-D	479.8	454
^{13}C-D	471.2	442
Transverse (E^-)		
^{12}C-H	670.8	740
^{13}C-H	652.9	725
^{12}C-D	656.6	684
^{13}C-D	635.3	662
Transverse (E^+)		
^{12}C-H	ND	559
^{13}C-H	ND	551
^{12}C-D	ND	437
^{13}C-D	ND	436

[a] Pritchard et al., 1994a.
[b] Wagner et al., 1995b.
[c] Jones et al., 1994.

by *ab initio* calculations (Pritchard *et al.*, 1994b) that showed that except for the case of four Al nearest neighbors, the C_{As}-H-Ga configuration has a total energy ~ 0.24 eV lower than the C_{As}-H-Al configuration. The calculations showed that the C_{As}-Ga bonds at the isolated carbon site are ~ 0.006 Å longer than the C_{As}-Al bonds and therefore can more easily accommodate a hydrogen atom.

The fact that hydrogen is incorporated preferentially between carbon and gallium atoms was used in a novel study of carbon diffusion in $Al_xGa_{1-x}As$ (Zhou *et al.*, 1997). $Al_xGa_{1-x}As$:C samples grown by MOMBE and annealed in hydrogen gas at 450°C for 1 hour showed IR absorption peaks corresponding to ^{12}C-H stretch modes with different numbers of Al nearest neighbors, as discussed earlier. Peaks arising from $AlGa_2C_{As}$-H-Ga, Al_2GaC_{As}-H-Ga, and Al_3C_{As}-H-Ga complexes exhibited a strong polarization dependence, suggesting a growth-induced alignment of the Al atoms along a specific [110] direction. To study carbon diffusion, $Al_xGa_{1-x}As$:C samples were annealed in N_2 at temperatures ranging from 500–625°C and then annealed in H_2 in order to re-form the C-H complexes. These annealed samples showed a loss in the dichroism of the stretch-mode peaks, indicating a diffusion jump of the C_{As} that destroyed the alignment of the complex. By measuring the rate of the dichroism loss, Zhou et al. (1997) obtained a carbon diffusion activation energy of 2.87 eV, in good agreement with previous measurements performed at higher temperatures.

c. Si—H Complexes

To date, the only observed donor-hydrogen complexes in AlAs are Si_{Al}—H complexes, which have hydrogen stretch- and wag-mode frequencies of 1609 and 891 cm^{-1}, respectively (Tuncel et al., 1992). These frequencies are similar to those of Si_{Ga}—H complexes in GaAs, which have stretch and wag modes at 1717 and 897 cm^{-1}, respectively (Chevallier et al., 1991). The Si-H stretch-mode peaks in GaAs and AlAs are plotted for several different temperatures in Fig. 15. As in the case of GaAs, the hydrogen is believed to attach directly to the Si donor, in an antibonding orientation along the [111] direction (Fig. 16). First-principles calculations (Pavesi, 1992) showed that the antibonding orientation minimizes the total energy, with the silicon relaxing toward the plane of arsenic atoms. The temperature dependence of the stretch and wag modes is well described by a model in which the Si—H complex interacts with a single phonon (see Sec. II.1.f) (Tuncel and Sigg, 1993).

d. Si—H Complexes in AlGaAs

In hydrogenated $Al_{0.2}Ga_{0.8}As$:Si alloys, the Si—H wag mode splits into several peaks lying between 878 to 912 cm^{-1} (Chevallier et al., 1991). The splitting arises from different configurations of the Al atoms, which are next-nearest neighbors to the Si atom. The stretch mode is presumably too broad to observe. Using SIMS, Chevallier et al. (1992) measured the deuterium concentration profiles for $Al_xGa_{1-x}As$:Si alloys exposed to a deuterium plasma. For $x < 0.055$, the diffusion profile closely resembles a complementary error function, suggesting little interaction with the Si donors. For $x \geq 0.055$, however, the diffusion profile matches the Si doping level, suggesting the formation of Si—H pairs. From these results, the authors concluded that hydrogen in GaAs has an acceptor level slightly above the conduction-band minimum. For $Al_xGa_{1-x}As$:Si alloys with

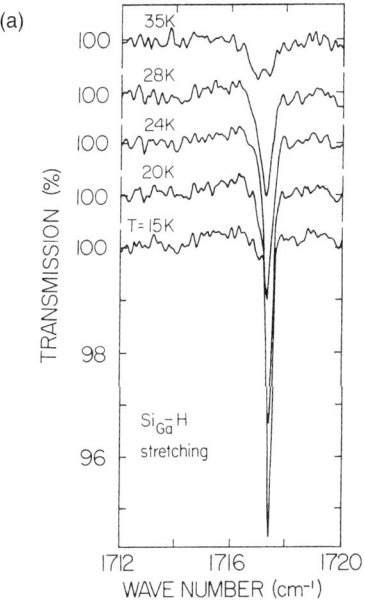

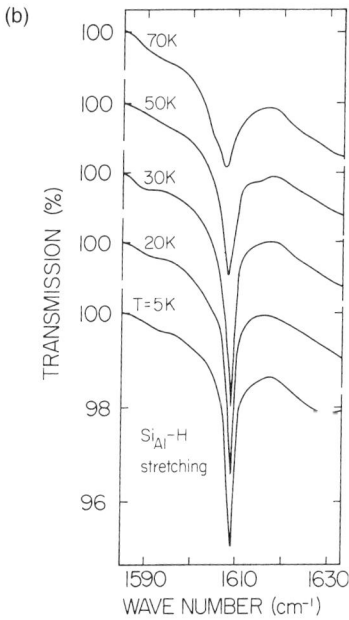

FIG. 15. Infrared absorption spectra of Si—H stretch modes in (a) GaAs and (b) AlAs. (From Tuncel and Sigg, 1993.)

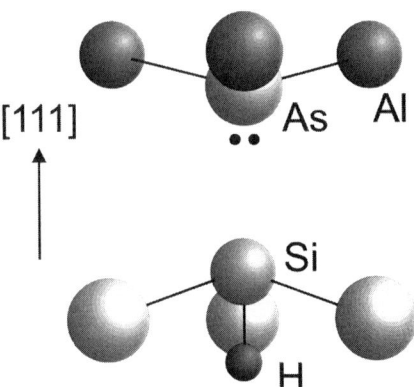

FIG. 16. Antibonding model of Si—H complexes in AlAs.

$x \geq 0.055$, the conduction-band minimum is high enough in energy that the hydrogen acceptor level lies within the bandgap.

3. HYDROGEN IN InP

a. *Group II Acceptor-Hydrogen Complexes*

Zn-H complexes were observed in bulk samples grown by the liquid encapsulated Czochralski (LEC) method (Chevallier et al., 1991), with the primary source of hydrogen contamination originating from wet boric oxide encapsulant. Recently, the stretch-mode frequencies of Mg-H and Mg-D complexes have been found to be 2366 and 1720 cm^{-1}, respectively (Bouanani-Rahbi et al., 1997). The isotopic frequency ratio is given by $r = v_H/v_D = 1.3756$. An extensive study of LVMs arising from group II acceptor-hydrogen complexes in InP was performed by Darwich et al. (1993). InP:Be, InP:Zn, and InP:Cd samples exposed to hydrogen plasma showed sharp stretch-mode absorption peaks at 2236, 2288, and 2332 cm^{-1}, respectively. In addition to these fundamental transitions, overtones ($N = 0 \rightarrow N = 2$) were also observed. The integrated peak areas of the overtones are less than 1% those of the fundamental modes. The overtone frequencies are slightly less than twice the fundamental frequency. The frequencies, widths, and intensity ratios are listed in Table IV.

The ratio of the overtone to fundamental frequency provides a measure

TABLE IV
FREQUENCIES AND WIDTHS OF HYDROGEN STRETCH-MODE FUNDAMENTALS (v_1) AND OVERTONES (v_2)

Complex	v_1 (cm^{-1})	FWHM (cm^{-1})	v_2 (cm^{-1})	FWHM (cm^{-1})	I_2/I_1	x_e
InP:Be,H	2236.49	0.43	ND			
InP:Be,D	1630.85	0.2	ND			
InP:Zn,H	2287.71	0.23	4487.80	0.56	0.009	0.0184
InP:Zn,D	1664.52	0.08	3283.34	0.14	0.007	0.0134
InP:Cd,H	2332.42	0.12	4580.20	1.0	0.006	0.0175
InP:Cd,D	1695.40	0.10	ND			

Note: ND = not discovered. I_2/I_1 is the overtone-fundamental intensity ratio, and x_e is the anharmonicity parameter.
From Darwich *et al.*, 1993.

of the anharmonicity of the hydrogen potential. The Morse potential (Morse, 1929), for example, is given by

$$V(z) = D_e[\exp(-\beta z) - 1]^2 \quad (3)$$

where D_e is the binding energy. For small z, the Morse potential approximates a harmonic potential, with a spring constant $k = 2D_e\beta^2$. The energy eigenvalues of the Morse potential are given by

$$E_n = \hbar\omega_e(n + \tfrac{1}{2})[1 - x_e(n + \tfrac{1}{2})] \quad (4)$$

where

$$\omega_e = \beta(2D_e/\mu)^{1/2} \quad (5)$$

and

$$\omega_e x_e = \frac{\hbar\beta^2}{2\mu} \quad (6)$$

The anharmonicity x_e is given by

$$x_e = \frac{2v_1 - v_2}{2(3v_1 - v_2)} \quad (7)$$

where v_1 and v_2 are the fundamental and overtone frequencies, respectively.

From Eq. (4) it can be seen that the overtone-to-fundamental-frequency ratio is less than 2.

Using Eq. (7), Darwich et al. (1993) experimentally derived x_e for acceptor-hydrogen complexes in InP. They found that x_e decreases as the acceptor mass increases from Zn to Cd (Table IV). This decrease in the anharmonicity is attributed to the greater interaction between hydrogen and the acceptor. In Section II.4.b we discuss this interaction in terms of bond compression.

The anharmonicity also affects the isotopic frequency ratio $r = \omega_H/\omega_D$. As pointed out by Newman (1990), the low r values of Sn-H and Si-H complexes in GaAs are more a result of anharmonicity than of the finite mass of the donor. The hydrogen has a larger vibrational amplitude than the deuterium, and its wave function samples more of the anharmonic part of the potential. From Eq. (4), the first excited state is given by

$$\Delta E = E_1 - E_0 = \hbar\omega_e - 2\hbar\omega_e x_e \tag{8}$$

The anharmonic term x_e is inversely proportional to the square root of the reduced mass. Therefore, the anharmonic term is greater for hydrogen than for deuterium, and the isotopic frequency ratio $r = \omega_H/\omega_D$ is reduced.

Similar to the case of GaAs, in group II acceptor-hydrogen complexes in InP, the hydrogen attaches to a phosphorus atom in a BC orientation, along the [111] direction adjacent to the acceptor (Fig. 17). This model is strongly supported by the following observations: First, the vibrational frequencies are similar to the P-H stretch mode of phosphine (2328 cm^{-1}) but much higher than, for example, the Zn-H bond-stretching frequency of 1600 cm^{-1}.

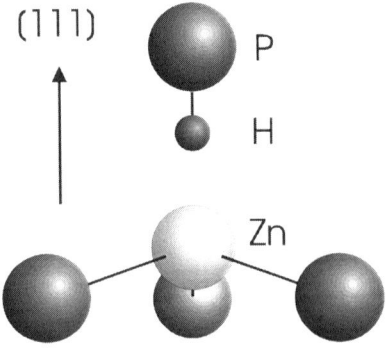

FIG. 17. Bond-centered model of group II acceptor-hydrogen complexes in InP and GaP.

Second, the vibrational frequencies vary by $\sim 100\,\text{cm}^{-1}$ from Be to Cd, indicating a fairly strong coupling to the acceptor. Third, polarized IR spectroscopy on samples under uniaxial stress verified the trigonal (C_{3v}) symmetry of the BC model (Darwich et al., 1993). Finally, ab initio local density cluster calculations (Ewels et al., 1996) showed that the BC orientation minimizes the total energy. The theoretical estimate of the InP:Be,H stretch mode is $2288\,\text{cm}^{-1}$, which is fairly close to the experimental value of $2237\,\text{cm}^{-1}$. The calculations showed that the Be relaxes so that it is coplanar with the plane of three neighboring P atoms.

Perturbed angular correlation (PAC) experiments have been performed on undoped InP implanted with ^{111m}Cd ions ($t_{1/2} = 48\,\text{min}$) with an energy of 60 keV and a dose of $1-3 \times 10^{12}\,\text{cm}^{-2}$ (Forkel-Wirth et al., 1995). After annealing at a temperature of 850°C for 10 sec, approximately 80% of the probe Cd atoms reside in a substitutional site with tetrahedral (T_d) symmetry. Electrically activated samples that were subsequently exposed to a hydrogen plasma showed Cd-H complexes with trigonal (C_{3v}) symmetry, in agreement with the BC model discussed earlier.

b. Vacancy-Hydrogen Complexes

In LEC-grown InP, hydrogen-decorated In vacancies ($V_{In}H_4$) give rise to an absorption peak at $2315\,\text{cm}^{-1}$. Polarized IR spectroscopy on samples under uniaxial stress verified the tetrahedral (T_d) symmetry of this complex (Darwich et al., 1993). InP samples implanted with protons or deuterons showed peaks at 2315 and $1683\,\text{cm}^{-1}$, respectively (Fischer et al., 1992), lending support to the claim that the peak at $2315\,\text{cm}^{-1}$ is due to a vacancy-hydrogen complex as opposed to a dopant-hydrogen complex. Zach (1994) observed that the peak at $2315\,\text{cm}^{-1}$ is correlated with the donor concentration in InP:Fe. The four hydrogen atoms in the $V_{In}H_4$ complex contribute a total of four electrons to the complex (Fig. 18). Since they effectively replace one group III atom, there exists an extra electron and therefore a donor level. Ab initio local density cluster calculations (Ewels et al., 1996) showed that the $V_{In}H_4$ donor level lies in the upper part of the bandgap and can compensate Fe^{3+} donors. The calculations predicted a P-H stretch-mode frequency of $2388\,\text{cm}^{-1}$, in good agreement with the observed frequency of $2315\,\text{cm}^{-1}$.

The $V_{In}H_4$ model received further support from IR absorption measurements performed on InP material grown with heavy water (D_2O) added to the boric oxide encapsulant (Bliss et al., 1995; Zach et al., 1996). The deuterated InP showed P-D stretch modes at 1684.5 and $1686.3\,\text{cm}^{-1}$ and

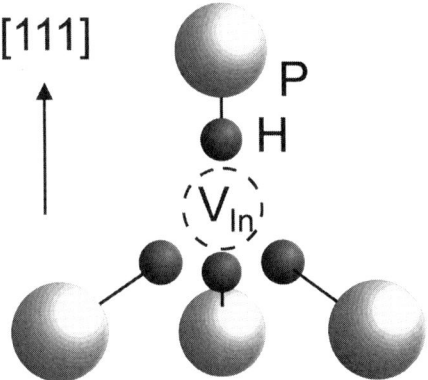

FIG. 18. Model of the hydrogen-decorated indium vacancy in InP.

P-H stretch modes at 2315.6, 2316.1, and 2316.6 cm^{-1} (Fig. 19). The splittings of the P-D and P-H modes arise from vacancies that contain both hydrogen and deuterium. The observation of these splittings clearly established that hydrogen-decorated vacancies in InP contain multiple hydrogen atoms.

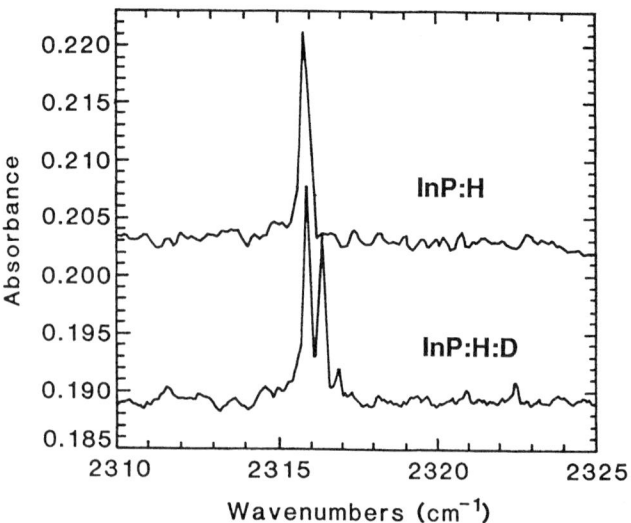

FIG. 19. Infrared absorption spectra of InP crystals grown with and without deuterium doping. (From Zach et al., 1996.)

c. Temperature Dependence of LVMs

As discussed in Section II.1.f, the temperature dependence the LVM frequencies of acceptor-hydrogen complexes in InP can be described by an interaction between the LVM and a single-phonon mode. The experimental data and fits to the single-phonon model are shown in Fig. 20. The temperature-dependent shifts and broadenings are greater for Cd-H complexes than for Zn-H complexes (Darwich et al., 1994), due to the fact that the interaction between Cd and H is stronger than the interaction between Zn and H (see Sec. II.4.b). The stronger interaction leads to an increased coupling to the acoustic phonon modes, which causes the LVM to broaden and shift to lower frequencies with increasing temperature.

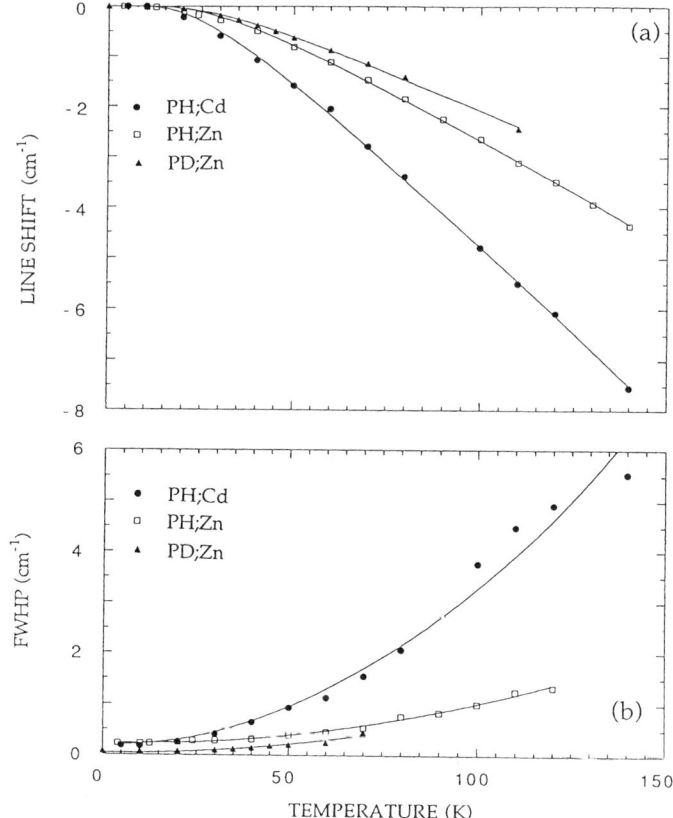

FIG. 20. Temperature dependence of hydrogen stretch mode (a) frequencies and (b) line widths of acceptor-hydrogen complexes in InP. (From Darwich et al., 1994.)

d. C-H Complexes

InP:C epilayers grown by MOCVD at a low growth temperature (500°C) and doped using a CCl_4 source are semi-insulating and contain concentrations of carbon and hydrogen in the range of 10^{18}–10^{20} cm^{-3} (Gardener et al., 1995). To explain the high resistivity of this material, Davidson et al. (1998) have proposed that hydrogen passivates carbon acceptors during growth. An IR peak was observed at 2703 cm^{-1} and is attributed to the C_P-H stretch mode. Lower-frequency modes were observed at 413.5 and 521 cm^{-1}. By comparison with C-H complexes in GaAs (Sec. II.1.a) and other semiconductors, the lower-frequency modes were assigned A_1 and E symmetries, respectively.

4. Hydrogen in GaP

a. C-H Complexes

As in many III-V semiconductors, carbon is also a common p-type dopant in GaP. In GaP samples grown by the LEC technique, hydrogen from the wet boric oxide encapsulant can passivate carbon acceptors (Clerjaud et al., 1991), forming neutral C-H complexes. The ^{12}C-H and ^{13}C-H stretch-mode frequencies are 2660.2 and 2652.6 cm^{-1}, respectively (Fig. 21). The ratio of intensities of the ^{12}C-H to ^{13}C-H stretch modes matches the ratio of the natural abundances of ^{12}C and ^{13}C, suggesting that the hydrogen attaches directly to the carbon. In addition, the shift in the vibrational frequency can be explained by a diatomic C-H model. Since these frequencies are similar to those found in GaAs (see Sec. II.1.a), it is assumed that in GaP the hydrogen also resides in a BC position. By adding heavy water (D_2O) to the boric oxide, Clerjaud et al. formed ^{12}C-D complexes with a stretch mode at 1980.8 cm^{-1}, clearly establishing that wet boric oxide is a source of hydrogen contamination in LEC-grown GaP. To date, no low-frequency C-H modes have been reported in GaP.

b. Group II Acceptor-Hydrogen Complexes

Hydrogenated and deuterated GaP:Be samples showed IR absorption peaks at 2292.2 and 1669.8 cm^{-1}, respectively, at a temperature of 10 K (Fig. 22a) (McCluskey et al., 1995a). The isotopic ratio of these frequencies is $r = 1.3727$. Neither peak was seen in GaP:Be that was not H- or D-plasma exposed. These values are similar to the corresponding values in InP:Be, which has a P-H bond-stretching mode at 2236.5 cm^{-1} and isotopic

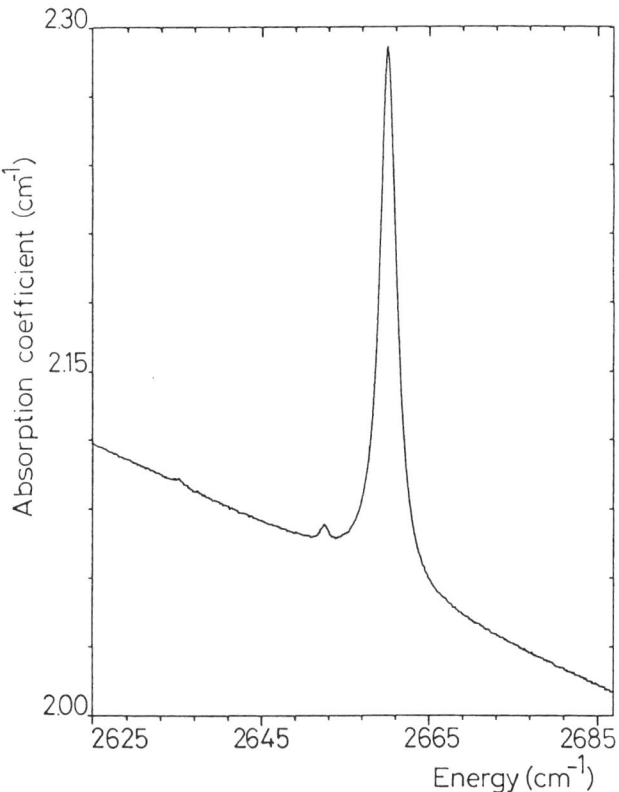

FIG. 21. Infrared absorption spectrum of ^{12}C-H (large peak) and ^{13}C-H (small peak) stretch modes in GaP. (From Clerjaud et al., 1991.)

frequency ratio $r = 1.3714$. It is therefore assumed that the absorption peaks arise from a P-H complex, oriented in a BC direction, adjacent to the beryllium acceptor.

Hydrogenated and deuterated GaP:Zn samples gave rise to IR absorption peaks at 2379.0 and 1729.4 cm^{-1}, respectively (Fig. 22b) (McCluskey et al., 1994a). The isotopic ratio of these frequencies, $r = v_H/v_D$, is 1.3756. By way of comparison, hydrogenated InP:Zn has a bond-stretching mode at 2287.7 cm^{-1} and an isotopic frequency ratio $r = 1.3744$ (Darwich et al., 1993). The bond-stretching mode has been attributed to a P-H complex oriented along a [111] BC direction, adjacent to the zinc acceptor, with the zinc relaxed into the plane of the phosphorus atoms (Fig. 17). Since the LVMs and the r factor for GaP:Zn are similar to the corresponding values for InP:Zn, it is assumed that the structures are the same. The P-H model

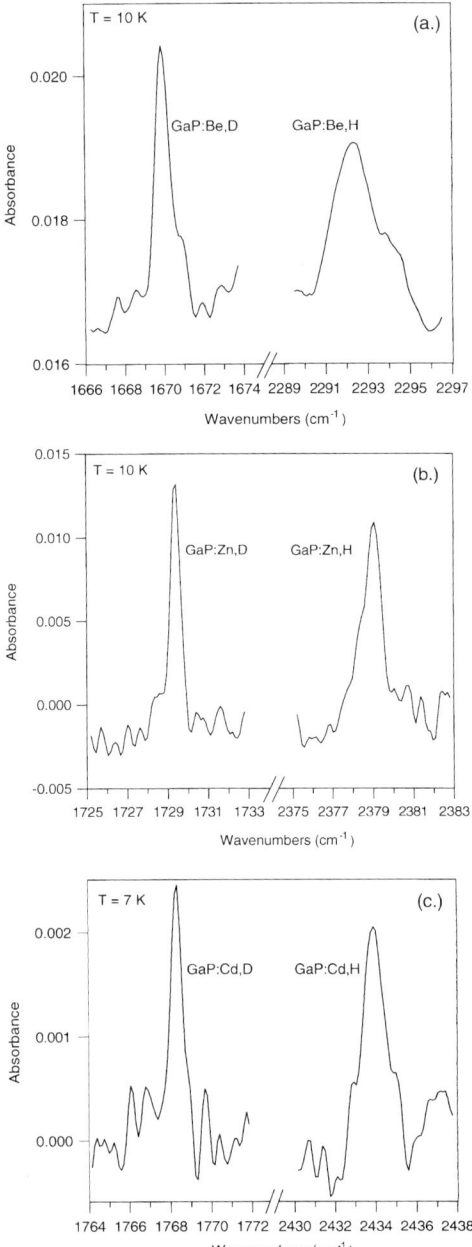

FIG. 22. Infrared absorption spectra of hydrogenated and deuterated (a) GaP:Be, (b) GaP:Zn, and (c) GaP:Cd. (From McCluskey *et al.*, 1995b.)

receives further support from the observation that the Zn-H bond-stretching frequency is 1600 cm^{-1}, far lower than the P-H bond-stretching mode of phosphine, which is 2328 cm^{-1}.

Hydrogenated and deuterated GaP:Cd samples showed IR absorption peaks at 2434.0 and 1768.3 cm^{-1}, respectively, at a temperature of 7 K (Fig. 22c) (McCluskey et al., 1995b). The isotopic ratio of these frequencies is $r = 1.3765$. Once again, these values are similar to the corresponding values for InP:Cd, which has a P-H bond-stretching mode at 2332.4 cm^{-1} and isotopic frequency ratio $r = 1.3757$. It therefore appears that for all group II acceptor-hydrogen complexes in GaAs (Sec. II.1.c), InP (Sec. II.3.a), and GaP, the hydrogen binds to the host anion in a [111] BC orientation.

Perturbed angular correlation (PAC) experiments have been performed on undoped InP implanted with ^{111m}Cd ions ($t_{1/2} = 48$ min) with an energy of 60 keV and a dose of $1-3 \times 10^{12}$ cm^{-2} (Forkel-Wirth et al., 1995). After annealing at a temperature of 1000°C for 10 sec, approximately 80% of the probe Cd atoms resided in a substitutional site with tetrahedral (T_d) symmetry. Electrically activated samples that were subsequently exposed to a hydrogen plasma showed Cd-H complexes with trigonal (C_{3v}), in agreement with the BC model discussed earlier.

The positions and FWHM of the observed IR absorption peaks are listed in Table V. The FWHMs of the P-D peaks are smaller than those of the P-H peaks. This narrowing effect has been observed in all group II acceptor-hydrogen complexes in III-V semiconductors and is correlated with the smaller vibrational amplitude of the deuterium as compared with the hydrogen. It has been suggested (Chevallier et al., 1991) that the smaller vibrational amplitude of deuterium leads to a weaker coupling with the lattice and thus an increase in the lifetime. Other hydrogen/deuterium-related complexes do not follow this trend, however, so there is still an open question.

TABLE V

FREQUENCIES AND FWHM OF P-H AND P-D LVM PEAKS IN GROUP II ACCEPTOR-HYDROGEN COMPLEXES IN GaP

	P-H stretch mode		P-D stretch mode		
Compound	Peak (cm^{-1})	FWHM (cm^{-1})	Peak (cm^{-1})	FWHM (cm^{-1})	$r = v_H/v_D$
GaP:Be	2292.2	2.7	1669.8	0.8	1.3727
GaP:Zn	2379.0	1.1	1729.4	0.5	1.3756
GaP:Cd	2434.0	1.2	1768.3	0.6	1.3765

In addition to the change in line width, several trends are immediately apparent. First, the frequencies of the P-H modes in GaP are higher than the corresponding P-H modes in InP (Fig. 23a). This is due to the fact that GaP has a smaller lattice constant than InP. Second, as the atomic number of the group II acceptor increases, the frequency of the P-H mode increases. The significant upward shift in frequency is evidence for hydrogen residing in a BC rather than an antibonding position. As the acceptor atomic number increases from Be to Cd, the acceptor-hydrogen bond is com-

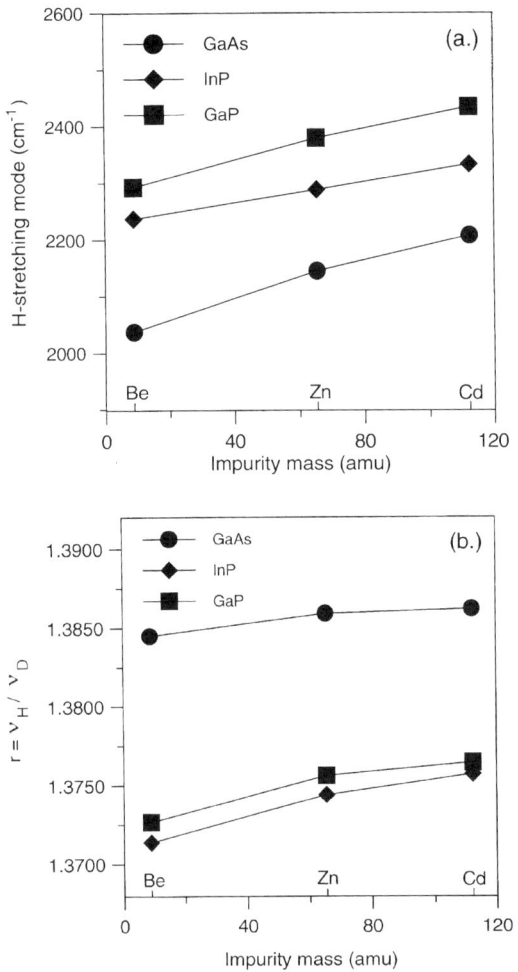

FIG. 23. Plot of (a) frequencies and (b) isotopic frequency ratios of group II acceptor-hydrogen complexes in InP, GaP, and GaAs as a function of acceptor mass. (From McCluskey et al., 1995b.)

pressed, increasing the LVM frequency (see discussion below). Finally, the isotopic ratio $r = v_\text{H}/v_\text{D}$ increases with increasing acceptor mass (Fig. 23b). Darwich *et al.* (1993) noted that the anharmonicity decreases from Be to Cd, resulting in an increase in the value of r (see Sec. II.3.a).

The increase in the hydrogen LVM frequency can be described empirically by considering the equilibrium bond lengths of the diatomic molecules BeH, ZnH, and CdH. The *compression factor* is defined

$$\Delta = d(X\text{-H}) + d(Y\text{-H}) - d_{nn} \quad (9)$$

where X is Be, Zn, or Cd, Y is P or As, and $d(X\text{-H})$ and $d(Y\text{-H})$ are equal to the molecular bond lengths (Table VI). d_{nn} is the nearest-neighbor lattice distance, given by

$$d_{nn} = \frac{\sqrt{3}\,a}{4} \quad (10)$$

where a is the lattice constant. This simple model does not account for the distribution of charge between the X-H and Y-H bonds or the influence of other atoms in the lattice.

Δ is a relative measure of how much the bonds are compressed. As Δ increases, the LVM frequency and r value increase. Figure 24 shows the LVM frequencies and r values as a function of Δ for acceptor-hydrogen complexes in GaAs, GaP, and InP. For GaAs and GaP, the LVMs vary linearly with Δ. The r values for GaP and InP lie on the same curve, since the complexes are so similar. The compression of dopant-hydrogen bonds

TABLE VI

EQUILIBRIUM BOND LENGTHS OF FREE MOLECULES
(ROSEN, 1970) AND SEMICONDUCTORS
(LANDOLT-BÖRNSTEIN, 1987)

Bond	Length (Å)
Free molecules	
P-H	1.42
As-H	1.52
Be-H	1.30
Zn-H	1.59
Cd-H	1.76
Semiconductors	
GaP	2.36
InP	2.54
GaAs	2.45

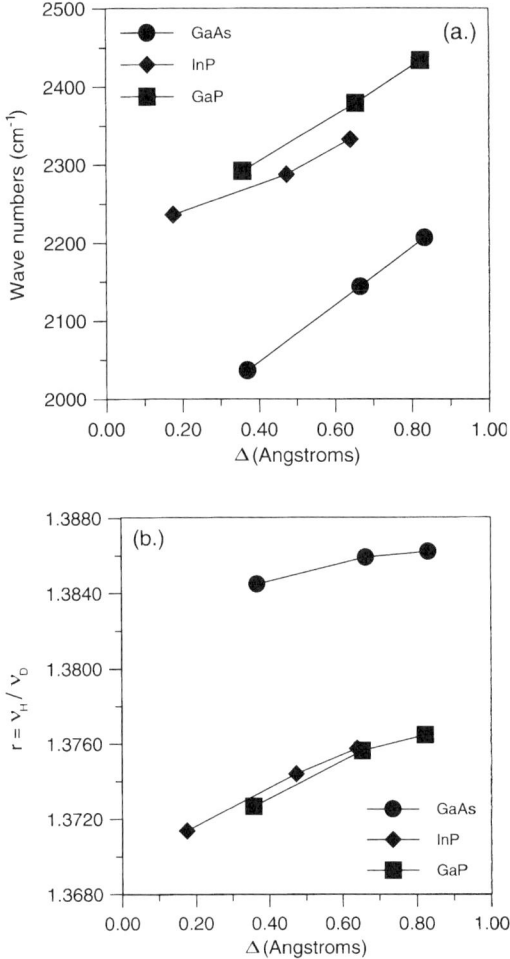

FIG. 24. Plot of (a) frequencies and (b) isotopic frequency ratios of group II acceptor-hydrogen complexes in InP, GaP, and GaAs as a function of bond compression (see text).

also has been studied through the application of hydrostatic pressure (see Sec. II.1.g).

The temperature dependence of the hydrogen-related LVMs in GaP:Be and GaP:Zn was studied between 7 and 150 K (Figs. 25 and 26). The line-width broadening and shift to lower energy with increasing temperature have been observed in numerous systems and are believed to be caused by anharmonic coupling between the localized mode and the extended-lattice

9 Hydrogen in III-V and II-VI Semiconductors

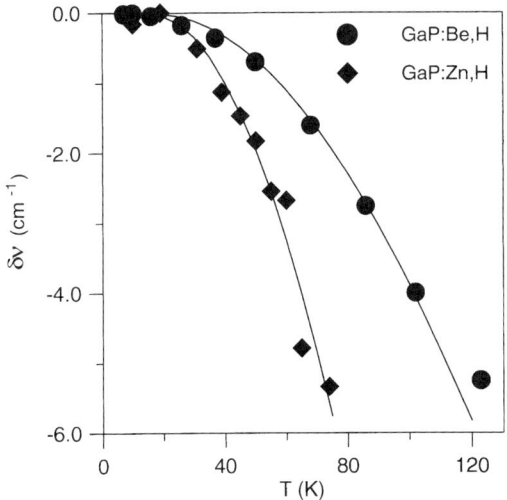

FIG. 25. Temperature-dependent shift in the LVM frequencies of Be-H and Zn-H stretch modes in GaP. The solid lines are fits to Eq. (11). (From McCluskey et al., 1995b.)

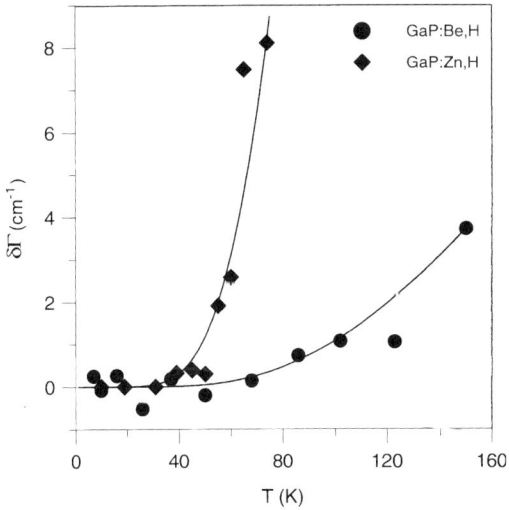

FIG. 26. Temperature-dependent shift in the line widths of Be-H and Zn-H stretch modes in GaP. The solid lines are fits to Eq. (12). (From McCluskey et al., 1995b.)

phonons. The temperature dependence of hydrogen LVMs in GaAs and InP has been explained with a model that assumes that the LVM interacts with a *single*-phonon mode (see Sec. II.1.f). In the present case, however, it is assumed that the LVM interacts with *all* the phonons and does not couple preferentially to any one mode (McCluskey et al., 1995b). Given this assumption, the LVM shift is proportional to the thermal lattice energy $U(T)$ (Elliot et al., 1965):

$$\delta(\hbar\omega) = \frac{\beta}{N_A} U(T) \qquad (11)$$

where $U(T)$ is given in units of energy per mole, N_A is Avagadro's number, and β is a dimensionless constant. Roughly speaking, β is the fraction of thermal energy that is transferred to the hydrogen's vibrational motion from its neighboring atoms.

Figure 25 shows the experimental data with fits to Eq. (11). The values of $U(T)$ were obtained by numerically integrating the reported experimental values of the specific heat $C_V(T)$ (Irwin and LaCombe, 1974), neglecting the zero-temperature energy. The data were approximated by linear least-squares fits, with coefficients $\beta_{Be-H} = -0.05$ and $\beta_{Zn-H} = -0.15$. The magnitudes of the shifts suggest that the zinc acceptor has more influence than the beryllium on the LVM frequency.

In the Debye approximation, elastic scattering by acoustical phonons leads to a temperature-dependent line width of the LVM given by

$$\delta\Gamma = \Gamma(T) - \Gamma(0) = A \left(\frac{T}{\theta_C}\right)^7 \int_0^{\theta_C/T} \frac{z^6 e^z}{(e^z - 1)^2} dz \qquad (12)$$

where $k\theta_C/\hbar$ is an effective cutoff frequency and A is an empirical constant. These two parameters have been adjusted to give reasonable fits to the data (Fig. 26). The value of $\theta_C = 400$ K is used, which is physically reasonable, since the Debye temperature for GaP ranges from 300–500 K as the sample is warmed from 10–150 K. Again, it can be seen that the GaP:Zn,H LVM is more sensitive to temperature variations than the GaP:Be,H LVM.

c. *Nitrogen-Dihydrogen Complexes*

GaP:N samples grown by the LEC technique with wet boric oxide encapsulant can form ^{14}N-2H and ^{15}N-2H complexes (Clerjaud et al., 1992b), resulting in strong stretch-mode peaks at 2885.5 and 2879.7 cm^{-1},

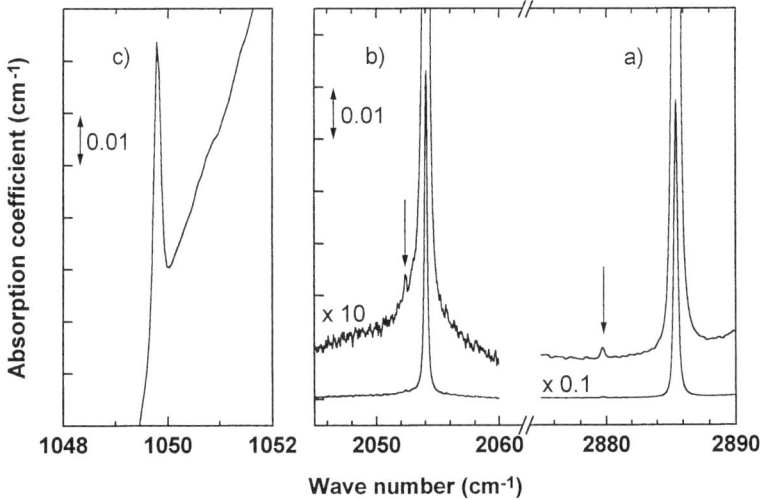

FIG. 27. Infrared absorption spectra of nitrogen-dihydrogen complexes in GaP. The peaks correspond to the (a) antisymmetric stretch mode, (b) symmetric stretch mode, and (c) transverse mode. The two peaks marked by arrows are ^{15}N-2H modes. (From Clerjaud et al., 1996.)

respectively (Fig. 27). The ratio of intensities of the ^{14}N-2H to ^{15}N-2H stretch modes matches the ratio of the natural abundances of ^{14}N and ^{15}N. It has been proposed (Clerjaud et al., 1996) that one hydrogen resides in a BC position and one resides in an antibonding position (Fig. 28). The antisymmetric stretch mode, in which the hydrogen atoms in the H-N-H complex oscillate out of phase with the nitrogen but in phase with each other, gives rise to the peak at 2885.5 cm^{-1}. The symmetric stretch mode, in which the hydrogen atoms oscillate out of phase with each other, gives rise to a weaker peak at 2054.1 cm^{-1}. In a symmetric H-N-H molecule, the symmetric stretch mode is not infrared active. The GaP:N,2H complex has trigonal (C_{3v}) symmetry, however, so the symmetric mode is weakly infrared active. In addition, a transverse mode was observed at 1049.8 cm^{-1}.

Clerjaud et al. (1992b) made the argument that since C-H and N-2H complexes are practically never observed in the same sample, neutral hydrogen atoms form N-2H complexes, whereas protons form C-H complexes. By monitoring the transition from C-H to N-2H complexes as a function of Fermi level, they concluded that the hydrogen donor level H^0/H^+ is located 0.3 eV above the valence-band maximum.

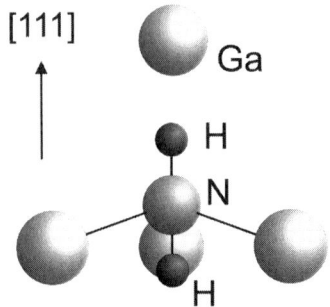

FIG. 28. Model of the nitrogen-dihydrogen complex in GaP.

d. *Vacancy-Hydrogen Complexes*

GaP grown by the LEC technique showed hydrogen-related peaks at 2072.7, 2175.1, 2204.3, 2226.0, 2244.0, and 2245.7 cm^{-1} (Dischler et al., 1991). These peaks are attributed to hydrogen-decorated vacancies with up to four hydrogen atoms.

5. HYDROGEN IN AlSb

a. *DX-Hydrogen Complexes*

In most semiconductors, the DX state of donors is the ground state only when the semiconductor is an alloy (Nelson, 1977; Lang and Logan, 1977) or when hydrostatic pressure is applied (Mizuta et al., 1985). In AlSb, however, the ground states of Se (Becla et al., 1995) and Te (Jost et al., 1994) donors are DX states without alloying or the application of pressure, making them convenient candidates for spectroscopic studies. The first IR spectroscopic study of DX-hydrogen complexes was performed with AlSb (McCluskey et al., 1996a).

Hydrogen can be introduced into a semiconductor sample by boiling in water, electrolysis, implantation, exposure to a hydrogen plasma, or exposure to a hydrogen atmosphere during the growth process (Haller, 1994). It has been demonstrated that annealing in hydrogen passivates shallow acceptors (Veloarisoa et al., 1991; McQuaid et al., 1993) and platinum (Williams et al., 1993) in silicon. Annealing of heavily doped epitaxial GaAs:C layers in a hydrogen ambient was shown to passivate carbon acceptors (Kozuch et al., 1993). McCluskey et al. (1996a) found that annealing bulk AlSb:Se or AlSb:Te in a hydrogen atmosphere at temperatures as low as 700°C followed by rapid quenching leads to the formation

9 Hydrogen in III-V and II-VI Semiconductors

of DX-hydrogen complexes. In addition, hydrogen passivation was achieved by annealing in methanol (CH_3OH) vapor.

AlSb:Se samples that were annealed in H_2 or CH_3OH at 900°C for 1 hour showed IR absorption peaks at 1608.6 and 1615.7 cm^{-1} at a temperature of 10 K (Fig. 29) (McCluskey et al., 1996a). The peaks at 1608.6 and 1615.7 cm^{-1} are attributed to hydrogen stretch modes. Since the bond-stretching mode of the free diatomic molecule AlH is 1624 cm^{-1} (Rosen, 1970), while the hydrogen stretch modes of H_2Se and SbH_3 are 2345 and 1891 cm^{-1}, respectively (Shimanouchi, 1972), the data suggest that the hydrogen binds to an aluminum atom. AlSb:Se samples that were annealed in D_2 or CD_3OD at 900°C for 1 hour showed only one stretch mode at 1173.4 cm^{-1} at a temperature of 10 K. The origin of the anomalous splitting for the hydrogen complexes is discussed in Section II.5.b.

The fundamental transitions of the Se-D and Se-H wag modes were not observed, since the spectral regions where they are expected contain significant phonon absorption features. However, the second, third, and fourth harmonics of the Se-D and Se-H wag modes have been observed (Table VII). The splittings of the peaks are a result of the threefold symmetry of the complex, as explained below. The energy-level spacings are approximately 240 and 330 cm^{-1} for the Se-D and Se-H wag modes, respectively.

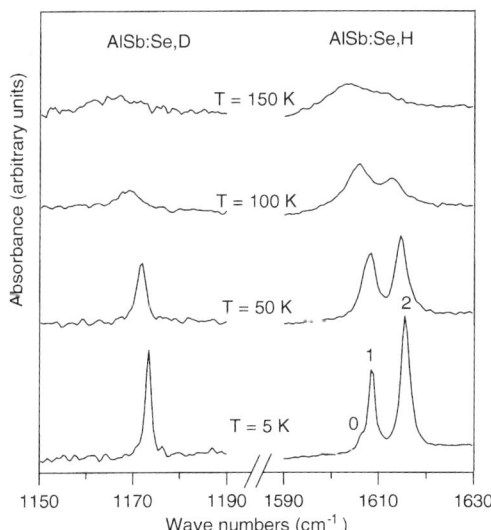

FIG. 29. Infrared absorption spectra of hydrogenated and deuterated AlSb:Se. The anomalous splitting of the Se-H mode is discussed in Section II.5.b. (From McCluskey et al., 1998.)

TABLE VII

THEORETICAL AND EXPERIMENTAL VALUES OF Se-H AND Se-D
WAG MODES IN AlSb

N	Symm.	\multicolumn{4}{c}{$0 \to N$ transition (cm^{-1})}			
		H (theory)	H (expt)	D (theory)	D (expt)
0	Γ_1				
1	Γ_3	342	ND	247	ND
2	Γ_1	666	666	484	478
2	Γ_3	689	692	496	497
3	Γ_1	1040	1032	748	742
3	Γ_2	1040	*	748	*
3	Γ_3	994	993	724	718
4	Γ_1	1304	1316	955	948
4	$\Gamma_3^{(1)}$	1327	1333	967	957
4	$\Gamma_3^{(2)}$	1396	ND	1002	ND
5	Γ_1	1665	ND	1212	ND
5	Γ_2	1665	*	1212	*
5	$\Gamma_3^{(1)}$	1629	ND	1193	ND
5	$\Gamma_3^{(2)}$	1735	ND	1247	ND

*IR inactive.
ND = not discovered.
From McCluskey et al., 1997b.

The splittings of the wag harmonics are consistent with a complex that possesses C_{3v} symmetry. In the plane perpendicular to the [111] axis, the C_{3v} potential is given by (Newman, 1969; Sciacca et al., 1995)

$$V(x, y) = \tfrac{1}{2}k(x^2 + y^2) + B(xy^2 - x^3/3) + C(x^2 + y^2)^2 + \cdots \quad (13)$$

where x and y are parallel to the [1$\bar{1}$0] and [11$\bar{2}$] crystallographic axes, respectively. For simplicity, the wag-stretch coupling terms have been omitted. The anharmonic terms in Eq. (13) lift the threefold degeneracy of the wave functions for $N = n_x + n_y > 1$. The dipole-allowed transitions are the $\Gamma_1 \to \Gamma_1$ and $\Gamma_1 \to \Gamma_3$ transitions. The higher harmonics give rise to weaker peaks, since they require higher-order anharmonic terms in Eq. (13). By adjusting the parameters k, B, and C and using second-order perturbation theory (McCluskey, 1997b) to calculate the energy levels, a good theoretical fit to the wag-mode harmonics was obtained (Table VII).

In addition to LVMs, hydrogenation affects the Se electronic spectrum. At temperatures below 90 K, AlSb:Se exhibits a large photoinduced persistent optical absorption (Becla et al., 1995). When AlSb:Se samples are exposed to light of energy 1 eV or more, the Se donors are transformed from a deep DX state to a metastable hydrogenic state. The hydrogenic absorption spectrum extends from 0.1–1.5 eV and is due to the excitation of the electron from the ground state to the X_1 and X_3 conduction bands (Ahlburn and Ramdas, 1968).

The persistent optical absorption of AlSb:Se samples annealed in a D_2 atmosphere was measured for annealing temperatures ranging from 700–950°C. Although only a fraction of the Se_{DX} centers are transferred into their hydrogenic states under typical illumination, the strength of the photoinduced absorption nonetheless provides a relative measure of the Se_{DX} concentration. The persistent absorption decreased with increasing annealing temperature, while the height of the Se-D stretch mode increased. The correlation between the LVM increase and the persistent absorption decrease indicates that the deuterium passivated a significant fraction of the Se DX centers.

AlSb:Te samples that were annealed in H_2 or D_2 atmospheres at 900°C for 1 hour showed stretch modes at 1599.0 and 1164.4 cm^{-1} and second harmonic wag modes at 665.0 and 478.2 cm^{-1}, respectively (Fig. 30). Like Se, Te also exhibits a DX-like bistability in AlSb (Jost et al., 1994). The fact that the hydrogen stretch and wag modes of AlSb:Se and AlSb:Te have similar vibrational frequencies and r values provides evidence that the hydrogen attaches to an aluminum atom in an antibonding rather than a bonding orientation (Fig. 31). As described in Section II.1.b, the antibonding model also applies to group VI donor-hydrogen complexes in GaAs, as well as group V donors in Si. Since the hydrogen is weakly coupled to the donor, the LVM frequency does not depend strongly on the donor species.

Theoretical studies (Chang et al., 1992) of the DX-hydrogen complex of GaAs:S under hydrostatic pressure suggested that two neutral hydrogen atoms can passivate a positively ionized donor and a negatively ionized DX center, resulting in two neutral complexes. The structures of the "DX-hydrogen" and "donor-hydrogen" complexes are identical, with the hydrogen in an antibonding [111] orientation. The results for DX-hydrogen complexes in AlSb lend further support to this antibonding model.

b. Resonant Interaction Between LVMs and Phonons

The temperature dependence of the Se-H and Se-D stretch modes is shown in Fig. 29. Temperature-dependent broadening causes peaks 0 and 1 to overlap such that they are not resolved for temperatures greater than 40 K. For variable-temperature measurements, therefore, the superposition of peaks 0 and 1 is referred to as "peak 1." As the temperature increases, the area of peak 1 increases, while the area of peak 2 decreases. The sum of the areas remains constant to within the experimental error.

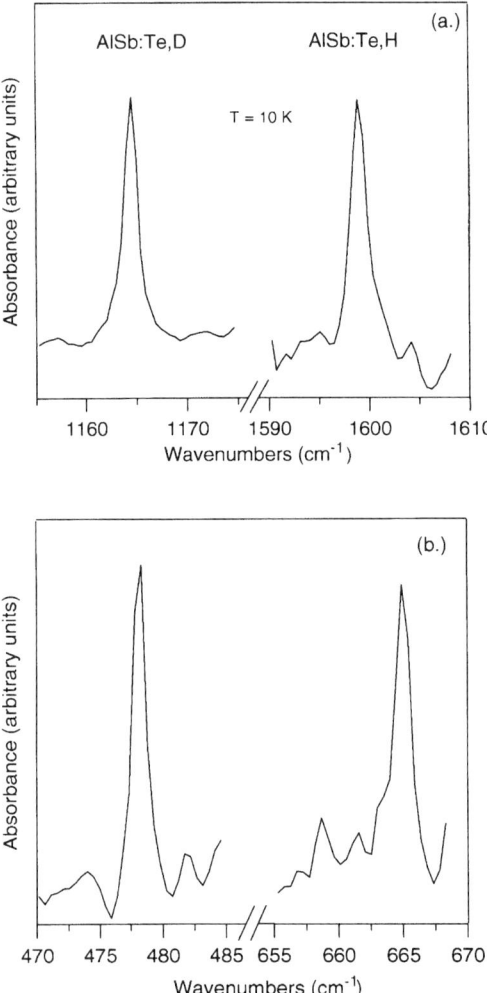

FIG. 30. Infrared absorption spectra of (a) stretch and (b) second harmonic wag modes of Te-H and Te-D complexes in AlSb. (From McCluskey et al., 1996a.)

To explain these observations, a model is proposed in which the Se-H stretch mode of frequency ω_s interacts with a nearly degenerate mode of frequency ω_1 that is the combination of two phonons and a third harmonic of the Se-H wag mode (McCluskey et al., 1998). The calculated AlSb optical phonon density of states has a sharp peak at $\omega_p = 290 \text{ cm}^{-1}$ (Giannozzi et

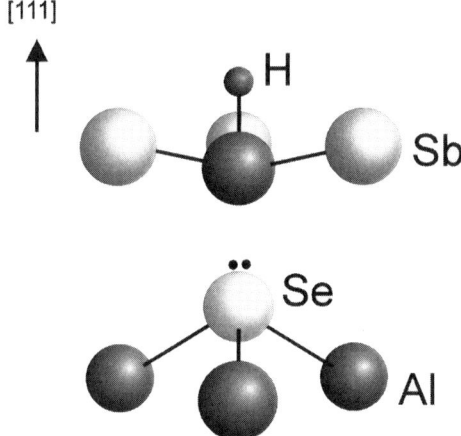

FIG. 31. Antibonding model of DX-hydrogen complexes in AlSb. (From McCluskey *et al.*, 1996a.)

al., 1991), and the experimentally measured frequency of the $N = 3$ (Γ_1) wag-mode harmonic is $\omega_{\text{wag},3} = 1032 \text{ cm}^{-1}$ (Table VII). The combination mode is therefore given by

$$\omega_1 = 2\omega_p + \omega_{\text{wag},3} \sim 1612 \text{ cm}^{-1} \qquad (14)$$

which is nearly degenerate with the Se-H stretch mode. The interaction between the stretch mode and the combination mode is similar to the Fermi resonance between wag modes and stretch modes in donor-hydrogen complexes in silicon (Zheng and Stavola, 1996).

The Hamiltonian of this system is represented by the matrix

$$H = \begin{bmatrix} \omega_s & A \\ A & \omega_1 \end{bmatrix} \qquad (15)$$

where $A = \langle N_s = 1 | H_{\text{int}} | N_p = 2, N_w = 3 \rangle$ and N_s, N_p, and N_w are the stretch-, phonon-, and wag-mode quantum numbers, respectively. The eigenvalues of this Hamiltonian are given by

$$\omega_{\pm} = \tfrac{1}{2}[\omega_s + \omega_1 \pm \sqrt{(\omega_s - \omega_1)^2 + 4A^2}] \qquad (16)$$

The corresponding wave functions are linear combinations of localized and extended lattice phonons:

$$|\psi\rangle = a|N_s = 1\rangle + b|N_p = 2, N_w = 3\rangle \quad (17)$$

The combination mode involves a total of five vibrational quanta and therefore has a negligible optical absorption cross section. The strength of the absorption peaks is determined solely by the stretch-mode contribution to the wave function in Eq. (17), given by

$$|a_\pm|^2 = \frac{A^2}{(\omega_s - \omega_\pm)^2 + A^2} \quad (18)$$

Experimentally, $|a_-|^2$ represents the normalized area of peak 1:

$$|a_-|^2 = A_1/(A_1 + A_2) \quad (19)$$

where A_1 and A_2 are the areas of peaks 1 and 2, respectively.

The temperature dependence of the unperturbed stretch mode is given by

$$\omega_s = 1612.9 - 0.034 U(T) \quad (20)$$

where $U(T)$ is the mean vibrational energy of the lattice in cal/mol, and ω_s is given in cm^{-1}. This temperature dependence is similar to that observed in GaP:Be,H and GaP:Zn,H (see Sec. II.4.b). The frequency of the combination mode can be approximated by a temperature-independent constant

$$\omega_{\text{phonon}} = 1611.1 \text{ cm}^{-1} \quad (21)$$

where the parameters in Eqs. (20) and (21) were adjusted to fit the data. As the temperature increases, the area of peak 1 increases as it becomes more "LVM-like" (Fig. 32a). Conversely, the area of peak 2 decreases as it becomes more "phonon-like." Using a value of $A = 3.45$ cm^{-1}, the temperature dependence of the peak positions (Fig. 32a) as well as the relative absorption strengths of the peaks (Fig. 32b) can be explained.

This model also explains the absence of splitting in the Se-D mode. The small interaction energy of $A = 3.45$ cm^{-1} means that a local mode must lie within a few wave numbers of the combination mode to show an appreciable splitting. The Se-D stretch mode at 1173.4 cm^{-1} is not degenerate with any sharp phonon-wag combination modes. The same is true for the Te-H stretch mode at 1599.0 cm^{-1}, which also does not show a splitting.

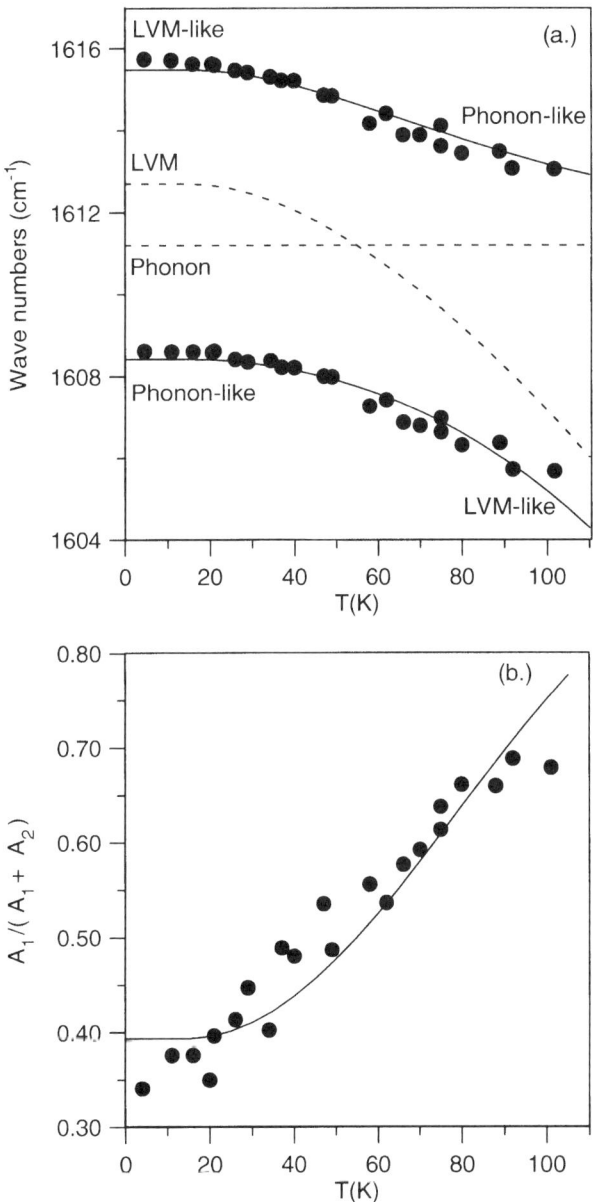

FIG. 32. (a) Se-H stretch mode frequencies as a function of temperature. The dashed lines are the unperturbed stretch and phonon-wag combination modes (Eqs. 20 and 21), and the solid lines are the perturbed modes (Eq. 16). (b) Normalized area of Se-H peak 1 (lower-frequency peak). The solid line is a plot of the theoretical model (Eq. 18). (From McCluskey et al., 1998.)

To further probe the properties of this interaction, hydrostatic pressure was used to change the resonance conditions between the local and extended modes. Varying the pressure has an advantage over varying the temperature in that the lines do not broaden, so all three peaks are resolved. Peak 0, which is negligibly small at ambient pressure, increases rapidly at the expense of peaks 1 and 2 with the application of pressure. At pressures above 4.5 kbar, only peak 0 can be detected. The integrated absorption for all the peaks remains constant to within experimental error.

To explain the existence of three peaks, one must consider an additional mode of frequency ω_2 that is the combination of one phonon and the fourth harmonic of the Se-H wag mode. The experimentally measured frequency of the $N=4$ (Γ_1) wag mode is $\omega_{\text{wag},4} = 1316 \text{ cm}^{-1}$ (Table VII). The combination-mode frequency is given by

$$\omega_2 = \omega_p + \omega_{\text{wag},4} \approx 1606 \text{ cm}^{-1} \qquad (22)$$

The pressure dependence of the peaks can be understood qualitatively as follows: The stretch mode interacts primarily with combination mode 1 and splits into two branches. The low-frequency branch then interacts with phonon 2, with a smaller coupling energy. The anticrossing between the three modes yields three infrared-active peaks at pressures of ~ 2 kbar. For higher pressures, only the lowest branch, peak 0, is "LVM-like."

To estimate the pressure dependence of the phonon-wag combination modes, the pressure dependence of the second harmonic ($N = 0 \rightarrow N = 2$) wag mode was measured. At liquid helium temperatures and atmospheric pressure, this mode has a frequency of 665.5 cm^{-1} (Table VII). As shown in Fig. 33, the pressure dependence of the frequency is linear, with a least-squares fit given by

$$\omega_{\text{wag},2} = 665.5 + 0.64P \qquad (23)$$

where P is the pressure in kbar and $\omega_{\text{wag},2}$ is in units of cm^{-1}. The second harmonic wag-mode peak was measured at the same time as the stretch-mode peaks. The fact that the stretch mode splits whereas the wag mode does not supports the claim that the splitting of the stretch mode is due to a resonant interaction, as opposed to a second configuration of the complex.

The measured pressure dependence of optical phonons in AlSb is given by $d\omega/dP \approx 0.55$ cm^{-1}/kbar (Ves et al., 1985). Using this value with Eq. (23), the pressure dependence of the combination modes were estimated:

$$\omega_1 = 1611.1 + 2.06P \qquad (24)$$

$$\omega_2 = 1605.6 + 1.83P \qquad (25)$$

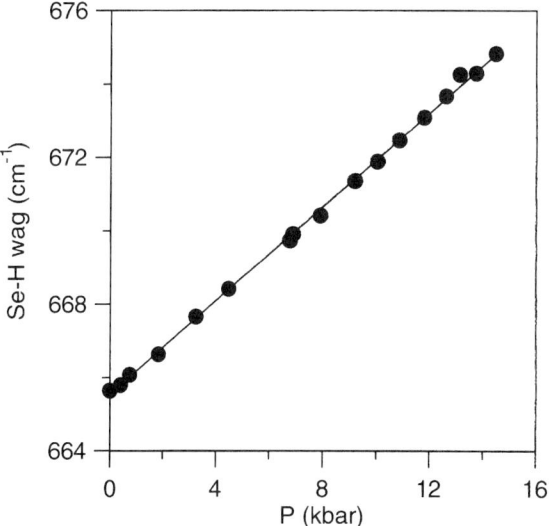

FIG. 33. Se-H second harmonic wag mode as a function of pressure. The solid line is a linear fit given by Eq. (23). (From McCluskey et al., 1998.)

where P is the pressure in kbar, and the frequencies are in units of cm^{-1}. The pressure dependence of the stretch mode was determined by measuring peak 0 at high pressures:

$$\omega_s = 1612.9 + 0.07P \qquad (26)$$

Note that the zero-pressure values are the same as those given in the variable-temperature analysis.

The three-level Hamiltonian is given by

$$H = \begin{bmatrix} \omega_s & A & B \\ A & \omega_1 & 0 \\ B & 0 & \omega_2 \end{bmatrix} \qquad (27)$$

where $A = \langle N_s = 1|H_{\text{int}}|N_p = 2, N_w = 3\rangle$, $B = \langle N_s = 1|H_{\text{int}}|N_p = 1, N_w = 4\rangle$, and for simplicity, the interaction between the combination modes is neglected. The values of the parameters are $A = 3.45$ cm^{-1}, as before, and $B = 0.9$ cm^{-1}. The eigenvalues of the Hamiltonian (Eq. 27) were calculated numerically and show good agreement with experiment (Fig. 34).

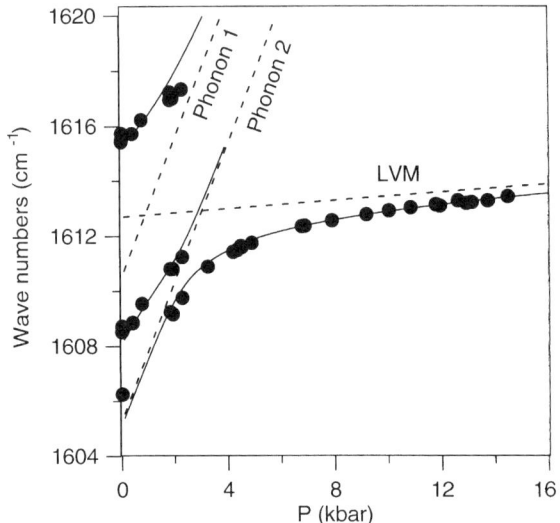

FIG. 34. Se-H stretch mode peaks as a function of pressure. The dashed lines are the unperturbed stretch (Eq. 26) and phonon-wag combination modes (Eqs. 24 and 25), and the solid lines are plots of the three-level theory (Eq. 27). (From McCluskey et al., 1998.)

It is unlikely that the observed anticrossing results from a Fermi resonance between the stretch mode and a wag-mode harmonic. From perturbation theory, the $N = 5$ (Γ_1) wag mode has a predicted frequency of 1665 cm^{-1} (Table VII), which is too far above the stretch mode (1610 cm^{-1}) to strongly interact. In addition, in C_{3v} symmetry there exists only one $N = 5$ (Γ_1) wag mode, while two "unknown" modes are observed. It is possible, however, that the stretch mode resonantly interacts with a local vibrational mode of the Se-H complex that has not yet been discovered.

6. HYDROGEN IN GaN

The development of blue light-emitting diodes (LEDs) (Nakamura et al., 1995) and laser diodes (LDs) (Nakamura et al., 1996) has focused a great deal of research activity on GaN-based III-V nitrides. The bandgaps of $In_xGa_{1-x}N$ alloys cover a wide spectral range, from red (InN) to ultraviolet (GaN), making this alloy system ideal for numerous optoelectronic applications (e.g., Ponce and Bour, 1997).

a. Mg-H Complexes

MOCVD is the dominant growth technique for III-V nitride devices, with Mg the most common p-type dopant. As a result of hydrogen passivation during growth, as-grown GaN:Mg is semi-insulating. It was shown empirically that low-energy electron-beam irradiation (LEEBI) (Amano *et al.*, 1989) or thermal annealing at temperatures above 600°C in N_2 ambient (Nakamura *et al.*, 1992) were required to activate the Mg acceptors. Since thermal annealing had been shown to dissociate acceptor-hydrogen complexes in numerous other semiconductors (e.g., Johnson *et al.*, 1985; Pankove and Johnson, 1991), it was assumed that Mg-H complexes were formed in GaN as well. It required IR spectroscopy, however, to positively identify the Mg-H complexes (Götz *et al.*, 1996). In GaN grown by MBE, the lack of hydrogen enables one to grow p-type GaN:Mg without LEEBI or thermal annealing (Moustakas and Molnar, 1993).

Activated GaN:Mg samples that were exposed to a remote deuterium plasma at a sample temperature of 600°C showed a decrease in the net acceptor concentration by a factor of 2 and an increase in the mobility, indicating passivation rather than compensation (Götz *et al.*, 1995). SIMS profiles show that the deuterium concentration closely tracks the Mg concentration ($\sim 10^{19}\,\mathrm{cm}^{-3}$). For n-type GaN, however, deuterium is not incorporated at concentrations greater than $10^{16}\,\mathrm{cm}^{-3}$. The inability to passivate n-type GaN is attributed to the extremely low diffusion coefficient of H^- in GaN (Neugebauer and Van de Walle, 1996).

First-principles total-energy calculations showed that hydrogen in GaN is a negative-U defect, with $U \approx 2.4\,\mathrm{eV}$ (Neugebauer and Van de Walle, 1995). Unlike the case in GaAs and Si (Van de Walle *et al.*, 1989; Pavesi and Gianozzi, 1992), in which H^+ occupies BC positions, calculations showed that isolated H^+ in GaN prefers an antibonding site near a nitrogen atom (Neugebauer and Van de Walle, 1996). Hydrogen attaches to a nitrogen atom in an antibonding orientation in the Mg-H complex (Fig. 35), in stark contrast with acceptor-hydrogen complexes in nonnitride semiconductors. This difference can be attributed to the ionicity of the Ga-N bond, in which there is no local maximum of the charge density at the bond center. In the strongly covalent Ga-As and Si-Si bonds, on the other hand, there is a maximum in the charge density that attracts the proton to the BC location.

The predicted stretch-mode frequency of hydrogen in the Mg-H complex was $3360\,\mathrm{cm}^{-1}$ (Neugebauer and Van de Walle, 1996), which is similar to that of NH_3 ($3444\,\mathrm{cm}^{-1}$). This prediction contrasted with earlier observations of IR and Raman peaks in GaN:Mg at 2168 and $2219\,\mathrm{cm}^{-1}$ (Brandt

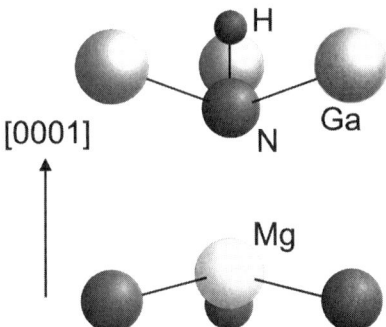

FIG. 35. Antibonding model of the Mg-H complex in GaN.

et al., 1994) that had been attributed to Mg-H LVMs. However, the Mg-D mode was not observed, and the LVMs did not correlate with p-type conductivity, leading the authors to suggest that the modes might arise from other hydrogen-related defects.

The "correct" stretch-mode frequency was observed at $3125\,\text{cm}^{-1}$ in 4-μm-thick epilayers of MOCVD-grown GaN:Mg (Götz et al., 1996). This frequency agrees quite well with the theoretical prediction of $3360\,\text{cm}^{-1}$ and is similar to the stretch-mode frequencies of N-H complexes in GaAs (Sec. II.1.d), GaP (Sec. II.3.c), and ZnSe (Sec. III). On annealing, the peak at $3125\,\text{cm}^{-1}$ decreased by a factor of 2 and was accompanied by an increase in the conductivity (Fig. 36). Annealed samples that were exposed to a remote deuterium plasma showed a deuterium stretch-mode peak at $2321\,\text{cm}^{-1}$. The isotopic frequency ratio is $r = v_H/v_D = 1.346$, which is very similar to that of NH_3 ($r = 1.342$), lending further support to the N-H model. Whether the hydrogen resides in an antibonding or BC position, however, has not yet been determined experimentally.

b. Ca-H Complexes

Although Mg is currently the preferred p-type dopant in GaN, Ca implantation has been shown to yield p-type conductivity (Zolper et al., 1996). Exposure to a hydrogen plasma at a sample temperature of 250°C led to a reduction of the sheet carrier concentration from 1.6×10^{12} to $1.8 \times 10^{11}\,\text{cm}^{-2}$ and an increase in the mobility from 6 to $18\,\text{cm}^2/\text{V} \cdot \text{sec}$ (Lee et al., 1996). The initial carrier concentration was restored after annealing at a temperature of 500°C for 1 min. This behavior strongly suggests Ca-H complex formation, but so far no spectroscopic signatures have been found for such a center.

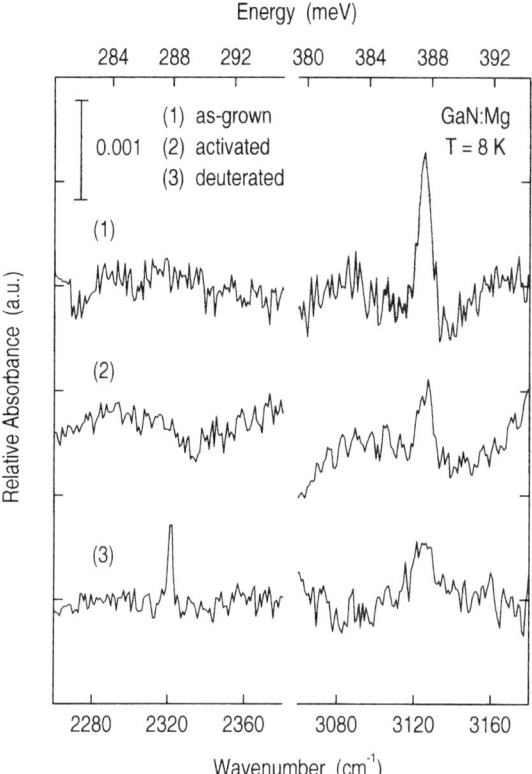

FIG. 36. Infrared absorption spectra of GaN:Mg grown by MOCVD. Spectra are shown for (1) as-grown material, (2) after RTA activation of the p-type conductivity, and (3) after deuteration. (From Götz et al., 1996.)

c. Cd-H Complexes

Perturbed angular correlation (PAC) experiments have been performed on n-type GaN implanted with ^{111m}Cd ions ($t_{1/2} = 48$ min) with an energy of 350 keV and a dose of 3×10^{12} cm^{-2} (Burchard et al., 1997a). After annealing at a temperature of 1030°C for several minutes, approximately 50% of the probe Cd atoms resided in a substitutional site, with threefold symmetry about the wurtzite c axis. Electrically activated samples that were subsequently implanted with low-energy (100-eV) protons showed evidence of two different Cd-H configurations, both of which were aligned along the c axis.

7. HYDROGEN IN OTHER III-V SEMICONDUCTORS

a. *InAs*

Electrical measurements have shown that p-type InAs:Be and InAs:Zn are passivated after exposure to a hydrogen or deuterium plasma (Polyakov et al., 1993). The deuterium concentration profiles measured by SIMS match the acceptor concentrations, providing evidence for the formation acceptor-hydrogen pairs.

Perturbed angular correlation (PAC) experiments have been performed on undoped InAs implanted with ^{111m}Cd ions ($t_{1/2} = 48$ min) with an energy of 60 keV and a dose of 5×10^{11} cm^{-2} (Burchard et al., 1997b). The samples were annealed at a temperature of 650°C for 10 min to activate the Cd acceptors and implanted with low-energy (100-eV) protons. The PAC spectra show evidence of two different complexes, oriented along [111] and [110] directions. The complexes oriented along the [111] directions are attributed to Cd-H pairs, whereas the complexes oriented along the [110] directions may involve an additional defect or more than one hydrogen. The [110]-oriented complexes transmute into [111]-oriented Cd-H pairs, with a time constant of approximately 2 hours at room temperature.

b. *GaSb*

Spreading-resistance and capacitance-voltage measurements have shown that atomic hydrogen passivates shallow acceptors (Zn and Si) and donors (Te) in GaSb (Polyakov et al., 1992). For hydrogenation temperatures below 250°C, the deuterium concentration near the surface is much higher than the concentration of acceptors, suggesting the formation of H_2 molecules.

III. Hydrogen in II-VI Semiconductors

1. HYDROGEN IN ZnSe

The interest in developing blue light-emitting diodes and diode lasers has focused a great deal of research on the growth and doping of wide-bandgap semiconductors. Continuous-wave ZnSe-based laser diodes have been fabricated from epilayers grown by molecular beam epitaxy (MBE), with high p-type doping achieved using a radiofrequency (rf) plasma nitrogen source (Park et al., 1990; Ohkawa et al., 1992). Epilayers grown by metalorganic chemical vapor deposition (MOCVD), however, have proved resistant to p-type doping (Morimoto and Fujino, 1993; Nishimura et al., 1993). In the following two subsections, the role of hydrogen passivation is discussed for the p-type dopants nitrogen and arsenic.

a. N-H Complexes

In ZnSe epilayers grown by photoassisted MOCVD on GaAs substrates and doped with nitrogen from NH_3, SIMS measurements revealed N concentrations above 3×10^{18} cm^{-3}, while capacitance-voltage measurements showed a net acceptor concentration of less than 10^{15} cm^{-3} (Wolk et al., 1993). IR absorption measurements showed peaks at 3193.6 and 783.0 cm^{-1} (Fig. 37), which are attributed to N-H stretch and wag modes, respectively (Wolk et al., 1993). The frequency of the N-H stretch mode is similar to that found in GaAs (see Sec. II.1.d) and is slightly lower than the N-H stretch mode in NH_3 at 3444 cm^{-1}. Polarized Raman spectroscopy was used to show that the symmetry of the N-H complex is C_{3v}, but it is not known whether the hydrogen resides in a BC or an antibonding position (Fig. 38).

Similar results were obtained by Kamata et al. (1993), who investigated N-H complexes in ZnSe epilayers grown by MOCVD on GaAs substrates

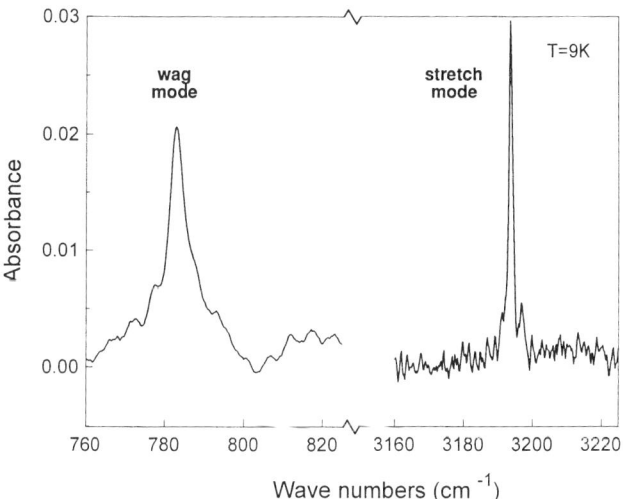

FIG. 37. Infrared absorption spectra of N-H wag and stretch modes in ZnSe. (From Wolk et al., 1993.)

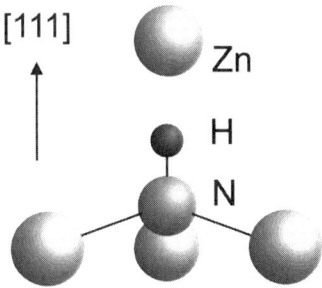

FIG. 38. Bond-centered model for N-H complexes in ZnSe. The hydrogen may, however, reside in the antibonding position.

and doped with nitrogen from NH_3. A SIMS profile clearly shows that the hydrogen and nitrogen concentrations closely match one another at a level of nearly 10^{20} cm^{-1} for a thickness of 3 µm. IR absorption measurements showed a N-H stretch mode at 3193.0 cm^{-1}.

The N-D stretch mode was observed at 2368 cm^{-1} in ZnSe layers grown on GaAs substrates by MBE and doped with nitrogen from an rf plasma atomic nitrogen source (Yu et al., 1996). Atomic hydrogen or deuterium was introduced during the growth by a thermal cracker using hydrogen or deuterium gas in the MBE chamber. In addition to the N-H stretch and wag modes, Yu et al. (1996) observed a N-D stretch mode at 2368 cm^{-1} (Fig. 39). The isotopic frequency ratio is $r = v_H/v_D = 1.348$, which is very similar to that of NH_3 ($r = 1.342$), lending further support to the N-H model.

Yu et al. (1996) also observed the substitutional N_{Se} LVM at 553 cm^{-1}, which had been identified previously by Stein (1994) in N-implanted ZnSe. For identical N concentrations, ZnSe layers grown with hydrogen showed smaller N_{Se} peaks than those grown without hydrogen, indicating passivation. For samples grown without hydrogen, the intensity of the N_{Se} peak increased sublinearly with N concentration, suggesting the formation of other nitrogen complexes such as N-N pairs.

b. *As-H Complexes*

Although arsenic-doped bulk ZnSe has only deep-level photoluminescence peaks (Watts et al., 1971), there is evidence that arsenic has a shallow acceptor level in ZnSe epilayers grown by MBE (Li et al., 1994). The incorporation of hydrogen in arsenic- and nitrogen-doped MOCVD-grown ZnSe has been studied by secondary ion mass spectrometry (SIMS) (Bourret-

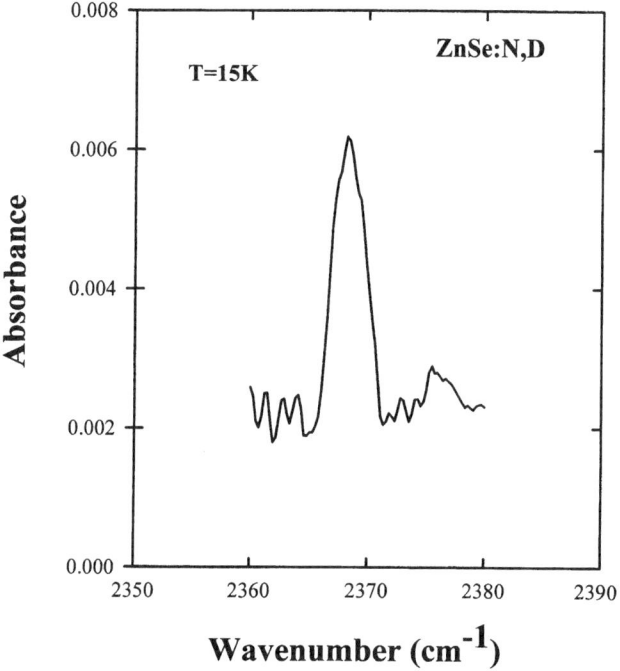

FIG. 39. Infrared absorption spectrum of the N-D stretch mode in ZnSe. (From Yu et al., 1996.)

Courchesne, 1996). The hydrogen incorporation increases strongly when hydrogen is used as a carrier gas instead of nitrogen, as discussed below.

A sample that was grown with hydrogen as a carrier gas showed an IR absorption peak at 2165.6 cm^{-1} at a sample temperature of 7 K (Fig. 40) (McCluskey et al., 1996b). When nitrogen was used as a carrier gas, the same peak was observed, but its area was reduced by a factor of 14, in good agreement with SIMS measurements (Bourret-Courchesne, 1996), which showed that the sample grown with hydrogen had [H] ~ 1.5 × 10^{19} cm^{-3}, while the sample grown with nitrogen had [H] = 1 × 10^{18} cm^{-3}. Under these growth conditions, the hydrogen most likely came from the metal-organic molecules. The sample that was grown with deuterium as a carrier gas showed an absorption peak at 1557.1 cm^{-1}, along with the hydrogen-related peak at 2165.6 cm^{-1} (Fig. 40). The isotopic ratio is $r = v_H/v_D$ = 1.3908. The peak positions, widths, areas, and r values of the LVMs are given in Table VIII. The area of the hydrogen-related peak is approximately 3 times that of the deuterium-related peak. Previous SIMS measurements of

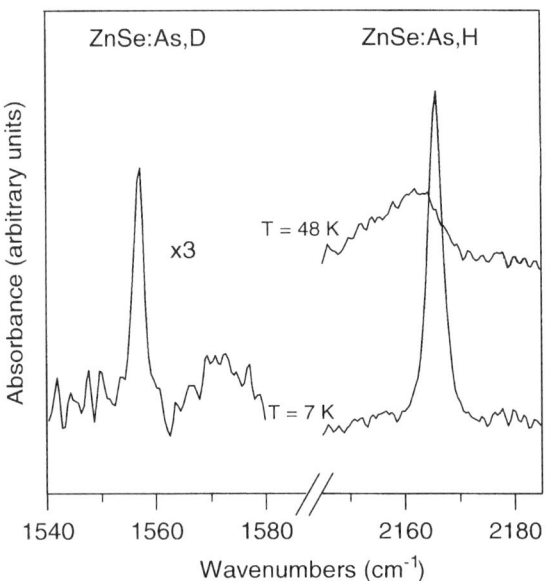

FIG. 40. Infrared absorption spectra of As-H and As-D stretch modes in ZnSe. (From McCluskey et al., 1996b.)

the samples showed $[H] = 6 \times 10^{18}$ cm^{-3} and $[D] = 1 \times 10^{18}$ cm^{-3} (Bourret-Courchesne, 1996). These results indicate that most of the hydrogen incorporation comes from by-products of reactions involving the hydrogen carrier gas and the metalorganic molecules. A sample that was grown at a lower temperature (360°C) contained high concentrations of hydrogen and arsenic ($[H] = 3 \times 10^{20}$ cm^{-3} and $[As] = 1.8 \times 10^{21}$ cm^{-3}) but did not

TABLE VIII

PEAK POSITIONS, WIDTHS, AND ISOTOPIC FREQUENCY RATIOS OF As-H AND As-D LVMs IN GaAs:Zn AND ZnSe:As

Compound	As-H stretch mode		As-D stretch mode		
	Peak (cm^{-1})	FWHM (cm^{-1})	Peak (cm^{-1})	FWHM (cm^{-1})	$r = v_H/v_D$
GaAs:Zn*	2146.9	1.8	1549.1	0.9	1.3860
ZnSe:As	2165.6	2.8	1557.1	1.9	1.3908

*See Chevallier et al., 1991.

show the hydrogen-related peak. At lower growth temperatures, hydrogen may be incorporated in forms that are infrared inactive, such as interstitial H_2 molecules.

The hydrogen bond-stretching-mode frequencies of the free molecules H_2Se, AsH_3, and ZnH are 2345, 2116 (Shimanouchi, 1972), and 1553 cm^{-1} (Rosen, 1970), respectively. Since the frequency of the ZnSe:As,H mode is 2165.6 cm^{-1}, it is likely that the hydrogen binds directly to the arsenic acceptor. In several respects, the As-H complex in ZnSe is similar to the Zn-H complex in GaAs (Chevallier et al., 1991). In GaAs, zinc is an acceptor that occupies a substitutional gallium site. Hydrogen passivates zinc by attaching to a host arsenic atom, in a BC orientation, adjacent to the zinc acceptor. In ZnSe:As the hydrogen attaches to the arsenic acceptor, in a BC orientation, adjacent to the host zinc atom (Fig. 41). The stretch mode of the GaAs:Zn,H complex is 2146.0 cm^{-1} at a temperature of 6 K, and the isotopic ratio is $r = 1.3860$ (Table VIII). The fact that the isotopic ratios and LVM frequencies of the two complexes are very similar lends further support to the BC model.

c. *Temperature Dependence*

The temperature-dependent behavior of the ZnSe:As,H LVM is shown in Figs. 42 and 43. As explained in Section II.4.b, to first order the LVM shift δv is proportional to the lattice thermal energy $U(T)$. The values of $U(T)$

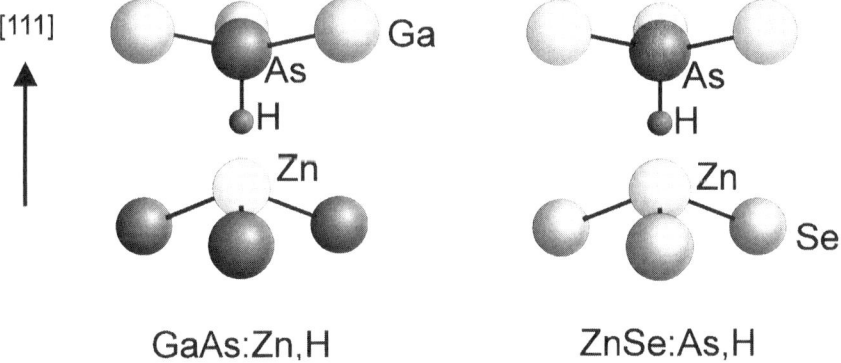

FIG. 41. Bond-centered models for GaAs:Zn,H and ZnSe:As,H complexes. In both examples, the hydrogen resides in a BC location between zinc and arsenic atoms. (From McCluskey et al., 1996b.)

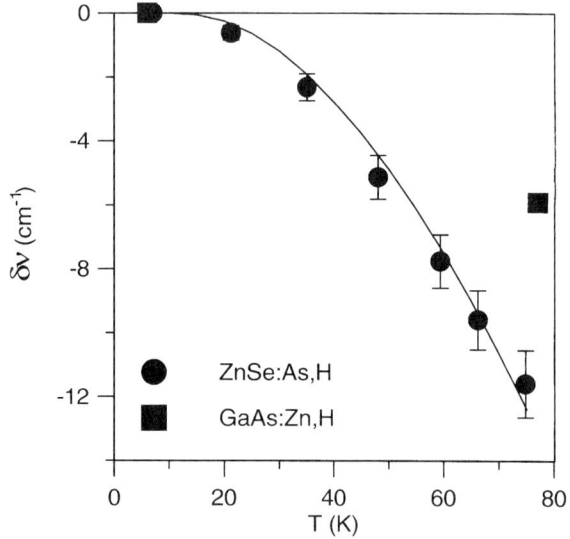

FIG. 42. Temperature-dependent shift in the frequencies of ZnSe:As,H and GaAs:Zn,H stretch modes. The solid line is a fit to Eq. (11). (From McCluskey et al., 1996b.)

are obtained by numerically integrating the experimental values of the specific heat $C_V(T)$ reported by Irwin and LaCombe (1974), neglecting the zero-temperature energy. The data can be approximated by a linear least-squares fit to Eq. (11), with $\beta = -0.17$. The temperature-dependent shift and the fit are shown in Fig. 42. At 77 K, the shift of the ZnSe:As,H mode is approximately twice that of the GaAs:Zn,H mode.

As discussed in Section II.4.b, elastic phonon scattering leads to a temperature-dependent line width

$$\delta\Gamma = \Gamma(T) - \Gamma(0) = A\left(\frac{T}{\theta_C}\right)^7 \int_0^{\theta_C/T} \frac{z^6 e^z}{(e^z - 1)^2} dz \qquad (28)$$

where $k\theta_C/h$ is the effective cutoff frequency, and A is an empirical constant. For high temperatures, Eq. (28) reduces to

$$\delta\Gamma = \alpha T^2 \qquad (29)$$

Elliot et al. (1965) point out that Eq. (29) is a good approximation even when T is a fraction of θ_C. Equation (29) is used to obtain a fit to the data, with $\alpha = 4 \times 10^{-3}\,\text{cm}^{-1}/\text{K}^2$. The temperature-dependent line width and the fit are plotted in Fig. 43.

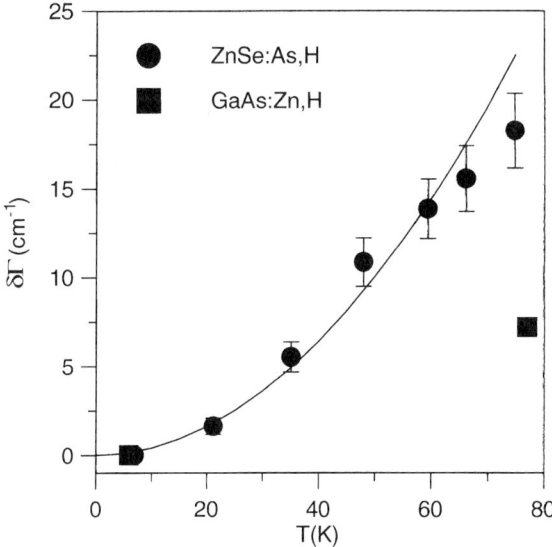

FIG. 43. Temperature-dependent shift in the line widths of ZnSe:As,H and GaAs:Zn,H stretch modes. The solid line is a fit to Eq. (29). (From McCluskey et al., 1996b.)

The ZnSe:As,H mode has a slightly higher frequency, higher r factor, and stronger temperature dependence than the GaAs:Zn,H mode. These observations suggest that the coupling between the zinc and the hydrogen is slightly weaker in GaAs than in ZnSe. The effect of the zinc can be modeled as a repulsive potential that confines the hydrogen atom. The potential increases the frequency and the r factor, the latter because hydrogen has a larger amplitude than deuterium and overlaps the potential more. The temperature-dependent shift of the frequency and line width are caused primarily by coupling between the hydrogen and the thermal motion of the zinc atom. Greater coupling leads to an LVM with a more pronounced temperature dependence. The cause of this greater coupling is currently not understood. It may be related to the fact that ZnSe is more ionic than GaAs.

2. HYDROGEN IN CdTe

It has been shown experimentally that arsenic shallow acceptors are passivated by hydrogen in MOCVD-grown CdTe epilayers (Clerjaud et al., 1993). In CdTe:As epilayers grown on GaAs substrates, SIMS measurements indicated $[H] = 3.5 \times 10^{18}\,\text{cm}^{-3}$ and $[As] = 1 \times 10^{19}\,\text{cm}^{-3}$. A sharp

IR peak was observed at 2022 cm^{-1} and was attributed to an As-H stretch mode. After annealing at a temperature of 550°C for 30 sec, the intensity of the IR peak was reduced to below the detection limit, indicating nearly complete dissociation of As-H complexes. In samples with [As] = 8 × 10^{16} cm^{-3}, nearly 100% of the dopants were activated after annealing (Svob et al., 1996). For more heavily doped samples, a maximum postanneal hole concentration of only 1 × 10^{17} cm^{-3} was attained. The authors suggested that the low hole concentration may be due to the formation of arsenic precipitates or a deep-level configuration of arsenic. The formation of compensating native defects during annealing may also play an important role in suppressing the hole concentration.

IV. Summary and Discussion

In this chapter on hydrogen in III-V and II-VI semiconductors, we have reviewed the properties of hydrogen-related complexes, with an emphasis on the microscopic structure of these defects and their interaction with the host lattice. Infrared and Raman spectroscopy have proved to be extremely useful tools in the study of hydrogen-related complexes. *Ab initio* calculations have provided detailed models as well as checks on the experimental data (and vice versa). With respect to hydrogen, III-V semiconductors have thus far received much more attention than II-VI semiconductors.

The structures of hydrogen-related complexes appear to follow two general trends. In donor-hydrogen complexes, hydrogen resides in an antibonding position, a result of the repulsion of the H$^-$ ion by the high electron density at the bond center. In acceptor-hydrogen complexes, hydrogen resides in a BC position, since it is attracted by the electron density in the covalent bond. GaN, unlike the nonnitride semiconductors, has a strongly ionic bond. For the Mg-H complex, theoretical studies predict that the hydrogen binds to a host nitrogen in an antibonding configuration. Although experimental results show conclusively that hydrogen binds to nitrogen, it is not known whether the hydrogen resides in an antibonding or BC position. One way to settle this question may be through the application of hydrostatic pressure, which should yield a sub- or superlinear LVM pressure derivative for antibonding or BC positions, respectively. Another approach is to measure the stretch modes of other acceptor-hydrogen complexes in GaN to determine the dependence of the LVM frequencies on the acceptor species. Given the technological importance of GaN and its alloys, the study of hydrogen in GaN undoubtedly will prove to be an active area of research.

Hydrogen-related defects interact with the host lattice in a number of ways. Group II acceptor-hydrogen complexes in GaP have higher stretch-mode frequencies than their counterparts in InP, a result of the compression of the P-H bond. In both the GaAs:Zn,H and ZnSe:As,H complexes, hydrogen binds to an arsenic atom in a BC position adjacent to a zinc atom. The subtle differences between these two complexes may result from the different ionicity of GaAs and ZnSe.

Lattice-defect interactions may be probed through variable-pressure or variable-temperature spectroscopy. Variable-pressure spectroscopy is a relatively new method that may help determine the structure of hydrogen-related complexes. We have described two models that explain the temperature-dependent frequency shift and line-width broadening of LVMs as a consequence of anharmonic coupling to thermally populated acoustical phonons. It would be interesting to develop a model to explain not only the magnitude but also the sign of the frequency shift.

A new interaction between LVMs and extended-lattice phonons was discovered in AlSb. The resonant interaction may result from a near degeneracy of the Se-H stretch mode with a wag-phonon combination mode. By varying the temperature or pressure, the resonance conditions were tuned such that a distinct anticrossing was observed. We expect that more examples of resonant LVM-phonon interactions will be found in other materials.

In closing, we can state confidently that hydrogen has remained a topic of great interest and will continue to attract attention, especially in the group III nitrides and their alloys.

ACKNOWLEDGMENTS

We would like to thank B. Clerjaud, W. Götz, N. M. Johnson, R. C. Newman, B. Pajot, M. Stavola, C. G. Van de Walle, W. Walukiewicz, J. Wolk, and P. Y. Yu for many enlightening discussions. We are indebted to all the authors who generously contributed figures for this chapter. This work was supported in part by U.S. NSF Grant DMR-94 17763 and in part by the Director, Office of Energy Research, Office of Basic Energy Sciences, Materials Sciences Division, of the U.S. Department of Energy under Contract No. DE-AC03-76SF00098. The work at Xerox was supported by DARPA under Agreement No. MDA972-96-3-0014.

References

Ahlburn, B. T., and Ramdas, A. K. (1968). *Phys. Rev.*, **167**, 717.
Amano, H., Kito, M., Hiramatsu, K., and Akasaki, I. (1989). *Jpn. J. Appl. Phys.*, **28**, L2112.
Becla, P., Witt, A., Lagowski, J., and Walukiewicz, W. (1995). *Appl. Phys. Lett.*, **67**, 395.
Bliss, D. F., Bryant, G. G., Gabbe, D., Iseler, G., Haller, E. E., and Zach, F. X. (1995). In *Proceedings of the 7th International Conference on InP and Related Materials*. New York: IEEE, p. 678.
Bonapasta, A. Amore. (1993). *Phys. Rev. B*, **48**, 8771.
Bouanani-Rahbi, R., Pajot, B., Ewels, C. P., Öberg, S., Goss, J., Jones, R., Nissim, Y., Theys, B., and Blaauw, C. (1997). In *Shallow-Level Centers in Semiconductors*, C. A. J. Ammerlaan and B. Pajot, eds. Singapore: World Scientific, p. 171.
Bourret-Courchesne, E. D. (1996). *Appl. Phys. Lett.*, **68**, 2481.
Brandt, M. S., Ager, J. W., III, Götz, W., Johnson, N. M., Harris, J. S., Molnar, R. J., and Moustakas, T. D. (1994). *Phys. Rev. B*, **49**, 14758.
Breuer, S. J., Jones, R., Briddon, P. R., and Öberg, S. (1996). *Phys. Rev. B*, **53**, 16289.
Burchard, A., Deicher, M., Forkel-Wirth, D., Haller, E. E., Magerle, R., Prospero, A., Stöltzer, A., and the ISOLDE Collaboration. (1997a). In *Proceedings of the 19th International Conference on Defects in Semiconductors (ICDS-19)*. *Mater. Sci. Forum*, **258-263**, 1099.
Burchard, A., Correia, J. G., Deicher, M., Forkel-Wirth, D., Magerle, R., Prospero, A., Stöltzer, A., and the ISOLDE Collaboration. (1997b). In *Proceedings of the 19th International Conference on Defects in Semiconductors (ICDS-19)*. *Mater. Sci. Forum*, **258-263**, 1223.
Chang, K.-J. (1991). *Solid State Commun.*, **78**, 273.
Chang, K.-J., Cheong, B. H., and Park, C. H. (1992). *Solid State Commun.*, **84**, 1005.
Cheng, Ying, Stavola, M., Abernathy, C. R., Pearton, S. J., and Hobson, W. S. (1994). *Phys. Rev. B*, **49**, 2469.
Chevallier, J., Dautremont-Smith, W. C., Tu, C. W., and Pearton, S. J. (1985). *Appl. Phys. Lett.*, **47**, 108.
Chevallier, J., Clerjaud, B., and Pajot, B. (1991). In *Semiconductors and Semimetals*, **34**, J. I. Pankove and N. M. Johnson, eds. New York: Academic Press.
Chevallier, J., Machayekhi, B., Grattepain, C. M., Rahbi, R., and Theys, B. (1992). *Phys. Rev. B*, **45**, 8803.
Clerjaud, B., Gendron, F., Krause, M., and Ulrici, W. (1990). *Phys. Rev. Lett.*, **65**, 1800.
Clerjaud, B., Côte, D., Hahn, W.-S., and Ulrici, W. (1991). *Appl. Phys. Lett.*, **58**, 1860.
Clerjaud, B., Côte, D., Gendron, F., Hahn, W.-S., Krause, M., Porte, C., and Ulrici, W. (1992a). In *Proceedings of the 16th International Conference on Defects in Semiconductors (ICDS-16)*. *Mater. Sci. Forum*, **83-87**, 563.
Clerjaud, B., Côte, D., Hahn, W.-S., Wasik, D., and Ulrici, W. (1992b). *Appl. Phys. Lett.*, **60**, 2374.
Clerjaud, B., Côte, D., Svob, L., Marfaing, Y., and Druilhe, R. (1993). *Solid State Commun.*, **85**, 167.
Clerjaud, B., Côte, D., Hahn, W.-S., Lebkiri, A., Ulrici, W., and Wasik, D. (1996). *Phys. Rev. Lett.*, **77**, 4930.
Clerjaud, B., Côte, D., Hahn, W.-S., Lebkiri, A., Ulrici, W., and Wasik, D. (1997). *Phys. Stat. Sol. (a)*, **159**, 121.
Darwich, R., Pajot, B., Rose, B., Robein, D., Theys, B., Rahbi, R., Porte, C., and Gendron, F. (1993). *Phys. Rev. B*, **48**, 17776.
Darwich, R., Song, C., Rahbi, R., and Pajot, B. (1994). In *Proceedings of the 17th International Conference on Defects in Semiconductors (ICDS-17)*. *Mater. Sci. Forum*, **143-147**, 927.

Davidson, B. R., Newman, R. C., Bullough, T. J., and Joyce, T. B. (1993). *Phys. Rev. B*, **48**, 17106.
Davidson, B. R., Newman, R. C., Kaneko, T., and Naji, O. (1994). *Phys. Rev. B*, **50**, 12250.
Davidson, B. R., Newman, R. C., and Bachem, K. H. (1995). *Phys. Rev. B*, **52**, 5179.
Davidson, B. R., Newman, R. C., Joyce, T. B., and Bullough, T. J. (1996). *Semicond. Sci. Technol.*, **11**, 455.
Davidson, B. R., Newman, R. C., and Button, C. C. (1998). *Phys. Rev. B*, **58**, 15609.
Dischler, B., Fuchs, F., and Seelewind, H. (1991). *Physica B*, **170**, 245.
Elliot, R. J., Hayes, W., Jones, G. D., MacDonald, H. F., and Sennet, C. T. (1965). *Proc. R. Soc. Lond.*, **A289**, 1.
Ewels, C. P., Öberg, S., Jones, R., Pajot, B., and Briddon, P. R. (1996). *Semicond. Sci. Technol.*, **11**, 502.
Fischer, D. W., Manasreh, M. O., and Matous, G. (1992). *J. Appl. Phys.*, **71**, 4805.
Forkel-Wirth, D., Achtziger, N., Burchard, A., and Correia, J. C. (1995). *Solid State Commun.*, **93**, 425.
Gardener, N. F., Hartmann, Q. J., Baker, J. E., and Stillman, G. E. (1995). *Appl. Phys. Lett.*, **67**, 3004.
Giannozzi, P., de Gironcoli, S., Pavone, P., and Baroni, S. (1991). *Phys. Rev. B*, **43**, 7231.
Götz, W., Johnson, N. M., Walker, J., Bour, D. P., Amano, H., and Akasaki, I. (1995). *Appl. Phys. Lett.*, **67**, 2666.
Götz, W., Johnson, N. M., Bour, D. P., McCluskey, M. D., and Haller, E. E. (1996). *Appl. Phys. Lett.*, **69**, 3725.
Haller, E. E. (1994). In *Handbook on Semiconductors*, **3b**, S. Mahajan, ed. Amsterdam: North-Holland, p. 1515.
Irwin, J. C., and LaCombe, L. (1974). *J. Appl. Phys.*, **45**, 567.
Johnson, N. M., Burnham, R. D., Street, R. A., and Thornton, R. L. (1985). *Phys. Rev. B*, **33**, 1102.
Johnson, N. M., Herring, C., and Chadi, D. J. (1986). *Phys. Rev. Lett.*, **56**, 769.
Jones, R., and Öberg, S. (1991). *Phys. Rev. B*, **44**, 3673.
Jones, R., Goss, J., Ewels, C., and Öberg, S. (1994). *Phys. Rev. B*, **50**, 8378.
Jost, W., Kunzer, M., Kaufmann, U., and Bender, H. (1994). *Phys. Rev. B*, **50**, 4341.
Kamata, A., Mitsuhashi, H., and Fujita, H. (1993). *Appl. Phys. Lett.*, **63**, 3353.
Kozuch, D. M., Stavola, M., Pearton, S. J., Abernathy, C. R., and Lopata, J. (1990). *Appl. Phys. Lett.*, **57**, 2561.
Kozuch, D. M., Stavola, M., Pearton, S. J., Abernathy, C. R., and Hobson, W. S. (1993). *J. Appl. Phys.*, **73**, 3716.
Landolt-Börnstein, New Series III. (1987). In *Semiconductors: Intrinsic Properties of Group IV Elements and III-V, II-VI and I-VII Compounds*, **22a**, O. Madelung, ed. Berlin: Springer-Verlag.
Lang, D. V., and Logan, R. A. (1977). *Phys. Rev. Lett.*, **39**, 635.
Lee, J. W., Pearton, S. J., Zolper, J. C., and Stall, R. A. (1996). *Appl. Phys. Lett.*, **68**, 2102.
Lee, S.-G., and Chang, K. J. (1996). *Phys. Rev. B*, **54**, 8522.
Li, M. Ming, Strachan, D. J., Ritter, T. M., Tamargo, M., and Weinstein, B. A. (1994). *Phys. Rev. B*, **50**, 4385.
McCluskey, M. D. (1997). Ph.D. thesis, University of California, Lawrence Berkeley National Laboratory Report No. LBNL-40451.
McCluskey, M. D., Haller, E. E., Walker, J., and Johnson, N. M. (1994a). *Appl. Phys. Lett.*, **65**, 2191.
McCluskey, M. D., Haller, E. E., Walker, J., and Johnson, N. M. (1995a). *Inst. Phys. Conf. Proc. Ser. No. 141*, Great Britain: Institute of Physics, p. 287.

McCluskey, M. D., Haller, E. E., Walker, J., and Johnson, N. M. (1995b). *Phys. Rev. B*, **52**, 11859.
McCluskey, M. D., Haller, E. E., Walukiewicz, W., and Becla, P. (1996a). *Phys. Rev. B*, **53**, 16297.
McCluskey, M. D., Haller, E. E., Zach, F. X., and Bourret-Courchesne, E. D. (1996b). *Appl. Phys. Lett.*, **68**, 3476.
McCluskey, M. D., Haller, E. E., Walker, J., Johnson, N. M., Vetterhöffer, J., Weber, J., Joyce, T. B., and Newman, R. C. (1997a). *Phys. Rev. B*, **56**, 6404.
McCluskey, M. D., Haller, E. E., Walukiewicz, W., and Becla, P. (1997b). In *Proceedings of the 19th International Conference on Defects in Semiconductors (ICDS-19)*. *Mater. Sci. Forum*, **258–263**, 1247.
McCluskey, M. D., Haller, E. E., Walukiewicz, W., and Becla, P. (1998). *Solid State Commun.*, **106**, 587.
McQuaid, S. A., Binns, M. J., Newman, R. C., Lightowlers, E. C., and Clegg, J. B. (1993). *Appl. Phys. Lett.*, **62**, 1612.
Mizuta, M., Tachikawa, M., Kukimoto, H., and Minomura, S. (1985). *Jpn. J. Appl. Phys.*, **24**, L143.
Morimoto, K., and Fujino, T. (1993). *Appl. Phys. Lett.*, **63**, 2384.
Morse, P. M. (1929). *Phys. Rev. Lett.*, **34**, 57.
Moustakas, T. D., and Molnar, R. (1993). *Mater. Res. Soc. Symp. Proc.*, **281**, 753.
Nakamura, S., Mukai, T., Senoh, M., and Iwasa, N. (1992). *Jpn. J. Appl. Phys.*, **31**, L139.
Nakamura, S., Senoh, M., Iwasa, N., and Nagahama, S. (1995). *Jpn. J. Appl. Phys.*, **34**, L797.
Nakamura, S., Senoh, M., Nagahama, S., Iwasa, N., Yamada, T., Matsushita, T., Kiyoko, H., and Sugimoto, Y. (1996). *Jpn. J. Appl. Phys.*, **35**, L74.
Nakayama, M. (1969). *J. Phys. Soc. Japan*, **27**, 636.
Nelson, R. J. (1977). *Appl. Phys. Lett.*, **31**, 351.
Neugebauer, J., and Van de Walle, C. (1995). *Phys. Rev. Lett.*, **75**, 4452.
Neugebauer, J., and Van de Walle, C. (1996). *Appl. Phys. Lett.*, **68**, 1829.
Newman, R. C. (1969). *Adv. Phys.*, **18**, 545.
Newman, R. C. (1990). *Semicond. Sci. Technol.*, **5**, 911.
Nishimura, K., Nagao, Y., and Sakai, K. (1993). In *Proceedings of the 6th International Conference on II-VI Compounds and Related Optoelectronic Materials, Journal of Crystal Growth*, **138**, 114.
Ohkawa, K., Karasawa, T., and Mitsuyu, T. (1992). *Jpn. J. Appl. Phys.*, **30**, L152.
Pajot, B., and Song, C. (1992). *Phys. Rev. B*, **45**, 6484.
Pankove, J. I. (1991). In *Semiconductors and Semimetals*, **34**, J. I. Pankove and N. M. Johnson, eds. New York: Academic Press.
Pankove, J. I., and Johnson, N. M. (1991). *Semiconductors and Semimetals*, **34**. New York: Academic Press.
Park, R. M., Troffer, M. B., Rouleau, C. M., DePuydt, J. D., and Haase, M. A. (1990). *Appl. Phys. Lett.*, **57**, 2127.
Pavesi, L. (1992). *Solid State Commun.*, **83**, 317.
Pavesi, L., and Gianozzi, P. (1992). *Phys. Rev. B*, **46**, 4621.
Polyakov, A. Y., Pearton, S. J., Wilson, R. G., Rai-Chadhury, P., Hillard, R. J., Bao, X. J., Stam, M., Milnes, A. G., Schlesinger, T. E., and Lopata, J. (1992). *Appl. Phys. Lett.*, **60**, 1318.
Polyakov, A. Y., Ye, M., Pearton, S. J., Wilson, R. G., Milnes, A. G., Stam, M., and Erickson, J. (1993). *J. Appl. Phys.*, **73**, 2882.
Ponce, F. A., and Bour, D. P. (1997). *Nature*, **386**, 351.
Pritchard, R. E., Davidson, B. R., Newman, R. C., Bullough, T. J., Joyce, T. B., Jones, R., and Öberg, S. (1994a). *Semicond. Sci. Technol.*, **9**, 140.

Pritchard, R. E., Newman, R. C., Wagner, J., Fuchs, F., Jones, R., and Öberg, S. (1994b). *Phys. Rev. B*, **50**, 10628.
Rahbi, R., Pajot, B., Chevallier, J., Marbeuf, A., Logan, R. C., and Gavand, M. (1993). *J. Appl. Phys.*, **73**, 1723.
Rahbi, R., Theys, B., Jones, R., Pajot, B., Öberg, S., Somogyi, K., Fille, M. L., and Chevallier, J. (1994). *Solid State Commun.*, **91**, 187.
Rosen, B. (1970). *International Tables of Selected Constants*, **17**. New York: Pergamon Press.
Sciacca, M. Dean, Mayur, A. J., Shin, N., Miotkowski, I., Ramdas, A. K., and Rodriguez, S. (1995). *Phys. Rev. B*, **51**, 6971.
Shimanouchi, T. (1972). *Tables of Molecular Vibrational Frequencies Consolidated*, **1**. Washington: National Bureau of Standards.
Song, C. (1992). Ph.D. thesis, University of Paris.
Stavola, M., and Pearton, S. J. (1991). In *Semiconductors and Semimetals*, **34**, J. I. Pankove and N. M. Johnson, eds. New York: Academic Press.
Stavola, M., Kozuch, D. M., Abernathy, C. R., and Hobson, W. S. (1992). In *Advanced III-V Compound Semiconductor Growth, Processing and Devices*, S. J. Pearton, D. K. Sadana, and J. M. Zavada, eds. *Mat. Res. Soc. Proc.*, **240**, 75.
Stein, H. J. (1994). *Appl. Phys. Lett.*, **64**, 1520.
Svensson, J. H., and Weber, J. (1993). *Mater. Sci. Forum*, **177–178**, 304.
Svob, L., Marfaing, Y., Clerjaud, B., Côte, D., Lebkiri, A., and Druilhe, R. (1996). *J. Crystal Growth*, **159**, 72.
Tuncel, E., and Sigg, H. (1993). *Phys. Rev. B*, **48**, 5225.
Van de Walle, C. G., Denteneer, P. J. H., Bar-Yam, Y., and Pantelides, S. T. (1989). *Phys. Rev. B*, **39**, 10791.
Veloarisoa, I. A., Stavola, M., Kozuch, D. M., Peale, R. E., and Watkins, G. D. (1991). *Appl. Phys. Lett.*, **59**, 2121.
Veloarisoa, I. A., Kozuch, D. M., Stavola, M., Peale, R. E., Watkins, G. D., Pearton, S. J., Abernathy, C. R., and Hobson, W. S. (1992). In *Proceedings of the 16th International Conference on Defects in Semiconductors (ICDS-16)*. *Mater. Sci. Forum*, **83–87**, 111.
Veloarisoa, I. A., Stavola, M., Cheng, Y. M., Uftring, S., Watkins, G. D., Pearton, S. J., Abernathy, C. R., and Lopata, J. (1993). *Phys. Rev. B*, **47**, 16237.
Ves, S., Strössner, K., and Cardona, M. (1986). *Solid State Commun.*, **57**, 483.
Vetterhöffer, J., Svensson, J. H., Weber, J., Leitch, A. W. R., and Botha, J. R. (1994). *Phys. Rev. B*, **50**, 2708.
Vetterhöffer, J., and Weber, J. (1996a). *Phys. Rev. B*, **53**, 12835.
Vetterhöffer, J., Wagner, J., and Weber, J. (1996b). *Phys. Rev. Lett.*, **77**, 5409.
Wagner, J., Maier, M., Lauterback, Th., Bachem, K. H., Fischer, A., Ploog, K., Mörsch, G., and Kamp, M. (1992). *Phys. Rev. B*, **45**, 9120.
Wagner, J., Bachem, K. H., Davidson, B. R., Newman, R. C., Bullough, T. J., and Joyce, T. B. (1995a). In *Proceedings of 22nd International Conference on the Physics of Semiconductors*, **3**, D. J. Lockwood, ed. Singapore: World Scientific, p. 2323.
Wagner, J., Pritchard, R. E., Davidson, B. R., Newman, R. C., Bullough, T. J., Joyce, T. B., Button, C., and Roberts, J. S. (1995b). *Semicond. Sci. Technol.*, **10**, 639.
Watts, R. K., Holton, W. C., and de Wit, M. (1971). *Phys. Rev. B*, **3**, 404.
Williams, P. M., Watkins, G. D., Uftring, S., and Stavola, M. (1993). *Phys. Rev. Lett.*, **70**, 3816.
Wolk, J. A., Ager, J. W., III, Duxstad, K. J., Haller, E. E., Taskar, N. R., Dorman, D. R., and Olego, D. J. (1993). *Appl. Phys. Lett.*, **63**, 2756.
Woodhouse, K., Newman, R. C., deLyon, T. J., Woodall, J. M., Scilla, G. J., and Cardone, F. (1991). *Semicond. Sci. Technol.*, **6**, 330.
Yu, Z., Buczkowski, S. L., Hirsch, L. S., and Myers, T. H. (1996). *J. Appl. Phys.*, **80**, 6425.

Zach, F. X. (1994). *J. Appl. Phys.*, **75**, 7894.
Zach, F. X., Haller, E. E., Gabbe, D., Iseler, G., Bryant, G. G., and Bliss, D. F. (1996). *J. Electr. Mater.*, **25**, 331.
Zavada, J. M., and Wilson, R. G. (1994). In *Proceedings of the 17th International Conference on Defects in Semiconductors (ICDS-17)*. *Mater. Sci. Forum*, **148–149**, 189.
Zheng, J.-F., and Stavola, M. (1996). *Phys. Rev. Lett.*, **76**, 1154.
Zhou, J. A., Song, C. Y., Zheng, J.-F., Stavola, M., Abernathy, C. R., and Pearton, S. J. (1997). In *Proceedings of the 19th International Conference on Defects in Semiconductors (ICDS-19)*. *Mater. Sci. Forum*, **258–263**, 1293.
Zolper, J. C., Wilson, R. G., Pearton, S. J., and Stall, R. A. (1996). *Appl. Phys. Lett.*, **68**, 1945.

CHAPTER 10

The Properties of Hydrogen in GaN and Related Alloys

S. J. Pearton

DEPARTMENT OF MATERIALS SCIENCE AND ENGINEERING
UNIVERSITY OF FLORIDA
GAINESVILLE, FLORIDA

J. W. Lee

PLASMATHERM
ST. PETERSBURG, FLORIDA

I. INTRODUCTION 441
II. HYDROGEN IN AS-GROWN NITRIDES 442
   1. Doped Material 442
   2. Sources of Hydrogen 445
   3. Diffusion 446
III. DOPANT PASSIVATION 450
   1. Calcium 451
   2. Carbon 454
   3. Summary 455
IV. DIFFUSION AND REACTIVATION MECHANISMS 456
   1. Alloys 456
   2. In-Containing Nitrides 461
   3. Mechanisms 462
   4. Heterostructures 466
V. ROLE OF HYDROGEN DURING PROCESSING 469
   1. Implant Isolation 470
   2. Wet Processing 470
   3. Deposition and Etching 471
VI. THEORY OF HYDROGEN IN NITRIDES 472
VII. SUMMARY AND CONCLUSIONS 476
   REFERENCES 477

I. Introduction

Hydrogen is a component of most of the gases and liquids used in the growth, annealing, and processing of semiconductors, and a great deal of eattention has been focused on the effects of hydrogen incorporation in Si, GaAs, SiC, and other materials (Pearton et al., 1987; Myers et al., 1992). It is fairly well established that H_2 (or larger aggregates) are basically electrically and optically inactive in all semiconductors and that the diatomic species has low diffusivity once formed inside the semiconductor. By contrast, atomic hydrogen (which may exist as H^0, H^+, or H^- depending on the position of the Fermi level) diffuses rapidly even at low temperatures (25–250°C) and can attach to dangling or defective bonds associated with point or line defects and also can form neutral complexes with dopants, that is,

$$D^+ + H^- \to (DH)^0 \tag{1}$$

$$A^- + H^+ \to (AH)^0 \tag{2}$$

where D^+ and A^- are ionized shallow donor and acceptor dopants, respectively. These reactions manifest themself as increases in resistivity of the semiconductor and an increase in carrier mobility as ionized impurity scattering is reduced. A typical signature of an unintentional hydrogen passivation process is a reduction in doping density in the near-surface region (≤ 1 µm) due to indifussion of atomic hydrogen from the liquid or gas in which the sample is immersed.

In GaN and related materials (AlN, InN, InGaN, InAlN, AlGaN) it has been found that hydrogen is present in relatively high concentrations in as-grown samples, especially p-type GaN, and that hydrogen is readily incorporated during many device process steps such as plasma etching, plasma-enhanced chemical vapor deposition (PECVD), solvent boiling, and wet chemical etching (Pearton, 1997).

In this chapter we will detail the effects of hydrogen in as-grown nitrides, its incorporation during processing, the theory of hydrogen and hydrogen-related defects, and its possible effects on dopant activation.

II. Hydrogen in As-Grown Nitrides

1. Doped Material

For many years it was not possible to achieve p-type doping in GaN, due partially to the often high residual n-type background (resulting from nitrogen vacancies or Si or O impurities) and the high ionization energy level of most acceptor dopants. However, even in GaN(Mg), in which the Mg concentration was easily sufficient to produce p-type conductivity, the material generally was resistive.

Amano et al. (1989) found that when the electron beam in a scanning electron microscope was focused on these samples, blue emission was evident, and the resistivity of the exposed area had dramatically decreased. Nakamura et al. (1991, 1992a) later found low p-type conductivity in as-grown GaN(Mg). Further treatment in a low-energy electron-beam irradiation facility reduced the resistivity from 4×10^4 to $\sim 3 \, \Omega \cdot$cm in the top 0.5 µm of the surface, as shown in Fig. 1. A maximum hole concentration of 7×10^{18} cm^{-3} and a mobility of $3 \, \text{cm}^2/\text{V} \cdot$sec were obtained by this method.

Subsequently, Nakamura et al. (1991, 1992b) showed that low-resistivity p-type GaN with uniform carrier densities throughout their entire thickness could be obtained by postgrowth thermal annealing in N_2 at temperatures $\geq 700°$C. Once again, the resistivity was observed to drop from $\sim 10^6$ to $2 \, \Omega \cdot$cm. Deep-level emission at ~ 750 nm sharply decreased with annealing, whereas the blue (450-nm) emission was maximized at 700°C. To elucidate the mechanism responsible for these changes, Nakamura et al. (1992b)

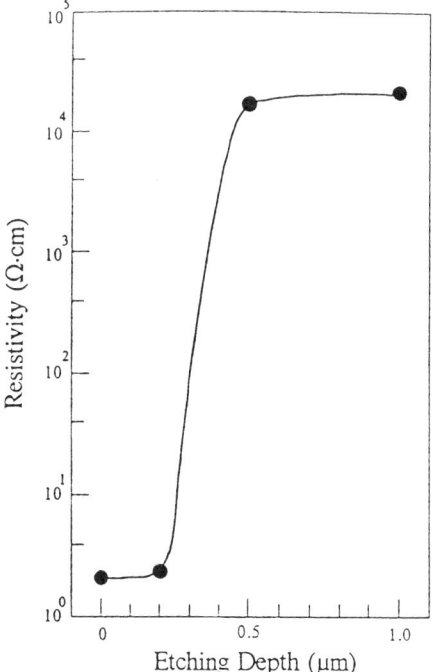

FIG. 1. Resistivity change of GaN(Mg) layer after low-energy electron-beam irradiation as a function of depth from the surface. (After Nakamura, 1991.)

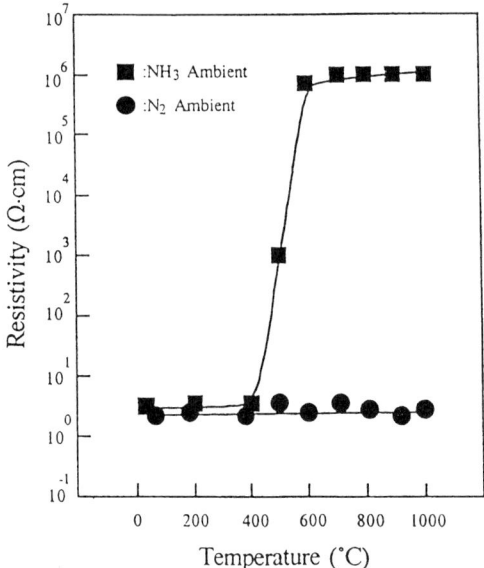

FIG. 2. Resistivity of GaN(Mg) layers as a function of annealing temperature in either NH_3 or NH_2 ambients. (After Nakamura, 1992b.)

annealed p-type GaN(Mg) in either N_2 or NH_3. As shown in Fig. 2, while there was no change in the resistivity with N_2 annealing, above $\sim 500°C$ the GaN returned to high resistivity when NH_3 was the annealing ambient. Subsequent annealing in N_2 returned these films to their low-resistivity condition, and these changes correlated well with changes in the photoluminescence spectrum from the samples (Fig. 3). NH_3 decomposes above $\sim 200°C$ to form N_2 and H_2 (99.9% dissociation at 800°C), and dissociation of the hydrogen can be catalyzed by the presence of the GaN surface. Thus there is a sizable flux of atomic hydrogen available for indiffusion into the GaN film, where it can form neutral complexes with the Mg acceptors through the reactions

$$Mg^- + H^+ \to (Mg - H)^0 \qquad (3)$$

This produces compensation of the holes from the acceptors and leads to high resistivity. The reaction can be driven in the reverse direction by either annealing or injection of minority carrier (electrons) through the low-energy electron-beam irradiation, by shining above-bandgap light onto the sample, or by forward biasing of a p-n junction structure. One usually observes a decrease in blue emission from samples annealed above $\sim 700°C$, unless

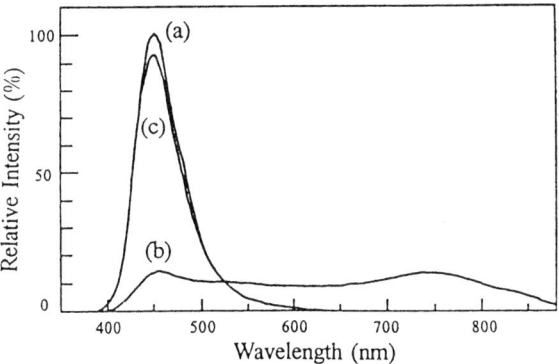

FIG. 3. PL spectra of GaN(Mg) layers after sequential (a) N_2 ambient annealing at 800°C, (b) subsequent NH_3 ambient annealing at 800°C, and (c) subsequent N_2 ambient annealing at 800°C. (After Nakamura, 1992b.)

special precautions are taken, because of the onset of surface dissociation. Amano et al. (1988) reported that the intensity of Zn-related blue emissions in GaN(Zn) was enhanced by electron injection, suggesting that $(Zn-H)^0$ complexes also form in this material.

2. SOURCES OF HYDROGEN

In as-grown GaN there are numerous potential sources of hydrogen, since the growth reaction by metal organic chemical vapor deposition (MOCVD) proceeds as

$$(CH_3)_3Ga + NH_3 \rightarrow GaN + CH_4\uparrow + H_2\uparrow$$

In some cases H_2 is also employed as a carrier gas for the trimethylgallium. Previous work in other semiconductor systems grown with metal-organic precursors has shown that all three of the components of the growth chemistry, that is, group III source, group V source, and carrier gas, can contribute to dopant passivation. Even the dopant sources (Cp_2Mg and Si_2H_6 are the most common for GaN) are potential supplies of hydrogen in the GaN films. There is a direct correlation between H and Mg concentrations in MOCVD GaN, suggesting that there is trapping of hydrogen at the acceptors (Ohba and Hatano, 1994); this result is found in other p-type semiconductors grown by gas-source techniques. Further implication of hydrogen passivation as the hole-reduction mechanism comes from the fact that p-type GaN has been achieved by Mg doping in molecular beam epitaxy material without postgrowth annealing (Moustakas et al., 1993). In this case N_2 is derived from a plasma source, and solid Ga is the group III source; hence no hydrogen is present in the growth environment.

3. DIFFUSION

In most III-V semiconductors it is found that atomic hydrogen diffuses more rapidly in p-type material, where it is likely in a positive charge state (H^+), and bonds more strongly to acceptors than it does to donors (Pankove and Johnson, 1991; Pearton et al., 1992). Figure 4 shows a SIMS depth profile of GaN(Mg) grown on Al_2O_3 by MOCVD; the as-grown material contains $\sim 5.5 \times 10^{19}$ Mg cm^{-3}, and this remains stable on annealing at 700°C. The hole concentration in the material was $\sim 6 \times 10^{17}$ cm^{-3}. Note that there is $\sim 10^{18}$ cm^{-3} hydrogen in the as-grown sample, more than enough to compensate all the holes. Postgrowth annealing at 700°C for 60 min reduces the hydrogen concentration to $1-5 \times 10^{17}$ cm^{-3}, which is below the hole concentration, and hence the p-type conductivity returns. An

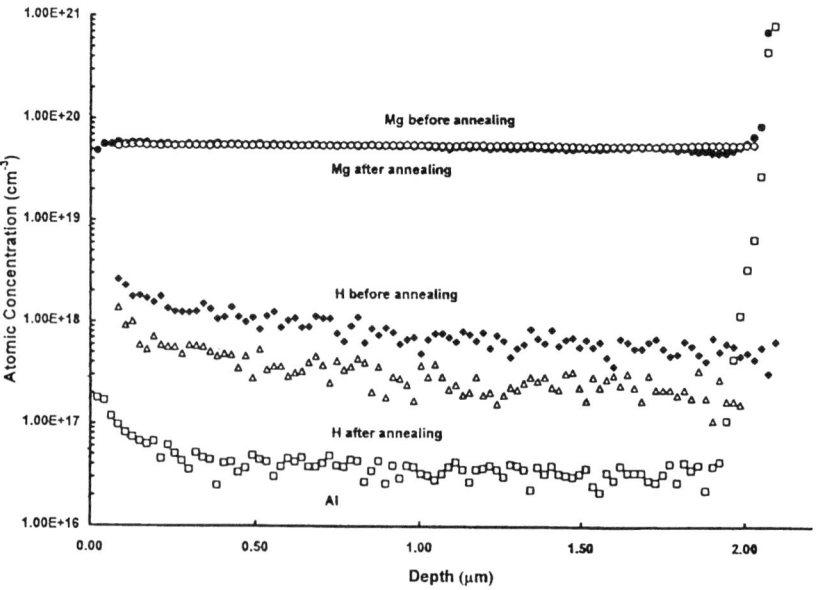

FIG. 4. SIMS depth profile of MOCVD-grown GaN(Mg) before and after 700°C, 60-min anneal in N_2.

interesting point from Fig. 4 is that a substantial amount of hydrogen remains in the material, probably bound at defects or internal surfaces. This residual hydrogen may give rise to effects such as current gain drift in transistors or a dependence of apparent material resistivity on the measurement current in Hall measurements because of minority-carrier reactivation of passivated acceptors.

Figure 5 shows the hydrogen concentration in undoped, lightly Mg-doped, and heavily Mg-doped GaN as a function of annealing time at 700°C under a N_2 ambient (Li et al., 1996). The hydrogen concentration is reduced with time but saturates at $\sim 2 \times 10^{17}$ cm^{-3}; it is not clear if this represents the background sensitivity of the SIMS apparatus or is a real concentration. After the anneal, the heavily Mg-doped ($Mg = 6 \times 10^{19}$ cm^{-3}) sample remained highly resistive. A problem with continuing to increase the Mg concentration is the onset of cracking of the GaN films and conversion of the conductivity to n-type. The reasons for this are as yet unclear, and the presence of Mg interstitials acting as shallow donors or formation of Mg-defect complexes with donor nature are just two of the possibilities. As found by Nakamura et al. (1991, 1992a, 1992b), the changes in resistivity are accompanied by strong changes in the PL spectra. Figure 6 shows 5-K spectra in heavily Mg-doped ($Mg = 6 \times 10^{19}$ cm^{-3}) GaN before and after different annealing times at 700°C. The peak at 3.285 eV has been ascribed to a free-to-bound Mg transition, corresponding to a binding energy for Mg of 155 meV, while that at 3.455 eV has been identified as a transition involving an exciton bound to Mg. Annealing for 20 min at 700°C shifts this

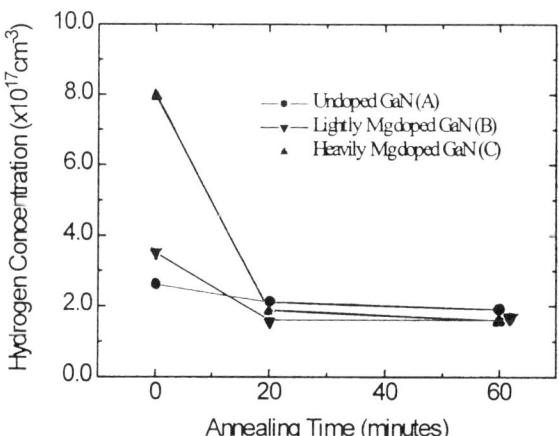

FIG. 5. Hydrogen concentration in undoped or Mg-doped GaN as a function of annealing time at 700°C in a N_2 ambient. (After Li et al., 1996.)

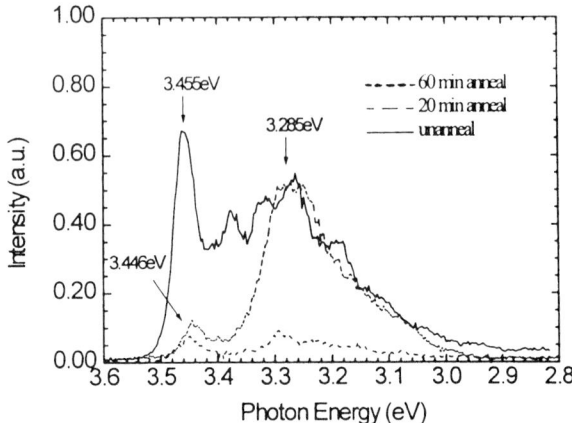

FIG. 6. Low-temperature (5 K) PL spectra from heavily Mg-doped (Mg = 6×10^{19} cm^{-3}) GaN as a function of annealing time at 700°C in a N_2 ambient. (After Li et al., 1996.)

latter peak to 3.446 eV. Li et al. (1996) attributed the 3.45-eV peak to a transition associated with an exciton bound to the Mg acceptor. For 60 min annealing, observation of a 3.465-eV peak was found in lightly Mg-doped GaN, consistent with a previous assignment to an exciton bound to a nitrogen vacancy. Therefore, loss of N_2 and compensation of the p-type doping by introduction of the shallow donor N_V states appear to be another reason it is difficult to get strong p-type conductivity.

Piner et al. (1997) found that the hydrogen concentration in MOCVD $In_xGa_{1-x}N$ was a strong function of H_2 and NH_3 flow rates during growth and also had a strong effect on the background C and O concentrations. Therefore, the overall compensation in the material is affected by the hydrogen flow. Figure 7 shows H, C, and O relative concentrations in InGaN grown at 730°C as a function of NH_3 flow rate. While the H clearly originates from the ammonia and the metalorganic sources [trimethylgallium, $(CH_3)_3Ga$; trimethylaluminum, $(CH_3)_3Al$; ethyldimethylindium, $(CH_3)_3C_2H_5In$] and the carbon also originates from the metalorganic sources, it appears that the purified NH_3 is the source of oxygen. The reduction in carbon with NH_3 flow probably results from H scavenging reactions. Note also in Fig. 8 that all three light-element impurities increase in concentration as the InN percentage composition increases, independent of H_2 flow, NH_3 flow, or growth temperature.

Johnson et al. (1997) performed a series of MBE growth experiments in which GaN(Mg) was grown with either pure N_2 plasmas or mixed N_2/H_2 plasmas in order to understand the effect of hydrogen under controlled

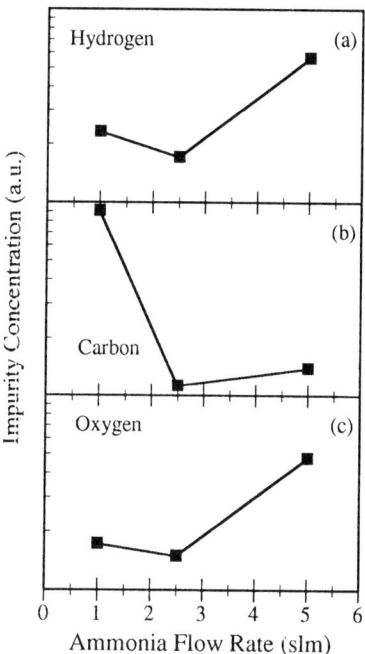

FIG. 7. H, C, and O relative concentrations as a function of NH_3 flow in MOCVD InGaN. (After Piner et al., 1997.)

conditions. However, PL examination showed only minor differences, in some cases due to Mg compensation. No measurements of H concentrations were reported, and it is likely under the particular growth conditions used that there was little H incorporation.

Soto (1996) reported that GaN grown by plasma-assisted MOCVD using triethylgallium and N radicals was highly resistive and attributed to the high density of C and H ($10^{19}-10^{20}$ cm^{-3}) incorporated into the films. The H signal intensity was linearly related to the C signal, suggesting complexing of the C and H. Annealing at 900°C did not affect the resistivity.

Lee and Yong (1997) grew hydrogenated AlN by radiofrequency (rf) magnetron sputtering in a $H_2/Ar/N_2$ gas mixture. The addition of H was found to reduce the oxygen background in the material, and the activation energy for the evolution of H_2 gas from the AlN:H flow was found to be 0.11 eV/atom, as determined by gas chromatography. A variety of N-H IR peaks were found between 3074–3532 cm^{-1}; these bonds may hinder formation of N-O bonds. The value of 0.11 eV/atom for hydrogen evolution is consistent with hydrogen bound at N atoms.

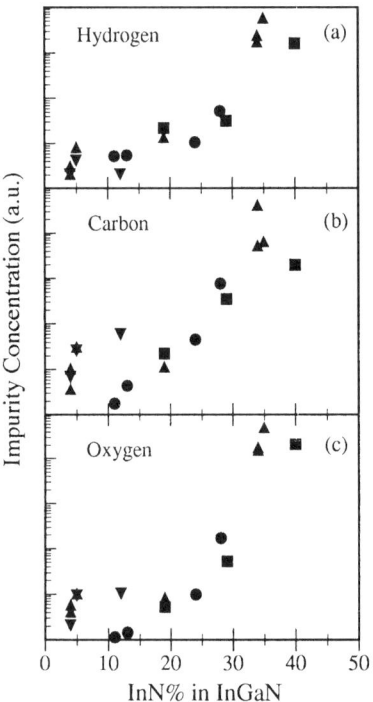

FIG. 8. H, C, and O relative concentrations as a function of InN composition in MOCVD InGaN. (After Piner et al., 1997.)

III. Dopant Passivation

We have already seen that atomic hydrogen passivates both Mg and Zn, but a number of reports have shown passivation of Ca (Lee et al., 1996), C (Pearton et al., 1994), and Cd (Burchard et al., 1997). Theoretical considerations initially suggested that Ca might be a shallower acceptor in GaN than Mg (Strite, 1994). Zolper et al. (1996) realized p-type doping of GaN using implantation of Ca^+ alone, or a coimplantation of Ca^+ and P^+, followed by rapid thermal annealing at $\geq 1100°C$. While the activation efficiency of Ca in both implant schemes was ~100%, temperature-dependent Hall measurements showed that the ionization level of Ca was ~168 meV, similar to that of Mg. The Ca atomic profile was thermally stable to temperatures up to 1125°C. Since Mg has a substantial memory effect in stainless steel epitaxial reactors (or in gas lines leading to quartz chamber systems), Ca may be a useful alternative p-dopant for epitaxial growth of laser diode or heterojunction bipolar transistor structures in which junction placement, and hence control of dopant profiles, is of critical importance.

1. CALCIUM

In considering Ca-doped GaN for device applications it is also necessary to understand the role of hydrogen, since there is always a ready supply of atomic hydrogen available from NH_3, the metalorganic group III source [typically $(CH_3)_3Ga$], or from the gaseous dopant source when using chemical vapor deposition techniques. We also have found that Ca acceptors in GaN are readily passivated by atomic hydrogen at low temperature (250°C), but they can be reactivated by thermal annealing at $\leq 500°C$ for 1 min in lightly doped ($3 \times 10^{17}\,cm^{-3}$) material. As the carrier density is restored by such annealing treatments, there is a true passivation and not just compensation of the Ca acceptors by the hydrogen.

Nominally undoped ($N < 3 \times 10^{16}\,cm^{-3}$) GaN was grown on double-side polished c-Al_2O_3 substrates prepared initially by $HCl/HNO_3/H_2O$ cleaning and an in situ H_2 bake at 1070°C. A GaN buffer $\leq 300\,\text{Å}$ thick was then grown at $\sim 500°C$ and crystallized by ramping the temperature to 1040°C, where trimethylgallium and ammonia are again used to grow the 2-μm-thick epitaxial layer. In brief, the double-crystal x-ray full widths at half maxima are ~ 300 arcsec, and the total defect density (threading dislocations, stacking faults) apparent in plan-view transmission electron microscopy was typically 2–$4 \times 10^9\,cm^{-2}$. The as-grown films are featureless and transparent and have strong band-edge (3.47 eV) luminescence.

^{40}Ca ions were implanted at 180 keV and a dose of $5 \times 10^{14}\,cm^{-2}$. In some cases, a coimplant of P^+ to the same dose at an energy of 130 keV was performed to try to enhance the substitutional fraction of Ca on subsequent annealing in analogy to the case of Mg implantation in GaN. For the case of Ca, we found there was little additional activation as a result of the coimplant. After rapid annealing at 1150°C for 15 sec under N_2 in a face-to-face geometry, we measured sheet carrier densities of $P \approx 1.6 \times 10^{12}\,cm^{-2}$ with a mobility at 500 K of $6\,cm^2/V \cdot sec$. Arrhenius plots of the hole density showed an ionization level of 169 meV for the Ca in GaN (Fig. 9). Samples with alloyed HgIn ohmic contacts were exposed to an electron cyclotron resonance (ECR) H_2 or He plasma (2.45 GHz) with 850 W forward power and a pressure of 10 mtorr. The exposure time was 30 min at 250°C, and the temperature was lowered to room temperature with the plasma on. The sheet carrier density and hole mobility at 300 K were obtained from Van der

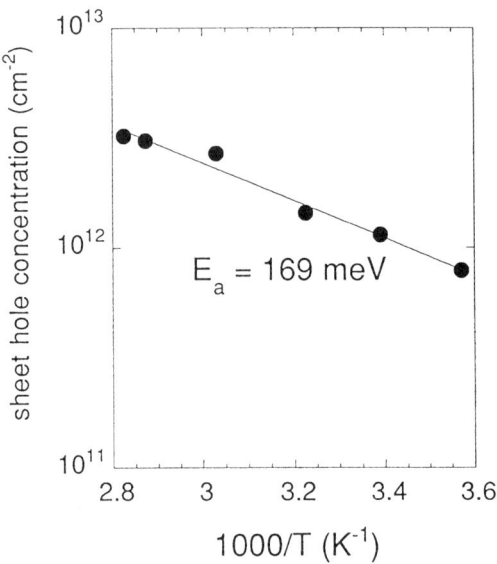

FIG. 9. Annealing plot of sheet hole density in Ca-implanted GaN. The activation energy is 169 ± 12 meV.

Pauw Hall measurements. Posthydrogenation annealing was performed at between 100 and 500°C for 60 sec under flowing N_2 with the ohmic contacts already in place.

The initial H_2 plasma exposure caused a reduction in sheet hole density of approximately an order of magnitude, as shown in Fig. 10. No changes in electrical properties were observed in the He-plasma-treated samples, showing that pure ion bombardment effects are insignificant and the chemical interaction of hydrogen with the Ca acceptors is responsible for the conductivity changes. Posthydrogenation annealing had no effect on the hole density up to 300°C, while the initial carrier concentration was essentially fully restored at 500°C. Assuming that the passivation mechanism is formation of neutral Ca-H complexes, then the hole mobility should increase upon hydrogenation. This is indeed the case, as shown in Fig. 11. Note that the mobility decreases to its initial value with posthydrogenation annealing. If the carrier reduction were due to introduction of compensating defects or impurities, then the hole mobility would decrease, which is not observed.

In other *p*-type III-V semiconductors it is generally accepted that atomic hydrogen is predominantly in a positive charge state, with the donor level being around midgap. If a similar mechanism exists in GaN, then the initial

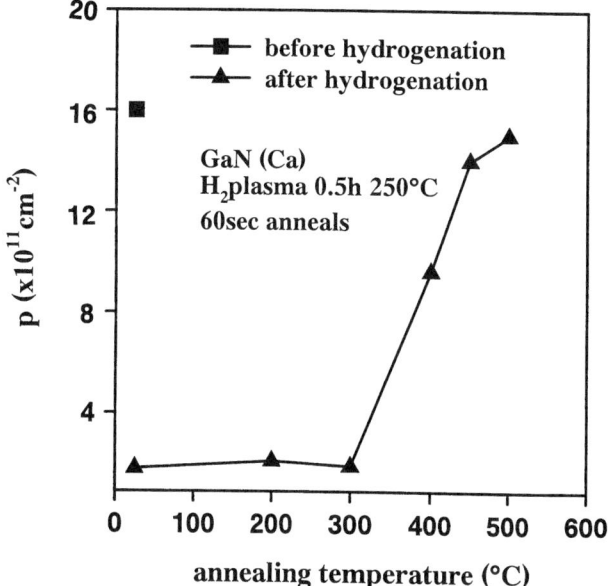

FIG. 10. Sheet hole density at 300 K in hydrogenated GaN(Ca) as a function of subsequent annealing temperature.

coulombic attraction between ionized acceptor and hydrogen leads to formation of a neutral close pair, that is,

$$Ca^- + H^+ \leftrightarrow (Ca\text{-}H)^0 \qquad (4)$$

The existence of the neutral complex should be verified by observation of a vibrational band, but to obtain the sensitivity needed for such a measurement will require a relatively thick epilayer of Ca-doped GaN. Our present implanted samples do not have a sufficient Ca density-times-thickness product to be suitable for IR spectroscopy.

If the dissociation of the Ca-H species is a first-order process, then the reactivation energy from the data in Fig. 10 is ~ 2.2 eV, assuming a typical attempt frequency of 10^{14} sec^{-1} for bond-breaking processes. This is similar to the thermal stability of Mg-H complexes in GaN, which we prepared in the same manner (implantation) with similar doping levels. In thicker, more heavily doped samples, the apparent thermal stability of hydrogen passivation is much higher because of the increased probability of retrapping of hydrogen at other acceptor sites. This is why for thick, heavily doped ($P > 10^{18}$ cm^{-3}) GaN(Mg), a postgrowth anneal of at least 700°C for 60 min

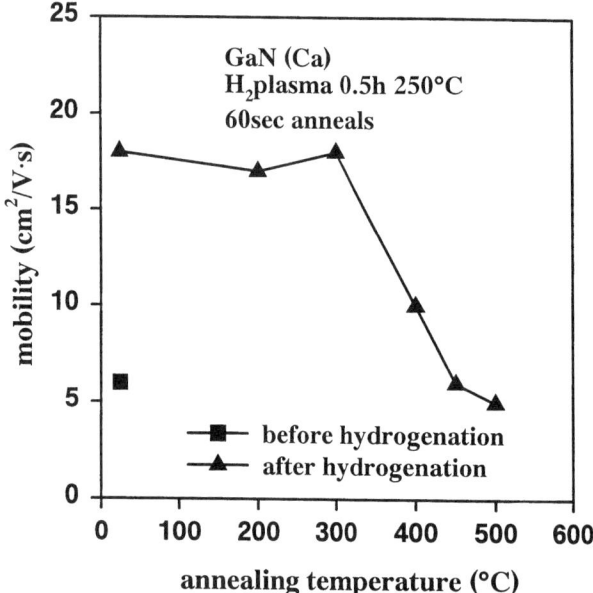

FIG. 11. Hole mobility at 300 K in hydrogenated GaN(Ca) as a function of subsequent annealing temperature.

is employed to ensure complete dehydrogenation of the Mg. True reactivation energies can only be determined in reserve-biased diode samples where the strong electrical fields present sweep the charged hydrogen out of the depletion region and minimize retrapping at the acceptors.

In conclusion, we have found that hydrogen passivation of acceptors in GaN occurs for several different dopant impurities and that postgrowth annealing also will be required to achieve full electrical activity in Ca-doped material prepared by gas-phase deposition techniques. The thermal stability of the passivation is similar for Ca-H and Mg-H complexes, with apparent reactivation energies of ~ 2.2 eV in lightly doped ($\sim 10^{17}$ cm^{-3}) material.

2. CARBON

Abernathy et al. (1995) reported carbon doping of GaN using CCl_4 in metal organic molecular beam epitaxy, with maximum hole concentrations of $\sim 3 \times 10^{17}$ cm^{-3}. The total carbon concentration in these films is somewhat higher than this (1–2 orders of magnitude), but the reason for the low doping efficiency is not known at this point. The activation energy of acceptor ionization was reported to be ~ 26 meV, but this may be due to impurity-band conduction. Exposure of the GaN(C) samples to an electron cyclotron resonance H_2 plasma at 250°C reduces the hole concentration by approximately a factor of 3, with an accompanying increase in hole mobility. Annealing of the hydrogenated material restored the initial hole concentration at ~ 450°C. It is well established in other dopant-hydrogen complexes that the apparent thermal stability is a strong function of the doping level and sample thickness because of hydrogen retrapping effects.

Burchard et al. (1997) studied the formation and properties of Cd-H pairs in GaN using radioactive ^{111}Cd and perturbed $\gamma\gamma$ angular correlation spectroscopy. The H was incorporated by 100-eV implantation, and formation of two different Cd-H complexes (different configurations) with dissociation energies of 1.1 and 1.8 eV, respectively, was found.

3. SUMMARY

Table I summarizes the information reported to date for hydrogen-acceptor complexes in GaN. To this point there have not been any direct observations of donor dopant passivation in any of the nitrides, although this is fairly typical of what has occurred previously in other semiconductor systems. For example, in Si, acceptor passivation was first reported in 1983 (Pankove and Johnson, 1991), and it was only in 1986 that reports of weak donor passivation were seen. Subsequently, in 1988, the first unambiguous observations by IR spectroscopy were made, but only after realizing

TABLE I

P-DOPANTS FOUND TO BE PASSIVATED BY ATOMIC HYDROGEN IN GaN

Dopant	Comments	References
Mg	Residual hydrogen in growth ambient leads to high resistivity in as-grown GaN(Mg)	Amano et al., 1989; Nakamura et al., 1991, 1992a, 1992b
Zn	Electron injection increases Zn-related emissions in GaN(Zn)	Amano et al., 1988
C	H_2 plasma exposure decreases hole density by a factor of 3, thermally reversible	Pearton et al., 1994
Ca	H_2 plasma exposure decreases hole density by a factor of 10, thermally reversible	Lee et al., 1996
Cd	Formation of Cd-H complexes seen by PAC, dissociates at <350°C	Burchard et al., 1997

that because of the lower binding energies for donor dopant-hydrogen complexes, the temperature at which the hydrogen was unincorporated should be lowered (Pearton et al., 1992). Once the plasma injection temperature was reduced (to 120°C from the usual 200°C for p-type Si), donor passivation efficiencies over 90% were realized. Another feature of dopant passivation by hydrogen is that it depends on the characteristics of the plasma used for injection, that is, the relative fluxes of H_2, H_2^+, H^0, H^+, and associated excited states, the average ion energy, and the Fermi level at the semiconductor surface. For example, we have seen that increasing the ion energy in a H_2 plasma increases the incorporation depth of hydrogen in p-type GaAs by up to 50%, and somewhat smaller effects have been observed in SiC.

IV. Diffusion and Reactivation Mechanisms

Typical SIMS profiles of ^2H in p-type GaN with [Mg] $\approx 2 \times 10^{19}$ cm^{-3} and hole concentration $\sim 1 \times 10^{17}$ cm^{-3} and in n-type GaN with [Si] $\approx 2.5 \times 10^{18}$ cm^{-3} and electron concentration 2×10^{18} cm^{-3} are shown in Figs. 12 and 13, respectively, for samples exposed to a ^2H plasma for 30 min at 250°C. Note in the p-type material the ^2H follows an in-diffusion profile, where ^2H is trapped at Mg or defects, and there is a very high concentration shallow feature extending to ~ 0.1 μm from the surface. This is deeper than the usual peak due to nonequilibrium sputtering effects in SIMS and may result from extended ^2H clusters or platelets. In the n-type material there is basically no measurable deuterium, suggesting that ^2H is not readily trapped at donor dopants. However, as pointed out earlier, this may simply be due to the nonoptimized plasma conditions employed thus far.

1. ALLOYS

A number of reports have shown that hydrogen can indeed pair with positively charged native donors in InN, InGaN, InAlN, and AlGaN (Polyakov et al., 1996). Frequently, undoped $Al_xGa_{1-x}N$ films with $x < 0.4$ show electron concentrations in the 10^{18}–10^{19} cm^{-3} range and mobilities on the order of 10 cm^2/V $\cdot$ sec. In addition, many groups have observed very strong tails at optical absorption near the band edge. High carrier concentrations and strong band tailing in AlGaN make it difficult to use such layers in most applications. Polyakov et al. (1996) found that hydrogen plasma exposure at 200°C for 1 hour led to a substantial decrease in carrier concentration

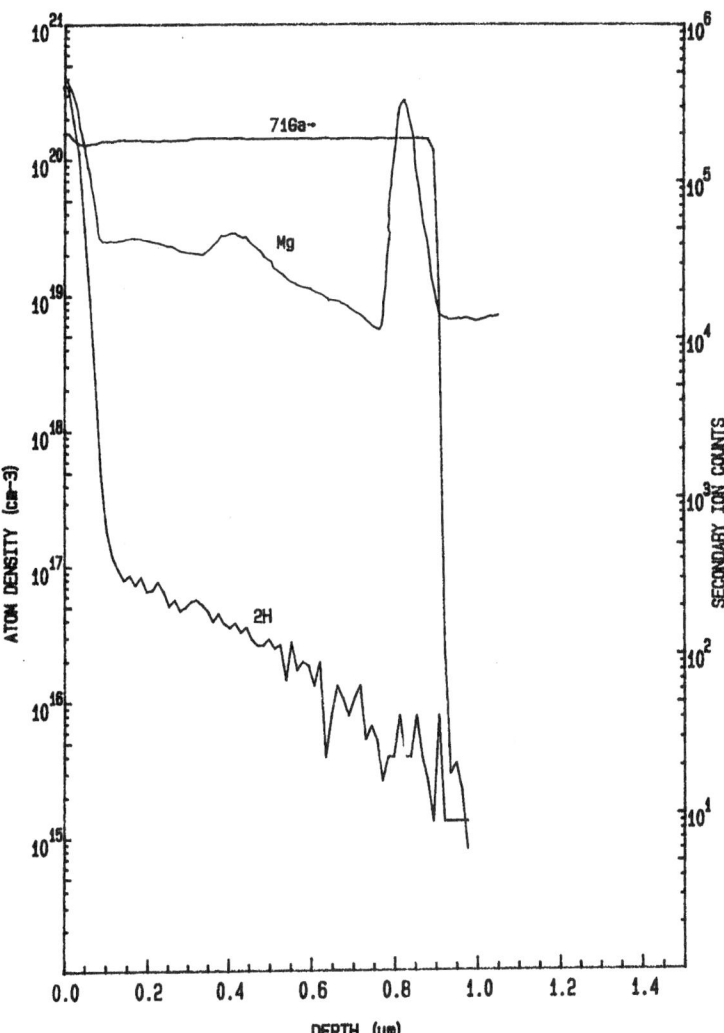

FIG. 12. SIMS profiles of ^{2}H and Mg in GaN(Mg) exposed to a ^{2}H plasma for 30 min at 250°C.

accompanied by an increase in electron mobility (see Table II). The passivation efficiency appears to be higher for AlGaN than for pure GaN, and there was little change in electrical properties for the closely compensated GaN sample with initial carrier concentration of 8×10^{14} cm^{-3}. By sharp contrast, changes in AlGaN carrier concentration still occur even

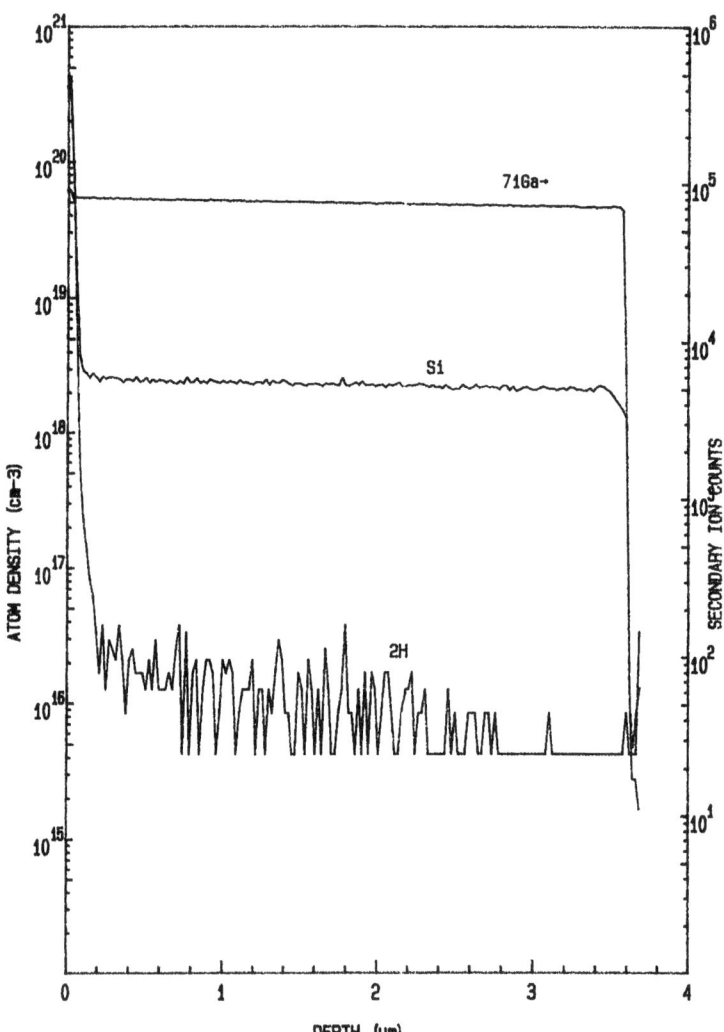

FIG. 13. SIMS profiles of ^{2}H and Si in GaN(Si) exposed to a ^{2}H plasma for 30 min at 250°C.

when the starting concentration is low. Polyakov et al. (1996) suggested that passivation of native donors in AlGaN proceeds via pairing with negatively charged hydrogen ions and that efficiency of passivation drops rapidly as the Fermi level crosses the hydrogen acceptor level. This would place the H$^-$ level in GaN somewhere above $E_c - 0.2$ eV and much deeper than that

TABLE II

ELECTRICAL PROPERTIES OF AlGaN LAYERS BEFORE AND AFTER HYDROGEN PLASMA PASSIVATION

Al content	Before treatment		After treatment	
	N, cm^{-3}	μ, cm^2/V·s	N, cm^{-3}	μ, cm^2/V·s
0	8×10^{17}	250	1.6×10^{17}	303
0	1.3×10^{18}	9.9	2.5×10^{16}	182
0	8×10^{14}	4.3	6.4×10^{14}	47.2
0.12	1.5×10^{18}	2.1	3.3×10^{10}	115.4
0.12	1.6×10^{11}	110.8	1.6×10^{10}	115.8
0.2	1.1×10^{19}	2.2	7.8×10^{17}	32.4
0.31	3.6×10^{18}	5.2	2.3×10^{17}	18.2
0.38	1.1×10^{19}	1.0	2.9×10^{18}	5
0.67	1.9×10^{14}	19.1	2.9×10^{13}	59.6

in AlGaN (e.g., in Al$_{0.12}$Ga$_{0.88}$N it should close to $E_c - 0.5$ eV). Measurements of the temperature dependence of carrier concentration in high-resistivity AlGaN samples after hydrogen treatment yield the same activation energies as before treatment (0.3 eV in Al$_{0.12}$Ga$_{0.88}$N, 0.22 eV in Al$_{0.67}$Ga$_{0.33}$N), indicating that no deeper compensating centers have been introduced and that the decreased carrier concentration is a result of passivation of the existing electrically active centers.

The effect of hydrogen plasma passivation on the absorption spectra near the fundamental absorption edge in AlGaN and AlN samples is shown in Fig. 14. Prior to passivation, the magnitude of the band tailing (manifested in deviation of the squared absorption coefficient versus photon energy from a straight line) is substantial. This band tailing is suppressed by the hydrogen plasma treatment. The origin of the band tails may be related to fluctuations in local electrical fields due to fluctuations in the density of charged defects. The hydrogenation treatment reduces the concentration of these defects, reducing the near-band-edge absorption. For AlN, no measurements of electrical properties could be performed because of its high resistivity. The near-band-edge absorption observed in that case could come from local fields associated with deep levels, and suppression of such absorption could be related to suppression of electrical activity of these deep centers. PL measurements on these samples showed a decrease in deep level emission and an increase in band-edge emission as a result of the hydrogenation.

The thermal stability of native donor-hydrogen complexes was measured by isochronal annealing, as shown in Fig. 15. The fraction of unannealed complexes is defined as $N_I - N_T/N_I - N_H$, where N_I is the electron concen-

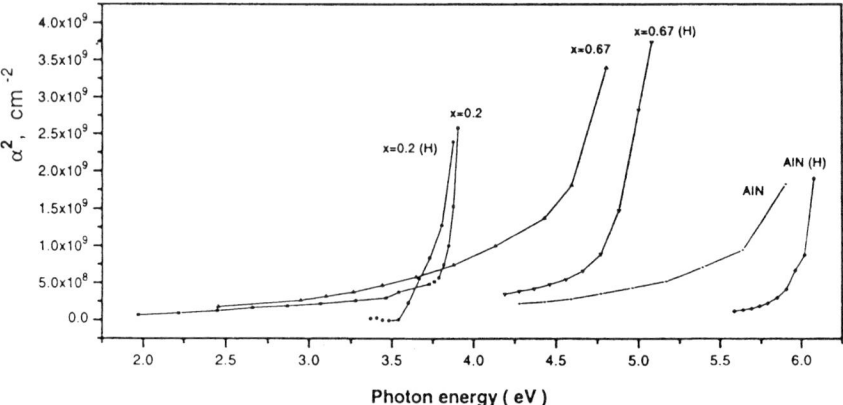

FIG. 14. Dependence of absorption coefficient squared on photon energy before and after hydrogenation of AlGaN samples of various Al contents.

tration before hydrogenation, N_H is the electron concentration after hydrogenation, and N_T is the electron concentration after hydrogenation and subsequent annealing at temperature T. The thermal stability of the donor-hydrogen complexes increases with AlN mole fraction, and this may play a

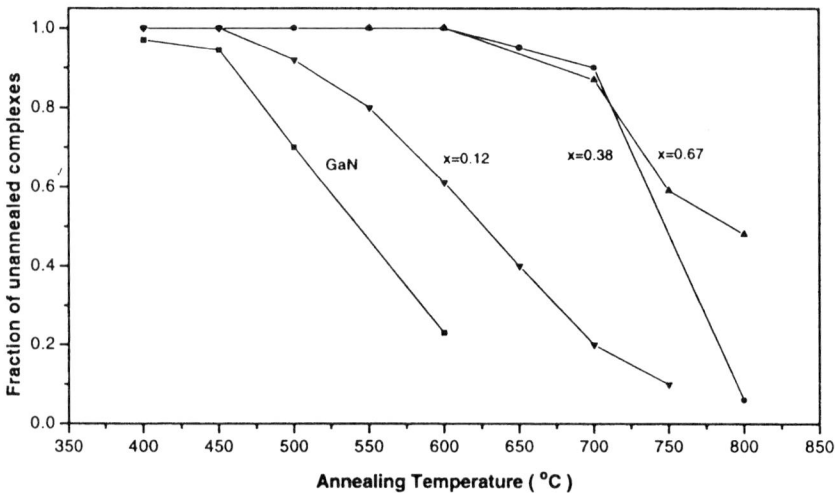

FIG. 15. Dependence of the fraction of unannealed native donor-hydrogen complexes in AlGaN.

role in the frequently observed decrease in carrier concentration in as-grown AlGaN layers with high Al mole fractions.

2. IN-CONTAINING NITRIDES

Lower thermal stabilities were found for In-containing nitrides. Figure 16 shows the fraction of passivated donors remaining in $In_{0.5}Al_{0.5}N$ and $In_{0.25}Al_{0.25}Ga_{0.5}N$ initially hydrogenated at 250°C for 30 min as a function of posthydrogenation annealing temperature. Both samples displayed a decrease in carrier concentration of approximately an order of magnitude after hydrogen plasma exposure. The passivated donors begin to reactivate around 400°C, and by 500°C, 78% of the lost carriers were restored in InAlN and 66% in the InAlGaN. The recovery of the donor activity occurred over a broader temperature range than generally observed for other passivated dopants and is consistent with the presence of a Gaussian distribution of activation energies. This may be due to nitrogen vacancies with different numbers of specific group III neighbors surrounding them (i.e., two In and two Al versus one In and three Al). Assuming a Gaussian

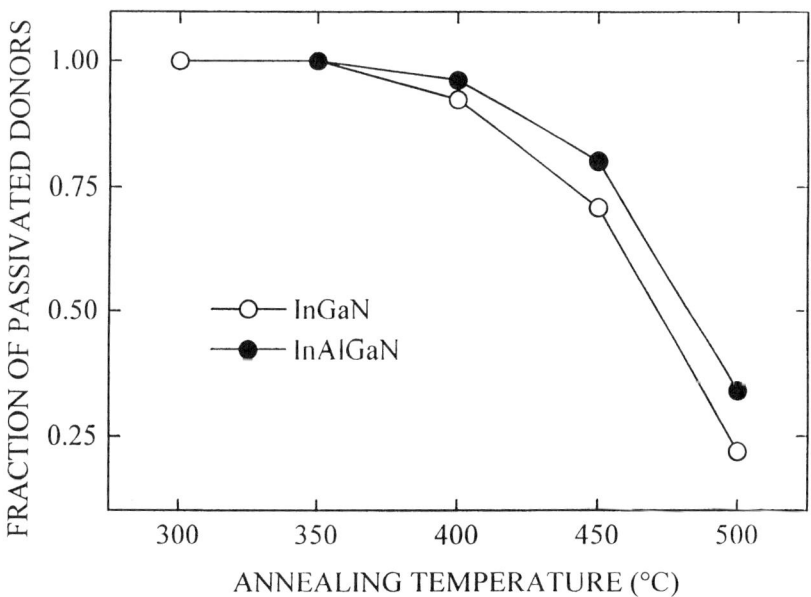

FIG. 16. Fraction of passivated donors remaining in InAlN or InGaN after deuteration at 250°C and subsequent annealing at different temperatures.

distribution of energies, we obtained values of 1.4 eV for the activation energy for donor reactivation, with a full width at half maximum of ~ 0.3 eV.

In AlN, hydrogenation at 250°C for 30 min was able to produce a uniform concentration of ^{2}H of $\sim 10^{21}$ cm^{-3} throughout a 1-μm-thick sample. Since this is well above the native donor concentration, the ^{2}H is presumably bound at structural defects in the material. Annealing at 800°C did not measurably affect the ^{2}H profile, but after 900°C, the ^{2}H plasma tends to form a high density of In droplets on its surface because of preferential loss of N as NH$_3$, whereas InGaN is more resistant to surface degradation. This is similar to the situation for InP and the related ternaries InGaAs and InAlAs.

3. MECHANISMS

In both Si and GaAs, injection of minority carriers either by forward biasing of a diode structure or illumination with above-bandgap light produces dissociation of neutral acceptor-hydrogen or donor-hydrogen complexes at temperatures at which they are normally thermally stable. While the details of the reactivation process are not clearly established, it is expected that for an acceptor A the reactions likely can be described by

$$(AH)^0 \leftrightarrow A^- + H^- \tag{5}$$

$$H^+ + e^- \leftrightarrow H^0 \tag{6}$$

The neutral hydrogen most likely forms diatomic or larger clusters with other neutral or charged hydrogen species. The mechanism for acceptor activation during the e-beam irradiation process has not been studied in detail to date. To establish that minority-carrier-enhanced debonding of Mg-H complexes in GaN is responsible for this phenomenon, we examined the effect of forward biasing in hydrogenated p-n junctions. We find that the reactivation of passivated acceptors obeys second-order kinetics and that the dissociation of the Mg-H complex is greatly enhanced under minority-carrier injection conditions.

The sample was grown on c-Al$_2$O$_3$ by MOCVD using a rotating-disk reactor. After chemical cleaning of the substrate in both acids (H$_2$SO$_4$) and solvents (methanol, acetone), it was baked at 1100°C under H$_2$. A thin (≤ 300 Å) GaN buffer was grown at 510°C, before growth of ~ 1-μm undoped material, 0.5 μm of GaN(Mg) with a carrier density of P $\approx 1.5 \times 10^{17}$ cm^{-3} after 700°C annealing and 0.3 μm GaN(Si) with a carrier density of 5×10^{18} cm^{-3}. Some of the sample was hydrogenated by annealing under NH$_3$ for 30 min at 500°C. This produces passivation of the Mg acceptors but has little effect on the Si donors.

Mesa p-n junction diodes were processed by patterning 500-μm-diameter TiAl ohmic contacts on the N-GaN by liftoff and then performing a self-aligned dry etch with an electron cyclotron resonance BCl$_3$/Ar plasma to expose the p-type GaN. E-beam evaporated NiAu was patterned by liftoff to make ohmic contact to the p-type material. The carrier profiles in the p-type layer were obtained from 10-kHz capacitance-voltage measurements at room temperature. Anneals were carried out in the dark at 175°C under two different types of conditions. In the first the diode was in the open-circuit configuration, while in the second the junction was forward biased at 9 mA to inject electrons into the p-type GaN. After each of these treatments the samples were returned to 300 K for measurement of the net electrically active acceptor profile in this layer.

Figure 17 shows a series of acceptor concentration profiles measured on the same p-n junction sample after annealing at 175°C under forward bias conditions. After the NH$_3$ hydrogenation treatment, the electrically active acceptor density decreased from 1.5×10^{17} to ~ 6–7×10^{16} cm^{-3}. If the

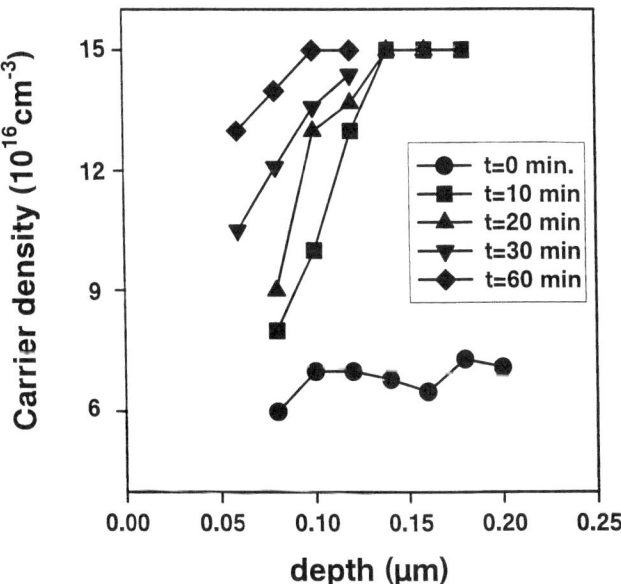

FIG. 17. Carrier concentration profiles in hydrogenated GaN(Mg) after annealing for various times at 175°C under forward-bias conditions.

subsequent annealing was carried in the open-circuit configuration, there was no change in the carrier profile for periods up to 20 hours at 175°C. By sharp contrast, Fig. 17 shows that for increasing annealing times under minority-carrier injections conditions there is a progressive reactivation of the Mg acceptors with a corresponding increase in the hole concentration. After 1 hour, the majority of these acceptors have been reactivated. Clearly, therefore, the injection of electrons has a dramatic influence on the stability of the MgH complexes. The Mg reactivation has a strong dependence on depth into the p-type layer, which may result from the diffusion distance of the injected electrons prior to recombination. We rule out heating of the sample during forward biasing as being a factor in the enhanced dissociation of the neutral dopant-hydrogen complexes. The samples were thermally bonded to the stainless steel stage, and the junction temperature rise is expected to be minimal ($\leq 10°C$). Moreover, from separate experiments we found that reactivation of the Mg did not begin until temperatures above $\sim 450°C$ under zero-bias conditions.

Previous experiments on minority-carrier-enhanced reactivation of hydrogen-passivated dopants in Si and GaAs have found that for long annealing times the kinetics can be described by a second-order equation:

$$d[N_A - N(t)]/dt = C[N_A - N(t)]^2 \tag{7}$$

where N_A is the uniform Mg acceptor concentration in the nonhydrogenated sample, $N(t)$ is the acceptor concentration in the hydrogenated GaN after forward-bias annealing for time t, and C is a second-order annealing parameter.

In order to quantitatively analyze the reactivation kinetics of the Mg-H complexes in GaN, we measured the inactive acceptor concentrations $N_A - N(t)$ using the capacitance-voltage measurements at a depth of 0.1 μm in the P-GaN layer. Figure 18 shows that there is a linear relationship between $[N_A - N(t)]^{-1}$ and annealing time t, confirming that the reactivation process can be described by a second-order equation with $C = 4 \times 10^{-20}$ cm^3 sec^{-1}. This value is consistent with those obtained in Si and GaAs, where minority-carrier-enhanced dopant reactivation also has been reported.

The rate of reactivation of passivated acceptors depends on the injected minority-carrier density. Moreover, for short annealing times it was found that the dopant reactivation occurred at a faster rate than predicted by the second-order equation for very short annealing times and that the annealing process was rate limited by the formation of stable, electrically inactive diatomic H species. At this point there have not been enough studies of the various states of hydrogen in GaN as determined by IR spectroscopy,

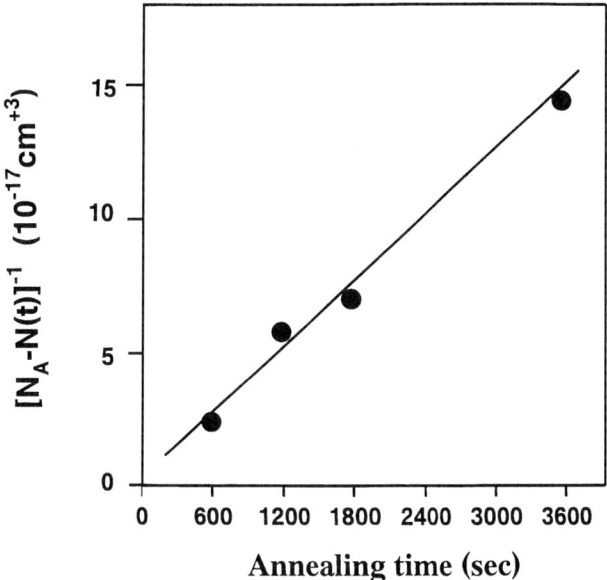

FIG. 18. Plot of inverse net active Mg concentration determined from Fig. 17 at a depth of 0.1 μm from the junction as a function of forward-bias annealing time.

channeling, or secondary-ion mass spectrometry for us to conclude anything about the ultimate fate of the atomic hydrogen once it has dissociated from the Mg-H complex, but it is likely that it then reacts with other hydrogen atoms to form diatomic or larger clusters. A strong dependence of reactivation rate on injected minority-carrier density would indicate the presence of a charge state for hydrogen and therefore influence the conversion of H^+ into the neutral state and then into the final hydrogen complexes.

The fact that the Mg-H complexes are unstable against minority-carrier injection has implications for several GaN-based devices. First, in a laser structure, the high level of carrier injection would rapidly dissociate any remaining Mg-H complexes and thus would be forgiving of incomplete removal of hydrogen during the postgrowth annealing treatment. In a heterojunction bipolar transistor, the lower level of injected minority carriers also would reactivate passivated Mg in the base layer, leading to an apparent time-dependent decrease in gain as the device was operated.

In summary, we have shown that hydrogen-passivated Mg acceptors in GaN may be reactivated at 175°C by annealing under minority-carrier injection conditions. The reactivation follows a second-order kinetics process in which the $(Mg-H)^0$ complexes are stable to $\geq 450°C$ in thin, lightly doped GaN layers. In thicker, more heavily doped layers where retrapping of hydrogen at the Mg acceptors is more prevalent, the apparent thermal stability of the passivation is higher, and annealing temperatures up to 700°C may be required to achieve full activation of the Mg. Our results suggest that the mechanism for Mg activation in e-beam-irradiated GaN is minority-carrier-enhanced debonding of the hydrogen.

4. HETEROSTRUCTURES

The diffusion and trapping behavior of hydrogen in device structures is more complicated than in single-layer structures. For example, light-emitting diodes or laser diodes contain both n- and p-type GaN cladding layers with one or more InGaN active regions. The first laser reported by Nakamura et al. (1996) contained 26 InGaN quantum wells. In other III-V semiconductors, the diffusivity of atomic hydrogen is a strong function of conductivity type and doping level, since trapping by acceptors is usually more thermally stable than trapping of hydrogen by donor impurities. Moreover, hydrogen is attracted to any region of strain within multilayer structures and has been shown to pile up at heterointerfaces in the GaAs/Si, GaAs/InP, and GaAs/AlAs materials systems. Therefore, it is of interest to investigate the reactivation of acceptors and trapping of hydrogen in double-heterostructure GaN/InGaN samples, since these are the basis for optical emitters. We find that the reactivation of passivated Mg acceptors also depends on the annealing ambient, with an apparently higher stability for annealing under H_2 rather than N_2. Hydrogen is found to redistribute to the regions of highest defect density within the structure.

The double-heterostructure sample consisted of 300-Å, low-temperature GaN buffer, 3.3 μm of N^+ (Si = 10^{18} cm^{-3}) GaN, 0.1 μm undoped InGaN, and 0.5 μm P^+ GaN (P = 3×10^{17} cm^{-3}, Mg-doped). In the immediate vicinity of the N–GaN/Al$_2$O$_3$ interface, the defect density was high but was reduced with increasing film thickness. However, after growth of the InGaN active layer, the threading dislocation density increased due to thermal decomposition of the top of the InGaN on raising the temperature to grow the P-GaN.

XTEM of the DH-LED showed dislocations as dark lines propagating in the direction normal to the substrate. Most of the dislocations appeared to traverse the entire double heterostructure, while some appeared to bend and follow the interface for a short distance before threading out to the surface. The nature of the threading dislocations was studied by conventional XTEM using the $\mathbf{g} \cdot \mathbf{b} = 0$ criteria. The dislocation will be invisible when $\mathbf{b}$ lies in the reflecting plane.

Some of the dislocations were invisible both in **g2** = (0002) and **g5** = (1$\bar{1}$01), and because **b** was common to both reflections, **b** was found to be 1/3[11$\bar{2}$0]. Assuming that the growth is the same as the translation vector of the dislocation, these defects would be pure edge type in nature. The average threading dislocation density also was found along the plane normal to the growth direction. The dislocation density was found to be $\sim 8 \times 10^{10}$ cm^{-2}.

The double-heterostructure sample was exposed to an electron cyclotron resonance plasma (500 W of microwave power, 10 mtorr pressure) for 30 min at 200°C. The hole concentration in the P-GaN layer was reduced from 3×10^{17} to $\sim 2 \times 10^{16}$ cm^{-3} by this treatment, as measured by capacitance voltage at 300 K. Sections from this material were then annealed for 20 min at temperatures from 500–900°C under an ambient of either N_2 of H_2 in a Heatpulse 410T furnace. Figure 19 shows the percentage of passivated Mg remaining after annealing at different temperatures in these two ambients. In the case of N_2 ambients, the Mg-H complexes show a lower apparent thermal stability (by ~ 150°C) than with H_2 ambients. This has been reported previously for Si donors in InGaP and AlInP and Be and Zn acceptors in InGaP and AlInP, respectively (Pearton, 1994) and most likely is due to in-diffusion of hydrogen from the H_2 ambients, causing a competition between passivation and reactivation. Therefore, an inert atmosphere is clearly preferred for the postgrowth reactivation anneal of P-GaN to avoid any ambiguity as to when the acceptors are completely active. Previous experimental results by Brandt *et al.* (1994) and total energy

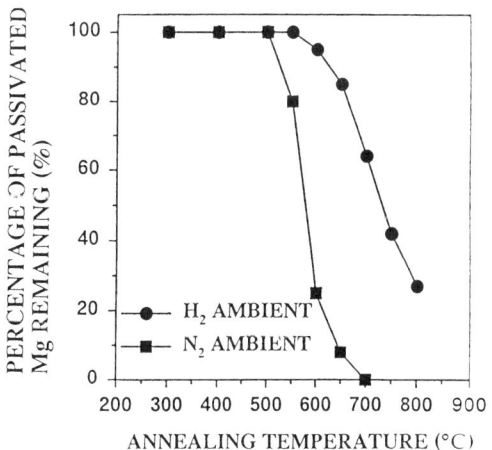

FIG. 19. Formation of passivated Mg acceptors remaining in hydrogenated *p*-type GaN after annealing for 20 min at various temperatures in either N_2 or H_2 ambients.

calculations by Neugebauer and Van de Walle (1995) suggest that considerable diffusion of hydrogen in GaN might be expected at ≤600°C.

Other sections of the double-heterostructure material were implanted with $^2H^+$ ions (50 keV, 2×10^{15} cm^{-2}) through an SiN$_x$ cap in order to place the peak of the implant distribution within the P$^+$GaN layer. Some of these samples were annealed at 90°C for 20 min under N$_2$. As shown in the secondary ion mass spectrometry (SIMS) profiles of Fig. 20, the ^2H diffuses out of the P$^+$GaN layer and piles up in the defective InGaN layer, which as we saw from the TEM results suffers from thermal degradation during growth of the P$^+$GaN. Note that there is still sufficient ^2H in the P$^+$GaN ($\sim 10^{19}$ cm^{-3}) to passivate all the acceptors present, but electrical measurements show that the p-doping level was at its maximum value of $\sim 3 \times 10^{17}$ cm^{-3}. These results confirm that as in other III-V semiconductors hydrogen can exist in a number of different states, including being

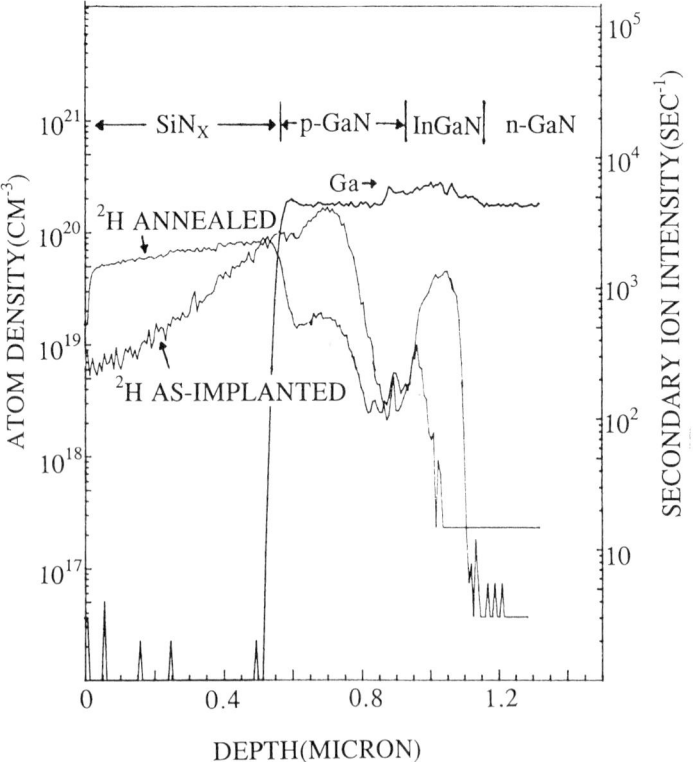

FIG. 20. SIMS profiles of ^2H in an implanted (50 keV, 2×10^{15} cm^{-2} through a SiN$_x$ cap) double-heterostructure sample before and after annealing at 900°C for 20 min.

bound at dopant atoms or in an electrically inactive form that is quite thermally stable. We expect that after annealing above 700°C all the Mg-H complexes have dissociated, and the electrical measurements show that they have not re-formed. In other III-V semiconductors, the hydrogen in p-type material is in a bond-centered position, forming a strong bond with a nearby N atom and leaving the acceptor threefold coordinated. Annealing breaks this bond and allows the hydrogen to make a short-range diffusion away from the acceptor, where it probably meets up with other hydrogen atoms, forming molecules or larger clusters that are relatively immobile and electrically inactive. This seems like a plausible explanation for the results of Figs. 19 and 20, where the Mg electrical activity is restored by 700°C but hydrogen is still present in the layer at 900°C. In material hydrogenated by implantation, there is almost certainly a contribution to the apparently high thermal stability by the hydrogen being trapped at residual implant damage, as is evident by the fact that the ^{2}H profile retains a Pearson IV type distribution even after 900°C annealing. The other important point from Fig. 20 is that as in other defective crystal systems, hydrogen is attracted to regions of strain, in this case the InGaN sandwiched between the adjoining GaN layers.

In conclusion, the apparent thermal stability of hydrogen-passivated Mg acceptors in P-GaN depends on the annealing ambient, as it does in other compound semiconductors. While the acceptors are reactivated at $\leq 700°$C for annealing under N_2, hydrogen remains in the material until much higher temperatures and can accumulate in defective regions of double-heterostructure samples grown on Al_2O_3. It will be interesting to compare the redistribution and thermal stability of hydrogen in homoepitaxial GaN in order to assess the role of the extended defects present in the currently available heteroepitaxial material.

V. Role of Hydrogen During Processing

Light-ion implantation is typically used in III-V technology to produce high-resistivity material through the introduction of point defects and complexes that are electron and/or hole traps. The thermal stability of this effect is an important consideration for designing the device processing sequence. High-resistivity material is produced by implantation of any ion, but annealing at a sufficiently high temperature will remove the trap states, and the resistivity of the material will revert to its original value. If the implanted species has a chemical deep level in the bandgap, then the material will remain resistive even at high annealing temperatures.

1. IMPLANT ISOLATION

A number of reports have discussed implant isolation in GaN using H, He, N, or O implantation. Binari et al. (1995) found that higher doses of implanted hydrogen were needed in comparison with He in order to isolate N^+GaN (Fig. 21). This is expected on the basis of the number of stable defects created by the two different species. The H-implanted material maintained high resistivity ($>10^9$ $\Omega\cdot$cm) only to $\sim 250°$C, and by 400°C, all samples had resistivity of only $\sim 10^3$ $\Omega\cdot$cm. In Fig. 22 we compare the thermal stability of proton implant isolation in N-GaN and GaAs. The stability is surprisingly higher in GaAs, but higher stabilities for GaN can be obtained with N^+, O^+ or F^+. The maximum incorporation depth of implanted protons in GaN is $\sim 2\,\mu$m for a typical 200-kV implanter. While the deep levels in H-implanted GaN anneal out at $\leq 200°$C, temperatures above 800°C are needed to cause most of the hydrogen to actually leave the material.

2. WET PROCESSING

In addition to direct implantation of hydrogen, it has been found that it can be unintentionally incorporated during many processing steps, including boiling in water, dry etching, chemical vapor deposition of dielectrics, and annealing in H_2 or NH_3 (Pearton et al., 1996). For example, GaN boiled in D_2O at 100°C for 30 min showed incorporation of ^{2}H up to

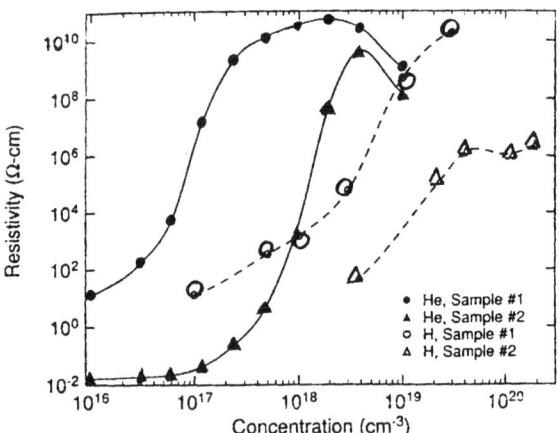

FIG. 21. As-implanted GaN resistivity for He- and H-implanted material as a function of implant volume concentration. (After Binari et al., 1995.)

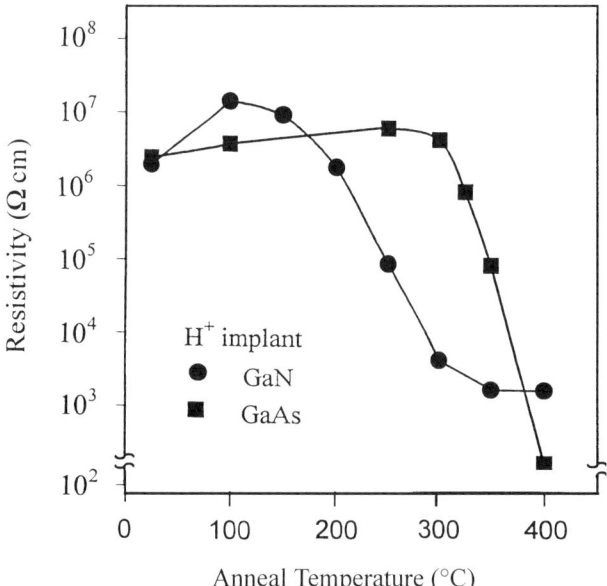

FIG. 22. Thermal stability of proton implant isolation in N-GaN or GaAs.

~1 µm in the material. It is not known how the columnar nature of the GaN growth affects the hydrogen incorporation, but comparisons with bulk crystals or homoepitaxial layers will be enlightening. Boiling in water has been shown previously to be an effective method of introducing hydrogen into Si, and it generally diffuses very rapidly under these conditions because the H flux is low energy to avoid the H-H pairing reactions seen in plasma-exposed material.

KOH-based solutions readily etch AlN selectively over GaN, with etch rates that are related to temperature and material quality. Exposure of undoped GaN to a solution of KOH/D_2O for 20 min leads to incorporation of ^2H at a concentration of $\sim 10^{17}$ cm^{-3} to a depth of ~ 0.5 µm. This shows that even at quite low processing temperatures (25°C), atomic hydrogen can be readily diffused into GaN.

3. DEPOSITION AND ETCHING

Another source of hydrogen incorporation is during PECVD of dielectrics such as SiN_x and SiO_2 for masking and surface passivation. It is known from GaAs work that PECVD of these dielectrics using SiN_x can cause extensive hydrogen passivation effects. We have previously reported that

ECR-CVD of SiN_x onto GaN at a rate of 250 Å/min at 250°C produced ^2H incorporation depths of $\sim 0.7\,\mu m$ using SiD_4/N_2. Postdeposition annealing of the SiN_x/GaN structure at 500°C for 20 min did not produce any further in-diffusion of ^2H from the SiN_x, indicating that the source of the initial incorporation is the exposure of the GaN surface to the silane discharge. Hydrogen incorporation in GaAs during similar ECR-CVD of silicon nitride can be reduced by employing an initially fast deposition rate to encapsulate the surface as quickly as possible.

There are a variety of dry-etch chemistries for GaN and related alloys, including Cl_2, BCl_3, $SiCl_4$, CCl_2F_2, CH_4/H_2, HBr, HI, ICl, and IBr. Additives such as Ar, N_2, and H_2 are often included to enhance sputter-assisted desorption of the etch products or to balance removal of Ga and N etch products. In particular, H_2 addition is effective in removing N as NH_3. ECR $CH_4/H_2/Ar$ discharges caused N-GaN to become more insulating and insulating films to become conducting. However, these effects may have been due to a combination of ion-induced defects (since the mobility decreased in N-GaN) and preferential loss of N from the near-surface (100 Å). Annealing at 800°C for 30 sec restored most of the original electrical properties. If elevated sample temperatures are employed to enhance desorption of In from In-containing nitrides, then hydrogen-indiffusion can be very substantial, with incorporation depths of $0.5-2\,\mu m$ in 40 sec at 170°C, that is, diffusivities of $10^{-9}-6 \times 10^{-11}\,cm^2/V \cdot sec$. Similar results were found in AlN and GaN in addition to $In_xGa_{1-x}N$. As mentioned earlier, it is not clear what role the extensive defect density in GaN heteroepitaxial samples plays, but it likely enhances the hydrogen diffusion substantially. The effect of H in dry etching of GaN is therefore twofold: It may diffuse into the bulk of the sample and passivate dopants or defects, leading to an increased resistivity of the material, or it may produce preferential loss of N right at the surface, leading to a thin N^+ layer ~ 100 Å thick. In some cases if the loss of N is not too severe, annealing at 400–800°C may be able to restore the surface stoichiometry, but in In-containing nitrides the vast difference in volatility of In and N products usually means that In droplets form, and the surface cannot be restored by annealing without first removing the droplets in HCl. Room-temperature etching minimizes indiffusion of hydrogen from the plasma.

VI. Theory of Hydrogen in Nitrides

Brandt et al. (1994) reported a new PL line at 3.35 eV after hydrogenation of p-type and unintentionally n-type GaN, suggesting the introduction of a hydrogen-related donor level. However, these same workers created some confusion by reporting local vibrational modes around 2200 cm^{-1} for Mg-H complexes in GaN. This led to a series of conflicting theoretical work that was guided by incorrect experimental data.

Okamoto *et al.* (1996) used local density approximation of the density functional theory to calculate the stable positions of Mg and Mg-H in GaN. Mg was found to occupy a Ga substitutional site, producing a shallow acceptor level. They found that the bond-centered (BC) site between Mg-N was the most stable site for hydrogen by ~ 0.1 eV relative to an antibonding (AB) site near the Mg atom. The N and intervening H were found to form a strong bond with the calculated bond length (1.02 Å) close to that of the NH_3 molecule (1.01 Å) and no bond charge between Mg and H. The distance between Mg and H was 1.96 Å, caused by displacement of the Mg atom from the substitutional site by 0.64 Å in the out-bonding direction (away from the position of the hydrogen). It was to be energetically favorable for the H to form the Mg-H complex rather than remain an interstitial. The calculated vibrational frequency was ~ 3490 cm^{-1}, similar to H in NH_3 molecules (3444 cm^{-1}).

Bosin *et al.* (1996) used *ab initio* calculations performed within local density functional theory to understand the energetics and geometry of H and its complexes in GaN. They found H to exhibit negative-U behavior, with thermal ionization energies for H_{III}^- and H_V^+ to be $E^{0/-} = 0.8$ eV and $E^{+/0} = 2.0$ eV, respectively. They calculated Be_{Ga} to be a relatively shallow impurity level, while C_N had a deep acceptor level ($E_V + 0.40$ eV). Mg-H complexes were calculated to have vibrational modes at ~ 3000 cm^{-1} for AB_V sites for H and at 3600–3900 cm^{-1} for BC-site hydrogen. It was found that Be_{Ga}, C_N, Ca_{Ga}, and Zn_{Ga} acceptors would all be susceptible to hydrogen passivation.

Estreicher and Maric (1996) employed molecular cluster calculations to examine hydrogen in cubic GaN. They considered three possibilities for the effect of hydrogen:

1. Passivation (an electrical level moves from the gap to a band)
2. Activation (an electrical level moves from a band to the gap)
3. Level shifting (a level shifts within the gap, e.g., a shallow level becomes deep)

The latter case may be important in wide-gap semiconductors because energy levels have to shift a lot to disappear from the gap. They found that H^+ is bound to N in a BC-like position, H^0 is at a BC site with H bound to Ga (and an activation energy for diffusion of <1 eV), while H^- is at the AB_{Ga} site (and an activation energy for diffusion of 1.5 eV). H_2 molecules were found to be stable at the T_{Ga} site.

Neugebauer and Van de Walle (1996) performed density functional calculations and found the following:

1. Isolated interstitial hydrogen has a large negative-U energy (2.5 eV), suggesting that H^0 is unstable.
2. The hydrogen donor level is near the conduction band, and the acceptor level is deep ($E_V + 0.9$ eV). For $E_F < 2$ eV, the stable state is H^+, while above that, the stable state is H^-. For $E_F = 2$ eV, H_2 molecules might form by association of H^+ and H^-.
3. H^+ is at the AB_N site, with N-H ≈ 1 Å bond length, but the BC configuration is only 0.1 eV higher in energy. The activation energy for diffusion of H^+ is 0.7 eV.
4. H^- is at the AB_{Ga} site, with an activation energy for diffusion of 3.4 eV.
5. H-Mg complexes have H at the AB_N site (Mg$\cdots$N-H), with a N-H stretching frequency of 3360 cm^{-1} and a pair dissociation energy of 1.5 eV. The energetics are summarized in Fig. 23.

Therefore, among theorists, there is agreement on the following:

1. H^- is at the AB_{Ga} site, with a large activation energy for diffusion.
2. H^- is within ~ 1 Å of N.
3. H is strongly bound to the N atom in the Mg-H pair.

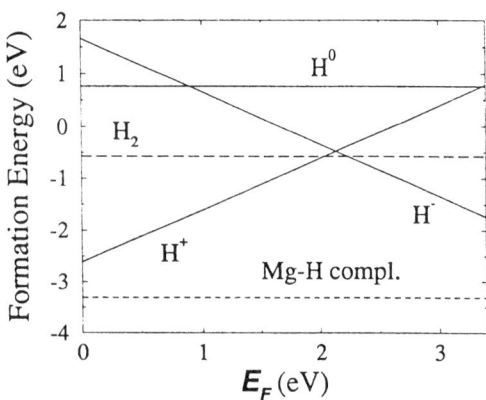

FIG. 23. Formation energy of H^+, H^-, and H^0 as a function of Ferris level position. Results are shown from Mg-H complexes and H_2 molecules. (After Neugebauer et al., 1995.)

There is still some disagreement on the following issues:

1. Whether H^0 is at the AB_{Ga} or BC site
2. Whether H^+ is at the AB_N site or at the BC site
3. Whether in the Mg-H complexes the H is at the AB_N side of N with a high N-H stretching frequency or at the BC site with a lower frequency

Van Vechten et al. (1992) suggested that hydrogen enables p-type doping by suppression of native defects and suggested the incorporation (and subsequent removal) of hydrogen as a method for improving doping in wide-bandgap semiconductors. This was expanded on by Neugebauer and Van de Walle (1996), who suggested that for this method to work, hydrogen would need to be the dominant compensating center, the dissociated hydrogen must have a high diffusion coefficient, and the energy needed to dissociate H-impurity complexes and remove the H would need to be lower than the formation energy of native defects.

Gotz et al. (1996) corrected their earlier work covering LVMs of Mg-H complexes in GaN, reporting a value of $3125\,cm^{-1}$ for Mg-H and $2321\,cm^{-1}$ for Mg-^{2}H, which were attributed to stretch modes of these complexes. Their previous work, reporting LVMs at $\sim 2200\,cm^{-1}$, presumably resulted from other defects. As-grown material was highly resistive ($10^{10}\,\Omega \cdot m$ at 400 K), while annealing at 800°C reduced this to $2\,\Omega \cdot cm$. Subsequent hydrogenation at 600°C increased the resistivity to $10^5\,\Omega \cdot cm$ at 400 K, due mainly to the fact that this is too high a temperature to prevent substantial dissociation of the Mg-H complexes. There was a relatively poor correlation of IR signal intensity with active Mg concentration. The IR stretch frequency is close to that of H in NH_3 ($3444\,cm^{-1}$) and similar to the value of N-H complexes in ZnSe ($3195\,cm^{-1}$).

Schematics of the possible configurations of hydrogen-dopant complexes in GaN are shown in Fig. 24. For donor dopants on either the Ga site (e.g., Si) or the N site (e.g., S), the hydrogen is in an AB position either attached to the dopant in the case of group IV donors or attached to the Ga neighbor in the case of a group VI donor. At this stage no reports of donor dopant passivation have been made for the reasons outlined earlier. In acceptor dopants, the hydrogen may be at a BC site bonded predominantly either to the acceptor or to an N neighbor, respectively, depending on whether the acceptor is from column IV or II of the periodic table, or in the AB position, creating an N-H bond. Current thinking favors the latter for Mg-H, as outlined earlier, but more work is needed to definitively establish this configuration and to see if the BC site is favored for the other acceptors.

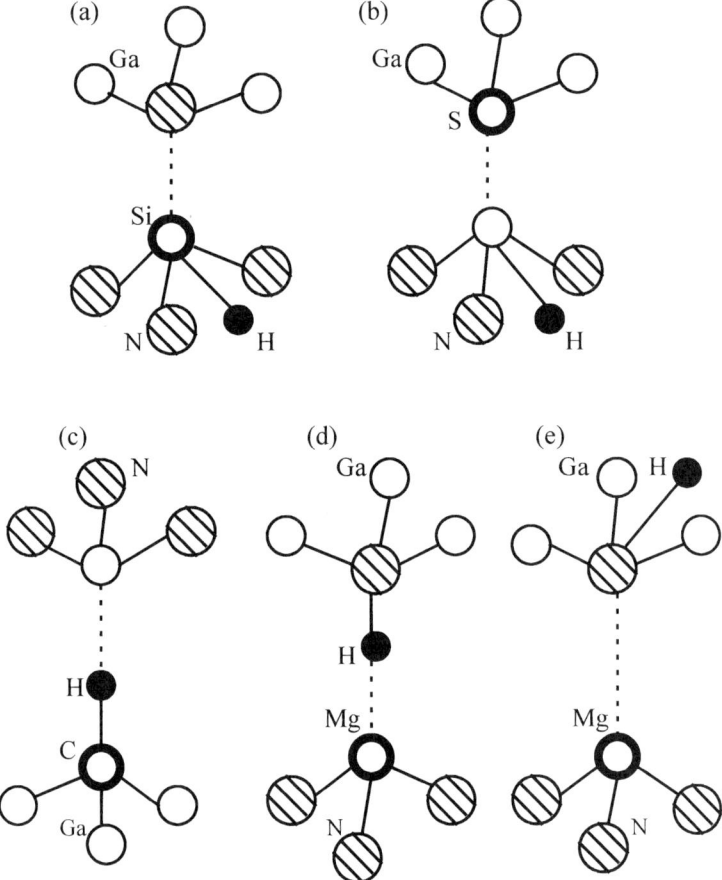

FIG. 24. Schematic representations of H-dopant complexes in GaN. The H occupies either an antibonding (AB) or a bond-centered (BC) site.

VII. Summary and Conclusions

Hydrogen plays a particularly prominent role in p-type GaN because of its ability to passivate acceptors, requiring postgrowth annealing of MOCVD material in order to electrically activate the acceptors. Donor passivation is more elusive but probably occurs under the right conditions. The use of deuterated gases has shown that hydrogen readily enters GaN and related materials during low-temperature processes such as CVD of dielectrics, plasma etching, wet etching, boiling in solvents, and contact sintering (metals may act as catalysts for H_2 dissociation). Therefore, one faces the situation that hydrogen may be incorporated unintentionally into GaN at many stages of a device-fabrication process, particularly when p-type layers are uppermost in the structure (i.e., LEDs and lasers). The out-diffusion and hydrogen retrapping behavior is more complex in multilayer (heterostructure) samples. Reactivation of passivated acceptors may be accomplished by thermal annealing or minority carrier injection to dissociate the dopant-hydrogen complexes. It will be very interesting to look at hydrogen incorporation in GaN bulk crystals with controlled doping levels (usually most of these samples are degenerately n-type, $N > 10^{19}\,cm^{-3}$) in order to establish the role of gain and column boundaries in heteroepitaxial material in assisting hydrogen diffusion.

ACKNOWLEDGMENTS

The contributions of C. R. Abernathy, C. B. Vartuli, J. M. Zavada, R. G. Wilson, R. J. Shul, J. C. Zolper, F. Ren, K. S. Jones, and S. Bendi to this work are gratefully acknowledged. This work was supported in part by the NSF (DMR-9421169) and a DARPA grant (A. Husain) monitored by the Air Force Office of Scientific Research (AFOSR).

REFERENCES

Abernathy, C. R., MacKenzie, J. D., Pearton, S. J., and Hobson, W. S. (1995). *Appl. Phys. Lett.*, **66**, 1969.
Amano, H., Kito, M., Hiramatsu, K., and Akasaki, I. (1989). *Jpn. J. Appl. Phys.*, **28**, L2112.
Amano, H., Akasaki, I., Kozawa, T., Sawaki, N., Ikeda, K., and Ishii, J. (1988). *J. Lumin.*, **4**, 121.
Binari, S. C., Dietrich, H. B., Kelner, G., Rowland, L. B., Doverspike, K., and Wickenden, D. K. (1995). *J. Appl. Phys.*, **78**, 3008.
Bosin, A., Fiorentini, V., and Vanderbilt, D. (1996). *Mater. Res. Soc. Symp. Proc.*, **395**, 503.
Brandt, M. S., Ager, J. W., Gotz, W., Johnson, N. M., Harris, J. S., Molnar, R. J., and Moustakas, T. D. (1994). *Phys. Rev. B*, **49**, 14758.
Burchard, A., Deicher, M., Forkel-Wirth, D., Haller, E. E., Magerle, R., Prospero, A., and Stotzler, R. (1997). *Mater. Res. Soc. Symp.*, **449**, 961.
Estreicher, S. K., and Maric, D. M. (1996). *Mater. Res. Soc. Symp. Proc.*, **423**, 613.
Gotz, W., Johnson, N. M., Bour, D. P., McCluskey, M. D., and Haller, E. E. (1996). *Appl. Phys. Lett.*, **69**, 3725.
Johnson, M. A. L., Yu, Z., Boney, C., Hughes, W. C., Cook, J. C., Schetzina, J. F., Zhao, H., Skromme, B. J., and Edmond, J. A. (1997). *Mater. Res. Soc. Symp. Proc.*, **449**, 215.

Lee, J. W., Pearton, S. J., Zolper, J. C., and Stall, R. A. (1996). *Appl. Phys. Lett.*, **68**, 2102.
Lee, J. Y., and Yong, Y.-J. (1997). *Mater. Res. Soc. Symp. Proc.*, **449**, 999.
Li, Y., Lu, Y., Shen, H., Wraback, M., Hwang, C.-Y., Schurman, M., Mayo, W., Salagaj, T., and Stall, R. A. (1996). *Mater. Res. Soc. Symp. Proc.*, **395**, 369.
Moustakas, T. D., Lei, T., and Molnar, R. J. (1993). *Mater. Res. Soc. Symp. Proc.*, **281**, 753.
Myers, S. M., Baskes, M. I., Birnbaum, H. K., Corbett, J. W., DeLeo, G. G., Estreicher, S. K., Haller, E. E., Jena, P., Johnson, N. M., Kircheim, R., Pearton, S. J., and Stavola, M. (1992). *Rev. Mod. Phys.*, **64**, 559.
Nakamura, S., Senoh, M., and Mukai, T. (1991). *Jpn. J. Appl. Phys.*, **30**, L1708.
Nakamura, S., Senoh, M., and Mukai, T. (1992a). *Jpn. J. Appl. Phys.*, **31**, L139.
Nakamura, S., Iwasa, N., Senoh, M., and Mukai, T. (1992b). *Jpn. J. Appl. Phys.*, **31**, L1258.
Nakamura, S., Senoh, M., Nagahama, S., Iwasa, N., Yamada, T., Matsushita, T., Kyoku, H., and Sugimoto, Y. (1996). *Jpn. J. Appl. Phys.*, **35**, L74.
Neugebauer, J., and Van der Walle, C. G. (1995). *Phys. Rev. Lett.*, **75**, 4452.
Neugebauer, J., and Van der Walle, C. G. (1996). *Mater. Res. Soc. Symp.*, **423**, 619; *Appl. Phys. Lett.*, **68**, 1829.
Ohba, Y., and Hatano, A. (1994). *Jpn. J. Appl. Phys.*, **33**, L1367.
Okamoto, Y., Saito, M., and Oshiyama, A. (1996). *Jpn. J. Appl. Phys.*, **35**, L807.
Pankove, J. J., and Johnson, N. M., eds. (1991). *Hydrogen in Semiconductors*, **34**. San Diego: Academic Press.
Pearton, S. J. (1994). *Int. J. Mod. Phys. B*, **8**, 1247.
Pearton, S. J. (1997). In *Optoelectronic Properties of Semiconductors and Superlattices*, **2**, M. O. Manasreh, ed. New York: Gordon and Breach.
Pearton, S. J., Corbett, J. W., and Shi, T. S. (1987). *Appl. Phys. A.*, **43**, 153.
Pearton, S. J., Corbett, J. W., and Stavola, M. (1992). *Hydrogen in Crystalline Semiconductors*. Berlin: Springer-Verlag.
Pearton, S. J., Abernathy, C. R., and Ren, F. (1994). *Electron. Lett.*, **30**, 527.
Pearton, S. J., Abernathy, C. R., Vartuli, C. B., Lee, J. W., MacKenzie, J. D., Wilson, R. G., Shul, R. J., Ren, F., and Zavada, J. M. (1996). *J. Vac. Sci. Technol.*, **A14**, 831.
Piner, E. L., Behbehani, M. K., El-Masry, N. A., Roberts, J. E., McIntosh, F. C., and Bedair, S. M. (1997). *Appl. Phys. Lett.*, **65**, 3206.
Polyakov, A. Y., Shin, M., Pearton, S. J., Skowronski, M., Greve, D. W., and Freitas, J. A. (1996). *Res. Soc. Symp. Proc.*, **423**, 607.
Soto, M. (1996). *Appl. Phys. Lett.*, **68**, 935.
Strite, S. (1994). *Jpn. J. Appl. Phys.*, **33**, L699.
Van Vechten, J., Zook, J. D., Hornig, R. D., and Goldenberg, B. (1992). *Jpn. J. Appl. Phys.*, **31**, 362.
Zolper, J. C., Wilson, R. G., Pearton, S. J., and Stall, R. A. (1996). *Appl. Phys. Lett.*, **68**, 1945.

CHAPTER 11

Theory of Hydrogen in GaN

Jörg Neugebauer

FRITZ-HABER-INSTITUT BERLIN
BERLIN, GERMANY

Chris G. Van de Walle

XEROX PALO ALTO RESEARCH CENTER
PALO ALTO, CALIFORNIA

I. INTRODUCTION 479
II. METHOD 481
   1. *Defect Concentrations and Solubility* 481
   2. *Energetics, Atomic Geometries, and Electronic Structure* 482
III. MONATOMIC HYDROGEN IN GaN 483
   1. *Atomic Geometries and Stable Positions* 484
   2. *Migration Path and Diffusion Barriers* 486
   3. *Formation Energies and Negative-U Effect* 486
IV. HYDROGEN MOLECULES IN GaN 489
V. HYDROGEN-ACCEPTOR COMPLEXES IN GaN 489
   1. *The Mg-H Complex* 490
   2. *Other H-Acceptor Complexes* 492
VI. COMPLEXES OF H WITH NATIVE DEFECTS 492
   1. *Hydrogen Interacting with Nitrogen Vacancies* 493
   2. *Hydrogen Interacting with Gallium Vacancies* 494
VII. ROLE OF HYDROGEN IN DOPING GaN 495
   1. *Doping in the Absence of Hydrogen* 495
   2. *Doping in the Presence of Hydrogen* 497
   3. *Activation Mechanism of the Dopants* 498
VIII. GENERAL CRITERIA FOR HYDROGEN TO ENHANCE DOPING 499
IX. CONCLUSIONS AND OUTLOOK 500
   REFERENCES 501

I. Introduction

GaN exhibits some unique properties among semiconductors such as a large direct bandgap ($E^{\mathrm{gap}} = 3.5\,\mathrm{eV}$), a high thermal conductivity, and very strong chemical bonds. Because of these properties, the materials system consisting of GaN and its alloys with AlN and InN has been considered ideal for fabricating optoelectronic devices in the blue/ultraviolet region of the optical spectrum (Maruska and Tietjen, 1969). Furthermore, the large melting temperature and the high thermal conductivity make it a good choice for high-temperature and/or high-power devices (see, e.g., Morkoc *et al.*, 1994). Despite significant research in the 1970s and 1980s, severe doping problems hampered the development of GaN technology.

First, as-grown GaN was commonly *n*-type with large carrier background concentrations of up to 10^{20} cm^{-3}. The high background concentrations commonly were attributed to the presence of N vacancies (see, e.g., Maruska and Tietjen, 1969; Ilegems and Montgomery, 1973). However, recent theoretical (Neugebauer and Van de Walle, 1994b; Van de Walle and Neugebauer, 1997) and experimental (Götz *et al.*, 1996b, 1997; Wetzel *et al.*, 1997) investigations provide strong evidence that unintentional doping with O and Si causes the high carrier concentrations. In fact, modern growth techniques now allow the growth of GaN epilayers with background concentrations of $<10^{16}$ cm^{-3}.

Second, *p*-type doping had been difficult to achieve. Efforts to incorporate acceptors in GaN did not produce *p*-type material. In 1989, Amano *et al.* (1989) achieved a major breakthrough when they observed that GaN grown by MOCVD (metal-organic chemical vapor deposition) can be activated by low-energy electron-beam irradiation (LEEBI). Nakamura *et al.* (1992a) later showed that the Mg activation also can be achieved by thermal annealing at 700°C under N_2 ambient. Nakamura *et al.* (1992a) further observed that the process was reversible, with *p*-type GaN reverting to semi-insulating when annealed in a NH_3 ambient, revealing the crucial role played by hydrogen.

Since many of the techniques commonly used to grow GaN, such as MOCVD and HVPE (hydride vapor-phase epitaxy), introduce large quantities of hydrogen-containing species into the growth chamber, the presence of large concentrations of hydrogen in the as-grown material should not come as a surprise. The involvement of hydrogen in the activation of acceptor-doped MOCVD-grown GaN is also confirmed by the achievement of *p*-type GaN by Mg doping in MBE (molecular-beam epitaxy) without any postgrowth treatment (Molnar *et al.*, 1993; Lin *et al.*, 1993); MBE growth, of course, does not introduce hydrogen into the material.

In this chapter we will summarize the properties hydrogen exhibits in GaN and discuss how hydrogen affects doping in GaN. In particular, it will be shown that H in GaN exhibits unique features not observed in the more "traditional" semiconductors such as Si or GaAs. Examples are a very large negative-*U* effect ($U \approx 2.4$ eV), the instability of the bond-centered (BC) site for positively charged H, high energies for H_2 molecules, and an unusual geometry for the Mg-H complex. All these features are shown to be a consequence of distinctive properties of GaN, namely, the strongly ionic nature and the large bond strength of the Ga-N bond. The formation and dissociation of the Mg-H complex and the consequences for activating of *p*-type GaN will be discussed. Finally, it will be explained why hydrogen is *beneficial* for acceptor incorporation in GaN based on the principle of codoping; the limitations of the codoping process are also identified.

II. Method

First-principles methods have had a major impact on description of the properties of hydrogen in semiconductors. The first applications focused on obtaining defect wave functions and levels in the bandgap. With the advent of the capability to calculate total energies, it becomes possible to identify the stable configurations of hydrogen in the semiconductor lattice and their formation energy, migration paths, and diffusion barriers (Van de Walle *et al.*, 1989). Furthermore, the interaction of H with defects and impurities and the formation of complexes have been studied (Van de Walle *et al.*, 1993). For a comprehensive overview of theoretical work on hydrogen in semiconductors, we refer the reader to the overview article by Estreicher (1995). More recently, formalisms have been developed to use these data to calculate defect and impurity concentrations and solubility limits (Laks *et al.*, 1991).

1. Defect Concentrations and Solubility

The key to describing solubility and doping issues is the calculation of the equilibrium concentration of defects and impurities:

$$c = N_{sites} g \exp(-E^f/k_B T) \qquad (1)$$

where N_{sites} is the number of sites the defect or impurity can be incorporated on, k_B is the Boltzmann constant, T is temperature, and E^f is the formation energy. g is a degeneracy factor representing the number of possible configurations a defect can assume on a single site.

The formation energy depends on the various growth parameters. For example, the formation energy of interstitial H depends on the abundance of H atoms. The formation energy of a substitutional Mg acceptor on a Ga site is determined by the relative abundance of Mg, Ga, and N atoms. In a thermodynamic context these abundances are described by the chemical potentials μ_{Ga}, μ_N, and μ_{Mg}. If the defect or impurity is charged, the formation energy depends further on the Fermi level (E_F), which acts as a reservoir for electrons. To be more specific, let us consider the creation of an Mg acceptor. Forming a substitutional Mg acceptor requires the removal of one Ga atom and the addition of one Mg atom. The formation energy is therefore

$$E^f(\text{GaN:Mg}_{Ga}^q) = E_{tot}(\text{GaN:Mg}_{Ga}^q) - \mu_{Mg} + \mu_{Ga} + qE_F \quad (2)$$

where $E_{tot}(\text{GaN:Mg}_{Ga}^q)$ is the total energy derived from a calculation for substitutional Mg, and q is the charge state of the Mg acceptor. Similar expressions apply to the hydrogen impurity and to the various native defects.

In order to compare the stability for different structures, the *free energy* instead of the *total energy* has to be employed. The difference between both energies is the energy contribution $-TS$, which can be divided into vibrational and configurational entropy. For point defects, the configurational entropy is simply given by the number of sites at which the defect can be created; this contribution is already taken into account in Eq. (1). The vibrational entropies are, at the present stage, not explicitly included, which would be computationally very demanding. Such entropy contributions cancel to some extent (Qian et al., 1992) and are small enough not to affect any qualitative conclusions. However, more powerful computers and improved methods will make accurate calculations of vibrational entropies more feasible to be carried out for various systems in the near future.

Also, the pressure dependence of the energy is generally small compared with the absolute value. The *formation enthalpy* thus can be replaced by the *total energy*.

2. Energetics, Atomic Geometries, and Electronic Structure

In most of the theoretical studies, the total energy has been calculated employing density-functional theory, a plane-wave basis set, and a supercell geometry (Neugebauer and Van de Walle, 1995a; Bosin et al., 1996; Okamota et al., 1996). Only one study has been performed using a different approach: Estreicher and Maric (1996) performed Hartree-Fock–based cluster calculations. We will provide here only a brief overview of these techniques. A thorough discussion has been given by Estreicher (1994, 1995).

a. Density-Functional Theory Calculations

Density-functional theory (DFT) calculations based on pseudopotentials, a plane-wave basis set, and a supercell geometry are now regarded as standard for performing first-principles studies of defects in semiconductors. DFT in the local density approximation (LDA) (Hohenberg and Kohn, 1964; Kohn and Sham, 1965) allows a description of the many-body electronic ground state in terms of single-particle equations and an effective potential. The effective potential consists of the ionic potential due to the atomic cores, the Hartree potential describing the electrostatic electron-electron interaction, and the exchange-correlation potential that takes into account the many-body effects. This approach has proven to describe with high accuracy quantities such as formation energies, atomic geometries, charge densities, etc. One shortcoming of this method is its failure to produce accurate excited-states properties—the bandgap is commonly underestimated. In cases where the bandgap enters the calculations, a careful analysis of the results still allows extraction of reliable data. Such situations will be discussed where appropriate (see Neugebauer and Van de Walle, 1994a, 1995b).

b. Hartree-Fock–Based Cluster Calculations

Hartree-Fock–based models commonly employ quantum-chemistry approaches that have been applied successfully to atoms and molecules. The main problems with the technique are the neglect of correlation effects (resulting in too high bandgaps) and the computational demands; *ab initio* Hartree-Fock methods can only be applied to systems with small numbers of atoms. The problem is that these methods require the evaluation of a large number of multicenter integrals. Simpler semiempirical methods have been developed that either neglect or approximate some of these integrals. The accuracy and reliability of these semiempirical methods are hard to assess.

III. Monatomic Hydrogen in GaN

Several groups have calculated energetics, atomic geometries, and electronic structure for isolated hydrogen in a variety of different interstitial positions in GaN. Neugebauer and Van de Walle (1995a, 1995b) and Bosin *et al.* (1996) employed pseudopotential-density-functional theory calculations, and their conclusions are very similar. Some quantitative differences will be discussed where appropriate. Estreicher and Maric (1996) reported preliminary Hartree-Fock–based calculations on 44 atom clusters of zinc-blende GaN.

The calculation of the total energy surface for monatomic hydrogen in semiconductors is a powerful approach to get a very direct insight into basic properties such as stable geometries, formation energies, and migration behavior. For hydrogen in *zinc-blende* GaN, such calculations have been reported by Neugebauer and Van de Walle (1995a). Bosin *et al.* (1996) studied hydrogen in *wurtzite* GaN but restricted their calculations to sites with C_{3v} symmetry. Since the zinc-blende and wurtzite structures only differ beyond third nearest neighbors, the differences with respect to atomic and electronic structure of the hydrogen impurity are expected to be small.

1. ATOMIC GEOMETRIES AND STABLE POSITIONS

a. Hydrogen in the Positive Charge State

Hydrogen in the positive charge state favors positions on a sphere with a radius of ~ 1 Å centered on an N atom; that is, H^+ prefers only positions with N as a nearest neighbor. The calculated H-N bond length for all these positions is 1.02 ⋯ 1.04 Å, close to the experimental bond length in NH_3 ($d_{N\text{-}H} = 1.04$ Å). Among these positions, the nitrogen antibonding site (AB_N) is energetically most stable (see Fig. 1b). All other sites, where a H-N bond cannot be formed, are found to be energetically less favorable.

The preference for the N antibonding site is in striking contrast with the behavior of H^+ in Si or GaAs, where the bond-centered (BC) position (see Fig. 1c) was found to be energetically most stable (Van de Walle *et al.*, 1989; Pavesi and Gianozzi, 1992). This difference can be understood by noting the very different character of the chemical bond in GaN and, for example, GaAs. In GaAs or Si there is a pronounced maximum of the charge density at the bond center due to the strong covalent character of these materials. For the more ionic GaN, however, there is no local maximum at the bond center but an almost spherical distribution of the charge density around the strongly electronegative nitrogen atom. H^+, which is simply a proton, prefers a position where it obtains maximum screening; that is, it prefers sites with a high charge density. This explains why H^+ prefers the BC position in covalent semiconductors such as Si and GaAs, whereas in an ionic compound such as GaN, *all* positions around the N atom are low in energy. At the BC site an extra energy cost needs to be paid due to the strain energy involved in relaxing the Ga and N atoms outward. The Ga-N bond

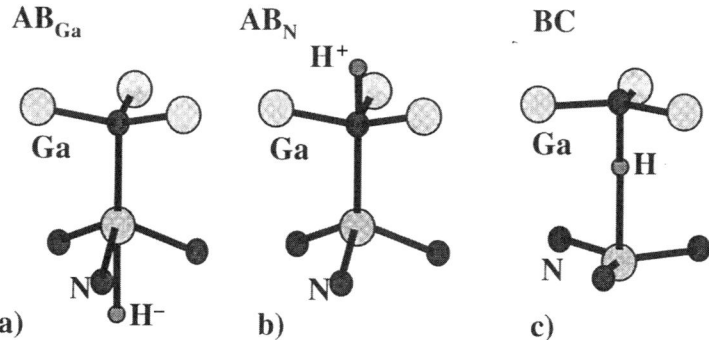

FIG. 1. Ball and stick model of configurations of interstitial hydrogen in GaN: (a) the Ga antibonding site (AB_{Ga}); (b) the N antibonding site (AB_N); (c) the bond-centered position (BC).

has to be stretched by more than 40% of the bond length. Since GaN has very strong interatomic bonds, this process is energetically very costly, rendering the BC site unfavorable.

Despite the high cost, the BC site is still only 0.1 eV higher in energy than the AB_N site, where virtually no relaxation is necessary. Estreicher and Maric (1996), in their Hartree-Fock–based calculations, actually found H^+ to prefer the BC site.

b. Hydrogen in the Neutral Charge State

For neutral hydrogen (H^0), much smaller energy differences (<1 eV) between different sites are found (except for regions within 1 Å of the nuclei, where a strong repulsion is found), indicating a rather flat total energy surface (Neugebauer and Van de Walle, 1995a). A similar behavior also has been found for H^0 in silicon (Van de Walle *et al.*, 1989). Neugebauer and Van de Walle (1995a) find the Ga antibonding site (AB_{Ga}) to be energetically preferred (see Fig. 1a). The H-Ga bond length for this site is only 0.1 Å smaller than the Ga-N bond length, causing the Ga antibonding site to nearly coincide with the position of the tetrahedral interstitial site (T_d^{Ga}). Bosin *et al.* (1996) find the BC site to have the lowest energy. The difference can be traced back to the difference in the bandgap. Because of the small value of the gap in the calculations by Bosin *et al.*, the lowest state to be occupied (for H^0 or H^-) was probably not a H-induced impurity level but a conduction-band state. Estreicher and Maric (1996) reported that the T_d^{Ga} and BC sites are almost degenerate in energy.

c. *Hydrogen in the Negative Charge State*

All calculations find that negatively charged hydrogen (H^-) prefers the Ga antibonding site (very close to the T_d^{Ga} site, as discussed earlier). At this site the distance to the neighboring Ga atoms is maximized, and the charge density of the bulk crystal has a local minimum. The Ga-H bond length is 1.76 Å (Neugebauer and Van de Walle, 1995b).

2. MIGRATION PATH AND DIFFUSION BARRIERS

Based on the total energy surfaces, the migrational path and diffusion barriers of H in GaN can be obtained immediately. In the *positive* charge state, intermittent diffusion occurs; the diffusion path is characterized by a basin of attraction around the N atoms with barriers smaller than 0.1 eV in which the H atom is localized until an intermittent jump to a nearby basin of attraction occurs. The diffusion barrier to jump from one basin to the next is 0.7 eV. This barrier is slightly larger than for H^+ in Si (Van de Walle *et al.*, 1989) but still sufficiently small to ensure high mobility even at temperatures only slightly above room temperature. Indeed, experiments on hydrogen diffusion by Götz *et al.* (1995) have confirmed this prediction.

For *negatively* charged H, a very different migration behavior has been found. H^- is strongly confined to the T_d^{Ga} site, giving rise to a very large migration barrier of ~ 3.4 eV. H^- is therefore virtually immobile in GaN.

3. FORMATION ENERGIES AND NEGATIVE-U EFFECT

Results for the formation energies of hydrogen in the positive, neutral, and negative charge state are shown in Fig. 2. For Fermi energies below ~ 2.1 eV, H^+ is the energetically most stable species. For Fermi levels higher in the gap, H^- is more stable. Since hydrogen has the lowest formation energies under extreme *n*- or *p*-type conditions, where it behaves as an acceptor or donor, respectively, it can be potentially a very efficient compensating center. We will come back to this point in Section VII.

The fact that hydrogen has a significantly lower formation energy under *p*-type conditions than under *n*-type conditions implies that the solubility of hydrogen is considerably higher in *p*-type than in *n*-type GaN. These theoretical predictions have been confirmed by the recent experimental observations of Götz *et al.* (1995) on plasma hydrogenation of MOCVD-grown GaN. H is readily incorporated in *p*-type GaN, consistent with a high diffusivity and solubility of H^+, whereas no detectable levels of H were

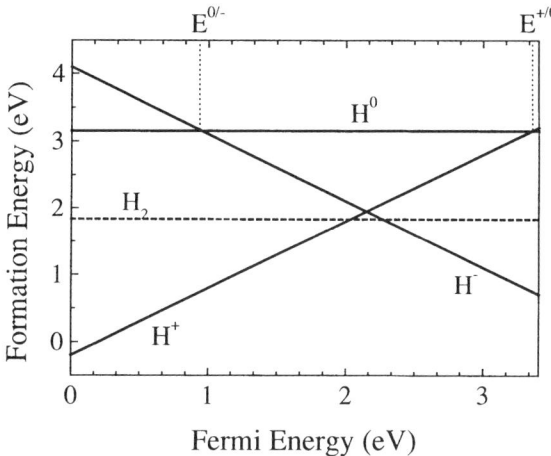

FIG. 2. Formation energy as a function of the Fermi level for H^+, H^0, and H^- (solid lines). Further included are the formation energies for a hydrogen molecule (dashed line) and for a magnesium-hydrogen complex (long-dashed line). $E_F = 0$ corresponds to the top of the valence band. The formation energy is referenced to the energy of a free H_2 molecule. (From Neugebauer and Van de Walle, 1995a.)

found in n-type GaN, consistent with a low diffusivity, as well as a lower solubility, of H^-.

The formation energies shown in Fig. 2 indicate that neutral hydrogen is never stable in GaN. This behavior is characteristic of a negative-U system. The value of U is given by the difference between acceptor $(0/-)$ and donor $(+/0)$ levels: $U = E^{0/-} - E^{+/0} \approx -2.4\,\text{eV}$. A negative-$U$ behavior also was found for H in Si ($U = 0.4\,\text{eV}$) (Van de Walle et al., 1989; Johnson et al., 1994). The value of $-2.4\,\text{eV}$, however, is unusually large and, to our knowledge, larger than any measured or predicted value for H in any other semiconductor.

The unusually large negative-U behavior can be understood by analyzing the total energy surface in more detail (Neugebauer and Van de Walle, 1995a). Figure 3 shows the variation in the hydrogen formation energy along a line between the AB_N site (preferred by H^+) and the AB_{Ga} site (preferred by H^0 and H^-). The results in Fig. 3 reveal some remarkable features: (1) for a *fixed* hydrogen position x, the transition levels $(0/-$ and $+/0)$ are nearly equal (within 0.1 eV), and (2) the total energy surface for neutral hydrogen is flat compared with both positively and negatively charged hydrogen. These observations apply not only to those positions depicted in Fig. 3 but also to *all* positions. The only exceptions are sites near the atomic cores, where the energy of H in any charge state is very high due

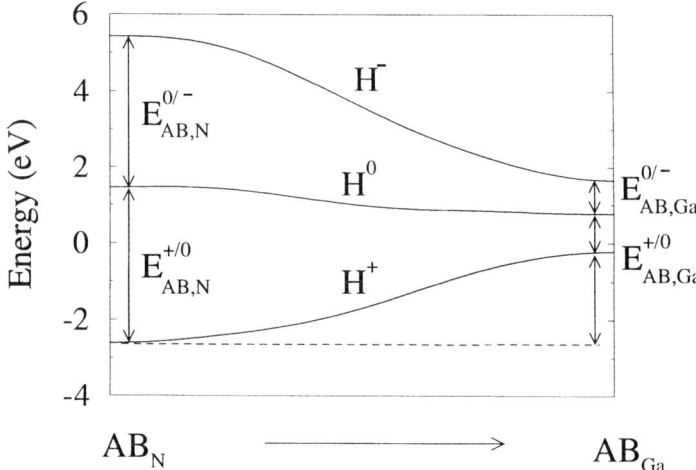

FIG. 3. Variation of the hydrogen formation energy for all three charge states along the shortest path between the AB_N and AB_{Ga} sites. The Fermi level is set to the top of the valence band. (From Neugebauer and Van de Walle, 1995a.)

to atomic repulsion. We note that these features are not specific to GaN but also apply to GaAs or Si (Van de Walle et al., 1989).

Both these features, as well as the large negative-U character, can be explained by analyzing the nature of the interaction between hydrogen and the semiconductor (Neugebauer and Van de Walle, 1995a). Neutral H consists of a proton and an electron. The *proton* prefers regions of high charge density, whereas the *electron* tends more toward regions where the charge density is low. This already explains the location of H^+ (a proton) in Si, GaAs, and GaN, as discussed earlier, and it explains why H^- (in which electrons dominate) prefers T_d or AB sites. Moving *neutral* hydrogen into regions with higher charge density causes the proton to gain about the same amount of energy as the electron loses. To first order, the total energy of H^0 is then independent of the charge density, which explains the flat total energy surface. Deviations occur because the electronic orbital (contrary to the proton) has a finite size and thus experiences variations in the charge density over a certain spatial extent. Nevertheless, this simple model works remarkably well, even for the BC site where the electron is known to reside in an antibonding combination of host-atom orbitals. For the preceding discussion, no assumptions about the specific nature of the semiconductor were made. Therefore, these features should be a general property of hydrogen in *any* semiconductor.

Based on the preceding discussion, a model can be constructed with the following assumptions: (1) the transition levels $E_x^{+/0}$ and $E_x^{0/-}$ are exactly equal for all positions x, and (2) the total energy surface for H^0 is completely flat. It then follows immediately that the total energy surface of the positive charge state is the exact mirror image of the negative charge state. Consequently, minima in the total energy surface for H^+ correspond to maxima in the charge density of H^- and vice versa (see Fig. 3). This immediately predicts that $U = E^{0/-} - E^{+/0}$, calculated for the most stable positions of each charge state, will be negative. As discussed earlier, for H^+, minima and maxima correspond to maxima and minima in the charge density. Consequently, in covalent systems (Si) or weakly ionic systems (GaAs), possible sites for H have a modest variation in charge density — hence the absolute value of U will be modest, although U will be negative. However, much larger variations in the charge density exist in ionic crystals — therefore, for this class of materials the largest negative-U values are expected.

IV. Hydrogen Molecules in GaN

Relatively little is known concerning hydrogen molecules in GaN. To our knowledge, there have been no experimental reports about the existence of H_2 molecules in GaN or its alloys. Hydrogen molecules have been studied theoretically using a standard first-principles method by Neugebauer and Van de Walle (1995a). The atomic geometry and formation energy for several symmetric as well as asymmetric configurations were considered. The value for the energetically most stable configuration has been included in Fig. 2. The H_2^* geometry, as proposed by Chang and Chadi (1990), also was investigated by Neugebauer and Van de Walle (1995b). Here one hydrogen atom is located at the BC site and the other at either the N or Ga antibonding site. It was found that H_2^* is even less favorable than the H_2 molecule geometry. As can be seen in Fig. 2, H_2 is unstable with respect to dissociation into monatomic hydrogen. The formation energy of $\sim 1.8\,eV$ is much higher than that of H_2 molecules in vacuum. Both features, the low stability of the hydrogen molecule and its unfavorably high formation energy, are distinct properties of GaN and again very different from the case of Si or GaAs.

V. Hydrogen-Acceptor Complexes in GaN

Hydrogen is well known to form complexes with shallow impurities in semiconductors. Here we focus on acceptor-hydrogen complexes since, as discussed in Section III.1.c, hydrogen has a low solubility in n-type GaN.

On formation of the acceptor-hydrogen complex, the acceptor level disappears into the valence band—the acceptor is passivated and electrically inactive. Hydrogen-acceptor complexes have been observed in a variety of semiconductors, and properties such as local vibrational modes, binding and activation energy, bonding geometry, etc. have been studied extensively (see, e.g., Pankove and Johnson, 1991).

1. THE Mg-H COMPLEX

The technologically most commonly used acceptor is magnesium, and consequently, most investigations focused on hydrogen interaction with Mg. The structure, energetics, and vibrational frequencies of the Mg-H complex have been studied by several groups for zinc-blende GaN (Neugebauer and Van de Walle, 1995a) and wurtzite GaN (Bosin et al., 1996; Okamoto et al., 1996).

Neugebauer and Van de Walle (1995a) and Bosin et al. (1996) find that the hydrogen atom does not form a bond with the Mg_{Ga} acceptor (as one could naively expect for a Mg-H complex) but prefers the AB site of one of the N neighbors (see Fig. 4). In cubic GaN, all four N neighbors are equivalent (Fig. 4a). Bosin et al. (1996) found that in wurtzite GaN a H-Mg complex with the H attached to the N neighbor along the c axis (Fig. 4b) is less stable than for a H atom on one of the three equivalent neighbors (the so-called a-axis orientation; see Fig. 4c). Okamoto et al. (1996) find the BC site to be slightly more favorable than the AB site, but only by 0.1 eV. This small quantitative difference may be due to their use of a smaller wurtzite supercell (16 atoms, as opposed to 32 atoms used by the other groups).

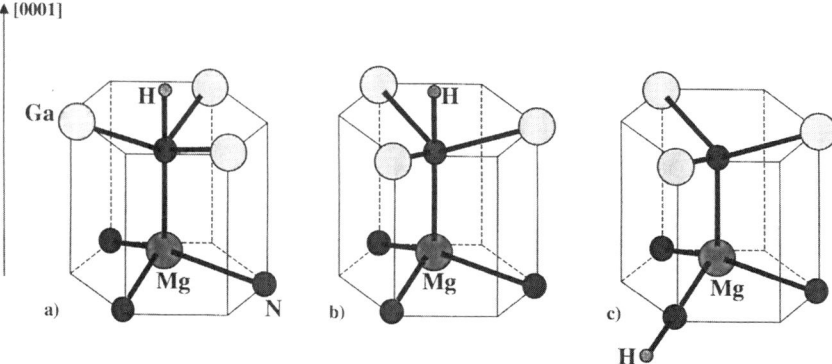

FIG. 4. Ball and stick model of the atomic structure of the Mg-H complex in cubic (a) and wurtzite (b, c) GaN with H on the N antibonding site. In wurtzite GaN two inequivalent configurations exist with the complex parallel to the z axis (b) and parallel to the a axis (c).

It is interesting to note that the atomic geometry (AB_N site) found for the Mg-H complex in GaN is very different from the well-established structure for acceptor-H complexes in other semiconductors where a BC configuration is preferred. The physical mechanism behind the unusual geometry of the Mg-H complex can be understood by analyzing the character of the acceptor level. First, it should be noted that the H atom donates its electron to the Mg acceptor level. The resulting H^+ prefers positions where (1) the charge density is high and (2) it is close to the donated electron (minimization of the electrostatic energy). Both conditions are closely related to the character of the Mg acceptor level. As can be seen in Fig. 5, this level is not characterized by Mg orbitals but by p-like orbitals located on the N atoms surrounding the Mg acceptor. The locations with the highest charge density are the BC position and the AB_N site. At the BC site, however, an additional relaxation energy has to be paid, explaining why H favors the AB_N site.

The calculated H stretch mode is $3360\,cm^{-1}$ for cubic GaN (Neugebauer and Van de Walle, 1995a) and $2939\,cm^{-1}$ for wurtzite GaN (Bosin et al., 1996). Both frequencies are close to the stretch mode of H in NH_3 ($\sim 3444\,cm^{-1}$), indicating that the Mg-H complex is mainly characterized by a H-N bond.

Experimentally, local vibrational modes (LVMs) in Mg-doped GaN had been observed for the first time by Brandt et al. (1994). Room-temperature frequencies of 2168 and $2219\,cm^{-1}$ were tentatively assigned to inequivalent configurations of hypothetical Mg-H complexes or to H-Ga bonds in the wurtzite lattice. Neither isotopic substitution nor correlations with activation/deactivation of the p-type conductivity were available in the study. Looking specifically in the region predicted by theory, Götz et al. (1996a) found a vibrational mode at $3125\,cm^{-1}$. There are strong experimental indications that the mode is a stretch mode from the Mg-H complex; activation of the Mg-dopants reduces the intensity of the local vibrational modes, and deuteration of the activated material shifts the mode to $2321\,cm^{-1}$.

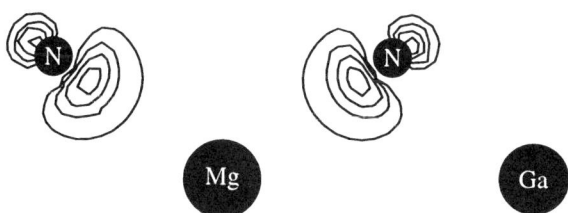

FIG. 5. Contour plot of the Mg acceptor level in GaN. The contour spacing is 0.005 bohr^{-3}. (From Neugebauer and Van de Walle, 1995a.)

The value calculated by Neugebauer and Van de Walle (1995a) of 3360 cm^{-1} is somewhat larger than the experimental value of 3125 cm^{-1}. It should be kept in mind that the calculated value does not include anharmonic effects, which are sizable in the case of N-H vibrations; in NH_3 the anharmonicity lowers the frequency by 170 cm^{-1} (Johnson et al., 1993).

The total energy surface for a H atom around a Mg_{Ga} acceptor in GaN has been calculated by Neugebauer and Van de Walle (1995a). From the total energy surface the dissociation barrier of the Mg-H complex can be estimated. Considering only a jump from the N antibonding site to a neighboring N atom, a value of 1.5 eV is found.

2. OTHER H-ACCEPTOR COMPLEXES

H complexes with other acceptors (Be_{Ga}, C_N, Ca_{Ga}, and Zn_{Ga}) have been studied by Bosin et al. (1996). However, the authors pointed out that their results were preliminary due to the limited size of the supercell (16 atoms) and an incomplete investigation of all possible sites. For each complex, two possible geometries have been calculated — the BC site and the N antibonding site, where the complex is parallel to the c-axis. Complexes oriented along the a axis have not been considered. For all complexes, the formation energy is exothermic. Thus at low temperatures we generally expect the formation of H-acceptor complexes and an efficient passivation of the acceptors. All complexes exhibit a competition between the BC and AB_N site; for Be and Zn, hydrogen prefers the BC site; for Ca and C, it prefers the AB_N site. One should keep in mind, however, that the a-axis orientation of the complex, which was the energetically most stable configuration for the Mg-H complex, had not been considered. Further experimental and theoretical studies are necessary to clarify the precise geometry of these complexes.

VI. Complexes of H with Native Defects

Besides interacting with impurities, hydrogen also can interact with native defects. By forming a complex, hydrogen modifies the character of the defect; defects can be passivated, but as found by Van de Walle (1997), defects also can be converted from acceptors to donors. Further, attaching hydrogen to defects can reduce their formation energy (resulting in higher concentrations) and affect their migration behavior.

Native point defects are known to play an important role in GaN.

Previous first-principles studies showed that the dominant native defects in GaN are vacancies. Other types of point defects (antisites or self-interstitials) are high in energy and therefore are unlikely to form in appreciable concentrations during GaN growth (Neugebauer and Van de Walle, 1994b; Boguslawski et al., 1995). We can therefore focus on the interaction of hydrogen with vacancies.

1. HYDROGEN INTERACTING WITH NITROGEN VACANCIES

The interaction of H with vacancies is often described in terms of the tying off of dangling bonds. This picture does not apply in the case of the nitrogen vacancy, which is surrounded by Ga atoms at a distance of 1.95 Å from the center of the vacancy. A typical Ga-H bond distance is too large for more than one hydrogen to fit inside the vacancy. It was previously pointed out (Neugebauer and Van de Walle, 1994b) that the dangling bonds on the Ga neighbors of the nitrogen vacancy strongly hybridize. This observation is consistent with the notion that H will not bond to any of the Ga neighbors in particular; rather, the H atom sits at the center of the vacancy, in a rather shallow potential well.

These observations are confirmed by explicit first-principles calculations. The V_NH complex behaves as a double donor. In the $2+$ charge state of the complex the hydrogen atom occupies a totally symmetric position, and the potential energy surface for displacements of the hydrogen atom is quite shallow; the corresponding vibrational frequency is less than 600 cm^{-1}. This frequency is obiously much smaller than frequencies associated with Ga-H bonds, consistent with the notion that no specific Ga-H bonds are being formed.

A binding energy of 1.56 eV was calculated for the V_NH complex, expressed with respect to interstitial H in the positive charge state. The formation energy of the complex (Van de Walle, 1997) is such that some concentration of V_NH complexes will be formed during high-temperature growth of p-type GaN. Note that formation of V_NH complexes *after* growth, by diffusion of a hydrogen atom toward V_N, is highly unlikely; since both H and V_N are donors and occur in the positive charge state (in p-type GaN), they repel one another.

It has been proposed (Van de Walle, 1997) that the V_NH complex is responsible for the disappearance and appearance of photoluminescence (PL) lines during postgrowth annealing of Mg-doped layers grown by MOCVD, as described by Götz et al. (1996c) and Nakamura et al. (1992b). A PL line at 3.25 eV is present in as-grown material. Annealing at temperatures above 500°C causes this line to "redshift." It was suggested that this shift actually corresponds to a decrease in intensity of the line at 3.25 eV (associated with the hydrogenated nitrogen vacancy), accompanied by an increase in intensity of a new line related to the isolated nitrogen vacancy. When the material is annealed, the hydrogenated vacancy complexes dissociate; the calculated removal energy of 1.56 eV is consistent with complex dissociation around 500°C. The resulting nitrogen vacancies have a level near the valence band, which may be responsible for the line around 2.9 eV.

2. Hydrogen Interacting with Gallium Vacancies

For the Ga vacancy it was found (Van de Walle, 1997) that one, two, three, or four hydrogen atoms can be accommodated in the vacancy, and levels are removed from the bandgap as more hydrogens are attached. Distinct N-H bonds are formed, with a bond length of about 1.02 Å and characteristic vibrational modes.

The isolated Ga vacancy is a triple acceptor; in the 3− charge state a triply degenerate defect level about 1 eV above the valence band is fully occupied. Introducing a H contributes an electron so that only two extra electrons are now required to fully occupy the defect levels; $V_{Ga}H$ is therefore a double acceptor. Similarly, the lowest-energy state of $V_{Ga}H_2$ is singly negatively charged, $V_{Ga}H_3$ is neutral, and $V_{Ga}H_4$ is actually positively charged (i.e., it behaves as a donor). All defect levels are removed from the gap in the latter configuration. The calculated formation energies (Van de Walle, 1997) indicate that these complexes may form during growth, depending on the likelihood of multiple hydrogen atoms being available at the time of incorporation. Capture of H at existing vacancies is unlikely; gallium vacancies are most likely to form in n-type GaN; hydrogen prefers the negative charge state here (Sec. III.1.a) and is repelled by the negatively charged vacancies. H^+ would be attracted by the gallium vacancies but is not stable for Fermi level positions above 2.1 eV (see Fig. 2).

The calculated frequency (in the harmonic approximation) for a hydrogen stretch mode in the $V_{Ga}H$ complex is 3100 cm^{-1}, somewhat lower than the value for an NH_3 molecule. For the $V_{Ga}H_4$ complex, the calculated frequency is 3470 cm^{-1}. The higher value reflects repulsion between H atoms.

Hydrogenated gallium vacancies may play a role in GaN materials and devices similar to that played by isolated gallium vacancies, that is, compensation of donors (Yi and Wessels, 1996) and as a source of the yellow luminescence (YL) (Neugebauer and Van de Walle, 1996c). The $V_{Ga}H$ and $V_{Ga}H_2$ complexes have levels in the bandgap only slightly lower (by 0.1–0.2 eV) than the isolated gallium vacancy; they therefore also may contribute to the YL, through transitions between shallow donors and a deep acceptor level. The slight shift in the positions of the levels between various complexes could contribute to the width of the luminescence line.

VII. Role of Hydrogen in Doping GaN

Having analyzed the properties of interstitial monatomic hydrogen and its interaction with acceptors, we will now focus on how the presence of hydrogen affects doping in GaN. As pointed out in the introduction, there is strong experimental evidence that H plays an active role during the growth and activation of p-type GaN, as shown in the work of Nakamura et al. (1992a).

Based on these observations Van Vechten et al. (1992) suggested that hydrogen enables p-type doping by suppressing compensation by native defects. These authors went on to propose the incorporation (and subsequent removal) of hydrogen as a general method for improving p-type as well as n-type doping of wide-bandgap semiconductors. The Van Vechten et al. model highlighted the important role of hydrogen but left various issues unexplained, such as the lack of hydrogen incorporation in n-type GaN and the success of p-type doping (without postgrowth treatments) in MBE (molecular-beam epitaxy).

In order to identify the role of hydrogen in achieving p-type GaN, we first analyze the case where hydrogen is absent, corresponding, for example, to MBE growth. We will then consider hydrogen-rich conditions that are characteristic for many of the high-temperature growth techniques such as MOCVD and HVPE.

1. DOPING IN THE ABSENCE OF HYDROGEN

Before discussing the role of hydrogen in p-type doping, let us briefly focus on the H-free case; that is, only the acceptor (we will here consider Mg) and native defects are present. Figure 6 shows the corresponding formation energies calculated by Neugebauer and Van de Walle (1996a) as a function of the Fermi energy. The dominant native defect under p-type conditions is the nitrogen vacancy; all other defects are higher in energy. The slope of the formation energies characterizes the charge state; a positive slope (as found for the nitrogen vacancy) indicates a positive charge state, corresponding to a donor. For Mg, the kink in the formation energy describes a change in the charge state from neutral to negative as characteristic for an acceptor. The position of the kink at $E_F \approx 0.2\,\text{eV}$ gives the

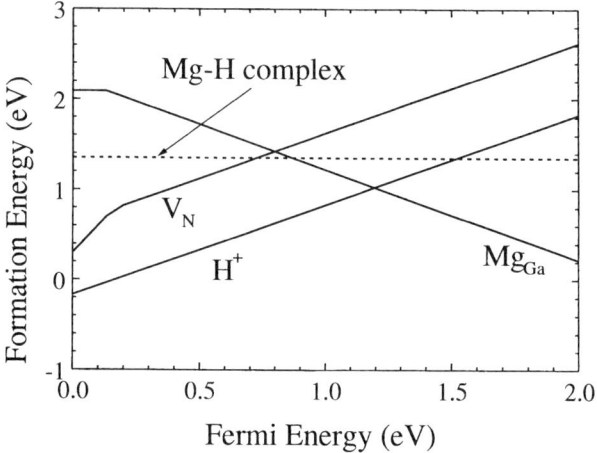

FIG. 6. Formation energy versus Fermi level for the Mg_{Ga} acceptor, the native defects, and interstitial hydrogen. Ga-rich conditions are assumed. (From Neugebauer and Van de Walle, 1996a. Copyright 1996 American Institute of Physics.)

position of the calculated acceptor level, which is close to the experimental value of 0.16 eV (Akasaki et al., 1991).

Using the calculated formation energies, and taking into account that the Fermi energy is fixed by the condition of charge neutrality, the equilibrium concentration (Eq. 1) for each defect can be calculated as a function of temperature. The results are shown in Fig. 7. As expected, the Mg concentration increases with increasing temperature. However, with increasing temperature, the nitrogen vacancy concentration also increases; at temperatures exceeding 1000 K, the Mg acceptors become increasingly compensated by nitrogen vacancies. Native defect compensation is therefore potentially an important concern for high-temperature growth techniques. Lower-temperature growth techniques such as MBE should suffer less from this problem. This conclusion is consistent with the fact that only in MBE-grown GaN p-type conductivity can be achieved without any postgrowth processing (Molnar et al., 1993; Lin et al., 1993). It should be kept in mind that the low growth temperatures employed in MBE might not be sufficient to bring the system into full thermodynamic equilibrium. In particular, a low diffusivity at these temperatures could efficiently suppress the formation of Mg_3N_2, thus shifting the solubility to higher Mg concentrations. Also, the Mg sticking coefficient is higher at lower temperatures. This might explain the high Mg concentrations ($\sim 10^{20}$ cm^{-3}) observed experimentally even at temperatures characteristic for MBE (between 1000 and 1100 K).

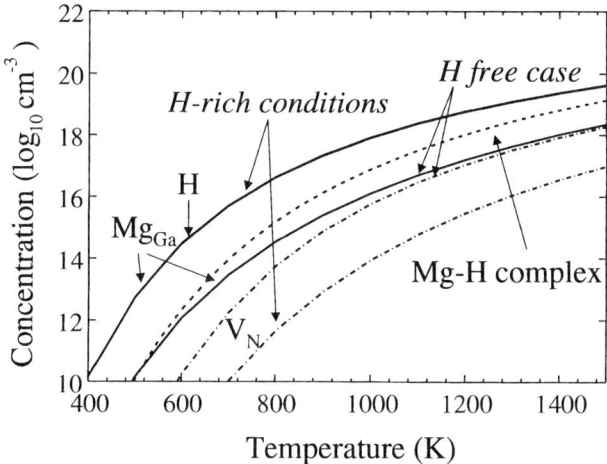

FIG. 7. Equilibrium concentrations of the Mg_{Ga} acceptors (solid line), nitrogen vacancies (dot-dashed line), and the Mg-H complexes (dashed line) as a function of the growth temperature. Two cases are assumed: H-free and H-rich conditions. Note that in the H-rich case the concentration of interstitial hydrogen is virtually identical with the Mg concentration. (From Neugebauer and Van de Walle, 1996a.)

2. DOPING IN THE PRESENCE OF HYDROGEN

We will now consider H-rich conditions that are characteristic for many of the high-temperature growth techniques such as MOCVD and HVPE. Figure 6 shows that under these conditions H becomes the dominant donor; the formation energy of H is always lower than the dominant native defect, the nitrogen vacancy. The calculated equilibrium concentrations are displayed in Fig. 7. The Mg and H concentration are for all temperatures virtually identical, indicating that H completely compensates the Mg acceptors. Further, compared with the H-free case, the concentration of Mg acceptors is increased and the defect concentration is decreased. Both effects contribute to increase doping levels.

What is the mechanism by which H changes the acceptor and defect concentration? In a plot such as Fig. 6, the Fermi-level position during growth can be roughly estimated to be near the crossing point between the acceptor and the dominant donor species. At this point their formation energies (and hence their concentrations) are equal, ensuring charge neutrality (ignoring the contributions from free carriers, which is small if E_{Fermi} is far from the band edges). By going from H-free conditions (only Mg and nitrogen vacancies present) to H-rich conditions, the crossing point shifts to higher Fermi energies (from ~ 0.8 to ~ 1.2 eV). An increase in the Fermi energy decreases the formation energy of acceptors and increases the formation energy of donors, thus resulting in a lowered donor defect concentration and an increased acceptor concentration. We note that this mechanism works only if H is able to significantly shift the Fermi energy, which is the case if (1) H is the dominant donor (i.e., its formation energy must be lower than that of all native defects) and (2) its formation energy must be comparable with that of the dopant impurity (a crossing point must exist in the bandgap).

It is interesting to note that condition (2) is not valid for n-type doping with, for example, Si. The reason is that under n-type conditions hydrogen has a higher formation energy than under p-type conditions; the Si donor has for *all* Fermi energies a lower formation energy than the H^- acceptor (Neugebauer and Van de Walle, 1996a).

3. ACTIVATION MECHANISM OF THE DOPANTS

We have thus established that growing under H-rich conditions enhances acceptor concentrations and suppresses defects. Figure 7 shows, however, that the Mg acceptors are almost completely compensated by the H impurities. In order to activate the Mg acceptors, postgrowth treatments are necessary to eliminate the compensation by hydrogen.

The H donors and Mg acceptors can form electrically neutral complexes with a binding energy of ~ 0.7 eV (see Sec. V.1). For the specific choice of chemical potentials made, this binding energy is low enough for the complexes to be dissociated at the growth temperature; however, the Mg and H will form pairs when the sample is cooled to room temperature, consistent with experimental observations (Götz *et al.*, 1995, 1996a).

The first step in the activation process is the dissociation of the Mg-H complex (see Fig. 8). The estimated dissociation barrier for the complex is 1.5 eV, calculated by considering a jump to a nearest-neighbor site; the total barrier may be slightly higher (Neugebauer and Van de Walle, 1995a). This barrier is low enough to be overcome at modest annealing temperatures (around 300°C). Experimental results show, however, that activation has to be carried out at much higher temperatures ($>600°C$). The reason is that dissociation alone is insufficient; in order to prevent the H from compensating the Mg acceptor, it has to be either removed from the p-type layer (to the surface or into the substrate) or neutralized (e.g., at an extended defect). Note that formation of H_2 molecules is not an option in GaN (see Sec. IV).

The calculated diffusion barrier for H^+ in GaN is low (~ 0.7 eV), indicating that H^+ is highly mobile and can easily migrate to the surface or

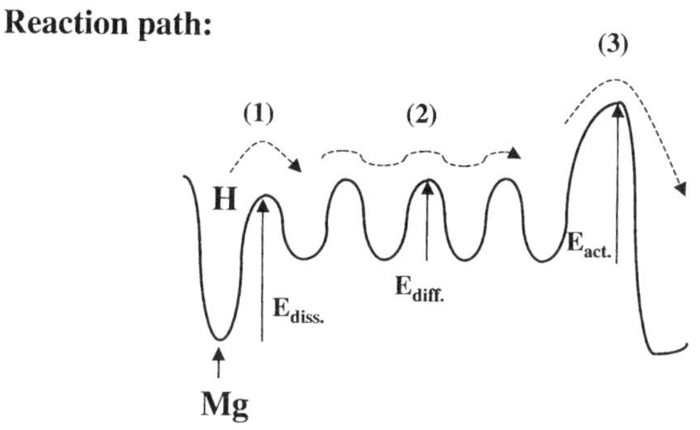

Reaction path:

FIG. 8. Schematic picture of the reaction path and characteristic energy barriers for the activation of the Mg acceptors. At low temperatures, the H forms a neutral complex with the Mg atom. With increasing temperature, the Mg-H complex dissociates (1), and the positively charged H can easily migrate through the GaN crystal (2). Increasing the temperature further allows the H to overcome the activation barrier (3) and become electrically inactive.

extended defects. The high temperature necessary to activate the Mg acceptors therefore reflects an activation barrier for eliminating H as a compensating center by incorporating it at extended defects (which typically occur in high concentrations in GaN; Lester *et al.*, 1994) or by removal of H through desorption at surfaces.

VIII. General Criteria for Hydrogen to Enhance Doping

The mechanism by which hydrogen enhances doping works only under specific conditions. First of all, hydrogen must be the dominant compensating defect (i.e., its formation energy must be lower than that of all native defects and comparable with the formation energy of the dopant impurity). Second, it must be possible to remove hydrogen from the doped layer after growth; an anneal at high temperature will work only in a specific temperature window. On the one hand, the temperature must be *high* enough to enable all three processes necessary to eliminate H as compensating center: complex dissociation, migration, and overcoming of the activation barrier (see Fig. 8). We call the lowest temperature where this mechanism occurs T_1. On the other hand, the temperature must be *low* enough ($< T_2$) to keep the dopant impurity immobile and to prevent the formation of other compensating defects; in the case of Mg doping of GaN, an anneal at too high a temperature would lead to Mg compensation by nitrogen vacancies, as shown in Fig. 7. Thus a necessary condition for this mechanism to work is $T_1 < T_2$.

This condition is met in p-type GaN, judging by the success of thermal annealing procedures (Nakamura et al., 1992a). We note that the preceding analysis is *general* and applies to dopants in any semiconductor. Whether the condition is realized, however, depends on the specific parameters of a system (migration barriers of impurities and defects, activation energies, etc.) and therefore must be addressed independently for each system. It is interesting to note that even for GaN (where the method works well for acceptor doping) it does not work for donor impurities. First of all, the formation energy of H is too high in n-type GaN for it to be incorporated in significant concentrations. And even if H *were* incorporated, the high diffusion barrier of H in n-type GaN (~ 3.4 eV) would render it immobile even at very high annealing temperatures, preventing the hydrogen from reaching the zones where it can be neutralized or removed.

Failure to meet these conditions is also the likely cause of the difficulties in obtaining p-type activity in MOCVD-grown nitrogen-doped ZnSe. Indeed, the N-H bond is very strong and would require high temperatures for dissociation; these temperatures exceed those at which the structural quality of the ZnSe crystal can be maintained.

IX. Conclusions and Outlook

First-principles total-energy calculations have had a major impact on exploring the properties of hydrogen and its interactions with dopant impurities and native defects in GaN. While many features are similar to the behavior of hydrogen in other semiconductors, a number of important differences occur. Computational studies revealed that H and acceptor-H complexes prefer unusual geometries and that hydrogen molecules are unstable. Furthermore, a huge negative-U effect is predicted for interstitial H. At this point, theory is leading experiment, and it would be extremely interesting to pursue these features experimentally.

Based on the computational studies, the mechanisms by which acceptors can be passivated and activated have been elucidated. Nevertheless, a number of aspects remain to be clarified. For instance, no conclusions have been reached regarding the mechanism by which hydrogen is removed from the p-type region, that is, whether it is brought to the surface (and thus disappears) or to extended defects (and stays in the sample). Also, the mechanism operative during the LEEBI treatment remains to be determined.

All theoretical investigations so far have focused on hydrogen in *bulk*

GaN. Experimentally, however, there is strong evidence that hydrogen at *surfaces* affects the growth rate and the crystalline quality (see, e.g., Ambacher *et al.*, 1997). Studies in this field could point the way toward improving growth, reducing defect densities, or enhancing dopant concentrations. We also note that investigations of the interaction between hydrogen and extended defects are still lacking.

Finally, it should be mentioned that all theoretical investigations and most of the experimental studies concerning hydrogen in the III-nitrides have focused on GaN. For device applications, however, ternary alloys such as AlGaN (as the cladding layer) and In GaN (as the active region) are crucial. Although qualitatively similar behavior of H in these alloys is expected, the quantitative results may turn out to be crucial. For instance, the larger bandgap and the stronger ionicity of AlGaN compared with GaN might severely affect the properties of interstitial H and its interaction with dopants. Studies of hydrogen at surfaces and in InGaN or AlGaN alloys are therefore called for.

ACKNOWLEDGMENTS

Thanks are due to W. Götz and N. Johnson for very productive collaborations and stimulating discussions. This work was supported in part by the Deutsche Forschungsgemeinschaft (DFG) and by DARPA under agreement number MDA972-96-3-014.

REFERENCES

Akasaki, I., Amano, H., Kito, M., and Hiramatsu, K. (1991). *J. Lumin.*, **48&49**, 666.
Amano, H., Kito, M., Hiramatsu, K., and Akasaki, I. (1989). *Jpn. J. Appl. Phys.*, **28**, L2112.
Ambacher, O., Dimitrov, R., Lentz, D., Metzger, T., Rieger, W., and Stutzmann, M. (1997). *J. Crystal Growth*, **170**, 335.
Boguslawski, P., Briggs, E. L., and Bernholc, J. (1995). *Phys. Rev. B*, **51**, 17255.
Bosin, A., Fiorentini, V., and Vanderbilt, D. (1996). *Mater. Res. Soc. Symp. Proc.*, **395**, 503.
Brandt, M. S., Ager, J. W., III, Götz, W., Johnson, N. M., Harris, J. S., Molnar, R. J., and Moustakas, T. D. (1994). *Phys. Rev. B*, **48**, 14758.
Chang, K. J., and Chadi, D. J. (1990). *Phys. Rev. B*, **42**, 2426.
Estreicher, S. K. (1994). *Mater. Sci. Forum*, **148–149**, 349.
Estreicher, S. K. (1995). *Mater. Sci. Engr. Reps.*, **14**, 319.
Estreicher, S. K., and Maric, D. M. (1996). *Mater. Res. Soc. Symp. Proc.*, **423**, 613.
Götz, W., Johnson, N. M., Walker, J., Bour, D. P., Amano, H., and Akasaki, I. (1995). *Appl. Phys. Lett.*, **67**, 2666.
Götz, W., Johnson, N. M., Bour, D. P., McCluskey, M. D., and Haller, E. E. (1996a). *Appl. Phys. Lett.*, **69**, 3725.

Götz, W., Johnson, N. M., Bour, Chen, C., Liu, H., Kuo, C., and Imler, W. (1996b). *Appl. Phys. Lett.*, **68**, 3114.
Götz, W., Johnson, N. M., Walker, J., Bour, and Street, R. A. (1996c). *Appl. Phys. Lett.*, **68**, 667.
Götz, W., Walker, J., Romano, L. T., and Johnson, N. M. (1997). *Proc. Mater. Res. Soc. Symp.*, **449**, 525.
Hohenberg, P., and Kohn, W. (1964). *Phys. Rev.*, **136**, B864.
Ilegems, M., and Montgomery, H. C. (1973). *J. Phys. Chem. Solids*, **34**, 885.
Johnson, B. G., Gill, P. M. W., and Pople, J. A. (1993). *J. Chem. Phys.*, **98**, 5612.
Johnson, N. M., Herring, C., and Van de Walle, C. G. (1994). *Phys. Rev. Lett.*, **73**, 130.
Kohn, W., and Sham, L. J. (1965). *Phys. Rev.*, **140**, A1133.
Laks, D. B., Van de Walle, C. G., Neumark, G. F., and Pantelides, S. T. (1991). *Phys. Rev. Lett.*, **66**, 648.
Lester, S. D., Ponce, F. A., Craford, M. G., and Steigerwald, D. A. (1994). *Appl. Phys. Lett.*, **66**, 1249.
Lin, M. E., Xue, C., Zhou, G. L., Greene, J. E., and Morkoc, H. (1993). *Appl. Phys. Lett.*, **63**, 932.
Maruska, H. P., and Tietjen, J. J. (1969). *Appl. Phys. Lett.*, **15**, 327.
Morkoc, H., Strite, S., Gao, G. B., Lin, M. E., Sverdlov, B., Burns, M. (1994). *J. Appl. Phys.*, **76**, 1363.
Molnar, R. J., Lei, T., and Moustakas, T. D. (1993). *Proc. Mater. Res. Soc. Symp.*, **281**, 753.
Nakamura, S., Iwasa, N., Senoh, M., and Mukai, T. (1992a). *Jpn. J. Appl. Phys.*, **31**, 1258.
Nakamura, S., Mukai, T., Senoh, M., and Iwasa, N. (1992b). *Jpn. J. Appl. Phys.*, **31**, L139.
Neugebauer, J., and Van de Walle, C. G. (1994a). *Mater. Res. Soc. Symp. Proc.*, **339**, 687.
Neugebauer, J., and Van de Walle, C. G. (1994b). *Phys. Rev. B*, **50**, 8067.
Neugebauer, J., and Van de Walle, C. G. (1995a). *Phys. Rev. Lett.*, **75**, 4452.
Neugebauer, J., and Van de Walle, C. G. (1995b). *Mater. Res. Soc. Symp. Proc.*, **378**, 503.
Neugebauer, J., and Van de Walle, C. G. (1996a). *Appl. Phys. Lett.*, **68**, 1829.
Neugebauer, J., and Van de Walle, C. G. (1996b). *Mater. Res. Soc. Symp. Proc.*, **423**, 619.
Neugebauer, J., and Van de Walle, C. G. (1996c). *Appl. Phys. Lett.*, **69**, 503.
Okamoto, Y., Saito, M., and Oshiyama, A. (1996). *Jpn. J. Appl. Phys.*, **35**, L807.
Pankove, J. I., and Johnson, N. M., eds. (1991). *Hydrogen in Semiconductors*, **34**. Boston: Academic Press.
Pavesi, L., and Giannozzi, P. (1992). *Phys. Rev. B*, **46**, 4621.
Pearton, S. J., Corbett, J. W., and Stavola, M. (1992). *Hydrogen in Crystalline Semiconductors*. Berlin: Springer-Verlag.
Qian, G.-X., Martin, R. M., and Chadi, D. J. (1992). *Phys. Rev. B*, **38**, 7649.
Van de Walle, C. G., Denteneer, P. J. H., Bar-Yam, Y., and Pantelides, S. T. (1989). *Phys. Rev. B*, **39**, 10791.
Van de Walle, C. G., Laks, D. B., Neumark, G. F., and Pantelides, S. T. (1993). *Phys. Rev. B*, **47**, 9425.
Van de Walle, C. G., and Neugebauer, J. (1997). *Mater. Res. Soc. Symp. Proc.*, **449**, 861.
Van de Walle, C. G. (1997). *Phys. Rev. B*, **56**, 10020.
Van Vechten, J. A., Zook, J. D., Horning, R. D., and Goldenberg, B. (1992). *Jpn. J. Appl. Phys.*, **31**, 3662.
Wetzel, C., Suski, T., Ager, J. W., Weber, E. R., Haller, E. E., Fischer, S., Meyer, B. K., Molnar, R. J., and Perlin, P. (1997). *Phys. Rev. Lett.*, **78**, 3923.
Yi, G.-C., and Wessels, B. W. (1996). *Appl. Phys. Lett.*, **69**, 3028.

Index

Numbers followed by the letter f indicate figures; numbers followed by the letter t indicate tables.

A

a-Si:H. *See* Hydrogenated amorphous silicon
Activation, 473
AlAs
 C-H complexes in, 390–392, 393
 Si-H complexes in, 394, 395
AlGaAs
 C-H complexes in, 393, 394
 Si-H complexes in, 394, 396
AlGaN, hydrogen behavior in, 9, 456, 457–459
AlN, hydrogen behavior in, 459–460
AlSb, 375
 DX-hydrogen complexes in, 412–415, 417
 resonant interaction between LVMs and phonons in, 415–422
Aluminum
 diffusivity of, 4
 hydrogen-enhanced diffusion of, 69–75
Amorphous silicon (a-Si)
 hydrogen interactions with, 248–271
 hydrogenated. *See* Hydrogenated amorphous silicon (a-Si:H)
As-grown nitrides
 hydrogen sources for, 445–446
 p-type doping in, 442–445

B

Bandgap problem, 247
Boron-doped poly-Si, 85
 dark electrical conductivity of, 137–140

C

Calcium, as dopant, 451–454, 455
Carbon, as dopant, 454–455
CdTe, hydrogen in, 433–434
Chemical vapor deposited diamond (CVD)
 covalent bonding environments in, 294–297
 dangling-bond hydrogen defects in, 302–304
 defects in, 284, 297–298
 deposition of, 284
 hydrogen concentrations in, 297–298
 hydrogen distribution in, 298–302, 304–305
 hydrogen properties in, 6, 284–286
 IR transmission in, 305–306
 thermal conductivity of, 306–308

D

Deep-level transient spectroscopy (DLTS), 3
Defects
 acceptor-like, 144–152
 EPR of, 47–48
 equilibrium concentrations of, 481–482
 grain-boundary, passivation of, 110–124
 hydrogen-intrinsic, 45–47
 light-induced, 125–132, 266
 overcoordination, 256–261
 paramagnetic, in CVD diamond, 302–304

Defects (*Continued*)
  passivation of, 1–2, 9, 26, 86, 244, 269–271, 373
  platelets, 2, 152–159
  uniaxial stress response of, 59–65
Density of states
  energy dispersion of, 210–212
  experimental determination of, 206–209
  model of diffusion, 102–110
Density-functional theory (DFT), 245–247, 483
Deuterated amorphous silicon, deuterium effusion in, 168–171, 172
Deuteration. *See* Hydrogenation
Deuterium
  diffusion from plasma source in LPCVD-grown poly-Si, 95–98
  diffusion from plasma source in solid-state poly-Si, 87–95
  diffusion in p-type silicon, 9
  diffusion of, 104–105
  passivation by, 269–271
  properties of, 5–6
Diffusion
  density of states model of, 102–110
  deviation from expected patterns of, 220–225
  and effusion, 226–229
  in heterostructures, 466–469
  of hydrogen through poly-Si, 87–98
  of hydrogen through Si, 13, 243–244
  of impurities, 69–75
  layer, 225–226
  mechanisms of, 462–466
  mechanisms in In-containing nitrides, 461–462
  mechanisms in III-V alloys, 456–461
  Meyer-Neldel rule of, 213–216
  plasma in-diffusion, 225–226
  source concentration dependence of, 209
  in III-V semiconductors, 446–449
  time dependence of, 217–220
Dopants
  activation mechanism of, 498–499
  hydrogen enhancement of, 499–500
  passivation of, 9, 86, 450–456

E

Effusion, 166–181
  compared to diffusion, 226–229
  related to IR absorption spectra, 229–230
Elastic recoil spectrometry (ERS), 293
Electron paramagnetic resonance (EPR), 3, 28, 290–293
  procedures for, 29

F

First-principles methods, 481
Formation energy, 481–482
Fourier transform spectroscopy (FTIR), 286–288

G

GaAs
  C-H complexes in, 376–381
  diatomic hydrogen in, 385
  group II acceptor-hydrogen complexes in, 375, 383–384, 407
  group VI donor-hydrogen complexes in, 382–383
  hydrogen complexes in, 374–375
  hydrogen passivation in, 373
  muonium transitions in, 359–364
  N-H complexes in, 384
  O-H complexes in, 384
  pressure effects on local vibrational modes, 389–390
  Si-H complexes in, 394
  temperature effects on local vibrational modes, 385–389
GaN
  Ca-H complexes in, 424
  Cd-H complexes in, 425
  characteristics of, 479–481
  doping of in the absence of hydrogen, 495–496
  doping of in the presence of hydrogen, 497–499
  gallium vacancies in, 494–495
  growing techniques for, 480
  hydrogen diffusion in, 456, 462–466
  hydrogen molecules in, 489
  hydrogen properties in, 8–9
  hydrogen solubility in, 486

hydrogen sources for, 445–446
hydrogen stability in, 487
hydrogen-acceptor complexes in, 489–492
interstitial configurations in, 485
Mg-H complexes in, 423–424, 490–492
monatomic hydrogen in, 483–486
muonium transitions in, 366
native defects in, 492–493
nitrogen vacancies in, 493–494
p-type doping in, 442–445, 480
GaP, 375
C-H complexes in, 402
group II acceptor-hydrogen complexes in, 398–402
nitrogen-dihydrogen complexes in, 411–412
vacancy-hydrogen complexes in, 412
GaSb
hydrogen in, 426
muonium transitions in, 365
Germanium, muonium transitions in, 354–359
Grain boundaries, 271–272
passivation of, 110–124, 272–273

H

Harris functional, calculations using, 247–248
Hartree-Fock-based cluster calculations, 483
Heterostructures, hydrogen behavior in, 466–469
Hydride gases, in semiconductor deposition, 1
Hydrogen
acceptor level in, Si, 20–21
acceptor-like defects induced in poly-Si by, 143–152
in As-grown nitrides, 443–449
behavior in heterostructures, 466–469
binding energy with silicon, 5
bond-centered, EPR spectrum of, 29–36
bond-centered, stress-induced alignment of, 37–42
charge states of, 3, 14, 17
chemical potential of, 106–110, 209, 212–220

compared with muonium, 6–7, 312–313, 366–369
complexes with impurities, 3–4
complexes with native defects, 492–493
complexes with Pt, 47–48
complexes with S, 47–48
configurations in Si, 245, 246
in CVD, 6, 284–286, 294–308
diffusion from plasma source in LPCVD-grown poly-Si, 95–98
diffusion from plasma source in solid-state poly-Si, 87–95
diffusion through Si, 13, 243–244
donor level in Si, 18–20
effect on IR transmission in CVD diamond, 305–306
effects induced in Si by, 51–75
effusion in a-Si:H, 167–171
enhancement of doping by, 499–500
equilibrium densities in Si, 21–22
extended complexes of, 245
facilitation of reactions by, 4
first-principles energies in Si, 249
formation energies of, 486–489
gap states induced by, 2
hydrogenation methods, 1, 50
hyperfine interactions in Si, 32–34
importance of, 2
interactions with a-Si, 241–242, 248–271
interactions with deep levels, 244
interactions with grain boundaries, 110–124, 272–273
interactions with overcoordination defects, 256–261
interactions with poly-Si, 241–242, 271–277
interactions with silicon dangling bonds, 5, 253–256, 269–271
interactions with vacancies, 493–495
interactions with weak Si Si bonds, 261–262
interstitial, in Si, 28–29
isolated interstitial, 242–243
molecular complexes of, 244–245
molecules of, 244
motion of, 248–252
negative charge state of, 3, 15, 17, 20–21, 486
neutral charge state of, 15, 16–17, 48
in nitrides, 472–477

Hydrogen (*Continued*)
  NRA analysis of, 293
  in oxygen thermal donor (NL10), 49–51
  passivation of defects by, 1–2, 9, 26, 244, 269–271, 373
  passivation of dopants by, 9, 86, 450–456
  platelets induced by, 2, 152–159
  positive charge state of, 14–16, 484–485
  potentials in Si, 248
  property differences from deuterium, 5–6
  reduction of defect concentration by, 165–166
  in Si-Si bonds, 6, 261–262
  solubility in a-Si:H, 192–199
  solubility in compact material, 231–232
  solubility in Si, 13
  sources of, 1
  stability of charge states of, 16, 17
  stable sites for, 3
  thermally activated disappearance in Si, 42–45
  in III-V semiconductors, 7–8, 374–501
  in II-VI semiconductors, 7–8, 375, 426–435
  void formation related to, 234–235
Hydrogen diffusion in a-Si:H, 131, 182–192
  defect motion models of, 201–202
  equilibrium energy band model of, 202–206
  hydrogen trapping model of, 199–201
  light-induced defects in, 130
Hydrogen diffusion in poly-Si
  in LPCVD-grown poly-Si, 95–98
  in solid-state poly-Si, 87–95
Hydrogen exchange, 210–212
Hydrogen traps, in poly-Si, 103–105
Hydrogenated amorphous networks, simulations of, 262–264
Hydrogenated amorphous silicon (a-Si:H), 5, 166–167
  deposition parameters of, 167
  hydrogen chemical potential in, 209
  hydrogen density of states distribution in, 206–209
  hydrogen diffusion in, 131, 182–192, 212–229
  hydrogen effusion in, 167–171, 226–230
  hydrogen solubility in, 192–199
  IR absorption data for, 174–181, 229–230
  mechanism of hydrogen exchange in, 210–212
  metastability in, 124
  structural characterization of, 172, 173, 174, 178–181
  void formation in, 234–235
Hydrogen-associated shallow donor (HSD), 51
  bistability of, 53–57
  formation of, 52–53
  in neutral charge state, 65–69
  role of oxygen in, 57–59

I

Impurities
  EPR analysis of, 290–293
  equilibrium concentrations of, 481–482
  ERS analysis of, 293
  Fourier transform analysis of, 286–288
  hydrogen-enhanced migration of, 69
  NMR analysis of, 288–290
  NRA analysis of, 293
InAlAs, hydrogen behavior in, 462
InAlN, hydrogen behavior in, 456, 461–462
InAs, hydrogen behavior in, 426
InGaAs, hydrogen behavior in, 462
InGaN, hydrogen behavior in, 456, 462
InN, hydrogen behavior in, 456
InP, 375
  C-H complexes in, 402
  group II acceptor-hydrogen complexes in, 396–399, 407
  hydrogen behavior in, 462
  temperature effects on local vibrational modes, 401
  vacancy-hydrogen complexes in, 399–400

L

Layer diffusion, 225–226
Level shifting, 473
Light, defects in poly-Si generated by, 125–132, 266
Longitudinal muon-spin relaxation, 323–325
Low-pressure chemical vapor deposition (LPCVD), 85, 95–98

M

Magnesium, as dopant, 450, 455
Material characterization, by hydrogen effusion, 166–181
Metastability, 124
  changes in electrical conductivity, 132–143
  energy distributions in, 266–267
  energy levels in, 264–266
  models of, 267–268
  in poly-Si, 124–143, 273–277
Meyer-Neldel rule, 213–216
Muon level-crossing resonance (LCR), 321–323
Muon-spin research (μSR), 316–318
  hyperfine decoupling curves in, 325–326
  longitudinal-field studies, 319–328
  relevance to hydrogen impurities, 366–369
  transverse-field studies, 318–319
Muon spin resonance, 320–321
Muonium, 26
  centers in semiconductors, 313–314
  characteristics of, 27, 312
  charged diamagnetic centers in, 334–340
  differences from hydrogen, 6–7, 312–313
  formation of, 312
  longitudinal-field studies of, 319–328
  neutral paramagnetic centers in, 329–333, 334
  states of, 314–315, 328–329
  state transitions in, 315, 341–366
  transverse-field studies of, 318–319

N

Negative-U center, 17
Negative-U effect, 486–489
NL10 center, 4
Nuclear magnetic resonance (NMR), 288–290
  in muonium research, 316–318
  longitudinal-field studies, 319–328
  transverse-field studies, 318–319
Nuclear reaction analysis (NRA), 293

O

Oxygen
  migration of, 4
  role in formation of HSD, 57–59

P

Passivation, 1–2, 9, 26, 86, 244, 269–271, 373, 450–456, 473
Phosphorus-doped poly-Si, 85
  dark electrical conductivity of, 134–136
Plasma in-diffusion, 225–226
Platelets, 2
  created in poly-Si, 152–159
Platinum, complexes with hydrogen, 47–48
Poly-Si
  acceptor-like defects in, 143–152, 277
  boron-doped, 85
  characteristics of, 83–84
  donor-like defects in, 273–277
  grain boundaries in, 271–273
  hydrogen diffusion from plasma source in, 87–98
  hydrogen diffusion from silicon layer in, 98–102
  hydrogen passivation in, 86, 110–124
  hydrogen properties in, 4–5, 85
  hydrogen traps in, 103–105
  hydrogenation of, 84
  light-induced defect generation in, 125–132
  metastability in, 125, 273–277
  metastable changes in electrical conductivity in, 132–143
  phosphorus-doped, 85
  platelets in, 152–159
  substrates for, 8
  temperature sensitivity of, 132–134
Polycrystalline chemical vapor deposited diamond. See Chemical vapor deposited diamond (CVD)
Polycrystalline silicon. See Poly-Si
Poole-Frenkel effect, 19
Pseudopotential-density-functional calculations, 245–248

R

Residual isotope effect, 36

S

Si-H bonds
  in amorphous environment, 254–255
  in crystalline environment, 253–254
  dissociation mechanisms of, 255–256

Silicon
amorphous. *See* Amorphous silicon (*a*-Si)

bonds with hydrogen at Si surface, 49
hydrogen acceptor level in, 20–21
hydrogen as contaminant of, 49–51
hydrogen donor level in, 18–20
hydrogen passivating dangling bonds in, 112–123, 269–271
hydrogen roles in, 241–248
hydrogen-enhanced migration of Al in, 69–75
hydrogen-induced effects in, 51–75
hydrogen-intrinsic defect complexes in, 45–47
hydrogenated amorphous. *See* Hydrogenated amorphous silicon (*a*-Si:H)
hyperfine interactions with hydrogen, 32–34
interstitial hydrogen in, 28–29
muonium states in, 342–354
muonium transitions in, 352
n-type, two-dimensional defects in, 2
polycrystalline. *See* Poly-Si
SIMS depth profiling, 182–184
Spin-lattice relaxation, 323
Stress-induced alignment
of AA1 spectrum, 59–65
in neutral charge state, 42
in positive charge state, 37–42
Sulfur, complexes with hydrogen, 47–48

T

Ternary alloys, hydrogen properties in, 9
Thermally activated annealing, 42–45
III-V semiconductors
deposition and etching of, 471–472
hydrogen in, 7–8, 374, 469
implant isolation in, 470
muonium transitions in, 364–366
wet processing of, 470–471
Tight-binding molecular dynamics, 248
II-VI semiconductors, hydrogen in, 7–8, 375

V

Void formation, hydrogen-related, 234–235

Z

Zero-field muon-spin relaxation, 327–328
Zinc, as dopant, 450, 455
ZnSe
As-H complexes in, 428–431
hydrogen behavior in, 375, 426–427
N-H complexes in, 375, 427–428
temperature effects on local vibrational modes, 431–433

Contents of Volumes in This Series

Volume 1 Physics of III–V Compounds

C. *Hilsum*, Some Key Features of III–V Compounds
F. *Bassani*, Methods of Band Calculations Applicable to III–V Compounds
E. O. *Kane*, The k-p Method
V. L. *Bonch-Bruevich*, Effect of Heavy Doping on the Semiconductor Band Structure
D. *Long*, Energy Band Structures of Mixed Crystals of III–V Compounds
L. M. *Roth and* P. N. *Argyres*, Magnetic Quantum Effects
S. M. *Puri and* T. H. *Geballe*, Thermomagnetic Effects in the Quantum Region
W. M. *Becker*, Band Characteristics near Principal Minima from Magnetoresistance
E. H. *Putley*, Freeze-Out Effects, Hot Electron Effects, and Submillimeter Photoconductivity in InSb
H. *Weiss*, Magnetoresistance
B. *Ancker-Johnson*, Plasma in Semiconductors and Semimetals

Volume 2 Physics of III–V Compounds

M. G. *Holland*, Thermal Conductivity
S. I. *Novkova*, Thermal Expansion
U. *Piesbergen*, Heat Capacity and Debye Temperatures
G. *Giesecke*, Lattice Constants
J. R. *Drabble*, Elastic Properties
A. U. *Mac Rae and* G. W. *Gobeli*, Low Energy Electron Diffraction Studies
R. Lee *Mieher*, Nuclear Magnetic Resonance
B. *Goldstein*, Electron Paramagnetic Resonance
T. S. *Moss*, Photoconduction in III–V Compounds
E. *Antoncik and* J. *Tauc*, Quantum Efficiency of the Internal Photoelectric Effect in InSb
G. W. *Gobeli and* I. G. *Allen*, Photoelectric Threshold and Work Function
P. S. *Pershan*, Nonlinear Optics in III–V Compounds
M. *Gershenzon*, Radiative Recombination in the III–V Compounds
F. *Stern*, Stimulated Emission in Semiconductors

Volume 3 Optical of Properties III–V Compounds

M. Hass, Lattice Reflection
W. G. Spitzer, Multiphonon Lattice Absorption
D. L. Stierwalt and R. F. Potter, Emittance Studies
H. R. Philipp and H. Ehrenveich, Ultraviolet Optical Properties
M. Cardona, Optical Absorption above the Fundamental Edge
E. J. Johnson, Absorption near the Fundamental Edge
J. O. Dimmock, Introduction to the Theory of Exciton States in Semiconductors
B. Lax and J. G. Mavroides, Interband Magnetooptical Effects
H. Y. Fan, Effects of Free Carries on Optical Properties
E. D. Palik and G. B. Wright, Free-Carrier Magnetooptical Effects
R. H. Bube, Photoelectronic Analysis
B. O. Seraphin and H. E. Bennett, Optical Constants

Volume 4 Physics of III–V Compounds

N. A. Goryunova, A. S. Borschevskii, and D. N. Tretiakov, Hardness
N. N. Sirota, Heats of Formation and Temperatures and Heats of Fusion of Compounds $A^{III}B^{V}$
D. L. Kendall, Diffusion
A. G. Chynoweth, Charge Multiplication Phenomena
R. W. Keyes, The Effects of Hydrostatic Pressure on the Properties of III–V Semiconductors
L. W. Aukerman, Radiation Effects
N. A. Goryunova, F. P. Kesamanly, and D. N. Nasledov, Phenomena in Solid Solutions
R. T. Bate, Electrical Properties of Nonuniform Crystals

Volume 5 Infrared Detectors

H. Levinstein, Characterization of Infrared Detectors
P. W. Kruse, Indium Antimonide Photoconductive and Photoelectromagnetic Detectors
M. B. Prince, Narrowband Self-Filtering Detectors
I. Melngalis and T. C. Harman, Single-Crystal Lead-Tin Chalcogenides
D. Long and J. L. Schmidt, Mercury-Cadmium Telluride and Closely Related Alloys
E. H. Putley, The Pyroelectric Detector
N. B. Stevens, Radiation Thermopiles
R. J. Keyes and T. M. Quist, Low Level Coherent and Incoherent Detection in the Infrared
M. C. Teich, Coherent Detection in the Infrared
F. R. Arams, E. W. Sard, B. J. Peyton, and F. P. Pace, Infrared Heterodyne Detection with Gigahertz IF Response
H. S. Sommers, Jr., Macrowave-Based Photoconductive Detector
R. Sehr and R. Zuleeg, Imaging and Display

Volume 6 Injection Phenomena

M. A. Lampert and R. B. Schilling, Current Injection in Solids: The Regional Approximation Method
R. Williams, Injection by Internal Photoemission
A. M. Barnett, Current Filament Formation

R. Baron and J. W. Mayer, Double Injection in Semiconductors
W. Ruppel, The Photoconductor-Metal Contact

Volume 7 Application and Devices
Part A

J. A. Copeland and S. Knight, Applications Utilizing Bulk Negative Resistance
F. A. Padovani, The Voltage-Current Characteristics of Metal-Semiconductor Contacts
P. L. Hower, W. W. Hooper, B. R. Cairns, R. D. Fairman, and D. A. Tremere, The GaAs Field-Effect Transistor
M. H. White, MOS Transistors
G. R. Antell, Gallium Arsenide Transistors
T. L. Tansley, Heterojunction Properties

Part B

T. Misawa, IMPATT Diodes
H. C. Okean, Tunnel Diodes
R. B. Campbell and Hung-Chi Chang, Silicon Junction Carbide Devices
R. E. Enstrom, H. Kressel, and L. Krassner, High-Temperature Power Rectifiers of $GaAs_{1-x}P_x$

Volume 8 Transport and Optical Phenomena

R. J. Stirn, Band Structure and Galvanomagnetic Effects in III–V Compounds with Indirect Band Gaps
R. W. Ure, Jr., Thermoelectric Effects in III–V Compounds
H. Piller, Faraday Rotation
H. Barry Bebb and E. W. Williams, Photoluminescence I: Theory
E. W. Williams and H. Barry Bebb, Photoluminescence II: Gallium Arsenide

Volume 9 Modulation Techniques

B. O. Seraphin, Electroreflectance
R. L. Aggarwal, Modulated Interband Magnetooptics
D. F. Blossey and Paul Handler, Electroabsorption
B. Batz, Thermal and Wavelength Modulation Spectroscopy
I. Balslev, Piezopptical Effects
D. E. Aspnes and N. Bottka, Electric-Field Effects on the Dielectric Function of Semiconductors and Insulators

Volume 10 Transport Phenomena

R. L. Rhode, Low-Field Electron Transport
J. D. Wiley, Mobility of Holes in III–V Compounds
C. M. Wolfe and G. E. Stillman, Apparent Mobility Enhancement in Inhomogeneous Crystals
R. L. Petersen, The Magnetophonon Effect

Volume 11 Solar Cells

H. J. Hovel, Introduction; Carrier Collection, Spectral Response, and Photocurrent; Solar Cell Electrical Characteristics; Efficiency; Thickness; Other Solar Cell Devices; Radiation Effects; Temperature and Intensity; Solar Cell Technology

Volume 12 Infrared Detectors (II)

W. L. Eiseman, J. D. Merriam, and R. F. Potter, Operational Characteristics of Infrared Photodetectors
P. R. Bratt, Impurity Germanium and Silicon Infrared Detectors
E. H. Putley, InSb Submillimeter Photoconductive Detectors
G. E. Stillman, C. M. Wolfe, and J. O. Dimmock, Far-Infrared Photoconductivity in High Purity GaAs
G. E. Stillman and C. M. Wolfe, Avalanche Photodiodes
P. L. Richards, The Josephson Junction as a Detector of Microwave and Far-Infrared Radiation
E. H. Putley, The Pyroelectric Detector — An Update

Volume 13 Cadmium Telluride

K. Zanio, Materials Preparations; Physics; Defects; Applications

Volume 14 Lasers, Junctions, Transport

N. Holonyak, Jr. and M. H. Lee, Photopumped III–V Semiconductor Lasers
H. Kressel and J. K. Butler, Heterojunction Laser Diodes
A Van der Ziel, Space-Charge-Limited Solid-State Diodes
P. J. Price, Monte Carlo Calculation of Electron Transport in Solids

Volume 15 Contacts, Junctions, Emitters

B. L. Sharma, Ohmic Contacts to III–V Compounds Semiconductors
A. Nussbaum, The Theory of Semiconducting Junctions
J. S. Escher, NEA Semiconductor Photoemitters

Volume 16 Defects, (HgCd)Se, (HgCd)Te

H. Kressel, The Effect of Crystal Defects on Optoelectronic Devices
C. R. Whitsett, J. G. Broerman, and C. J. Summers, Crystal Growth and Properties of $Hg_{1-x}Cd_xSe$ alloys
M. H. Weiler, Magnetooptical Properties of $Hg_{1-x}Cd_xTe$ Alloys
P. W. Kruse and J. G. Ready, Nonlinear Optical Effects in $Hg_{1-x}Cd_xTe$

Volume 17 CW Processing of Silicon and Other Semiconductors

J. F. Gibbons, Beam Processing of Silicon
A. Lietoila, R. B. Gold, J. F. Gibbons, and L. A. Christel, Temperature Distributions and Solid Phase Reaction Rates Produced by Scanning CW Beams

A. Leitoila and J. F. Gibbons, Applications of CW Beam Processing to Ion Implanted Crystalline Silicon
N. M. Johnson, Electronic Defects in CW Transient Thermal Processed Silicon
K. F. Lee, T. J. Stultz, and J. F. Gibbons, Beam Recrystallized Polycrystalline Silicon: Properties, Applications, and Techniques
T. Shibata, A. Wakita, T. W. Sigmon, and J. F. Gibbons, Metal-Silicon Reactions and Silicide
Y. I. Nissim and J. F. Gibbons, CW Beam Processing of Gallium Arsenide

Volume 18   Mercury Cadmium Telluride

P. W. Kruse, The Emergence of $(Hg_{1-x}Cd_x)Te$ as a Modern Infrared Sensitive Material
H. E. Hirsch, S. C. Liang, and A. G. White, Preparation of High-Purity Cadmium, Mercury, and Tellurium
W. F. H. Micklethwaite, The Crystal Growth of Cadmium Mercury Telluride
P. E. Petersen, Auger Recombination in Mercury Cadmium Telluride
R. M. Broudy and V. J. Mazurczyck, (HgCd)Te Photoconductive Detectors
M. B. Reine, A. K. Soad, and T. J. Tredwell, Photovoltaic Infrared Detectors
M. A. Kinch, Metal-Insulator-Semiconductor Infrared Detectors

Volume 19   Deep Levels, GaAs, Alloys, Photochemistry

G. F. Neumark and K. Kosai, Deep Levels in Wide Band-Gap III–V Semiconductors
D. C. Look, The Electrical and Photoelectronic Properties of Semi-Insulating GaAs
R. F. Brebrick, Ching-Hua Su, and Pok-Kai Liao, Associated Solution Model for Ga-In-Sb and Hg-Cd-Te
Y. Ya. Gurevich and Y. V. Pleskon, Photoelectrochemistry of Semiconductors

Volume 20   Semi-Insulating GaAs

R. N. Thomas, H. M. Hobgood, G. W. Eldridge, D. L. Barrett, T. T. Braggins, L. B. Ta, and S. K. Wang, High-Purity LEC Growth and Direct Implantation of GaAs for Monolithic Microwave Circuits
C. A. Stolte, Ion Implantation and Materials for GaAs Integrated Circuits
C. G. Kirkpatrick, R. T. Chen, D. E. Holmes, P. M. Asbeck, K. R. Elliott, R. D. Fairman, and J. R. Oliver, LEC GaAs for Integrated Circuit Applications
J. S. Blakemore and S. Rahimi, Models for Mid-Gap Centers in Gallium Arsenide

Volume 21   Hydrogenated Amorphous Silicon
Part A

J. I. Pankove, Introduction
M. Hirose, Glow Discharge; Chemical Vapor Deposition
Y. Uchida, di Glow Discharge
T. D. Moustakas, Sputtering
I. Yamada, Ionized-Cluster Beam Deposition
B. A. Scott, Homogeneous Chemical Vapor Deposition

F. J. Kampas, Chemical Reactions in Plasma Deposition
P. A. Longeway, Plasma Kinetics
H. A. Weakliem, Diagnostics of Silane Glow Discharges Using Probes and Mass Spectroscopy
L. Gluttman, Relation between the Atomic and the Electronic Structures
A. Chenevas-Paule, Experiment Determination of Structure
S. Minomura, Pressure Effects on the Local Atomic Structure
D. Adler, Defects and Density of Localized States

Part B

J. I. Pankove, Introduction
G. D. Cody, The Optical Absorption Edge of a-Si:H
N. M. Amer and W. B. Jackson, Optical Properties of Defect States in a-Si:H
P. J. Zanzucchi, The Vibrational Spectra of a-Si:H
Y. Hamakawa, Electroreflectance and Electroabsorption
J. S. Lannin, Raman Scattering of Amorphous Si, Ge, and Their Alloys
R. A. Street, Luminescence in a-Si:H
R. S. Crandall, Photoconductivity
J. Tauc, Time-Resolved Spectroscopy of Electronic Relaxation Processes
P. E. Vanier, IR-Induced Quenching and Enhancement of Photoconductivity and Photo luminescence
H. Schade, Irradiation-Induced Metastable Effects
L. Ley, Photoelectron Emission Studies

Part C

J. I. Pankove, Introduction
J. D. Cohen, Density of States from Junction Measurements in Hydrogenated Amorphous Silicon
P. C. Taylor, Magnetic Resonance Measurements in a-Si:H
K. Morigaki, Optically Detected Magnetic Resonance
J. Dresner, Carrier Mobility in a-Si:H
T. Tiedje, Information about band-Tail States from Time-of-Flight Experiments
A. R. Moore, Diffusion Length in Undoped a-Si:H
W. Beyer and J. Overhof, Doping Effects in a-Si:H
H. Fritzche, Electronic Properties of Surfaces in a-Si:H
C. R. Wronski, The Staebler-Wronski Effect
R. J. Nemanich, Schottky Barriers on a-Si:H
B. Abeles and T. Tiedje, Amorphous Semiconductor Superlattices

Part D

J. I. Pankove, Introduction
D. E. Carlson, Solar Cells
G. A. Swartz, Closed-Form Solution of I–V Characteristic for a a-Si:H Solar Cells
I. Shimizu, Electrophotography
S. Ishioka, Image Pickup Tubes

P. G. LeComber and *W. E. Spear*, The Development of the a-Si:H Field-Effect Transistor and Its Possible Applications
D. G. Ast, a-Si:H FET-Addressed LCD Panel
S. Kaneko, Solid-State Image Sensor
M. Matsumura, Charge-Coupled Devices
M. A. Bosch, Optical Recording
A. D'Amico and G. Fortunato, Ambient Sensors
H. Kukimoto, Amorphous Light-Emitting Devices
R. J. Phelan, Jr., Fast Detectors and Modulators
J. I. Pankove, Hybrid Structures
P. G. LeComber, A. E. Owen, W. E. Spear, J. Hajto, and W. K. Choi, Electronic Switching in Amorphous Silicon Junction Devices

Volume 22 Lightwave Communications Technology
Part A
K. Nakajima, The Liquid-Phase Epitaxial Growth of InGaAsP
W. T. Tsang, Molecular Beam Epitaxy for III–V Compound Semiconductors
G. B. Stringfellow, Organometallic Vapor-Phase Epitaxial Growth of III–V Semiconductors
G. Beuchet, Halide and Chloride Transport Vapor-Phase Deposition of InGaAsP and GaAs
M. Razeghi, Low-Pressure Metallo-Organic Chemical Vapor Deposition of $Ga_xIn_{1-x}As P_{1-y}$ Alloys
P. M. Petroff, Defects in III–V Compound Semiconductors

Part B
J. P. van der Ziel, Mode Locking of Semiconductor Lasers
K. Y. Lau and A. Yariv, High-Frequency Current Modulation of Semiconductor Injection Lasers
C. H. Henry, Special Properties of Semiconductor Lasers
Y. Suematsu, K. Kishino, S. Arai, and F. Koyama, Dynamic Single-Mode Semiconductor Lasers with a Distributed Reflector
W. T. Tsang, The Cleaved-Coupled-Cavity (C^3) Laser

Part C
R. J. Nelson and N. K. Dutta, Review of InGaAsP InP Laser Structures and Comparison of Their Performance
N. Chinone and M. Nakamura, Mode-Stabilized Semiconductor Lasers for 0.7–0.8- and 1.1–1.6-μm Regions
Y. Horikoshi, Semiconductor Lasers with Wavelengths Exceeding 2 μm
B. A. Dean and M. Dixon, The Functional Reliability of Semiconductor Lasers as Optical Transmitters
R. H. Saul, T. P. Lee, and C. A. Burus, Light-Emitting Device Design
C. L. Zipfel, Light-Emitting Diode-Reliability
T. P. Lee and T. Li, LED-Based Multimode Lightwave Systems
K. Ogawa, Semiconductor Noise-Mode Partition Noise

Part D

F. *Capasso*, The Physics of Avalanche Photodiodes
T. P. *Pearsall and M. A. Pollack*, Compound Semiconductor Photodiodes
T. *Kaneda*, Silicon and Germanium Avalanche Photodiodes
S. R. *Forrest*, Sensitivity of Avalanche Photodetector Receivers for High-Bit-Rate Long-Wavelength Optical Communication Systems
J. C. *Campbell*, Phototransistors for Lightwave Communications

Part E

S. *Wang*, Principles and Characteristics of Integrable Active and Passive Optical Devices
S. *Margalit and A. Yariv*, Integrated Electronic and Photonic Devices
T. *Mukai, Y. Yamamoto, and T. Kimura*, Optical Amplification by Semiconductor Lasers

Volume 23   Pulsed Laser Processing of Semiconductors

R. F. *Wood, C. W. White, and R. T. Young*, Laser Processing of Semiconductors: An Overview
C. W. *White*, Segregation, Solute Trapping, and Supersaturated Alloys
G. E. *Jellison, Jr.*, Optical and Electrical Properties of Pulsed Laser-Annealed Silicon
R. F. *Wood and G. E. Jellison, Jr.*, Melting Model of Pulsed Laser Processing
R. F. *Wood and F. W. Young, Jr.*, Nonequilibrium Solidification Following Pulsed Laser Melting
D. H. *Lowndes and G. E. Jellison, Jr.*, Time-Resolved Measurement During Pulsed Laser Irradiation of Silicon
D. M. *Zebner*, Surface Studies of Pulsed Laser Irradiated Semiconductors
D. H. *Lowndes*, Pulsed Beam Processing of Gallium Arsenide
R. B. *James*, Pulsed CO_2 Laser Annealing of Semiconductors
R. T. *Young and R. F. Wood*, Applications of Pulsed Laser Processing

Volume 24   Applications of Multiquantum Wells, Selective Doping, and Superlattices

C. *Weisbuch*, Fundamental Properties of III–V Semiconductor Two-Dimensional Quantized Structures: The Basis for Optical and Electronic Device Applications
H. *Morkoc and H. Unlu*, Factors Affecting the Performance of (Al,Ga)As/GaAs and (Al,Ga)As/InGaAs Modulation-Doped Field-Effect Transistors: Microwave and Digital Applications
N. T. *Linh*, Two-Dimensional Electron Gas FETs: Microwave Applications
M. *Abe et al.*, Ultra-High-Speed HEMT Integrated Circuits
D. S. *Chemla, D. A. B. Miller, and P. W. Smith*, Nonlinear Optical Properties of Multiple Quantum Well Structures for Optical Signal Processing
F. *Capasso*, Graded-Gap and Superlattice Devices by Band-Gap Engineering
W. T. *Tsang*, Quantum Confinement Heterostructure Semiconductor Lasers
G. C. *Osbourn et al.*, Principles and Applications of Semiconductor Strained-Layer Superlattices

Volume 25 Diluted Magnetic Semiconductors

W. Giriat and J. K. Furdyna, Crystal Structure, Composition, and Materials Preparation of Diluted Magnetic Semiconductors
W. M. Becker, Band Structure and Optical Properties of Wide-Gap $A^{II}_{1-x}Mn_xB_{IV}$ Alloys at Zero Magnetic Field
S. Oseroff and P. H. Keesom, Magnetic Properties: Macroscopic Studies
T. Giebultowicz and T. M. Holden, Neutron Scattering Studies of the Magnetic Structure and Dynamics of Diluted Magnetic Semiconductors
J. Kossut, Band Structure and Quantum Transport Phenomena in Narrow-Gap Diluted Magnetic Semiconductors
C. Riquaux, Magnetooptical Properties of Large-Gap Diluted Magnetic Semiconductors
J. A. Gaj, Magnetooptical Properties of Large-Gap Diluted Magnetic Semiconductors
J. Mycielski, Shallow Acceptors in Diluted Magnetic Semiconductors: Splitting, Boil-off, Giant Negative Magnetoresistance
A. K. Ramadas and R. Rodriquez, Raman Scattering in Diluted Magnetic Semiconductors
P. A. Wolff, Theory of Bound Magnetic Polarons in Semimagnetic Semiconductors

Volume 26 III–V Compound Semiconductors and Semiconductor Properties of Superionic Materials

Z. Yuanxi, III–V Compounds
H. V. Winston, A. T. Hunter, H. Kimura, and R. E. Lee, InAs-Alloyed GaAs Substrates for Direct Implantation
P. K. Bhattacharya and S. Dhar, Deep Levels in III–V Compound Semiconductors Grown by MBE
Y. Ya. Gurevich and A. K. Ivanov-Shits, Semiconductor Properties of Supersonic Materials

Volume 27 High Conducting Quasi-One-Dimensional Organic Crystals

E. M. Conwell, Introduction to Highly Conducting Quasi-One-Dimensional Organic Crystals
I. A. Howard, A Reference Guide to the Conducting Quasi-One-Dimensional Organic Molecular Crystals
J. P. Pouquet, Structural Instabilities
E. M. Conwell, Transport Properties
C. S. Jacobsen, Optical Properties
J. C. Scott, Magnetic Properties
L. Zuppiroli, Irradiation Effects: Perfect Crystals and Real Crystals

Volume 28 Measurement of High-Speed Signals in Solid State Devices

J. Frey and D. Ioannou, Materials and Devices for High-Speed and Optoelectronic Applications
H. Schumacher and E. Strid, Electronic Wafer Probing Techniques
D. H. Auston, Picosecond Photoconductivity: High-Speed Measurements of Devices and Materials
J. A. Valdmanis, Electro-Optic Measurement Techniques for Picosecond Materials, Devices, and Integrated Circuits.
J. M. Wiesenfeld and R. K. Jain, Direct Optical Probing of Integrated Circuits and High-Speed Devices
G. Plows, Electron-Beam Probing
A. M. Weiner and R. B. Marcus, Photoemissive Probing

Volume 29 Very High Speed Integrated Circuits: Gallium Arsenide LSI

M. Kuzuhara and T. Nazaki, Active Layer Formation by Ion Implantation
H. Hasimoto, Focused Ion Beam Implantation Technology
T. Nozaki and A. Higashisaka, Device Fabrication Process Technology
M. Ino and T. Takada, GaAs LSI Circuit Design
M. Hirayama, M. Ohmori, and K. Yamasaki, GaAs LSI Fabrication and Performance

Volume 30 Very High Speed Integrated Circuits: Heterostructure

H. Watanabe, T. Mizutani, and A. Usui, Fundamentals of Epitaxial Growth and Atomic Layer Epitaxy
S. Hiyamizu, Characteristics of Two-Dimensional Electron Gas in III–V Compound Heterostructures Grown by MBE
T. Nakanisi, Metalorganic Vapor Phase Epitaxy for High-Quality Active Layers
T. Nimura, High Electron Mobility Transistor and LSI Applications
T. Sugeta and T. Ishibashi, Hetero-Bipolar Transistor and LSI Application
H. Matsueda, T. Tanaka, and M. Nakamura, Optoelectronic Integrated Circuits

Volume 31 Indium Phosphide: Crystal Growth and Characterization

J. P. Farges, Growth of Discoloration-free InP
M. J. McCollum and G. E. Stillman, High Purity InP Grown by Hydride Vapor Phase Epitaxy
T. Inada and T. Fukuda, Direct Synthesis and Growth of Indium Phosphide by the Liquid Phosphorous Encapsulated Czochralski Method
O. Oda, K. Katagiri, K. Shinohara, S. Katsura, Y. Takahashi, K. Kainosho, K. Kohiro, and R. Hirano, InP Crystal Growth, Substrate Preparation and Evaluation
K. Tada, M. Tatsumi, M. Morioka, T. Araki, and T. Kawase, InP Substrates: Production and Quality Control
M. Razeghi, LP-MOCVD Growth, Characterization, and Application of InP Material
T. A. Kennedy and P. J. Lin-Chung, Stoichiometric Defects in InP

Volme 32 Strained-Layer Superlattices: Physics

T. P. Pearsall, Strained-Layer Superlattices
F. H. Pollack, Effects of Homogeneous Strain on the Electronic and Vibrational Levels in Semiconductors
J. Y. Marzin, J. M. Gerárd, P. Voisin, and J. A. Brum, Optical Studies of Strained III–V Heterolayers
R. People and S. A. Jackson, Structurally Induced States from Strain and Confinement
M. Jaros, Microscopic Phenomena in Ordered Superlattices

Volume 33 Strained-Layer Superlattices: Materials Science and Technology

R. Hull and J. C. Bean, Principles and Concepts of Strained-Layer Epitaxy
W. J. Schaff, P. J. Tasker, M. C. Foisy, and L. F. Eastman, Device Applications of Strained-Layer Epitaxy

S. T. Picraux, B. L. Doyle, and J. Y. Tsao, Structure and Characterization of Strained-Layer Superlattices
E. Kasper and F. Schaffer, Group IV Compounds
D. L. Martin, Molecular Beam Epitaxy of IV-VI Compounds Heterojunction
R. L. Gunshor, L. A. Kolodziejski, A. V. Nurmikko, and N. Otsuka, Molecular Beam Epitaxy of II-VI Semiconductor Microstructures

Volume 34 Hydrogen in Semiconductors

J. I. Pankove and N. M. Johnson, Introduction to Hydrogen in Semiconductors
C. H. Seager, Hydrogenation Methods
J. I. Pankove, Hydrogenation of Defects in Crystalline Silicon
J. W. Corbett, P. Deák, U. V. Desnica, and S. J. Pearton, Hydrogen Passivation of Damage Centers in Semiconductors
S. J. Pearton, Neutralization of Deep Levels in Silicon
J. I. Pankove, Neutralization of Shallow Acceptors in Silicon
N. M. Johnson, Neutralization of Donor Dopants and Formation of Hydrogen-Induced Defects in n-Type Silicon
M. Stavola and S. J. Pearton, Vibrational Spectroscopy of Hydrogen-Related Defects in Silicon
A. D. Marwick, Hydrogen in Semiconductors: Ion Beam Techniques
C. Herring and N. M. Johnson, Hydrogen Migration and Solubility in Silicon
E. E. Haller, Hydrogen-Related Phenomena in Crystalline Germanium
J. Kakalios, Hydrogen Diffusion in Amorphous Silicon
J. Chevalier, B. Clerjaud, and B. Pajot, Neutralization of Defects and Dopants in III-V Semiconductors
G. G. DeLeo and W. B. Fowler, Computational Studies of Hydrogen-Containing Complexes in Semiconductors
R. F. Kiefl and T. L. Estle, Muonium in Semiconductors
C. G. Van de Walle, Theory of Isolated Interstitial Hydrogen and Muonium in Crystalline Semiconductors

Volume 35 Nanostructured Systems

M. Reed, Introduction
H. van Houten, C. W. J. Beenakker, and B. J. van Wees, Quantum Point Contacts
G. Timp, When Does a Wire Become an Electron Waveguide?
M. Büttiker, The Quantum Hall Effects in Open Conductors
W. Hansen, J. P. Kotthaus, and U. Merkt, Electrons in Laterally Periodic Nanostructures

Volume 36 The Spectroscopy of Semiconductors

D. Heiman, Spectroscopy of Semiconductors at Low Temperatures and High Magnetic Fields
A. V. Nurmikko, Transient Spectroscopy by Ultrashort Laser Pulse Techniques
A. K. Ramdas and S. Rodriguez, Piezospectroscopy of Semiconductors
O. J. Glembocki and B. V. Shanabrook, Photoreflectance Spectroscopy of Microstructures
D. G. Seiler, C. L. Littler, and M. H. Wiler, One- and Two-Photon Magneto-Optical Spectroscopy of InSb and $Hg_{1-x}Cd_xTe$

Volume 37   The Mechanical Properties of Semiconductors

A.-B. Chen, A. Sher and W. T. Yost, Elastic Constants and Related Properties of Semiconductor Compounds and Their Alloys
D. R. Clarke, Fracture of Silicon and Other Semiconductors
H. Siethoff, The Plasticity of Elemental and Compound Semiconductors
S. Guruswamy, K. T. Faber and J. P. Hirth, Mechanical Behavior of Compound Semiconductors
S. Mahajan, Deformation Behavior of Compound Semiconductors
J. P. Hirth, Injection of Dislocations into Strained Multilayer Structures
D. Kendall, C. B. Fleddermann, and K. J. Malloy, Critical Technologies for the Micromachining of Silicon
I. Matsuba and K. Mokuya, Processing and Semiconductor Thermoelastic Behavior

Volume 38   Imperfections in III/V Materials

U. Scherz and M. Scheffler, Density-Functional Theory of sp-Bonded Defects in III/V Semiconductors
M. Kaminska and E. R. Weber, El2 Defect in GaAs
D. C. Look, Defects Relevant for Compensation in Semi-Insulating GaAs
R. C. Newman, Local Vibrational Mode Spectroscopy of Defects in III/V Compounds
A. M. Hennel, Transition Metals in III/V Compounds
K. J. Malloy and K. Khachaturyan, DX and Related Defects in Semiconductors
V. Swaminathan and A. S. Jordan, Dislocations in III/V Compounds
K. W. Nauka, Deep Level Defects in the Epitaxial III/V Materials

Volume 39   Minority Carriers in III–V Semiconductors: Physics and Applications

N. K. Dutta, Radiative Transitions in GaAs and Other III–V Compounds
R. K. Ahrenkiel, Minority-Carrier Lifetime in III–V Semiconductors
T. Furuta, High Field Minority Electron Transport in p-GaAs
M. S. Lundstrom, Minority-Carrier Transport in III–V Semiconductors
R. A. Abram, Effects of Heavy Doping and High Excitation on the Band Structure of GaAs
D. Yevick and W. Bardyszewski, An Introduction to Non-Equilibrium Many-Body Analyses of Optical Processes in III–V Semiconductors

Volume 40   Epitaxial Microstructures

E. F. Schubert, Delta-Doping of Semiconductors: Electronic, Optical, and Structural Properties of Materials and Devices
A. Gossard, M. Sundaram, and P. Hopkins, Wide Graded Potential Wells
P. Petroff, Direct Growth of Nanometer-Size Quantum Wire Superlattices
E. Kapon, Lateral Patterning of Quantum Well Heterostructures by Growth of Nonplanar Substrates
H. Temkin, D. Gershoni, and M. Panish, Optical Properties of $Ga_{1-x}In_xAs$/InP Quantum Wells

Volume 41 High Speed Heterostructure Devices

F. Capasso, F. Beltram, S. Sen, A. Pahlevi, and A. Y. Cho, Quantum Electron Devices: Physics and Applications
P. Solomon, D. J. Frank, S. L. Wright, and F. Canora, GaAs-Gate Semiconductor–Insulator–Semiconductor FET
M. H. Hashemi and U. K. Mishra, Unipolar InP-Based Transistors
R. Kiehl, Complementary Heterostructure FET Integrated Circuits
T. Ishibashi, GaAs-Based and InP-Based Heterostructure Bipolar Transistors
H. C. Liu and T. C. L. G. Sollner, High-Frequency-Tunneling Devices
H. Ohnishi, T. More, M. Takatsu, K. Imamura, and N. Yokoyama, Resonant-Tunneling Hot-Electron Transistors and Circuits

Volume 42 Oxygen in Silicon

F. Shimura, Introduction to Oxygen in Silicon
W. Lin, The Incorporation of Oxygen into Silicon Crystals
T. J. Schaffner and D. K. Schroder, Characterization Techniques for Oxygen in Silicon
W. M. Bullis, Oxygen Concentration Measurement
S. M. Hu, Intrinsic Point Defects in Silicon
B. Pajot, Some Atomic Configurations of Oxygen
J. Michel and L. C. Kimerling, Electical Properties of Oxygen in Silicon
R. C. Newman and R. Jones, Diffusion of Oxygen in Silicon
T. Y. Tan and W. J. Taylor, Mechanisms of Oxygen Precipitation: Some Quantitative Aspects
M. Schrems, Simulation of Oxygen Precipitation
K. Simino and I. Yonenaga, Oxygen Effect on Mechanical Properties
W. Bergholz, Grown-in and Process-Induced Effects
F. Shimura, Intrinsic/Internal Gettering
H. Tsuya, Oxygen Effect on Electronic Device Performance

Volume 43 Semiconductors for Room Temperature Nuclear Detector Applications

R. B. James and T. E. Schlesinger, Introduction and Overview
L. S. Darken and C. E. Cox, High-Purity Germanium Detectors
A. Burger, D. Nason, L. Van den Berg, and M. Schieber, Growth of Mercuric Iodide
X. J. Bao, T. E. Schlesinger, and R. B. James, Electrical Properties of Mercuric Iodide
X. J. Bao, R. B. James, and T. E. Schlesinger, Optical Properties of Red Mercuric Iodide
M. Hage-Ali and P. Siffert, Growth Methods of CdTe Nuclear Detector Materials
M. Hage-Ali and P. Siffert, Characterization of CdTe Nuclear Detector Materials
M. Hage-Ali and P. Siffert, CdTe Nuclear Detectors and Applications
R. B. James, T. E. Schlesinger, J. Lund, and M. Schieber, $Cd_{1-x}Zn_xTe$ Spectrometers for Gamma and X-Ray Applications
D. S. McGregor, J. E. Kammeraad, Gallium Arsenide Radiation Detectors and Spectrometers
J. C. Lund, F. Olschner, and A. Burger, Lead Iodide
M. R. Squillante, and K. S. Shah, Other Materials: Status and Prospects
V. M. Gerrish, Characterization and Quantification of Detector Performance
J. S. Iwanczyk and B. E. Patt, Electronics for X-ray and Gamma Ray Spectrometers
M. Schieber, R. B. James, and T. E. Schlesinger, Summary and Remaining Issues for Room Temperature Radiation Spectrometers

Volume 44 II–IV Blue/Green Light Emitters: Device Physics and Epitaxial Growth

J. Han and R. L. Gunshor, MBE Growth and Electrical Properties of Wide Bandgap ZnSe-based II–VI Semiconductors
S. Fujita and S. Fujita, Growth and Characterization of ZnSe-based II–VI Semiconductors by MOVPE
E. Ho and L. A. Kolodziejski, Gaseous Source UHV Epitaxy Technologies for Wide Bandgap II–VI Semiconductors
C. G. Van de Walle, Doping of Wide-Band-Gap II–VI Compounds—Theory
R. Cingolani, Optical Properties of Excitons in ZnSe-Based Quantum Well Heterostructures
A. Ishibashi and A. V. Nurmikko, II–VI Diode Lasers: A Current View of Device Performance and Issues
S. Guha and J. Petruzello, Defects and Degradation in Wide-Gap II–VI-based Structures and Light Emitting Devices

Volume 45 Effect of Disorder and Defects in Ion-Implanted Semiconductors: Electrical and Physiochemical Characterization

H. Ryssel, Ion Implantation into Semiconductors: Historical Perspectives
You-Nian Wang and Teng-Cai Ma, Electronic Stopping Power for Energetic Ions in Solids
S. T. Nakagawa, Solid Effect on the Electronic Stopping of Crystalline Target and Application to Range Estimation
G. Müller, S. Kalbitzer and G. N. Greaves, Ion Beams in Amorphous Semiconductor Research
J. Boussey-Said, Sheet and Spreading Resistance Analysis of Ion Implanted and Annealed Semiconductors
M. L. Polignano and G. Queirolo, Studies of the Stripping Hall Effect in Ion-Implanted Silicon
J. Stoemenos, Transmission Electron Microscopy Analyses
R. Nipoti and M. Servidori, Rutherford Backscattering Studies of Ion Implanted Semiconductors
P. Zaumseil, X-ray Diffraction Techniques

Volume 46 Effect of Disorder and Defects in Ion-Implanted Semiconductors: Optical and Photothermal Characterization

M. Fried, T. Lohner and J. Gyulai, Ellipsometric Analysis
A. Seas and C. Christofides, Transmission and Reflection Spectroscopy on Ion Implanted Semiconductors
A. Othonos and C. Christofides, Photoluminescence and Raman Scattering of Ion Implanted Semiconductors. Influence of Annealing
C. Christofides, Photomodulated Thermoreflectance Investigation of Implanted Wafers. Annealing Kinetics of Defects
U. Zammit, Photothermal Deflection Spectroscopy Characterization of Ion-Implanted and Annealed Silicon Films
A. Mandelis, A. Budiman and M. Vargas, Photothermal Deep-Level Transient Spectroscopy of Impurities and Defects in Semiconductors
R. Kalish and S. Charbonneau, Ion Implantation into Quantum-Well Structures
A. M. Myasnikov and N. N. Gerasimenko, Ion Implantation and Thermal Annealing of III-V Compound Semiconducting Systems: Some Problems of III-V Narrow Gap Semiconductors

Volume 47 Uncooled Infrared Imaging Arrays and Systems

R. G. Buser and M. P. Tompsett, Historical Overview
P. W. Kruse, Principles of Uncooled Infrared Focal Plane Arrays
R. A. Wood, Monolithic Silicon Microbolometer Arrays
C. M. Hanson, Hybrid Pyroelectric-Ferroelectric Bolometer Arrays
D. L. Polla and J. R. Choi, Monolithic Pyroelectric Bolometer Arrays
N. Teranishi, Thermoelectric Uncooled Infrared Focal Plane Arrays
M. F. Tompsett, Pyroelectric Vidicon
T. W. Kenny, Tunneling Infrared Sensors
J. R. Vig, R. L. Filler and Y. Kim, Application of Quartz Microresonators to Uncooled Infrared Imaging Arrays
P. W. Kruse, Application of Uncooled Monolithic Thermoelectric Linear Arrays to Imaging Radiometers

Volume 48 High Brightness Light Emitting Diodes

G. B. Stringfellow, Materials Issues in High-Brightness Light-Emitting Diodes
M. G. Craford, Overview of Device issues in High-Brightness Light-Emitting Diodes
F. M. Steranka, AlGaAs Red Light Emitting Diodes
C. H. Chen, S. A. Stockman, M. J. Peanasky, and C. P. Kuo, OMVPE Growth of AlGaInP for High Efficiency Visible Light-Emitting Diodes
F. A. Kish and R. M. Fletcher, AlGaInP Light-Emitting Diodes
M. W. Hodapp, Applications for High Brightness Light-Emitting Diodes
I. Akasaki and H. Amano, Organometallic Vapor Epitaxy of GaN for High Brightness Blue Light Emitting Diodes
S. Nakamura, Group III-V Nitride Based Ultraviolet-Blue-Green-Yellow Light-Emitting Diodes and Laser Diodes

Volume 49 Light Emission in Silicon: from Physics to Devices

D. J. Lockwood, Light Emission in Silicon
G. Abstreiter, Band Gaps and Light Emission in Si/SiGe Atomic Layer Structures
T. G. Brown and D. G. Hall, Radiative Isoelectronic Impurities in Silicon and Silicon-Germanium Alloys and Superlattices
J. Michel, L. V. C. Assali, M. T. Morse, and L. C. Kimerling, Erbium in Silicon
Y. Kanemitsu, Silicon and Germanium Nanoparticles
P. M. Fauchet, Porous Silicon: Photoluminescence and Electroluminescent Devices
C. Delerue, G. Allan, and M. Lannoo, Theory of Radiative and Nonradiative Processes in Silicon Nanocrystallites
L. Brus, Silicon Polymers and Nanocrystals

Volume 50 Gallium Nitride (GaN)

J. I. Pankove and T. D. Moustakas, Introduction
S. P. DenBaars and S. Keller, Metalorganic Chemical Vapor Deposition (MOCVD) of Group III Nitrides
W. A. Bryden and T. J. Kistenmacher, Growth of Group III-A Nitrides by Reactive Sputtering
N. Newman, Thermochemistry of III-N Semiconductors
S. J. Pearton and R. J. Shul, Etching of III Nitrides

S. M. Bedair, Indium-based Nitride Compounds
A. Trampert, O. Brandt, and K. H. Ploog, Crystal Structure of Group III Nitrides
H. Morkoc, F. Hamdani, and A. Salvador, Electronic and Optical Properties of III–V Nitride based Quantum Wells and Superlattices
K. Doverspike and J. I. Pankove, Doping in the III-Nitrides
T. Suski and P. Perlin, High Pressure Studies of Defects and Impurities in Gallium Nitride
B. Monemar, Optical Properties of GaN
W. R. L. Lambrecht, Band Structure of the Group III Nitrides
N. E. Christensen and P. Perlin, Phonons and Phase Transitions in GaN
S. Nakamura, Applications of LEDs and LDs
I. Akasaki and H. Amano, Lasers
J. A. Cooper, Jr., Nonvolatile Random Access Memories in Wide Bandgap Semiconductors

Volume 51A Identification of Defects in Semiconductors

G. D. Watkins, EPR and ENDOR Studies of Defects in Semiconductors
J.-M. Spaeth, Magneto-Optical and Electrical Detection of Paramagnetic Resonance in Semiconductors
T. A. Kennedy and E. R. Glaser, Magnetic Resonance of Epitaxial Layers Detected by Photoluminescence
K. H. Chow, B. Hitti, and R. F. Kiefl, μSR on Muonium in Semiconductors and Its Relation to Hydrogen
K. Saarinen, P. Hautojärvi, and C. Corbel, Positron Annihilation Spectroscopy of Defects in Semiconductors
R. Jones and P. R. Briddon, The Ab Initio Cluster Method and the Dynamics of Defects in Semiconductors

Volume 51B Identification of Defects in Semiconductors

G. Davies, Optical Measurements of Point Defects
P. M. Mooney, Defect Identification Using Capacitance Spectroscopy
M. Stavola, Vibrational Spectroscopy of Light Element Impurities in Semiconductors
P. Schwander, W. D. Rau, C. Kisielowski, M. Gribelyuk, and A. Ourmazd, Defect Processes in Semiconductors Studied at the Atomic Level by Transmission Electron Microscopy
N. D. Jager and E. R. Weber, Scanning Tunneling Microscopy of Defects in Semiconductors

Volume 52 SiC Materials and Devices

K. Järrendahl and R. F. Davis, Materials Properties and Characterization of SiC
V. A. Dmitriev and M. G. Spencer, SiC Fabrication Technology: Growth and Doping
V. Saxena and A. J. Steckl, Building Blocks for SiC Devices: Ohmic Contacts, Schottky Contacts, and p-n Junctions
M. S. Shur, SiC Transistors
C. D. Brandt, R. C. Clarke, R. R. Siergiej, J. B. Casady, A. W. Morse, S. Sriram, and A. K. Agarwal, SiC for Applications in High-Power Electronics
R. J. Trew, SiC Microwave Devices

J. Edmond, H. Kong, G. Negley, M. Leonard, K. Doverspike, W. Weeks, A. Suvorov, D. Waltz, and C. Carter, Jr., SiC-Based UV Photodiodes and Light-Emitting Diodes
H. Morkoç, Beyond Silicon Carbide! III–V Nitride-Based Heterostructures and Devices

Volume 53 Cumulative Subject and Author Index Including Tables of Contents for Volume 1–50

Volume 54 High Pressure in Semiconductor Physics I

W. Paul, High Pressure in Semiconductor Physics: A Historical Overview
N. E. Christensen, Electronic Structure Calculations for Semiconductors under Pressure
R. J. Neimes and M. I. McMahon, Structural Transitions in the Group IV, III-V and II-VI Semiconductors Under Pressure
A. R. Goni and K. Syassen, Optical Properties of Semiconductors Under Pressure
P. Trautman, M. Baj, and J. M. Baranowski, Hydrostatic Pressure and Uniaxial Stress in Investigations of the EL2 Defect in GaAs
M. Li and P. Y. Yu, High-Pressure Study of DX Centers Using Capacitance Techniques
T. Suski, Spatial Correlations of Impurity Charges in Doped Semiconductors
N. Kuroda, Pressure Effects on the Electronic Properties of Diluted Magnetic Semiconductors

Volume 55 High Pressure in Semiconductor Physics II

D. K. Maude and J. C. Portal, Parallel Transport in Low-Dimensional Semiconductor Structures
P. C. Klipstein, Tunneling Under Pressure: High-Pressure Studies of Vertical Transport in Semiconductor Heterostructures
E. Anastassakis and M. Cardona, Phonons, Strains, and Pressure in Semiconductors
F. H. Pollak, Effects of External Uniaxial Stress on the Optical Properties of Semiconductors and Semiconductor Microstructures
A. R. Adams, M. Silver, and J. Allam, Semiconductor Optoelectronic Devices
S. Porowski and I. Grzegory, The Application of High Nitrogen Pressure in the Physics and Technology of III-N Compounds
M. Yousuf, Diamond Anvil Cells in High Pressure Studies of Semiconductors

Volume 56 Germanium Silicon: Physics and Materials

J. C. Bean, Growth Techniques and Procedures
D. E. Savage, F. Liu, V. Zielasek, and M. G. Lagally, Fundamental Crystal Growth Mechanisms
R. Hull, Misfit Strain Accommodation in SiGe Heterostructures
M. J. Shaw and M. Jaros, Fundamental Physics of Strained Layer GeSi: Quo Vadis?
F. Cerdeira, Optical Properties
S. A. Ringel and P. N. Grillot, Electronic Properties and Deep Levels in Germanium-Silicon
J. C. Campbell, Optoelectronics in Silicon and Germanium Silicon
K. Eberl, K. Brunner, and O. G. Schmidt, $Si_{1-y}C_y$ and $Si_{1-x-y}Ge_xC_y$ Alloy Layers

Volume 57 Gallium Nitride (GaN) II

R. J. Molnar, Hydride Vapor Phase Epitaxial Growth of III-V Nitrides
T. D. Moustakas, Growth of III-V Nitrides by Molecular Beam Epitaxy
Z. Liliental-Weber, Defects in Bulk GaN and Homoepitaxial Layers
C. G. Van de Walle and N. M. Johnson, Hydrogen in III-V Nitrides
W. Götz and N. M. Johnson, Characterization of Dopants and Deep Level Defects in Gallium Nitride
B. Gil, Stress Effects on Optical Properties
C. Kisielowski, Strain in GaN Thin Films and Heterostructures
J. A. Miragliotta and D. K. Wickenden, Nonlinear Optical Properties of Gallium Nitride
B. K. Meyer, Magnetic Resonance Investigations on Group III-Nitrides
M. S. Shur and M. Asif Khan, GaN and AlGaN Ultraviolet Detectors
C. H. Qiu, J. I. Pankove, and C. Rossington, III-V Nitride-Based X-ray Detectors

Volume 58 Nonlinear Optics in Semiconductors I

A. Kost, Resonant Optical Nonlinearities in Semiconductors
E. Garmire, Optical Nonlinearities in Semiconductors Enhanced by Carrier Transport
D. S. Chemla, Ultrafast Transient Nonlinear Optical Processes in Semiconductors
M. Sheik-Bahae and E. W. Van Stryland, Optical Nonlinearities in the Transparency Region of Bulk Semiconductors
J. E. Millerd, M. Ziari, and A. Partovi, Photorefractivity in Semiconductors

Volume 59 Nonlinear Optics in Semiconductors II

J. B. Khurgin, Second Order Nonlinearities and Optical Rectification
K. L. Hall, E. R. Thoen, and E. P. Ippen, Nonlinearities in Active Media
E. Hanamura, Optical Responses of Quantum Wires/Dots and Microcavities
U. Keller, Semiconductor Nonlinearities for Solid-State Laser Modelocking and Q-Switching
A. Miller, Transient Grating Studies of Carrier Diffusion and Mobility in Semiconductors

Volume 60 Self-Assembled InGaAs/GaAs Quantum Dots

Mitsuru Sugawara, Theoretical Bases of the Optical Properties of Semiconductor Quantum Nano-Structures
Yoshiaki Nakata, Yoshihiro Sugiyama, and Mitsuru Sugawara, Molecular Beam Epitaxial Growth of Self-Assembled InAs/GaAs Quantum Dots
Kohki Mukai, Mitsuru Sugawara, Mitsuru Egawa, and Nobuyuki Ohtsuka, Metalorganic Vapor Phase Epitaxial Growth of Self-Assembled InGaAs/GaAs Quantum Dots Emitting at 1.3 μm
Kohki Mukai and Mitsuru Sugawara, Optical Characterization of Quantum Dots
Kohki Mukai and Mitsuru Sugawara, The Photon Bottleneck Effect in Quantum Dots
Hajime Shoji, Self-Assembled Quantum Dot Lasers
Hiroshi Ishikawa, Applications of Quantum Dot to Optical Devices
Mitsuru Sugawara, Kohki Mukai, Hiroshi Ishikawa, Koji Otsubo, and Yoshiaki Nakata, The Latest News

ISBN 0-12-752170-4